NUCLEI IN THE COSMOS III
THIRD INTERNATIONAL SYMPOSIUM ON NUCLEAR ASTROPHYSICS

Nuclei in the Cosmos III

Scientific Committee

M. Arnould (Belgium)
E. Bellotti (Italy)
V. Castellani (Italy)
C. Chiosi (Italy)
D. D. Clayton (USA)
F. Käppeler (Germany)
R. W. Kavanagh (USA)

D. L. Lambert (USA)
A. Omont (France)
A. Renzini (Italy)
C. Rolfs (Germany)
G. J. Wasserburg (USA)
S. E. Woosley (USA)
E. K. Zinner (USA)

Organizing Committee

C. M. Raiteri, M. Busso (Astronomical Observatory of Torino)
R. Gallino (University of Torino)
O. Straniero (Astronomical Observatory of Teramo)
M. I. Ferrero (INFN and University of Torino)
P. Monacelli (INFN-LNGS and University of L'Aquila)

Sponsored by

European Union—Directorate General XII
Istituto Nazionale di Fisica Nucleare
Osservatorio Astronomico di Torino
Università di Torino
Osservatorio Astronomico di Teramo
Gruppo Nazionale di Astronomia—Consiglio Nazionale delle Ricerche
Soc. Alenia Spazio

AIP CONFERENCE PROCEEDINGS 327

NUCLEI IN THE COSMOS III

THIRD INTERNATIONAL SYMPOSIUM ON NUCLEAR ASTROPHYSICS

ASSERGI, ITALY JULY 1994

EDITORS: **MAURIZIO BUSSO**
OSSERVATORIO ASTRONOMICO DI TORINO
ROBERTO GALLINO
UNIVERSITÀ DI TORINO
CLAUDIA M. RAITERI
OSSERVATORIO ASTRONOMICO DI TORINO

American Institute of Physics   New York

Authorization to photocopy items for internal or personal use, beyond the free copying permitted under the 1978 U.S. Copyright Law (see statement below), is granted by the American Institute of Physics for users registered with the Copyright Clearance Center (CCC) Transactional Reporting Service, provided that the base fee of $2.00 per copy is paid directly to CCC, 27 Congress St., Salem, MA 01970. For those organizations that have been granted a photocopy license by CCC, a separate system of payment has been arranged. The fee code for users of the Transactional Reporting Service is: 0094-243X/ 87 $2.00.

© 1995 American Institute of Physics.

Individual readers of this volume and nonprofit libraries, acting for them, are permitted to make fair use of the material in it, such as copying an article for use in teaching or research. Permission is granted to quote from this volume in scientific work with the customary acknowledgment of the source. To reprint a figure, table, or other excerpt requires the consent of one of the original authors and notification to AIP. Republication or systematic or multiple reproduction of any material in this volume is permitted only under license from AIP. Address inquiries to Series Editor, AIP Conference Proceedings, AIP, 500 Sunnyside Boulevard, Woodbury, NY 11797-2999.

L.C. Catalog Card No. 95-75492
ISBN 1-56396-436-8
DOE CONF-9407131

Printed in the United States of America.

Ante mare et terras et, quod tegit omnia, coelum
Unus erat toto naturae vultus in orbe,
Quem dixere chaos, rudis indigestaque moles
Nec quicquam nisi pondus iners congestaque eodum
Non bene junctarum discordia semina rerum.
   [*Ovidius: Metamorphoses, I, 5-9*]

"Ere land and sea and the all-covering sky
Were made, in the whole world the countenance
Of nature was the same, all one, well named
Chaos, a raw and undivided mass,
Naught but a lifeless bulk, with warring seeds
Of ill-joined elements compressed together"
  [*Translated by A. D. Melville, in World's Classics,
     (Oxford Univ. Press*: *Oxford), 1986*]

Contents

(Invited talks are in italics)

Preface xv
Acknowledgments xvi
Conference Photo xvii

SECTION I: OBSERVATIONAL ASTRONOMY

Surface Composition of Unevolved Stars 3
    R. G. Gratton
Isotopic Abundances in Interstellar Medium and Circumstellar Envelopes 19
    C. Kahane
The LUNA Project: Status and First Results 31
    C. Arpesella, C. A. Barnes, F. Bartolucci, E. Bellotti, C. Broggini,
    P. Corvisiero, G. Fiorentini, A. Fubini, G. Gervino, F. Gorris, U. Greife,
    C. Gustavino, M. Junker, R. W. Kavanagh, A. Lanza, G. Mezzorani,
    P. Prati, P. Quarati, W. S. Rodney, C. E. Rolfs, W. H. Schulte,
    H.-P. Trautvetter, and D. Zahnow
A Very Metal-Poor Field Giant with Extreme *r*-Process Abundance Enhancements 37
    C. Sneden, G. W. Preston, A. McWilliam, L. Searle, and J. J. Cowan
Abundance Patterns in Unevolved A Stars and in Blue Stragglers 41
    H. Holweger, M. Lemke, I. Rentzsch-Holm, and S. Stürenburg
Triple-Alpha Burning Products on the Surface of Peculiar Post-AGB Stars ... 45
    K. Werner, S. Dreizler, U. Heber, and T. Rauch
Abundances for Type I Planetary Nebulae: Evidence for Convective Envelope Burning?! 51
    R. L. Kingsburgh
Experimental Results on High Energy Cosmic Rays 55
    G. Navarra
Recent Results of the GALLEX Solar Neutrino Detector and their Implications 59
    W. Hampel
Mid-Infrared Imaging of AGB Star Envelopes 63
    M. Marengo, G. Silvestro, P. Persi, and L. Origlia
Mass and Radius of the Ellipsoidal Variable Barium Star HD 121447 ... 67
    A. Jorissen
Theoretical Modelling of Dust Shells Around AGB Stars 71
    E. Paravicini Bagliani, M. Marengo, and G. Silvestro
Is the Cosmic Ray Spectrum Formed by Splits of Nuclei? 75
    W. Tkaczyk
Cosmic Ray Tracking with Streamer Tubes 79
    U. Brandt, K. Daumiller, P. Doll, R. Gumbsheimer, H. Hucker,
    H. O. Klages, P. Kleinwächter, G. Kolb, and J. Lang

SECTION II: NUCLEAR ASTROPHYSICS

Charged-Particle Thermonuclear Reactions 83
    H.-P. Trautvetter, M. Möhle, and S. Schmidt

Experiments with Radioactive Beams 91
    W. Galster

New Experimental Data for the s-Process: Problems and Consequences 101
    F. Käppeler

Nuclear Properties Far from Stability and the r-Process 113
    K.-L. Kratz

The rp-Process: Rates from Direct and Indirect Reaction Studies 125
    M. Wiescher

Weak Interaction Rates 135
    E. Sugarbaker

First Direct Measurement of a $p(n,\gamma)d$ Reaction Cross Section at Stellar Energy 145
    T. S. Suzuki, Y. Nagai, T. Shima, M. Igashira, T. Kikuchi, H. Sato, and T. Kii

Search for Low-Energy Resonances in the $^3\text{He}(^3\text{He}, 2p)^4\text{He}$ Reaction 149
    P. Descouvemont

The PIAFE Project and Its Possible Implications in r-Process Studies 153
    G. Fioni and the PIAFE Collaboration

Measurements of (n,γ) Cross Sections with Small Samples 157
    P. E. Koehler and F. Käppeler

Neutron Capture in Barium and Gadolinium Isotopes: Implications for the s-Process 161
    K. Wisshak, F. Voss, and F. Käppeler

The Stellar Capture Rate of ^{208}Pb 165
    F. Corvi, P. Mutti, K. Athanassopulos, and H. Beer

Recent (n, p)- and (n,α)- Measurements of Relevance to Astrophysics 169
    C. Wagemans, S. Druyts, and R. Barthélémy

Recent Results of Measurements of the $^{14}\text{N}(n, p)^{14}\text{C}$, $^{35}\text{Cl}(n, p)^{35}\text{S}$, $^{36}\text{Cl}(n, p)^{36}\text{S}$ and $^{36}\text{Cl}(n, \alpha)^{33}\text{P}$ Reaction Cross Sections 173
    Yu. M. Gledenov, V. I. Salatski, P. V. Sedyshev, M. V. Sedysheva, P. E. Koehler, V. A. Vesna, and I. S. Okunev

Far from Stability: Evaluation of Properties of Nuclear Excited States 179
    A. Mengoni, G. Maino, and Y. Nakajima

Shell Effects in Neutron Capture on Pb 183
    T. Rauscher, R. Bieber, S. Lingner, and H. Oberhummer

G-T Strength Distribution in f-p Shell Nuclei and ν_e from Stellar Collapse 187
    F. K. Sutaria and A. Ray

Beta-Decay Studies of Nuclei Far from Stability Near N=28 191
O. Sorlin, R. Anne, L. Axelsson, D. Bazin, W. Böhmer, V. Borrel,
D. Guillemaud-Mueller, Y. Jading, H. Keller, K.-L. Kratz,
M. Lewitowicz, S. M. Lukyanov, T. Mehren, A. C. Mueller,
Yu. E. Penionzhkevich, B. Pfeiffer, F. Pougheon, M. G. Saint-Laurent,
V. S. Salamatin, S. Schoedder, and A. Wöhr

Measurement of the $\Gamma_\gamma/\Gamma_{tot}$ of the E_x=2.646 MeV State in ^{20}Na 195
M. A. Hofstee, J. C. Blackmon, A. E. Champagne, N. P. T. Bateman,
Y. Butt, P. D. Parker, S. Utku, K. Yildiz, M. S. Smith,
R. B. Vogelaar, and A. J. Howard

Is ^{16}O a Strong Neutron Poison? 201
Y. Nagai, M. Igashira, T. Shima, K. Masuda, T. Ohsaki, and H. Kitazawa

New Measurement of the ^{14}N(n, p)^{14}C Cross Section at Stellar Energy with a Drift Chamber 205
T. Shima, K. Watanabe, H. Sato, T. Kii, Y. Nagai, and M. Igashira

Charge-Changing Cross Sections Near the Δ-Isobar Resonance 209
P. Doll and H. J. Crawford

S-Factor in the Framework of the "Shadow" Model 213
A. Scalia

An Improved Experimental Approach to the ^{12}C$(\alpha,\gamma)^{16}$O Problem 217
G. Roters, C. Rolfs, and H.-P. Trautvetter

Approximate Penetration Factor for a Diffuse Edge Well with Coulomb Barrier. Astrophysical Applications 223
C. Grama, N. Grama, and I. Zamfirescu

The s-Process Cross Sections of the Tin Isotopes 227
Ch. Theis, K. Wisshak, K. Guber, F. Voss, F. Käppeler, L. Kazakov,
and N. Kornilov

The Stellar (n,γ) Cross Sections of Unstable s-Process Nuclei 231
S. Jaag and F. Käppeler

β-Delayed Neutron Decay in ^{17}B and ^{19}C 235
G. Raimann, A. Ozawa, R. N. Boyd, F. R. Chloupek, M. Fujimaki,
K. Kimura, T. Kobayashi, J. J. Kolata, S. Kubono, I. Tanihata, Y. Watanabe,
and K. Yoshida

Shapes of Nuclei in the Inner Crust of a Neutron Star 239
K. Oyamatsu and M. Yamada

Capture Reactions at Astrophysically Relevant Energies 243
P. Mohr, V. Kölle, S. Wilmes, U. Atzrott, F. Hoyler, C. Engelmann,
H. Grollmuss, G. Staudt, and H. Oberhummer

Solar Abundances and s-Process by Combined Burning of Two Neutron Sources 247
H. Beer, B. Spettel, and H. Palme

A Direct Measurement of the ^{19}Ne(p,γ)^{20}Na Reaction 253
R. D. Page, G. Vancraeynest, A. C. Shotter, M. Huyse, C. R. Bain,
F. Binon, R. Coszach, T. Davinson, P. Decrock, Th. Delbar, P. Duhamel,
M. Gaelens, W. Galster, P. Leleux, I. Licot, E. Liénard, P. Lipnik,
C. Michotte, A. Ninane, P. J. Sellin, Cs. Sükösd, P. Van Duppen,
J. Vanhorenbeeck, J. Vervier, M. Wiescher, and P. J. Woods

Neutron Producing Reactions in Stars 255
A. Denker, H. W. Drotleff, M. Grosse, H. Knee, R. Kunz, A. Mayer,
R. Seidel, M. Soiné, A. Wöhr, G. Wolf, and J. W. Hammer

The Reaction Rate for ^{31}S(p,γ)^{32}Cl and its Influence on the SiP-Cycle in Hot Stellar Hydrogen Burning 259
S. Vouzoukas, C. P. Browne, J. Görres, H. Herndl, C. Iliadis,
J. Meissner, J. G. Ross, L. van Wormer, M. Wiescher, A. Lefébvre,
P. Aguer, A. Coc, and J.-P. Thibaud

Sequential Proton Capture Reactions in the rp-Process 263
L. Van Wormer, J. Görres, H. Herndl, M. Wiescher, B. A. Brown,
and F.-K. Thielemann

Nuclear Recoil from Atomic Electrons and Resulting Effects on Low-Energy Fusion Cross Sections 267
T. D. Shoppa

Subthreshold Resonance Effect in the ^{6}Li(d,α)^{4}He Reaction 271
H. Bucka, K. Czerski, P. Heide, A. Huke, G. Ruprecht, and B. Unrau

The Reaction ^{70}Ge(α,γ)^{74}Se (p-Process) 277
Zs. Fülöp, Á. Z. Kiss, E. Somorjai, C. E. Rolfs, H.-P. Trautvetter,
T. Rauscher, and H. Oberhummer

$\beta-\beta$ Decay with Majorana Neutrino as Possible Reason for the Lack of Solar Neutrinos 281
V. I. Tretyak and V. V. Kobychev

The Search of 2β Decay of ^{116}Cd with ^{116}CdWO$_4$ Crystal Scintillators 285
F. A. Danevich, A. Sh. Georgadze, V. V. Kobychev, B. N. Kropivyansky,
V. N. Kuts, A. S. Nikolaiko, V. I. Tretyak, and Yu. Zdesenko

The keV Neutron Capture Cross Sections of ^{146}Nd, ^{148}Nd, and ^{150}Nd 291
K. A. Toukan, K. Debus, and F. Käppeler

Reaction Rate for ^{6}Li(p,α)^{3}He and ^{6}Li(d,α)^{4}He Reactions 295
J. Szabó

Analysis of the Triple-Alpha Process in the Potential Model 299
H. Krauss, K. Grün, H. Herndl, H. Oberhummer, P. Mohr, H. Abele,
and G. Staudt

Measurement of the Half-Life of ^{44}Ti 303
J. Meissner, J. Görres, H. Schatz, S. Vouzoukas, M. Wiescher,
L. Buchmann, D. Bazin, J. A. Brown, M. Hellström, J. H. Kelley,
R. A. Kryger, D. J. Morrissey, M. Steiner, K. W. Scheller, and R. N. Boyd

Measurement of the Neutron Capture Cross Sections of ^{15}N and ^{18}O 307
J. Meissner, H. Schatz, H. Herndl, M. Wiescher, H. Beer, and F. Käppeler

Determination of Alpha Widths in ^{19}F 311
      F. de Oliveira, A. Coc, P. Aguer, C. Angulo, G. Bogaert, S. Fortier,
      J. Kiener, A. Lefébvre, J. M. Maison, L. Rosier, G. Rotbard, V. Tatischeff,
      J.-P. Thibaud, and J. Vernotte

Electron Excitations in Nuclear Reactions at Astrophysically Relevant Energies 315
      K. Grün, H. Huber, J. Jank, H. Leeb, and H. Oberhummer

Classical Distributions for Interacting Particles 319
      G. Kaniadakis, A. Erdas, G. Mezzorani, and P. Quarati

Effect of Radiative Recombination in Thermonuclear Fusion Rates 323
      A. Erdas, G. Kaniadakis, G. Mezzorani, and P. Quarati

Quantum Statistical Distributions for Interacting Particles 327
      G. Kaniadakis and P. Quarati

The Rapid Hydrogen Burning Process and the Solar Abundances of ^{92}Mo, ^{94}Mo, ^{96}Ru and ^{98}Ru 331
      M. Hencheck, R. N. Boyd, B. S. Meyer, M. Hellström, D. J. Morrissey,
      M. J. Balbes, F. R. Chloupek, M. Fauerbach, C. A. Mitchell, R. Pfaff,
      C. F. Powell, G. Raimann, B. M. Sherrill, M. Steiner, J. Vandegriff,
      and S. J. Yennello

SECTION III: STELLAR MODELS AND NUCLEOSYNTHESIS

Stellar Evolution with Turbulent Diffusion 337
      C. Chiosi, A. Bressan, and L. Deng

Evolution and Mixing in Low and Intermediate Mass Stars 353
      J. C. Lattanzio

Nucleosynthesis and Supernovae in Massive Stars 365
      S. E. Woosley and T. A. Weaver

Nucleosynthesis in Core Collapse Supernovae 379
      F.-K. Thielemann, K. Nomoto, and M. Hashimoto

Sodium Enrichment and the Evolution of A-F Type Supergiants 393
      M. F. El Eid

Structure and Evolution of AGB Stars 399
      T. Blöcker and D. Schönberner

***s*-Process Calculations in Thermal Pulses: a Word of Caution** 403
      N. Mowlavi, A. Jorissen, M. Forestini, and M. Arnould

Radiative ^{13}C Burning in AGB Stars and *s*-Processing 407
      O. Straniero, R. Gallino, M. Busso, A. Chieffi, C. M. Raiteri, M. Salaris,
      and M. Limongi

Production of CNO Isotopes and ^{26}Al in Massive Single and Binary Stars 413
      N. Langer and C. Henkel

Presupernova Evolution and Explosive Nucleosynthesis in Massive Stars 419
      M. Hashimoto, K. Nomoto, and F.-K. Thielemann

Effect of the Equation of State on the Rapidly Rotating Neutron Star 425
    M. Hashimoto, K. Oyamatsu, Y. Eriguchi, and M. Kan
Explosion of a C-O Accreting White Dwarf 429
    E. Bravo, A. Tornambé, and J. Isern
CNO Anomalies on the Red Giant Branch: Tests for Classical Evolutionary Models of Low Mass Stars 433
    C. Charbonnel
Simulation of a White Dwarf Explosion with a 3D Particle Hydrocode 437
    E. Bravo, D. García, and N. Serichol
Neutrino Spallation Reactions on ^{4}He and the r-Process 441
    B. S. Meyer, S. E. Woosley, R. D. Hoffman, G. J. Mathews, and J. R. Wilson
Zero Age Main Sequence Stars: Structure and Neutrino Emission 447
    D. Hartmann, B. S. Meyer, D. D. Clayton, N. Luo, and T. Krishnan
ONeMg Novae: Nuclear Uncertainties on the ^{26}Al and ^{22}Na Yields 453
    A. Coc, R. Mochkovitch, Y. Oberto, J.-P. Thibaud, and E. Vangioni-Flam
^{26}Al and ^{22}Na Production in Neon Novae 457
    S. Wanajoh, M. Hashimoto, and K. Nomoto
Rare Neutron-Rich Nucleosynthesis in Type IA Supernovae 463
    S. E. Woosley, T. A. Weaver, and R. D. Hoffman
Hot Bottom Burning in Red Giants 469
    R. C. Cannon, C. A. Frost, J. C. Lattanzio, and P. R. Wood
Advanced Evolution of Massive Stars: Central Carbon Burning 473
    O. Aubert, I. Baraffe, and N. Prantzos
Nucleosynthesis in TP-AGB Stars 477
    P. Marigo
Stellar Models and Microstructural Investigations of Stardust 481
    A. C. Andersen, M.-L. Andersen, K. Glejbøl, and U. G. Jørgensen
Properties of Strange-Matter Stars 485
    F. Weber, Ch. Kettner, and N. K. Glendenning

SECTION IV: CHEMICAL EVOLUTION AND COSMOCHRONOLOGY

Light Elements: From Big Bang to Stars 491
    B. E. J. Pagel
Chemical Evolution of the Galaxy 501
    J. W. Truran and F. X. Timmes
Dating Methods from Stellar Evolution 513
    R. Buonanno and G. Iannicola
Gamma-Ray Line Astronomy and Interstellar Radioactivity 521
    M. D. Leising and D. D. Clayton
Galactic Evolution of D, ^{3}He, Li, Be, B 531
    N. Prantzos
Genesis and Evolution of LiBeB Isotopes II: Galactic Evolution 539
    M. Cassé, E. Vangioni-Flam, R. Lehoucq, and Y. Oberto

Galactic Chemical Evolution: Neutrino-Process Contributions 543
     F. X. Timmes, S. E. Woosley, and T. A. Weaver
The Orion Phenomenon: Particle Fluences in the Solar Nebula 549
     K. Marti and R. E. Lingenfelter
Distribution of Al-26 in the Galaxy 553
     N. Prantzos
The Main s-Component as a Superposition of Mean Neutron Exposures by TP-AGB Stars 557
     R. Gallino, M. Busso, and C. M. Raiteri
In Early Universe Primordial Molecules and Thermal Effects 561
     D. Puy

SECTION V: ISOTOPIC COMPOSITION OF METEORITES AND THE EARLY SOLAR SYSTEM

Interstellar Grains from Primitive Meteorites: New Constraints on Nucleosynthesis Theory and Stellar Evolution Models 567
     E. K. Zinner
Interstellar Graphite from the Murchison Meteorite 581
     S. Amari, E. K. Zinner, and R. S. Lewis
Oxygen-Rich Stardust in Meteorites 585
     L. R. Nittler, C. M. O'D. Alexander, X. Gao, R. M. Walker, and E. K. Zinner
The Abundance of ^{60}Fe in the Early Solar System 591
     G. W. Lugmair, A. Shukolyukov, and Ch. MacIsaac
Evidence in Meteorites for the Presence of ^{41}Ca in the Early Solar System 595
     J. N. Goswami and G. Srinivasan
The Astrophysical Site of the Origin of the Solar System Inferred from Extinct Radionuclides 599
     C. L. Harper Jr.
Chemically Fractionated Fission Xenon (CFF-Xe) on the Earth and in Meteorites 603
     A. P. Meshik, Yu. A. Shukolyukov, and E. K. Jessberger
List of Participants 607
Author Index 621

PREFACE

The Symposium **Nuclei in the Cosmos III** was held in the *Laboratori Nazionali del Gran Sasso* of the *Istituto Nazionale di Fisica Nucleare* (LNGS-INFN) at Assergi, near L'Aquila (Italy), from July 8 to July 13, 1994. It was intended as an international forum in which scientists working in different research branches, all in some way connected to Nucleosynthesis, could meet and discuss the most recent achievements in Nuclear Astrophysics. Not only in the title, but also in the content and in the wide interdisciplinary approach, the Symposium continued the tradition initiated since 1990 with the first Conference of the series, held in Baden (Vienna) and prosecuted with the second appointment of Karlsruhe (Germany) in 1992. Through its first three meetings, this series of Symposia encountered a steadily increasing attention from the scientific community, so that in Assergi we reached the number of 152 participants, from 19 countries.

The topics addressed spanned from laboratory work of Nuclear Astrophysics, especially on reaction rates, to models of Nucleosynthesis in single and double stars; from measurements of abundances and isotopic anomalies in pristine meteorites to the reconstruction of the early evolutionary phases of the Solar System; from optical and radio spectroscopic studies of the composition in stellar atmospheres and circumstellar envelopes to modeling of the global chemical evolution of galaxies. The place of the meeting (a top-level Laboratory in Underground Physics and Astrophysics) lead to an underlining of the topics related to Experimental Nuclear Astrophysics, which became the largest session both in the Conference and in this book. Also for the other sessions, however, the presentation should be quite complete, and suitable to a wide scientific audience.

For each topic, a thorough review is provided by introducing invited talks that should serve as a clear guide for interested researchers coming from other fields. For the readers directly working in the areas addressed by the book, the many research contributions can give a closer view of the most recent results, usually presented here before publication in specialized journals.

Acknowledgments

We are indebted, for their continuous help and advice, to all the members of the Scientific Committee. Thanks are due, for their warm welcome words and for the facilities provided to the Symposium, to P. Monacelli and A. Ferrari, directors of the LNGS and of the Observatory of Torino, respectively.

Several other people helped us in the secretarial work, both in the months before the Conference and during the Conference itself. For reasons of space, we cannot acknowledge in detail the efforts of all of them. However, a particular thanks must at least be paid, for their continuous and creative contribution, to Ms. Patrizia Santoro, of the Observatory of Torino, and to Mrs. Franca Masciulli, of the INFN-LNGS.

This Symposium was possible thanks to generous support from a number of Institutions, whose help is here gratefully recognized. A large fraction of the financial aid, more than 45% of the total, was given by the **European Union, Directorate General XII:** it was completely used for providing grants to the participants and was essential in order to allow many researchers, especially from less-favored countries, to participate. Among Italian sponsoring Institutions we must thank first of all the **INFN** (Istituto Nazionale di Fisica Nucleare), which offered not only economical support, but also active collaboration and a beautiful site for the Conference in the Gran Sasso Laboratory. Financial support from the **University of Torino,** the **Astronomical Observatory of Torino,** the **Astronomical Observatory of Teramo,** the **Consiglio Nazionale delle Ricerche (CNR-GNA),** and **Alenia Spazio** is also acknowledged.

The Editors:
Maurizio Busso
Roberto Gallino
Claudia M. Raiteri

Photograph by Massimo Villata.

SECTION I

OBSERVATIONAL ASTRONOMY

Surface Composition of Unevolved Stars

Raffaele G. Gratton

Osservatorio Astronomico di Padova

Abstract. This review is divided into two parts: first some of the mechanisms altering the surface composition of small mass single stars are considered, including pre-main sequence nuclear burning, dilution during the subgiant phase, first dredge-up and further mixing mechanisms along the RGB. A fairly good agreement exists between theoretical predictions and observations for stars in not too dense environments; departures from theoretical predictions observed for stars in the field and in open clusters might be attributed to the effects of some additional parameters (e.g. rotation), which is not included in standard models. The behaviour of stars in globular clusters is still not understood. On the whole, it may be safely concluded that surface abundances of most elements remain unchanged during the evolution of small mass stars up to the AGB-phase.

The aftermaths of some of the observed element-to-element abundance ratio runs in metal-poor stars are then examined. α-elements show a similar behaviour: there is a small but significant trend with [Fe/H] in the halo ([Fe/H]< −1) that can be attributed to a variation of the yields as a function of the mass for type-II SNe; the location of the change of slope (related to the onset of the contribution by type Ia SNe, and to the timescale of the halo collapse) is still not well established, being in the range −1.7 <[Fe/H]< −1 (corresponding to a timescale for the halo collapse between 3 and 1 Gyr). An early onset of the contribution by type Ia SNe is suggested by the analysis of the Cu/Fe trend. Recent results for the n-capture elements supports the predictions that the bulk of the $r-$process contribution is due to the less massive type II SNe, and that the $s-$process contribution is a primary-like process in intermediate and small mass star. Finally, in metal-poor stars there is a small excess of those elements which solar abundance is mainly attributed to the $s-$process, with respect to the predictions of a solar-scaled $r-$process distribution. This excess can be explained by a small contribution by the $s-$process, by too large solar $s-$fractions, or by a dependence of the $r-$process nucleosynthesis on metallicity.

Key words: Stars: abundances - Stars: evolution - Galaxy (The): chemical evolution - Nucleosynthesis

1. MECHANISMS ALTERING THE SURFACE COMPOSITION OF SMALL MASS SINGLE STARS

The abundances of individual elements in the surface layers of stars of different age (and hence overall metal abundance as represented by [Fe/H]) [1]

---

[1] I adopt two standard spectroscopic notations in this paper. First, $[X]=\log(X)_{star} - \log(X)_{\odot}$ for any abundance quantity X. Second, $\log n(X)=\log(N_X/N_H) + 12.0$ for absolute number density abundances.

provide a wealth of information about the chemical evolution of galaxies, and about the nucleosynthesis mechanisms (Wheeler et al 1989, Matteucci 1991). In these studies, it is usually assumed that the surface composition of stars reflects that of the interstellar medium (ISM) from which they formed. This is exactly true in virtually no case. However, there are at least two distinct groups of stars in which modifications are found to be rather small: the main-sequence B-stars, and the late-type single stars which have a radiative core and an outer convective envelope; a strong observational support to this last assertion is the good agreement existing between photospheric and meteoritic abundances (Anders & Grevesse 1990). Small mass stars are of paramount interest, because their long main sequence lifetime make them well suited for studying the composition of the ISM at different epochs, and then the galactic chemical evolution. Unfortunately, small mass main sequence stars have low luminosities; then evolved stars are often used. In the last years there has been a considerable progress in our understanding of the processes that cause modifications in the surface abundances during the evolution of single small mass stars both from theoretical and observational point of views; in this subsection, we will examine these processes following an evolutionary sequence, and will stop just before the third dredge-up phase on the asymptotic giant branch (AGB), which will be considered in the talk by Gustafsson. Furthermore, for lack of space we will skip the very interesting arguments related to the evolution of the surface Li and Be abundances in main sequence stars; recent reviews of this topic may be found in Michaud & Charbonneau (1991) and the volume edited by D'Antona (1991).

1.1. Pre main sequence nuclear burning

When the protostar contracts, the central temperature raises; roughly speaking, burning of ^{2}H, ^{6}Li, ^{7}Li, and ^{9}Be occur when the internal temperature reach respectively ~ 1, ~ 2, ~ 3, and $\sim 3.5 \times 10^6$ K (Bodenheimer 1965). For small mass stars these temperatures are reached at the base of the very deep outer convective envelope; surface depletion increases as mass decreases. Theoretical predictions (D'Antona & Mazzitelli 1994, Forestini 1994) are very sensitive on model details (nuclear burning rates, equation of state, opacity and convection). Note that in very small mass objects (brown dwarfs) central temperature never raises enough for nuclear burning. This may be a way to identify brown dwarfs (Magazzù et al 1993). Table 1 lists the minimum mass for nuclear burning (Pozio 1991, Nelson et al. 1993, D'Antona & Mazzitelli 1994); these values are rather model independent.

Observations of T Tau stars (Martín et al 1994), α Per (Balachandran 1993), the Pleiades (Soderblom et al 1993, Garcia Lopez et al 1994), and the Hyades (Thornburn et al 1993) show that (i) there is no PMS depletion for $M > 0.9$ M$\odot$; (ii) for $M < 0.9$ M$\odot$ there is a significant spread, that cannot be justified by observational errors alone; and (iii) this spread is likely related to rotation: fast rotators show smaller Li depletion. Possible explanations include the presence of magnetic fields (expected in fast rotating stars with deep convective envelopes) inhibiting turbulent diffusion (Schatzman 1993) and an inversion between the region of Li burning and the lower edge of the convective envelope.

Table 1: Minimum mass for nuclear burning

Isotope	M_{min} ($M_\odot$)
^{2}H	0.015
^{6}Li	0.059
^{7}Li	0.060-0.065
^{9}Be	0.085
^{11}B	0.092

1.2. Dilution during the subgiant phase

On the main-sequence, low mass stars retain Li only on the surface, about 1-2% by mass. Beneath the surface the temperature is too hot for Li to survive; Li nuclei react with protons to form Be. Low-mass stars leaving the main sequence when hydrogen is depleted develop surface convection zones, which deepen as the stars continues to evolve at lower surface temperature. The surface convection zone mixes the surface layer with deeper material in which the Li has been depleted (Iben 1965). This dilution of the surface layer causes the observed surface Li abundance to fall. In models with diffusion during the MS (Proffitt & Michaud 1991), Li settles below the surface, reducing the surface abundance, but does not fall far enough to reach a temperature where it will be burned. When surface convection begins, the Li rich layer is mixed to the surface, initially raising the surface Li; the Li abundance at even cooler temperatures falls as convection continues to deepen into the stars. Once the convective zone reaches its deepest extent, the dilution is complete, and the Li abundance should level off at a constant value. The process starts at the turn-off (TO) and it is completed at the base of the red-giant branch (RGB), where it is expected that the surface Li abundace is lowered by a factor of 60 at 3 $M_\odot$, and a factor of 18 at 1 $M_\odot$ with respect to the original value. Therefore, giants of nearly solar metallicity are expected to have $\log n(Li) \sim 1.3$ at 3 $M_\odot$ and 1.7 at 1 $M_\odot$. However, lower abundances are expected if part of the surface Li is burnt during the main sequence phase.

Observation shows that most (98%) field population I stars (Bonsack 1959, Lambert et al 1980, Brown et al 1989) have Li abundances much lower than the predicted maximum for a giant, suggesting that the bulk of them evolve from low-mass stars which have suffered an extra-Li depletion during the main-sequence phase. More insight is obtained from observations of selected groups of stars, where age (and masses) are better defined. Lambert et al (1980) obtained Li abundances for giants in the Hyades in rough agreement with expectations; similar results were obtained by Pilachowski et al (1988) for giants in NGC 7789 and NGC 752, while much lower abundances were obtained for stars in M 67, which result from the evolution of stars in the Li-dip (Boesgaard & Tripicco 1986). Gilroy (1989) surveyed Li abundances in giants of 20 open clusters; she found no strong correlation with TO masses, with Li abundances in general lower than theoretically predicted values. However, these analysis refer to stars in a later evolutionary phase, and the low Li abundances can be explained by some mass-loss or by the effect of the first dredge-up.

Pilachowsky et al (1993) observed a rather large number of metal-poor subgiant stars, resulting from the evolution of subdwarfs in the Spite & Spite (1982) plateau; those stars which are faint enough not to have experienced the first dredge up and the further mixing mechanism discussed in the next subsection show a good comparison with theoretical predictions both without and with diffusion during the MS (Proffitt & Michaud 1991), with some hints for a "diffusion bump"; however there is some scatter at the end of the dilution phase.

Finally, it must be mentioned that while the vast majority of giants are Li-poor, there are various groups of Li-rich giants (D'Antona 1991); most of them are chromospherically active, again suggesting that rotation may play an important role in the Li-depletion mechanisms.

1.3. First dredge-up and further mixing along the RGB

As the star evolves up the giant branch, the outer convective envelope expands inward and penetrates into the CN-cycle processed interior regions, and mixes approximately the outer 50% of the star by mass. The major elements to be mixed to the surface are ^{13}C and ^{14}N, both products of the CN cycle which occurs in the hydrogen shell outside the core. The convective mixing also transports the primordial ^{12}C and fragile, light elements like Li, Be, and B, from the surface to the interior. The net result of this so-called first dredge-up phase (Iben 1965) is to decrease the surface $^{12}C/^{13}C$ and $^{12}C/^{14}N$ ratios and the abundances of the light elements. Theoretical calculations (Iben 1967, Dearborn et al 1976, VandenBerg 1992, Charbonnel 1994) show that for solar composition, the $^{12}C/^{13}C$ ratios of first ascent giants will decrease from $^{12}C/^{13}C=90$ (the solar ratio) to ~ 25 in the luminosity range during the first dredge-up ($4 > M_V > 2$), the original ^{12}C abundances will drop by a factor of ~ 2, and the nitrogen abundances will rise by corresponding amount (the oxygen content remaining constant). Calculations of VandenBerg (1992) for metal-poor low-mass stellar models suggest that the $^{12}C/^{13}C$ decrease in the atmosphere of these stars should begin at brighter luminosities ($M_V = +2.5$) and should be completed at $M_V = +0.6$. Mixing is predicted to be slightly less effective in metal-poor than in population I stars.

The cleanest observational results are obtained for the $^{12}C/^{13}C$ isotopic ratio; results based on molecules may be affected by small offsets, while those from neutral lines may require non-LTE corrections. Theory and observations are in agreement for field population I stars (Lambert & Ries 1981, Kjaergaard et al 1982, Lambert et al 1980) and young open cluster giants (Gilroy 1989). Results for field old disk giants (Cottrell & Sneden 1986, Shetrone et al 1993) and metal-poor stars (Sneden et al 1986, Gratton et al 1994) shows that the first dredge-up occurs at the predicted luminosities ($M_V \sim 2$); however, mixing in bright giants is much more extreme than predicted by evolutionary models (VandenBerg 1992): small mass stars have lower carbon isotope ratios yet higher C/N ratios than the young disk giants. Similar results are obtained from Li destruction in giants (Brown et al 1989, Pilachowski et al 1993). At prima facie, these results might suggest that turbulent mixing (Genova & Schatzman 1977) or meridional circulation (Dearborn et al 1976) during the main sequence may cause a larger mixing than expected from the standard first dredge-up model. However when accurate estimates of the luminosities (and then of the evolutionary phase) are available, it turns out that stars with very low $^{12}C/^{13}C$ ratios either are brighter than $M_V \sim 0$, or are in the core He-burning phase, while bona fide stars that have just completed the first dredge-up ($2 < M_V < 0$) do not show any evidence of excess mixing (Gilroy 1989, Gilroy & Brown 1991).

Hence, on the whole observations support theoretical computations for the first dredge-up. As noted by VandenBerg (1992) after the end of the dredge-up phase is reached, the convective envelope begins to recede, leaving behind a chemical discontinuity. This is subsequently contacted by the H-burning shell. Thereafter, since there will be no mean molecular weight gradient between the convective envelope and the near vicinity of the shell, it is possible that circulation currents could give rise to further mixing. However, before this contact is made, the H-burning shell is advancing through a region where there are appreciable composition gradients, which should inhibit any rotational induced mixing (Tassoul & Tassoul 1984). Charbonnel (1994) remarked the difference between the behavior of small mass stars ($M < 2$ M$\odot$), which develop a degenerate core, and larger mass stars which ignite He before the core becomes degenerate. In this last case the hydrogen-burning shell never reaches the region of constant molecular weight that has been homogenized by the convective envelope during the first dredge-up. Under this condition, the extra-mixing does not occur, as confirmed by observational data.

It must be remarked that there are a large variety of evidences for this further mixing during the late RGB evolution, which is not the same in all stars, and it depends on some parameters other than mass and age; in fact stars in globular clusters show a large variation in the strength of C and N features, which are correlated with the variations of lines of other elements too (Smith 1987). Clear evidences for this further mixing is provided by Na and O abundances in field and globular cluster giants which show a global anticorrelation of [Na/Fe] and [O/Fe] (Kraft et al 1993); the most plausible interpretation is that the decline of oxygen is the result of deep mixing, in which the ashes of O→N burning and quantities of ^{23}Na formed by proton captures on ^{22}Ne are dredged into the atmosphere (Denisenkov & Denisenkova 1990, Langer et al 1993). Although the effect appears related primarily to evolutionary state, it may well be controlled in any given star by the degree of internal angular momentum, since the best candidate for this further mixing is meridional circulation induced by core rotation (Sweigart & Mengel 1979). A support to this interpretation of the deep mixing in the brightest giants is provided by the observation that the brightest giants in M13 are far more oxygen depleted and sodium enriched than the brightest giants in M3; on the other hand, Peterson (1983) found that M13 horizontal branch stars rotate faster than M3 horizontal branch (HB) stars in the same range of color and magnitude.

The abundance pattern of CNO elements in globular cluster stars is however more intriguing than suggested by these observations alone: stars on the main sequence of 47 Tuc and NGC 6752 show evidence for star-to-star fluctuations in the strength of CN and CH bands that cannot be explained by deep mixing processes (Hesser 1978, Briley et al 1991, Suntzeff 1989). The same anti-correlation of Na and O is not observed in bright giants in metal-rich clusters (Brown et al 1990, Brown & Wallerstein 1992, Carretta & Gratton 1994, Sneden et al 1994); metal-poor field halo giants are all oxygen-rich, although they show a large variation in the [Na/Fe] ratio, which is related to the overall metallicity (Kraft et al 1993, Gratton et al 1994). A possible explanation for these observations is that the presence of deep mixing is influenced by metallicity and by the environment; this can be understood if we consider that the molecular weight barrier in the interiors of high metallicity giants is less penetrable than that of low metallicity stars (Sweigart & Mengel 1979); and if we accept e.g. that stars in cluster form with angular momenta on average larger than field halo stars. However, at least another mechanism is important in globular cluster stars, perhaps related to star-to-star interactions in the dense core or primordial star-to-star variations in the abundances of a few selected elements (Smith & Wirth 1991).

1.4. Conclusion

For most elements, the surface abundance in small mass stars remains unaltered at least up to the thermal-pulse phase on the AGB. Li, Be and B may be partially or entirely depleted by various mechanisms during the pre-main sequence and main sequence phases, and they are diluted during the red giant branch evolution. Abundances of ^{12}C, ^{13}C and ^{14}N are changed at the first dredge-up. Further mixing may occur in the last stages of the RGB evolution, causing a further reduction of ^{12}C and possibly of ^{16}O, and enhancements of ^{13}C and ^{14}N and possibly of Na.

2. ABUNDANCE PATTERNS OF UNEVOLVED STARS AND GALACTIC CHEMICAL EVOLUTION

2.1. Nucleosynthesis mechanisms

Table 4: Schematic review of nucleosynthesis mechanisms

Group	Mass range	Remnant	Mass loss mechanism	Contribution	Timescale
			BIG BANG		
				^{2}H,^{3}He,^{4}He ^{7}Li,^{9}Be	
			SINGLE STARS		
very low mass	< 0.5	He WDs	no	no	>15 Gyr
low mass	0.5-2.3	CO WDs	winds, PN	He,^{14}N, main $s-$ (AGB phase)	1-15 Gyr
intermediate	2.3-8	CO WDs	winds, PN	He,^{12}C,^{13}C ^{14}N,^{17}O main $s-$	0.1-1 Gyr
???	≈ 8	no	winds type 1 1/2 SNe?	He,^{12}C,^{13}C ^{14}N,^{17}O main $s-$ Fe-group?	0.1 Gyr
large mass	8-12	NS	winds type II-L SNe?	He	0.01-0.1 Gyr
	12-100	NS or BH	winds type II-P SNe? type Ib SNe?	O,Ne,Mg,Si,Ca Fe-group weak $s-$ $r-$	<0.01 Gyr
	> 100	no	winds pair inst. SNe	O	1 Myr
supermassive?	400-7.5E5	no	winds	He,^{15}N,^{7}Li	<1 Myr
	>7.5E5	BH			
			BINARIES		
C-deflagration or delayed C-detonation	1.4-8	no	type Ia SNe?	Fe-group Si, Ca	0.1-15 Gyr
He-detonation	>2.3	no	type Ib SNe?	Fe-group	1 Gyr

Table 4 gives a schematic list of the major nucleosynthesis mechanism (Matteucci 1991, Reeves 1992, Kajino 1992, Nomoto et al 1992, Thielemann et al 1992). In the following sections we briefly review a few comparisons between the results that can be drawn from galactic evolutionary models which considers these mechanisms and observations (they are mainly taken from Matteucci and coworkers). For lack of space, we will skip important arguments as the chemical evolution of the disk (Edvardsson et al 1993) and of the bulge (McWilliam & Rich 1994), and concentrate on results obtained for halo stars.

2.2. α-elements and the timescale for halo collapse

Recently, King (1994) examined existing data about O abundances looking for evidences of (i) a trend of [O/Fe] with [Fe/H] within the halo, and (ii) of a break at [Fe/H]~ -1.7 which will be the signature of an early onset of the contribution by type Ia SNe, which in turn would imply a rather long timescale for the halo collapse (Matteucci & François 1992: ~ 3 Gyr). King concluded that current data do not clearly set the break point; the only indication is that it is at [Fe/H]< -1 (i.e > 1 Gyr). However, O abundances are sensitive to details in the analysis; hence, the spread of [O/Fe] ratios at any given [Fe/H] is generally large. The abundances of other α-elements (which are still mainly produced in type II SNe) are much better defined (see Figure 1). The King analysis should then be repeated using e.g. Mg abundances; a look at Figure 1 suggests that the change of slope might occur at [Fe/H]~ -1, although any value between $-1.7 <$[Fe/H]< -1 may perhaps be accepted. Here, we performed a simple exercise, deriving the least square linear regression coefficients for relations of the type [A/Fe]$=a$[Fe/H]$+b$ into three metallicity bins in the Galaxy, from a compilation of high quality data (Magain 1989, Zhao & Magain 1990, Edvardsson et al 1993, Nissen et al 1994, Gratton & Sneden 1991, Gratton et al 1994: see Table 2). While doing this exercise from a simple compilation of literature data increases the scatter given by individual studies, we think that the larger error bars better represents the uncertainties related e.g. to the temperature scale. Once this will be better established, systematic corrections could be applied, in order to avoid biases possibly present in our estimates. Note that for O we considered only abundances from [OI] lines; similar results are obtained from OH bands, while the analysis of the OI lines is very sensitive to the adopted T_{eff} scale.

Inspection of the results of Table 2 indicates:

- small but significant slopes are obtained among halo stars for all α-elements, the average value being -0.08; this is not unexpected, since the yields by type II SNe likely depends on precursor mass (Thielemann et al 1992); a similar slope is predicted by models of the chemical evolution of the Galaxy (Matteucci & François 1992). We get larger values of the slope for O, Mg and Ti; the trends for Si and Ca are flatter, suggesting that the bulk of Si and Ca is produced in less massive stars than the other elements

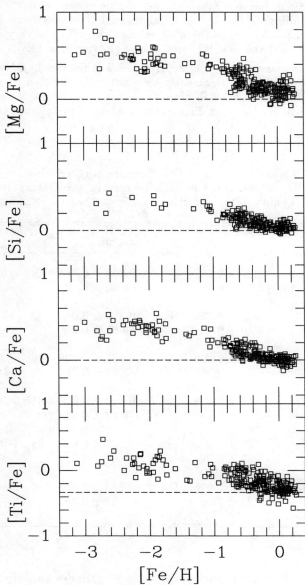

Figure 1: Runs of the abundance ratios of α−elements to Fe with [Fe/H] from a compilation of literature data (Magain 1989, Zhao & Magain 1990, Edvardsson et al 1993, Nissen et al 1994, Gratton & Sneden 1991, Gratton et al 1994). Panel a: [Mg/Fe]; panel b: [Si/Fe]; panel c: [Ca/Fe]; panel d: [Ti/Fe].

- the relations become much steeper in the range $-1.0 <$[Fe/H]< -0.2, corresponding to the kynematical transition between the halo and the disk (thick disk), where the slope is close to -0.3. The very small spreads indicates that the ISM was well mixed (Edvardsson 1993).
- slopes decreases again for [Fe/H]> -0.2 (thin disk); Edvardsson et al (1993) suggests that this nearly constant ratio between α-elements and Fe is an indication of infall of metal-poor material from the disk; this also agrees with the lack of a discernible gradient of metal abudance with age.

Table 2: Linear regression coefficients for relations of the type [A/Fe]$=a$[Fe/H]$+b$ into three metallicity bins in the Galaxy from a compilation of high quality data. a is given on the first line, b on the second.

Element Phase		[Fe/H]< -1.0 Halo		$-1.0 <$[Fe/H]< -0.2 Thick disk		[Fe/H]> -0.2 Thin Disk
[O/Fe]	42	-0.087 ± 0.062 $+0.25 \pm 0.15$	53	-0.323 ± 0.074 -0.00 ± 0.12	48	-0.091 ± 0.133 $+0.02 \pm 0.13$
[Mg/Fe]	42	-0.085 ± 0.029 $+0.30 \pm 0.10$	103	-0.360 ± 0.045 $+0.03 \pm 0.08$	80	$+0.020 \pm 0.060$ $+0.10 \pm 0.07$
[Si/Fe]	14	-0.040 ± 0.028 $+0.23 \pm 0.06$	108	-0.191 ± 0.027 $+0.03 \pm 0.05$	86	-0.002 ± 0.035 $+0.04 \pm 0.04$
[Ca/Fe]	43	-0.057 ± 0.024 $+0.24 \pm 0.08$	106	-0.250 ± 0.030 -0.03 ± 0.05	84	-0.089 ± 0.036 $+0.01 \pm 0.04$
[Ti/Fe]	36	-0.087 ± 0.030 $+0.13 \pm 0.09$	103	-0.233 ± 0.047 $+0.03 \pm 0.09$	84	-0.106 ± 0.065 $+0.05 \pm 0.07$
[α/Fe]	43	-0.082 ± 0.018 $+0.21 \pm 0.06$	108	-0.262 ± 0.208 $+0.01 \pm 0.05$	86	-0.045 ± 0.034 $+0.05 \pm 0.04$

2.3. Cu/Fe: balancing SN and s-process contributions

Observational data (Sneden & Crocker 1988, Sneden et al 1991) shows that while Zn abundances scale as the Fe ones, Cu is deficient relative to Fe in metal-poor stars. In order to interprete these results, Matteucci et al

(1993) considered that the solar abundances of Cu and Zn are due to contributions by various processes: the weak component of the $s-$process in massive stars (secondary-like); explosive nucleosynthesis in massive stars (type II SNe, primary-like), which for Cu is due to explosive Ne-burning (with some contribution by hydrostatic C-burning) and for Zn by $\alpha-$rich freeze-out of nuclear statistical equilibrium; and explosive nucleosynthesis in binary systems (type Ia SNe, delayed primary-like). The $s-$process yields must be normalized using the solar abundance ^{80}Kr. The predictions from type II SNe depend on the mass cut, while those from type Ia SNe are quite uncertain. The model which best fit the observational data have both Cu yields from type Ia and II SNe increased: in this model most solar Cu and Zn are produced in type Ia SNe. Note that in this model the contribution by type Ia SN occurs already at [Fe/H]< -1 (else we would have a constant [Cu/Fe] in metal-poor stars, at variance with observations).

Table 3: Relative $s-$ and $r-$ fractions in the Solar System

Element	log n	$s-$main	$s-$weak	$r-$
Rb	2.40	0.43	0.05	0.52
Sr	2.93	0.65	0.18	0.17
Y	2.22	0.81	0.15	0.04
Zr	2.61	0.66	0.07	0.27
Ba	2.21	0.89	0.01	0.10
La	1.20	0.75	0.01	0.24
Ce	1.61	0.78	0.01	0.21
Pr	0.71	0.46	0.00	0.54
Nd	1.47	0.45	0.00	0.55
Sm	0.97	0.30	0.00	0.70
Eu	0.54	0.07	0.00	0.93
Dy	1.15	0.12	0.00	0.88

2.4. n-capture elements

The heavy elements observed in the Sun are attributed to $s-$ and $r-$processes (see Table 3); these occur in very different astrophysical sites. The main component of the $s-$process (we will not consider here the weak component, which is relevant mainly for lighter elements) is usually related with the AGB phase of small and intermediate mass stars (Iben & Renzini 1982, Hollowell & Iben 1989, Gallino et al 1988, Kappeler et al 1990), while the astrophysical site of the $r-$process is still not certain (Mathews & Cowan 1990, Mathews et al 1992), although the favorite candidate are the lower mass end of type II SNe (Woosley & Weaver 1986, Pinto & Woosley 1988).

Spite & Spite (1978) first found clear evidences that [Ba/Fe] declines below [Fe/H]~ -1.5; the seminal work by Spite & Spite was followed by several others: a summary of previous Ba/Fe determinations and a few new ones have been presented by François (1992); an exam of this material suggests that the downward turn possibly occurs at slightly lower metallicity, and that the scatter is rather large. Eu abundances are discussed by Gilroy et al (1988), who found that Eu is overabundant in metal-poor stars, although it may be underabundant in extremely deficient stars ([Fe/H]< -2.5); the scatter in these determinations is also rather large. A possible explanation is that in stars with [Fe/H]< -2 we are seeing the contribution by the $r-$process alone (Truran 1981), while the flat portion seen for [Fe/H]> -1.5 (i.e. a primary-like behaviour) could be explained if the main neutron source for the $s-$process is ^{13}C (Raiteri et al 1992). A (difficult) analysis of Ba isotopes by Magain & Zhao (1993) indeed suggests that the $r-$process contribution is larger in very metal-poor stars; however Magain (1989) pointed out that this hypothesys does not seem to agree with the observed variation with metallicity which show a rise of e.g. [Ba/Fe] at the lowest metallicity.

Gratton & Sneden (1994) derived the abundances of several neutron-capture elements in 19 field stars with $-2.8 <$[Fe/H]< 0. The main results are:

- those elements whose solar abundances are mainly attributed to the $s-$process are overdeficient in extremely metal-poor stars ([Fe/H]< -2) with respect to those elements whose solar abundances are mainly attributed to the $r-$process. There is no clear evidence for a plateau in abundance ratios like [Ba/Eu] at these low values of [Fe/H].

- Eu itself begins to decline (with respect either to Fe and Mg) in the most metal-poor stars, with a sharp drop in stars with [Fe/H]< -2.5. If the $r-$process mainly occurs in SN explosions of massive stars, then the abundances of the ejecta are a function of initial stellar mass and/or metallicity, in agreement with some nucleosynthesis prediction.

- the abundance pattern of n-capture elements in metal-poor stars show clear differences with respect to scaled solar system $r-$process distribution: there is a relative excess of all elements that in the Sun are mainly attributed to the $s-$process. This pattern may be explained if the contribution by the $s-$process in the Sun is overestimated, or the relative production of different elements by the $r-$process changes with metallicity, or there was an early onset of the contribution of the main component of the $s-$process to the galactic chemical evolution.

REFERENCES

Anders, E., & Grevesse, N. 1990, Geochim. Cosmochim. Acta, 53, 197

Balachandran, S. 1993, in IAU Colloq. 137, Inside the Stars, ed. A. Baglin & W. Weiss (ASP Con. Ser.; San Francisco: ASP), 333

Bodenheimer, P. 1965, ApJ, 142, 451

Boesgaard, A.M., & Tripicco, M.J. 1986, ApJ, 302, L49

Bonsack, W.K. 1959 ApJ, 130, 843

Briley, M.M., Hesser, J.E., & Bell, R.A. 1991, ApJ, 373, 482

Brown, J.A., & Wallerstein, G. 1992, AJ, 104, 1818

Brown, J.A., Sneden, C., Lambert, D.L., & Dutchover, E.Jr. 1989, ApJS, 71, 293

Brown, J.A., Wallerstein, G., & Oke, J.B. 1990, AJ, 104, 1818

Carretta, E., & Gratton, R.G. 1994, in preparation

Charbonnel, C. 1994, A&A, 282, 811

Cottrell, P.L., & Sneden, C. 1986, A&A, 161, 314

D'Antona, F., ed. 1991, Mem. SAIt, 62

D'Antona, F., & Mazzitelli, I. 1994, ApJS, 90, 467

Dearborn, D.S.P., Eggleton, P.P., & Schramm, D.N. 1976, ApJ, 203, 455

Denisenkov, P.A., & Denisenkova, S.N. 1990, Sov. Astron. Lett., 16, 275

Edvardsson, B., Andersen, J., Gustafsson, B., Lambert, D.L., Nissen, P.E., & Tomkin, J. 1993, A&A, 275, 101

Forestini 1994, A&A, 285, 473

François, F. 1992, in Elements in the Cosmos, edited by M.G. Edmunds & R. Terlevich, Cambridge Un. Press, Cambridge, p. 137

Gallino, R., Busso, M., Picchio, G., Raiteri, C.M., & Renzini, A. 1988, ApJL, 334, L45

Garcia Lopez, R.J., Rebolo, R., Martín, E.L. 1994, A&A, 282, 518

Genova, F., & Schatzman, E. 1979, A&A, 78, 323

Gilroy, K.K. 1989, ApJ, 347, 835

Gilroy, K.K., & Brown, J.A. 1991, ApJ, 371, 578

Gilroy, K.K., Sneden, C., Pilachowski, C.A., & Cowan, J.J. 1988, ApJ, 327, 298

Gratton, R.G., & Sneden, C. 1991, A&A, 241, 501

Gratton, R.G., & Sneden, C. 1994, A&A, 287, 927

Gratton, R.G., Carretta, E., & Sneden, C. 1994, in preparation

Hesser, J.E. 1978, ApJ, 223, L117

Hollowell, D.E., & Iben, I.Jr. 1989, ApJ, 340, 966

Iben, I.Jr. 1965, ApJ, 142, 1447

Iben, I.Jr. 1967, ARAA, 5, 571

Iben, I.Jr., & Renzini, A. 1982, ApJL, 259, L79

Kajino, T. 1992 in Elements in the Cosmos, edited by M.G. Edmunds & R. Terlevich, Cambridge Un. Press, Cambridge, p. 8

Kappeler, F., Gallino, R., Busso, M., Picchio, G., & Raiteri, C.M. 1990, ApJ, 354, 630

King, J.R. 1994, AJ, 107, 350

Kjaergaard, P., Gustafsson, B., Walker, G.A.H., & Hultqvist, L. 1982, A&A, 115, 145

Kraft, R.P., Sneden, C., Langer, G.E., & Shetrone, M.D. 1993, AJ, 106, 1490
Lambert, D.L.. & Ries, L.M. 1981, ApJ, 248, 228
Lambert, D.L., Dominy, J.F., & Sivertsen S. 1980, ApJ, 235, 114
Langer, G.E., Hoffman, R., & Sneden, C. 1993, PASP, 105, 301
Magain, P. 1989, A&A, 209, 211
Magain, P., & Zhao, G. 1993, A&A, 268, L27
Magazzu', A., Martín, E.L.,& Rebolo, R. 1993, ApJ, 404, L17
Martín, E.L., Rebolo, R., Magazzù, A., Pavlenko, Y.V. 1994, A&A, 282, 503
Mathews, G.J., & Cowan, J.J. 1990, Nature, 345, 491
Mathews, G.J., Bazan, G., & Cowan, J.J. 1992, ApJ, 391, 719
Matteucci, F. 1991, in Frontiers of Stellar Evolution, edited by D.L. Lambert, ASP Conf. Ser. 20, p. 539
Matteucci, F., & François, P. 1992, A&A, 262, L1
Matteucci, F., Raiteri, C.M., Busso, M., Gallino, R., & Gratton, R.G. 1993, A&A, 272, 421
Michaud, G., & Charbonneau, P. 1991, Space Sci. Rev., 57, 1
McWilliam, A., & Rich, R.M. 1994, preprint
Nelson, L.A., Rappaport, S.,& Chiang, E. 1993, ApJ, 413, 364
Nissen, P.E., Gustafsson, B., Edvardsson, B., & Gilmore G. 1994, A&A, 285, 440
Nomoto, K., Tsujimoto, T., Yamaoka, H., Kumagai, S., & Shigeyama, T. 1992 in Elements in the Cosmos, edited by M.G. Edmunds & R. Terlevich, Cambridge Un. Press, Cambridge, p. 55
Peterson, R.C. 1983, ApJ, 275, 737
Pilachowski, C.A., Saha, A., & Hobbs, L.M. 1988, PASP, 100, 626
Pilachowski, C.A., Sneden, C., & Booth, J. 1993, ApJ, 407, 699
Pinto, P., & Woosley, S.E. 1988, ApJL, 331, L101
Pozio, F. 1991, Mem. SAIt, 62, 171
Proffitt, C.P.,& Michaud, G. 1991, ApJ, 371, 584
Raiteri, C.M., Gallino, R., & Busso, M. 1992, ApJ, 387, 263
Reeves, H. 1992 in Elements in the Cosmos, edited by M.G. Edmunds & R. Terlevich, Cambridge Un. Press, Cambridge, p. 1
Schatzman, E. 1993, A&A, 271, L29
Shetrone, M.D., Sneden, C., & Pilachowski, C.A. 1993, PASP, 105, 337
Soderblom, D.R., Jones, B.F., Balachandran, S., Stauffer, J.R., Duncan, D.K., Fedele, S.B., & Hudon, J.D. 1993, AJ, 106, 1059
Smith, G.H. 1987, PASP, 99, 67
Smith, G.H., & Wirth, G.D. 1991, PASP, 103, 1158
Sneden, C., & Crocker, D. 1988, ApJ, 335, 406
Sneden, C., Pilachowski, C.A., & VandenBerg, D.A. 1986, ApJ, 311, 826
Sneden, C., Gratton, R.G., Crocker, D.A. 1991, A&A, 246, 354
Sneden, C., Kraft, R.P., Langer, G.E., Prosser, C.F., & Shetrone, M.D. 1994, AJ, 107, 1773
Spite, F., & Spite, M. 1978, A&A, 67, 23

Spite, M., & Spite, F. 1982, A&A, 115, 357

Suntzeff, N. 1989, in The Abundance Spread Within Globular Clusters, edited by G. Cayrel de Strobel, M. Spite, and T.L. Evans (Obs. de Paris, Paris)

Sweigart, A.V., & Mengel, J.G. 1979, ApJ, 229, 624

Tassoul, M., & Tassoul, J.-L. 1984, ApJ, 279, 384

Thielemann, F.-K., Nomoto, K., Shigeyama, T., Tsujimoto, T., & Hashimoto, M. 1992 in Elements in the Cosmos, edited by M.G. Edmunds & R. Terlevich, Cambridge Un. Press, Cambridge, p. 68

Thornburn, J.A., Hubbs, L.M., Deliyannis, C.P., & Pinsonneault, M.H. 1993, ApJ, 415, 150

Truran, J.W. 1981, A&A, 97, 391

VandenBerg, D.A. 1992, ApJ, 391, 685

Wheeler, J.C., Sneden, C., & Truran, J.W. 1989, ARAA, 27, 279

Woosley, S.E., & Weaver, T.A. 1986, ARAA, 24, 205

Zhao, G., & Magain, P. 1990 A&A, 238, 1990

Isotopic abundances in interstellar medium and circumstellar envelopes

Claudine Kahane

Observatoire de Grenoble, BP 53, 38041 Grenoble Cedex 9, FRANCE

Abstract. This talk reviews the measurements of isotopic abundances in the galactic and extragalactic interstellar medium and in the circumstellar envelopes, based on radio molecular lines observations. The derivation of isotopic abundance ratios from molecular lines is discussed in terms of accuracy and reliability. The observational data are compared with theoretical results on galactic chemical evolution and stellar processing.

Key words: abundances — ISM — circumstellar matter — radio lines

1. INTRODUCTION

Nucleosynthesis in stars produces all elements heavier than lithium (1). Elemental and isotopic abundances measured in various locations (solar vicinity, interstellar or circumstellar medium, galactic center or galactic disk, external galaxies) are thus essential to our knowledge of nuclear processing. Elemental abundances based on optical observations of atomic lines (towards stellar atmospheres (2), HII regions (3), Planetary Nebulae (4)) are limited to low extinction regions. Atomic radio recombination lines observations of HII regions reach more heavily obscured regions (3). In both cases, elemental abundances are obtained after careful modeling and are potentially affected by depletion (i.e. condensation of certain elements onto grains), in particular in HII regions. Elemental abundances cannot be derived in molecular gas due to depletion and complex chemistry. In contrast, derivation of isotopic abundance ratios is generally simpler because most uncertainties similarly affect both isotopes and cancel out. Except for the lighter elements, the isotope shifts of the atomic lines are smaller than the thermal linewidths so that isotopic abundances are mainly derived from molecular lines, for which isotope shifts are easily resolved. We report in Table 1 all the nuclei presently observed in the galactic and extragalactic molecular gas.

TABLE 1. Nuclei observed in molecular gas.

Element	Nuclei	Location[a]
hydrogen	H, D	interstellar
carbon	^{12}C, ^{13}C	interstellar, circumstellar, extragalactic
nitrogen	^{14}N, ^{15}N	interstellar, circumstellar
oxygen	^{16}O, ^{17}O, ^{18}O	interstellar, circumstellar, extragalactic
silicon	^{28}Si, ^{29}Si, ^{30}Si	interstellar, circumstellar
sulfur	^{32}S, ^{33}S, ^{34}S, ^{36}S	interstellar, circumstellar
chlorine	^{35}Cl, ^{37}Cl	circumstellar

© 1995 American Institute of Physics

Molecules exist both in circumstellar envelopes and in interstellar clouds. The surface composition of unevolved and late-type stars, derived from optical observations, are reviewed in other talks at this conference. We will deal here only with the so-called "dusty envelopes" of evolved stars, which are too optically thick to allow optical star surface observations, but present intense molecular radio emission. In the cold and dense interstellar medium, our knowledge of the gas composition mostly rely on the radio observations of the molecular rotational spectra.

Isotopic abundances measured in the molecular clouds are representative both of large-scale and long-term average of nuclear evolution : the observed enrichments are integrated over different stellar types and successive generations of stars. In our Galaxy, however, the spatial resolution provided by the radio telescopes allows to study the galactic distribution of isotopic abundances. It is then possible to compare processing in regions dominated by stellar populations with different mass distributions and/or star formation rates, i.e. showing various nucleosynthesis processes and material recycling rates.

In external galaxies, isotopic abundances are mostly determined for a single position and correspond to averages over huge fractions of the galaxy. Their interest lays in the comparison of the mean isotopic composition of galaxies showing various metallicities, stellar masses distributions and/or star formation rates.

In contrast, abundances in the circumstellar envelopes of evolved stars trace the nuclear processing by a single star. It thus represents a powerful tool for testing nucleosynthesis and stellar evolution models.

Several detailed reviews have been devoted to abundances in the interstellar medium in the recent past (see in particular 2, 5, 6, 7) so that the data and the discussions presented here are partially a summary of these much longer reviews.

2. FROM MOLECULAR RADIO LINES TO ELEMENTAL ISOTOPIC RATIOS

2.1. Basic assumptions

The intensity of a radio molecular line, expressed in terms of brightness temperature, is given by

$$T_B = (1 - e^{-\tau})F(T_{ex}) \tag{1}$$

where τ is the line optical depth and T_{ex} the excitation temperature of the transition. If the line is optically thin, the intensity of the line is proportionnal to the column density N of the molecule and to a factor which depends on the line excitation, i.e.

$$T_B = NG(T_{ex}) \tag{2}$$

When the same transition of two isotopic species of the same molecule, ^aXY and ^bXY, (called *isotopomers* in the following) is observed, the line intensity ratio is an accurate measurement of the molecular abundance ratio provided that (i) both lines are optically thin, (ii) both transitions have the same excitation temperature, (iii) both isotopomers show the same spatial distribution along the line of sight. The next step to derive an isotopic abundance ratio is to assume that (iv) the isotopomer abundance ratio $[^aXY]/[^bXY]$ faithfully reflects the elemental isotopic ratio $[^aX]/[^bX]$. (In the following, for the molecular species, the superscript of the normal isotopes is often omitted.)

2.2. Required cautions

Several effects may prevent to derive an accurate value of the elemental isotopic ratio from the observed molecular line intensity ratio. Being aware of them may help to choose the molecules to use for a safer determination, to predict the signe of the correction which should be applied to the ratio, and to be cautious for any interpretation of measured ratios, since presently unknown systematic biases may reveal in the future.

2.2.1. Excitation

As soon as isotopomers showing very different abundances are compared, several excitation effects are likely to affect the derivation of the abundance ratio (for detailed discussions see 5, 6, 8, 9). The most obvious difficulty is that the strongest line may be optically thick and equation (2) is no longer valid (for instance the $^{12}CO/^{13}CO$ intensity ratio is generally useless to derive the $^{12}C/^{13}C$ ratio). Even if both observed lines are optically thin, their excitation may involve optically thick transitions and the assumption of identical excitation conditions for the isotopomers is no longer valid (see the discussion for H_2CO in 9). For large abundance ratios, the line intensity ratio is likely to underestimate the relative abundance of the most abundant isotopomer.

To avoid, or at least minimize, excitation effects, one should try

1. to make sure that both lines are optically thin (it often means using weak lines ; in some cases, such as the circumstellar envelopes, the line shape is of a great help to check the line opacity (10))

2. to compare isotopomers showing comparably low abundances, such as $C^{18}O$ and $C^{17}O$, or ^{13}CS and $C^{34}S$ (it may result in the measurement of a double isotopic ratio).

3. if only isotopomers showing a large abundance ratio can be used, to observe several transitions of both isotopomers to get more informations about the lines excitation.

Otherwise, a detailed modeling of the lines excitation is necessary, or the line intensity ratio must be considered as a lower limit to the column density ratio of the main/rare isotopomers.

2.2.2. Selective photodissociation

For abundant molecules which are dissociated by spectral line radiation, the external layers of the molecular cloud protect the inner regions against dissociation. As the protection is more efficient for more abundant species, selective photodissociation favors the more abundant isotopomer, so that the line intensity ratio overestimates the abundance of the more abundant species. The (H_2/HD) ratio is very sensitive to selective photodissociation (11). Its effect has been more recently measured and calculated for CO and its isotopomers (see for instance 12, 13) and appears much weaker. Other molecules are not likely to be affected.

2.2.3. Chemical fractionation

Two isotopomers share the same chemical properties. However, the difference in nuclear mass is responsible for a difference in chemical behaviour, called chemical fractionation, which favors the formation of one isotopomer. The fractionation reactions likely to be important are exothermic ion-molecule reactions (the neutral-neutral reactions are too slow), so that only the carbon and hydrogen isotopes are concerned by chemical fractionation since the fractional ionization of the other elements is too small (14,15). All deuterated molecules are favored by fractionation reactions, so that any molecular abundance ratio overestimates the actual (D/H) ratio (14). For carbon-bearing molecules, the effect of chemical fractionation is more complex (15) : it does not operate in warm (T > 35 K) and dense (not ionized) molecular cores and it results in an enhancement in ^{13}CO and a depletion in other ^{13}C-bearing molecules. The conclusion is that the effect of chemical fractionation on the derivation of the ^{12}C/^{13}C ratio depends on the observed cloud and on the observed molecule.

2.2.4. Measurements reliability and accuracy

As discussed above, care is needed to derive actual isotope ratios from line intensity ratios. One way to check that systematic effects are not too large is to compare the ratios derived from different molecules. In addition the accuracy of the ratios depends on the accuracy of the line intensity calibration. Due to several uncertainties (receiver gain, atmospheric and antenna losses,...) the absolute calibration of a line intensity is hardly more accurate than 5 to 10 %. When possible, simultaneous observations of two lines (for instance H^{13}CN and HC^{15}N) with the same receiver, reduces significantly the uncertainties on the calibration of the line ratio. In addition, when very weak lines are observed, the chance of accidental blending with another weak line is non negligible (see the discussion in (10) for the C^{17}O and C^{18}O lines and in (16) for ^{13}C^{18}O). For all these reasons (added to the interpretation uncertainties mentionned above), isotopic ratios are hardly determined with an absolute accuracy better than 10%.

3. AN OVERVIEW OF THE MAIN MOLECULAR MEASUREMENTS

3.1. Galactic interstellar medium

The large scale distribution of CO emission in our Galaxy shows that the molecular gas is located in two main components (17) : the Galactic Center, which concentrates 10 % of the galactic molecular gas in a thin nuclear disk of diameter 500 pc, and the Molecular Ring (also called Galactic Disk in the following) lying between 4 and 8 kpc from the Galactic Center. Most of the isotopic ratios reported in this section belong to surveys aimed to compare the composition of the Galactic Center, the Molecular Ring, the Local Interstellar Medium and the Solar System, since these regions are likely to present different nuclear processing histories.

3.1.1. Deuterium

The (D/H) ratio is the only primordial isotopic ratio likely to be measured by molecular line observations and is of central importance to determine the conditions in the Big Bang. Unfortunately, its derivation suffers from almost all the difficulties mentionned above : the (D/H) ratio is large, chemical fractionation and self shielding are significant. Numerous observations of deuterated molecules (see (6) for a summary) lead to the same conclusion : all species show a large enrichment in D, compared to the (D/H) ratio derived from atomic lines. However molecular observations are the only way to estimate (D/H) in distant regions. In particular, they tend to indicate a D/H ratio roughly 4 times smaller towards the galactic center than in the disk (18).

3.1.2. Carbon

Since the end of the seventies, many hours of telescope time have been devoted to surveys of $^{12}C/^{13}C$ ratio in the galaxy. The main molecules used for this purpose are CO, H_2CO, CS, HCO^+ and HCN. For the formaldehyde, with a detailed and careful excitation modeling, derivation of $^{12}C/^{13}C$ from the observation of several transitions of $H_2^{12}CO$ and $H_2^{13}CO$ seems to be a relatively secure method (9, 20, 21, 22). For the other species, only the rare isotopomers are used. With the improvement of the receivers sensitivity, it has become possible in the recent past to derive directly the carbon isotopic ratio from a comparison of the $^{12}C^{18}O$ and $^{13}C^{18}O$ lines (16, 19). This method seems to be free of excitation and selective photodissociation effects. We will not consider measurements which imply double isotopic ratios (such as $^{13}CO/C^{18}O$ or $H^{13}CN/HC^{15}N$) because both isotopic ratios are likely to present variations accross the Galaxy. The data are reported in Fig. 1a.

The 2 molecular species indicate the same trend : a decrease of the $^{12}C/^{13}C$ ratio from the outer to the inner regions of the Molecular Ring, a smaller value in the Galactic Center than in the Molecular Ring, most of the ratios being smaller than the Solar System value of 89, in particular in the solar vicinity.

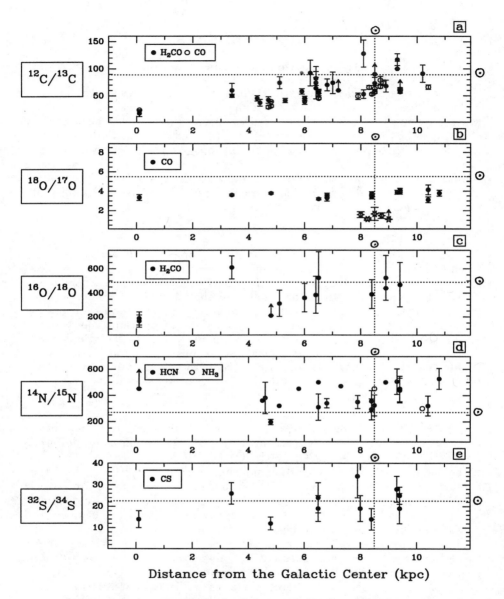

FIGURE 1. Isotopic ratios measured in the Galaxy, as a function of the distance from the Galactic Center. (a), $^{12}C/^{13}C$ ratio : H_2CO data from (9,20,21,22), CO data from (16,19). (b), $^{18}O/^{17}O$ ratio : interstellar CO data from (22), the stars are CO data in circumstellar C-rich envelopes (10). (c) $^{16}O/^{18}O$ ratio : H_2CO data from (24) combined with average $^{12}C/^{13}C$ ratio derived from Fig. 1a. (d) $^{14}N/^{15}N$ ratio : HCN data from (26,27) combined with average $^{12}C/^{13}C$ ratio derived from Fig. 1a and NH_3 data from (25). (e) $^{32}S/^{34}S$ ratio : CS data from (31) combined with average $^{12}C/^{13}C$ ratio derived from Fig. 1a.

3.1.3. Oxygen

The $^{18}O/^{17}O$ ratio which implies two rare isotopes is very accurately measured with the $C^{18}O$ and $C^{17}O$ lines (23). It appears remarkably constant throughout the Galaxy, with an average value of 3.5, significantly smaller than the Solar System ratio of 5.5 (see Fig. 1b).

Derivation of the $^{16}O/^{18}O$ ratio mainly relies on $H_2^{13}CO/H_2C^{18}O$ observations (24) and requires the knowledge of the $^{12}C/^{13}C$ ratio. The main measurements, combined with the $^{12}C/^{13}C$ ratios derived from Fig. 1a, are reported in Fig. 1c. Towards the Galactic Center, assuming an $^{12}C/^{13}C$ ratio of ~ 20, the $^{16}O/^{18}O$ ratio appears more than two times smaller than the Solar System value of 490. In the Galactic Disk, most of the measurements lead to values smaller than 490. A decrease of the ratio towards the inner regions cannot be ruled out, but it may also be an artefact due to the $^{12}C/^{13}C$ ratio.

3.1.4. Nitrogen

The nitrogen isotopic ratio is derived either from a model of NH_3 excitation (25) or from the double isotopic ratio $HC^{13}CN/HC^{15}N$ (26,27) and the $^{12}C/^{13}C$ ratio. The values reported in Fig. 1d show that in the Galactic Disk, the nitrogen ratio is higher than the Solar System value and seems to present a gradient. Towards the Galactic Center, only a lower limit can be derived from HCN and the NH_3 observations indicate an increase by at least a factor 4 compared to the Solar System ratio.

3.1.5. Heavy elements

Much less work has been devoted to the measurements of heavy elements isotopic ratios in the molecular gas. The only silicon bearing molecule detected in the galactic interstellar gas is SiO. The isotopic ratios $^{28}Si/^{29}Si$ and $^{29}Si/^{30}Si$ derived from optically thin lines, appear in agreement with the terrestrial ratios (28,29). The sulfur $^{34}S/^{33}S$ ratio has been derived from the CS isotopomers towards two clouds (in the Disk and in the Galactic Center) and appears in both locations compatible with the Solar System value of 5.5 (30). A survey of $^{13}CS/C^{34}S$ accross the Galaxy (31), combined with the average $^{12}C/^{13}C$ ratios derived from Fig. 1a, leads to the results plotted in Fig. 1e : the sulfur $^{32}S/^{34}S$ ratio is constant in the Galaxy and compatible with the Solar System ratio (22.5). (It can be noticed that he $C^{36}S$ isotopomer has been detected in the Galactic Disk and towards the Galactic Center (32)).

3.2. Circumstellar envelopes

Most of the measurements of elemental and isotopic ratios in the envelopes of evolved stars come from optical observations. However, radio observations of molecular lines are the only way to measure isotopic ratios in the very dusty envelopes of high mass loss Asymptotic Giant Branch stars. In addition, due to the simple geometry and velocity field of these objects, modeling of the radio line intensity is quite simple and secure, so that the isotopic ratios are likely to be derived with a good accuracy. On the other hand, due to the weakness of the optically thin lines, isotopic ratios are derived from radio lines only in a small number of envelopes, most of them carbon-rich.

3.2.1. Carbon

The first estimates of $^{12}C/^{13}C$ rely on CO observations in about fifteen bright envelopes (33). With the improvement of telescope sensitivity, many more envelopes have been observed since (34,35,36) but, unless an excitation model is used, the ratios derived from CO isotopomers are lower limits due to the main line opacity (the four J-type stars observed by (37) represent probably an exception because the $^{12}C/^{13}C$ ratio is very small and radio measurements are in good agreement with optical data). The $^{12}CO/^{13}CO$ intensity ratios range from about 5 to more than 100. The C-rich envelopes tend to show somewhat higher ratios than the O-rich ones, but both average values are significantly lower than the Solar System ratio. A more reliable $^{12}C/^{13}C$ ratio $\sim$ 44 has been derived from optically thin lines of various species in the envelope of the nearby C-rich star IRC+10216 (38,39). Analogous observations of the carbon star IRAS 15194-5115 lead to a much lower ratio of $\sim$ 9 (40). The double isotopic ratio $C^{34}S/^{13}CS$, combined with a sulfur $^{32}S/^{34}S$ ratio of 22.5 (see below), has been used to derive $^{12}C/^{13}C$ in four additional C-rich envelopes (10) leading to ratios from 30 to more than 65.

3.2.2. Oxygen

The oxygen isotopic ratios have been measured in 5 carbon-rich envelopes (10,36). The $^{18}O/^{17}O$ ratio is accurately measured from the CO isotopomers and appears considerably smaller than the terrestrial ratio and than the value measured troughout the Galaxy (see Fig. 1b). The $^{16}O/^{17}O$ and $^{16}O/^{18}O$ ratios require the knowledge of the $^{12}C/^{13}C$ ratio to be derived from $^{13}CO/C^{17}O$ and $^{13}CO/C^{18}O$. The $^{16}O/^{17}O$ ratio ranges from 250 to 850, significantly less than the terrestrial ratio of 2750. The $^{16}O/^{18}O$ ratio shows a huge dispersion, from 300 to more than 1300, and this range includes the Solar System value of 490.

3.2.3. Nitrogen

Measurements of $^{14}N/^{15}N$ rely on the observations of $H^{13}CN$ and $HC^{15}N$. Due to the $H^{13}CN$ line opacity (and in some cases to the uncertainties on the carbon isotopic ratio), the nitrogen ratios derived by (41) in eight carbon-rich envelopes are lower limits. They are all in excess of the terrestrial value, ranging from a lower limit of 300 to a lower limit of 5300 (for IRC+10216).

3.2.4. Heavy elements

Measurements of sulfur and silicon isotopic ratios in the C-rich envelope of IRC+10216, based on optically thin lines of various molecular species (CS, SiS, SiO, SiCC) lead to values in agreement with terrestrial ratios (38). (The observations of ^{29}SiO and ^{30}SiO maser emission in circumstellar envelopes (42) cannot be used to derive a silicon isotopic ratio because of the complexity of line excitation). The chloride ^{35}Cl/^{37}Cl ratio has also been measured towards IRC+10216, using NaCl and AlCl lines (43), and was found compatible with the Solar System value.

3.3. Extragalactic interstellar gas

Most of the carbon and oxygen rare isotopomers have been observed within the last four years towards the nuclei of active galaxies. The results have been recently presented and discussed in detail by (7), so that this section is only a brief summary of this review.

3.3.1. Carbon isotopic ratio

In most of the sources, only the ^{12}CO/^{13}CO intensity ratio is available. As discussed above, it provides most probably a lower limit to the actual ^{12}C/^{13}C ratio. In the compilation given by (7), the values range from 5 to 55. In two nuclear regions (NGC253 and NGC4945) the ^{12}C/^{13}C ratio has been derived from a detailed analysis of the line intensities from several molecular species (44,45) and is respectively ~ 40 and ~ 50, i.e. significantly higher than the ratio measured towards the Galactic Center. There are very few studies of the radial behaviour of ^{12}CO/^{13}CO in external galaxies and the variations proposed by (46) have not been confirmed (47).

3.3.2. Oxygen isotopic ratios

The ^{18}O/^{17}O ratio has been measured from the CO isotopomers in three active nuclear regions (45,48) and shows very high values (~ 8), i.e. much higher than anywhere in the Galaxy. In contrast, the ^{16}O/^{18}O ratio (~ 200) is somewhat lower than the galactic ratios (44,45).

4. COMPARISON WITH THEORETICAL PREDICTIONS

The aim of this section is not to review the isotopes production schemes nor the chemical evolution models. These topics are extensively discussed by (2), (5) and in these Conference Proceedings. We will rather try to point out the isotopic ratio behaviours which seem at least roughly understood, those which remain controversial and those which are still unexplained. Following (5), we will call *primary elements* those produced from the primordial elements

TABLE 2. Isotopic ratios derived from radio molecular observations.

Isotopic Ratios	Solar[a] System	Local ISM	Gradient[b]	Galactic Center	Carbon Stars	External Galaxies	Products[c] Behaviour
$^{12}C/^{13}C$	89	~ 65	+	~ 20	≥ 10	~ 40	I/II
$^{16}O/^{18}O$	490	~ 400	(+)	~ 200	300–1300	~ 200	I/II
$^{18}O/^{17}O$	5.5	3.5	no	3.5	0.6–0.9	~ 8	II/II
$^{14}N/^{15}N$	270	~ 450	(+)	≥ 1000	≥ 5300	?	II/I
sulfur		solar	no	solar	solar	?	I/I
silicon		solar	no	solar	solar	?	I/I

[a] Isotopic ratios from (49)
[b] + means a ratio increasing with galactic radius ; (+) means a controversial gradient
[c] I means primary product and II secondary product

H and He, on time scales short compared to the age of the Galaxy. In contrast, elements due to processing of nuclei heavier than He or on time scales comparable to the Galaxy age, are called *secondary elements*. With such a definition, the relative abundance of secondary products compared to primaries should increase with time and with processing rate. The observed isotopic ratios and the nature (primary or secondary) of the isotopes are summarized in Table 2.

The carbon isotopic ratio behaviour is consistent with nucleosynthesis and chemical evolution models : ^{13}C is produced by CNO processing during red giant phase so that a (I/II) behaviour is expected. The solar system ratio being likely to represent the local composition 4.5 Gyr ago, the present local ISM value is consistent with moderate processing in the solar vicinity. The galactic gradient is predicted by chemical evolution models (5) and the low value towards the Galactic Center is consistent with active stellar processing in this region. However, the nucleosynthesis of ^{13}C is not fully understood, and more observational data would be of great interest, in particular towards the Galactic Center, in the O-rich dusty envelopes and in external galaxies.

The behaviour of the oxygen ratios is much more puzzling. The present local $^{16}O/^{18}O$ ratio is uncertain : (5) and data in Fig. 1c indicate no change or a slight decrease compared to the solar value, whereas (6), including unpublished data not taken into account in Fig. 1c, indicates a significant increase. The first behaviour, together with a galactic gradient and the low value at the Galactic Center would be consistent with a I/II behaviour. However, the production schemes of ^{18}O are still quite uncertain. The behaviour of the $^{18}O/^{17}O$ ratio throughout the Galaxy is also somewhat mysterious. Its value is too low to be explained by an enrichment by intermediate mass red giants (10). Its remarkable constancy is also puzzling since the production schemes of ^{17}O and ^{18}O are likely to be quite different. The high $^{18}O/^{17}O$ ratios observed in active galactic nuclei are interpreted by (41) as indicative of ejection of ^{18}O by massive stars and of ^{17}O by intermediate mass stars. With such an assumption, they also suggest that the present Solar system $^{18}O/^{17}O$ ratio could result from processing by massive stars from an initial ratio of 3.5, as observed anywhere else in the Galaxy. Obviously, additional data (in the solar vicinity, towards O-rich dusty envelopes, in active and quiescent galaxies) are required and a better understanding of the processing of ^{18}O is also needed.

The nitrogen ^{14}N/^{15}N ratio also raises several questions : how to concile the local and the Galactic Center values, which would indicate a II/I behaviour, and the possible decrease (or at least absence of increase) of the ratio from the outer to the inner disk ? How high is the depletion in ^{15}N in evolved carbon stars, where only a lower limit to ^{14}N/^{15}N could be determined ? Do evolved oxygen-rich envelopes show such a high ratio ? Can this ratio be measured in external galaxies ?

The isotopic ratios of heavy elements measured in the galactic interstellar gas and in a single C-rich circumstellar envelope seem to be in agreement with the Solar System values. However, the accuracy of these measurements could be improved and other heavy elements are likely to be observed, at least in circumstellar envelopes.

The increase in sensitivity and spatial resolution provided, since the last ten years, by the radiotelescopes have allowed significant progress in the measurements of isotopic ratios. However, a number of key questions remain open and their solution will come both from theoretical and observational efforts.

ACKNOWLEDGEMENTS

It is a pleasure to thank M. Forestini, T. Forveille and M. Guélin for useful comments on the manuscript.

REFERENCES

1. Boesgaard, A.M., and Steigman, G., *Ann. Rev. Astron. Astrophys.* **23**, 319–78 (1985).
2. Matteucci, F., "Chemical enrichment of the interstellar medium", in *Chemistry in space*, 1–41, 1991, J.M. Greenberg and V. Pironello eds., Kluwer Academic Publishers.
3. Shields, G., *Ann. Rev. Astron. Astrophys.* **28**, 525–60 (1991).
4. Clegg, R.E.S., "Abundances in Planetary Nebulae", in *Planetary Nebulae*, 139–56, 1989, Torres-Peimbert, S., ed., Kluwer, Dordrecht.
5. Wilson, T.L., and Matteucci, F., *Astron. Astrophys. Rev.* **4**, 1–33 (1992).
6. Wilson, T.L., and Rood, R.T., *Ann. Rev. Astron. Astrophys.* **32**, in press (1994).
7. Henkel, C., and Mauersberger, R., *Astron. Astrophys.* **274**, 730–42 (1993).
8. Penzias, A.A., *Science* **208**, 663–69 (1980).
9. Henkel, C., Walmsley, C.M., Wilson, T.L., *Ap. J.* **82**, 41–47 (1980).
10. Kahane, C., Cernicharo, J., Gómez-González, J., Guélin, M., *Astron. Astrophys.* **256**, 235–250 (1992).
11. Watson, W.D., *Ann. Rev. Astron. Astrophys.* **16**, 585–615 (1976).
12. Bally, J., and Langer, W.D., *Ap. J.* **255**, 143–48 (1982).
13. Glassgold, A.E., Huggins, P.J., Langer, W.D., *Ap. J.* **290**, 615–26 (1985).
14. Herbst, E., and Klemperer, W., *Ap. J.* **185**, 505–33 (1973).
15. Langer, W.D., Glassgold, A.E., Wilson, R.W., *Ap. J.* **277**, 581–604 (1984).

16. Langer, W.D., and Penzias, A.A, *Ap. J.* **357**, 477–92 (1990).
17. Combes, F., *Ann. Rev. Astron. Astrophys.* **29**, 195–237 (1991).
18. Walmsley, C.M., and Jacq, T., "Deuterium in Galactic Center molecular clouds", in *Atoms, Ions and Molecules*, 305–312, Haschick, A.D., and Ho, P.T.P., eds., (1991).
19. Langer, W.D., and Penzias, A.A., *Ap. J.* **408**, 539–47 (1993).
20. Henkel, C., Wilson, T.L., Bieging, J., *Astron. Astrophys.* **109**, 344–51 (1982).
21. Henkel, C., Güsten, R., Gardner, F.F., *Astron. Astrophys.* **143**, 148–52 (1985).
22. Güsten, R., Henkel, C., Batrla, W., *Astron. Astrophys.* **149**, 195–98 (1985).
23. Penzias, A.A., *Ap. J.* **249**, 518–23 (1981).
24. Gardner, F.F., and Whiteoak, J.B., *MNRAS* **194**, 37–41p (1981).
25. Güsten, R., and Ungerechts, H., *Astron. Astrophys.* **145**, 241–50 (1985).
26. Wannier, P.G., Linke, R.A., Penzias, A.A., *Ap. J.* **247**, 522–29 (1981).
27. Dahmen, G., Wilson, T.L., Matteucci, F., *in prep.* (1994).
28. Wolff, R.S., *Ap. J.* **242**, 1005–12 (1980).
29. Penzias, A.A., *Ap. J.* **249**, 513–17 (1981).
30. Wilson, R.W., Penzias, A.A., Wannier, P.G. Linke, R.A., *Ap. J.* **204**, L135–37 (1976).
31. Frerking, M.A., Wilson, R.W., Linke, R.A., Wannier, P.G., em Ap. J. **240**, 65–73 (1980).
32. Guélin, M., private communication.
33. Knapp, G.R., and Chang, K.M., *Ap. J.* **293**, 281–87 (1985).
34. Wannier, P.G., and Sahai, R., *Ap. J.* **319**, 367–82 (1987).
35. Sopka, R.J., Olofsson, H., Johansson, L.E.B., Nguyen-Q-Rieu, Zuckerman, B., *Astron. Astrophys.* **210**, 78–92 (1989).
36. Bujarrabal, V., Fuente, A., Omont, A., *Astron. Astrophys.* **285**, 247–71 (1994).
37. Jura, M., Kahane, C., Omont, A., *Ap. J.* **201**, 80–88 (1988).
38. Kahane, C., Gómez-González, J., Cernicharo, J., Guélin, M., *Astron. Astrophys.* **190**, 167–177 (1988).
39. Cernicharo, J., Guélin, M., Kahane, C., Bogey, M., Demuynck, C., Destombes, J.L., *Astron. Astrophys.* **246**, 213–20 (1990).
40. Nyman, L.-Å., Olofsson, H., Johansson, L.E.B., Booth, R.S., Carlström, U., Wolstencroft, R., *Astron. Astrophys.* **269**, 377–89 (1993).
41. Wannier, P.G., Andersson, B-G., Olofsson, H., Ukita, N., Young, K., *Ap. J.* **380**, 593–605 (1991).
42. Alcolea, J., and Bujarrabal, V., *Astron. Astrophys.* **253**, 475–86 (1992).
43. Cernicharo, J., and Guélin, M., *Astron. Astrophys.* **183**, L10–12 (1987).
44. Henkel, C., Mauersberger, R., Wiklind, T., Hüttemester, S., Lemme, C., Millar, T.J., *Astron. Astrophys.* **268**, L17–20 (1993).
45. Henkel, C., Whiteoak, J.B., Mauersberger, R., *Astron. Astrophys.* **284**, 17–27 (1994).
46. Rickard, L.J., and Blitz, L., *Ap. J.* **292**, L57–60 (1985).
47. Young, J.S., and Sanders, D.B., *Ap. J.* **302**, 680–92 (1986).
48. Sage, J.L., Mauersberger, R., Henkel, C., *Astron. Astrophys.* **249**, 31–35 (1991).
49. Anders, E., and Grevesse, N., *Geochim. Cosmochim. Acta* **53**, 197–214 (1989).

The LUNA Project:
Status and First Results

C. Arpesella(1), C.A. Barnes(2), F. Bartolucci(1), E. Bellotti(3),
C. Broggini(4), P. Corvisiero(5), G. Fiorentini(6), A. Fubini(7), G.
Gervino(8), F. Gorris(9), U. Greife(9), C. Gustavino(1), M. Junker(9), R.W.
Kavanagh(2), A. Lanza(5), G. Mezzorani(10), P. Prati(5), P. Quarati(11),
W.S. Rodney(12), C. Rolfs(9), W.H. Schulte(9), H.P. Trautvetter(9), D.
Zahnow(9)

(1) Laboratori Nazionali del Gran Sasso, LNGS, Assergi, (2) Kellog Radiation Laboratory, Caltech, Pasadena, (3) Dipartimento di Fisica and INFN, Milano, (4) INFN Padova, (5) Dipartimento di Fisica and INFN, Genova, (6) Dipartimento di Fisica and INFN, Ferrara, (7) ENEA, Frascati and INFN, Torino, (8) Dipartimento di Fisica and INFN, Torino, (9) Institut für Physik mit Ionenstrahlen, Ruhr-Universität Bochum, (10) Dipartimento di Fisica and INFN, Cagliari, (11) Politecnico, Torino and INFN, Cagliari, (12) Georgetown University, Washington

Abstract. LUNA is a pilot project initially focused on the $^3He(^3He,2p)^4He$ cross section measurement within the thermal energy region of the Sun (15-27 KeV). A compact high current 50 KV ion accelerator facility including a windowless gas target system, a beam calorimeter and four detector telescopes has been built, tested and installed underground at the Laboratori Nazionali del Gran Sasso. The sensitivity has been improved by more than four orders of magnitude, as compared to the previous experiment.
In particular, thanks to the cosmic ray suppression, we could attain a background level of less than 1 event per week, a rate similar to the one expected from $^3He(^3He,2p)^4He$ at the lower edge of the Sun thermal energy region.

1. INTRODUCTION

LUNA (Laboratory for Underground Nuclear Astrophysics) is a pilot project which has the purpose of measuring a few cross sections of the reactions which are important for the generation of energy and for the synthesis of elements in the stars.
These processes occur in the stars at energies which are far below the Coulomb barrier. In this region the cross section $\sigma(E)$ drops nearly esponentially with decreasing energy and its measurement becomes increasingly difficult. Indeed, it was not yet possible to measure $\sigma(E)$ within the thermal energy region in the stars. Instead, the observed energy dependence of $\sigma(E)$ at high energies had to be extrapolated to thermal energies, leading to substantial uncertainties. In particular, a possible resonance in the unmeasured region is not accounted for by the extrapolation, but it can completely dominate the reaction rate for low stellar temperatures. It is therefore compelling to extend the measurements in the low energy region.
This gives the possibility to study an additional effect: the electron screening. The beam and the target used in an experiment are usually made of ions and

neutral atoms, respectively. The electron clouds surrounding the interacting nuclides act as a screening potential, thus reducing the height of the Coulomb barrier and enhancing the cross section. The screening effect has to be measured and taken into account in order to derive the cross section for bare nuclei, which is the starting point for nuclear astrophysics calculations.

The LUNA project is initially focused on the $^3He(^3He,2p)^4He$ cross section measurement within the thermal energy region of the Sun (15-27 KeV).

A resonance in this region has been suggested [Fowler 1972] [Fetysov and Kopysov 1972] to explain the observed high energy solar neutrino flux, a factor between 2 and 3 lower than the expected one. The enhancement in the $^3He(^3He,2p)^4He$ cross section would decrease the relative contribution of the alternative reaction $^3He(\alpha,\gamma)^7Be$, which generates the branch responsible for high energy neutrino production in the Sun.

A resonance at energy far below 100 KeV has been recently suggested [Straniero 1994] to explain the galactic abundance of ^3He. It is known [Reeves et al. 1973] that big-bang nucleosynthesis alone generates enough ^3He to account for the observations. The ^3He production by the stars is not required: the resonance in the $^3He(^3He,2p)^4He$ cross section could provide the mechanism through which the produced ^3He is also destroyed inside the star.

2. THE EXPERIMENTAL SETUP

The $^3He(^3He,2p)^4He$ cross section has been measured down to the centre of mass energy of 24.5 KeV [Krauss et al. 1987], where it has the value of 7±2 $pbarn$. This point is just at the upper edge of the thermal energy region of the Sun.

In order to measure at lower energy, where the cross section is getting smaller and smaller, it is necessary to make several improvements. In particular the accelerator has to deliver a high intensity beam even at low energy, where the space charge repulsion within the beam ions is a severe problem; the detector efficiency has to be reasonably high and the background has to be minimized. The 50 KV LUNA accelerator setup (fig. 1) consists of a duoplasmatron ion source, a beam extraction-acceleration system and a double focusing analyzing magnet. The energy spread of the source is less than 20 eV and the acceleration voltage is know with an accuracy better than 10^{-4}.

The $^3He^+$ beam selected by the magnet enters the windowless gas target system and can interact in the detector chamber, inside which a constant ^3He pressure of 0.5 $mbar$ is kept. The $^3He^+$ current entering the detector chamber is of ~ 0.5 mA. Finally the beam is stopped in a calorimeter where, from the heat deposition, the beam intensity is derived with a 3% accuracy.

The detector setup consists of four $\Delta E - E$ silicon telescopes placed around the beam axis at the distance of 2.5 and 3.5 cm. Each detector is a square of 5 cm side and 140 μm thick (ΔE detector) and 1 mm thick (E detector).

A mylar foil and an Al foil (each of 1.5 μm thickness) are placed in front of each telescope to shield the detectors from intense elastic scattering yields as well as from beam induced light and heat. Moreover the detectors are cooled down to -20 °C in order to have a small leakage current and to mantain a good energy resolution.

With this setup, which has a ∼13% detection efficiency, we can distinguish a proton from an α and measure their energy. Thanks to this possibility one can suppress the background reactions $d(^3He,p)^4He$ and $^3He(d,p)^4He$, which were one of the most severe problems in previous investigations. The deuterium is contained either in the target or in the beam, as a molecule of HD^+. Because of the lower Coulomb barrier, even a tiny deuterium contamination is enough to produce a higher rate than the one of $^3He(^3He,2p)^4He$. However the protons coming out from the two different reactions have different energies: 14.7 MeV in $d+^3He$ and less than 10.7 MeV in $^3He+^3He$. Therefore, by measuring the proton energy it is possible to distinguish between the two processes.

In addition to the background due to deuterium contamination we have to consider the background due to the natural radioactivity of the detector itself and of the environment, to the cosmic rays and to the neutrons.

A passive shielding around the detectors can provide a reduction of gammas and neutrons from the environment, but it produces at the same time an increase of gammas and neutrons due to the cosmic ray interactions in the shielding itself. An active shielding can only partially reduce the problem, because some of the activation due to spallation or μ^- capture gives rise to a delayed background.

The best solution is to strongly suppress the cosmic ray flux by going underground. At the end of 1993 we installed our accelerator facility at the Laboratori Nazionali del Gran Sasso (LNGS), at a depth equivalent to ∼ 3600 meters of water. Here the muon flux is reduced by a factor 10^6 and the neutron flux by a factor 10^3. LUNA is in a dedicated room separated from the other experiments by at least 60 m of rock.

The background due to the natural radioactivity of the detector setup itself is suppressed by the coincidence requirement between the proton signals in the ΔE and E detectors of a telescope.

3. STATUS AND FIRST RESULTS

The 50 KV facility has been tested over a period of 3 months at the Bochum University and then moved to LNGS in late 1993.

During the test phase it was possible to obtain some interesting physics results. In particular we measured the $d(^3He,p)^4He$ cross section down to the centre of mass energy of 5.4 KeV, where we clearly saw the enhancement due to the electron screening. Of course we will repeat and improve the experiment underground, thus exploiting the strong background reduction. In particular we will measure in the lower energy region, where the screening effect is getting larger and larger and its determination is less dependent on the extrapolation from the high energy data. However the results already show the LUNA capability of performing cross section measurements at energies far below those of previous works.

Our facility can be easily connected to an accelerator of higher energy. We did so with the 450 KV Bochum accelerator and we measured the $^3He(^3He,2p)^4He$ cross section over the centre of mass energy region from 46 to 92 KeV. The results are in good agreement with the existing measurement [Krauss et al. 1987] both in the energy dependence as well as in the absolute scale.

We also verified that the $^3He(^3He,2p)^4He$ protons can be easily separated from the protons due to $d(^3He,p)^4He$ or $^3He(d,p)^4He$.

The LUNA facility is now installed underground at LNGS, where, first of all, we checked the background reduction. Background events are all the proton-like events in the energy region of the $^3He(^3He,2p)^4He$ protons. We counted less than 1 of them per week, with at least a factor 200 reduction as compared to the background we measured in Bochum.

We are now starting the measurement of the $^3He(^3He,2p)^4He$ cross section. First at the centre of mass energy of 25 KeV, an important point for the overlap with the previous work [Krauss et al. 1987], and where we expect the "huge" rate of $\sim$130 events per day. We will then decrease the energy to cover the thermal energy region of the Sun, thus approaching the low rate typical of many underground experiments: for instance $\sim$2 events per week at the centre of mass energy of 17 KeV (this rate has been calculated without taking into account the screening effect and a possible resonance).

4. CONCLUSION

LUNA is a pilot project initially focused on the $^3He(^3He,2p)^4He$ cross section measurement within the thermal energy region of the Sun.

To achieve this goal the experimental sensitivity has been improved, as compared to the previous experiment, by more than four orders of magnitude: a factor 3 in the beam current, a factor 20 in the detection efficiency and more than a factor 200 in the background reduction.

We are now starting the experiment with the 50 KV accelerator facility installed underground at LNGS.

The $^3He(^3He,2p)^4He$ cross section measurement will last for about one year, after which, we hope, LUNA will have shown that it is possible to explore new interesting regions in nuclear astrophysics by going underground and by using some of the typical techniques of underground physics.

REFERENCES

Fowler W.A. 1972, Nature, 238, 24

Fetysov V.N. and Kopysov Y.S. 1972, Phys. Lett., B40, 602

Straniero O. 1994, talk given at the Summer Institute on Nuclear Physics and Astrophysics, LNGS (Assergi), 27 June-7 July 1994

Reeves H. et al. 1973, ApJ, 179, 909

Krauss A. et al. 1987, Nucl. Phys. A467, 273

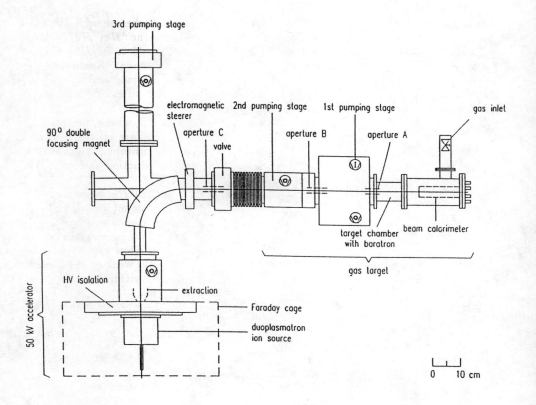

Fig. 1.— Schematic diagram of the 50 *KV* LUNA facility.

A Very Metal-Poor Field Giant with Extreme r-Process Abundance Enhancements

C. Sneden(1), G. W. Preston(2), A. McWilliam(2), L. Searle(2), J. J. Cowan(3)

(1) Department of Astronomy, University of Texas, Austin, Texas 78712
(2) Carnegie Observatories, 813 Santa Barbara St. Pasadena, California 91106-1292
(3) Department of Physics & Astronomy, University of Oklahoma, Norman, Oklahoma 73019

Abstract. The southern Galactic halo star CS 22892-052 is extremely metal-poor, but has very large enhancements of all elements formed by neutron capture nucleosynthesis. In this paper, we discuss a high resolution spectroscopic analysis of this star, showing that r-process neutron capture syntheses may account fully for the relative abundances of elements Ba through Er, and that probably a combination of the r-process and the weak-component s-process may match the abundances of the lighter elements Sr-Y-Zr.

Key words: Population II stars – Galactic halo – nucleosynthesis

1. INTRODUCTION

The chemical compositions of the lowest metallicity halo stars are records of very few prior nucleosynthesis generations in the Galaxy. Analysis of the spectra of large numbers of stars with overall metallicities $[Fe/H] < -2$ is necessary to address such issues as: the bulk yields of the first generation of supernovae; the mass range of these supernovae; the chemical homogeneity of the early Galaxy.

Beers et al. (1985, 1992) used a Ca II HK objective prism survey to identify many new very low metallicity stars. We have been conducting a high resolution spectroscopic abundance study of 20 HK survey giants (mean metallicity of our sample $<[Fe/H]> \sim -3.2$) and a control group of known low metallicity giants (which have $<[Fe/H]> \sim -2.8$). The high resolution ($R \sim 22,000$), modest signal-to-noise ($S/N \sim 35$) spectra for this work have been obtained with Shectman's Cassegrain echelle spectrograph at the Las Campanas 2.5m telescope. Basic spectroscopic data are given by Preston et al. (1994) and abundance results are discussed by McWilliam et al. (1994). The program star CS 22892-052 has a unique spectrum that is apparent at first glance: although it possesses the general weak-lined appearance of a very low metallicity K-giant, the features of CH and all neutron-capture elements ($Z > 30$) are extremely strong. Sneden et al. (1994) have completed a detailed analysis of the neutron capture elements in CS 22892-052; here we discuss the abundances of this star, comparing them to solar system neutron capture element r- and s-process abundance fractions.

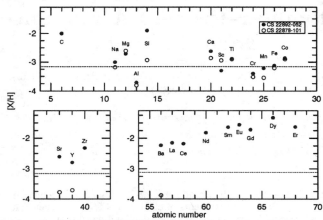

Fig. 1.— Abundances of CS 22892-052, rich in neutron-capture elements, and of the comparison star CS 22878-101.

2. ABUNDANCE ANALYSIS

We measured equivalent widths of some 200 lines of Fe-peak elements appearing on our spectrum of CS 22892-052. Then with an initial set of model atmosphere parameters estimated from the Beers et al. (1992) objective prism metallicity value and photometric data, we used a model atmosphere line analysis of our equivalent widths to derive parameters $(T_{\rm eff}/\log g/[{\rm Fe/H}]/v_t) =$ (4725 K/1.00/2.0 km s^{-1}/−3.12). We plot the relative abundances of elements with $Z < 30$ in the upper panel of Fig. 1, and for comparison we add points for the abundances of CS 22878-101 (McWilliam et al. 1994), a star with nearly identical atmosphere parameters (4750 K/1.20/2.2 km s^{-1}/−3.23). The relative abundances of these elements are very similar in the two stars, and are consistent with those of typical Pop II stars.

The similarity between CS 22892-052 and CS 22878-101 does not extend to the neutron capture elements. In the spectrum of CS 22878-101 only the strongest Sr, Y, and Ba lines are detectable, but in CS 22892-052, lines of many neutron-capture elements (along with CH features) are plentiful. In fact, many strong features in the blue spectral region of this star contain multiple components of neutron-capture elements and CH; this necessitates full spectrum syntheses of many features to extract reliable abundances of these elements. To illustrate this, in Fig. 2 we show small regions of spectrum around strong features of Sr II and Eu II in CS 22892-052 and CS 22878-101. The difference in neutron-capture element line strengths is obvious, and is especially acute for Eu II.

In the lower panels of Fig. 1 we plot the neutron-capture element abundances in CS 22892-052 (Sneden et al. 1994). As noted above, the extreme weakness of these elements in CS 22878-101 allowed derivation of abundances for only three neutron-capture elements in that star. Abundances of these

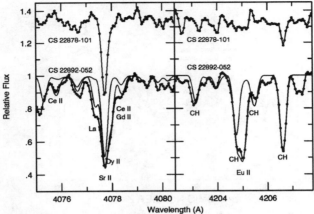

Fig. 2.— Observed spectra (heavy dotted lines) and synthetic spectra (thin lines) of CS 22892-052, along with the observed spectrum (shifted vertically) of CS 22878-101. The shallower synthetic spectrum assumes solar mixes of all elements, while the deeper synthetic spectrum has greatly enhanced CH and neutron-capture element abundances.

elements in CS 22892-052 are based on more transitions than is usual for metal-poor stars (e.g. six lines of Ba II, nine of Nd II, four of Eu II); not only are the neutron-capture element features strong but competing Fe-peak transitions are often extremely weak. The numbers of lines employed (and hence the abundance reliability) could easily be increased with acquisition of a higher S/N spectrum of this star. More elements with $Z > 30$ also await detection with renewed spectroscopic attention to this star.

3. THE PRODUCTION OF NEUTRON-CAPTURE ELEMENTS IN CS 22892-052

A significant clue to the origin of the neutron-capture elements is the slope of [X/H] with Z in the Ba–...–Er element group: the overabundances *increase* with increasing Z. In particular, compare Eu (made predominantly by the r-process) with Ba (mostly synthesized via the s-process): [Eu/Ba] $\simeq +0.7$. Other ratios yield similar results, and in Fig. 3 we replot the observed abundances for elements with $Z > 30$ along with curves that represent the r- and s-process fractions of solar system abundances of these elements. We have employed the Käppeler et al. (1989) division of solar system isotopic abundances into r- and s-process contributions to these two processes. We then have (arbitrarily) normalized each of these curves to the observed Nd abundance of CS 22892-052. The main result seems clear: an r-process neutron capture event created the elements Ba–...–Er for this star. Because it was an r-process, CS 22892-052 cannot have internally generated these elements, and instead a preceding local supernova probably seeded the halo ISM cloud that

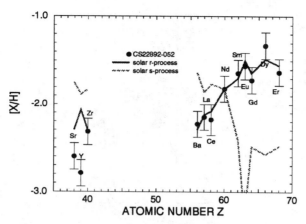

Fig. 3.— The observed abundances of CS 22892-052 compared with scaled solar-system r- and s-process abundance fractions. The error bars on the observed points are the approximate uncertainties in [X/H].

formed this star.

The case is less clear for the Sr–Y–Zr element group. For example, the relative Y abundance is much too low relative to Sr or Zr to match a pure r-process yield. Probably these elements were created with some combination of r-process and the so-called "weak component" s-process, most likely by the same progenitor star. We will explore matches between the abundance distribution of CS 22892-052 and theoretical r- and s-process production mechanisms more fully in a future paper.

ACKNOWLEDGEMENTS

We thank Debra Burris for assistance in the generation of the theoretical abundance curves. This work has been partially supported by NSF grants AST 91-15026 and 93-15068 to C.S and AST 93-14936 to J.J.C.

REFERENCES

Beers, T. C., Preston, G. W., and Shectman, S. A. 1985, AJ, 90, 2089
Beers, T. C., Preston, G. W., and Shectman, S. A. 1992, AJ, 103, 1987
Käppeler, F., Beer, H., and Wisshak, K. 1989, Rep. Prog. Phys., 52, 945
McWilliam, A., Preston, G. W., Sneden, C., and Searle, L. 1994, in preparation
Preston, G. W., McWilliam, A., Sneden, C., and Searle, L. 1994, in preparation
Sneden, C., Preston, G. W., McWilliam, A., and Searle, L. 1994, ApJL, in press

Abundance Patterns in Unevolved A Stars and in Blue Stragglers

Hartmut Holweger[1], Michael Lemke[2],
Inga Rentzsch-Holm[1] and Sven Stürenburg[1]

[1] *Institut für Theoretische Physik und Sternwarte,*
Universität Kiel, D-24098 Kiel, Germany
[2] *Institute of Astronomy,*
University of Cambridge, Cambridge CB3 0HA, U.K.

Abstract. On the basis of high-resolution spectrometry and NLTE analysis we investigate abundance patterns of non-nuclear origin in normal main-sequence A stars and in their metal-deficient counterparts, the Lambda Bootis stars. We try to identify and separate the effects of diffusion (gravitational settling, radiative levitation) and accretion (separation of gas and dust) on the surface composition. Two blue stragglers in the old open cluster M67 are compared with the normal A stars in an attempt to trace the signature of anomalous stellar evolution.

INTRODUCTION

Cosmic matter displayed at the surface of stars does not necessarily represent the unaltered yield of primordial, galactic, or *in situ* nucleosynthesis. Low-energy processes may occur in the low-density stellar environment or in the stars themselves that are capable of distorting the original nuclear abundance pattern in a complex way, sometimes even mimicking nuclear processes.

The chemically peculiar stars on the upper main sequence are drastic examples. Since the pioneering work of Michaud (1) it has become clear that in their largely static envelopes the interplay of *radiative pressure* and *gravitational forces* determines whether, and to what extent, a given element is driven to the surface or removed from it. On the other hand, the *separation of gas and dust* during star formation can in principle change the overall 'metal' content (2) because refractory elements have condensed onto grains. The existence of a metal-deficient class of main-sequence A stars, the λBoo stars, has been explained in this way by preferential accretion of depleted gas (3).

Is the occurrence of chemical separation exceptional, and how does it depend upon spectral type? In an ongoing project we are investigating these problems. Results are available for main-sequence A stars including the λBoo

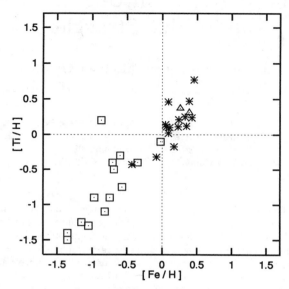

FIGURE 1. Titanium vs. iron in normal A stars (asterisks), Lambda Bootis stars (squares), and two blue stragglers in M67 (triangles). Dotted lines indicate solar abundances.

group; the current status is illustrated here. Details about the stellar sample, model-atmosphere analysis, and NLTE calculations have been published (4,5 and references therein). New NLTE calculations performed for nitrogen are taken into account; NLTE abundance corrections for N I are in the range −0.1 to −0.6 dex, depending on temperature and metallicity. In the present sample we also include two blue stragglers in M67 taken from the LTE analysis by Mathys (6), applying appropriate NLTE corrections.

RESULTS AND DISCUSSION

Figures 1-3 show logarithmic abundances with respect to the Sun for three characteristic cases. The common deficiency of *iron group* elements in λBoo stars (Fig. 1) supports the suggestion (3) that these stars have accreted gas depleted in refractory elements. The deviating stars near [Fe/H] = −0.8 have the lowest gravities of the sample, indicating evolutionary effects. The normal A stars appear to form the continuation of the sequence towards the metal-rich side - but see below. *Carbon,* on the other hand, behaves quite differently (Fig. 2); its abundance is close to solar even in the most metal-depleted objects. The same is true for *nitrogen* (cf. Fig. 5). This lends strong support to the accretion model: volatile elements like C and N are expected to stay with hydrogen in the gas phase (3,4). A group of 7 stars on the metal-rich side deviates by being *carbon deficient.* Remarkably, they also show the largest

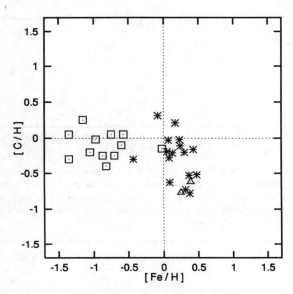

FIGURE 2. Carbon vs. iron. The symbols are the same as in Fig. 1.

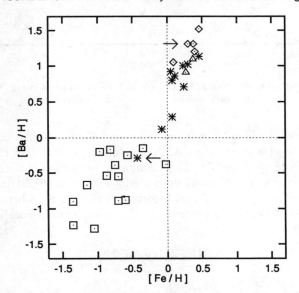

FIGURE 3. Barium vs. iron. Normal A stars deficient in carbon ([C/H]< −0.5) are shown as diamonds. Arrows: Sirius (top) and Vega (bottom). Other symbols are the same as in Fig. 1.

excess of *barium* (Fig. 3); the same holds for Sr. This is a clear signature of *diffusion*. Theory (7) predicts that carbon experiences too little radiative support to remain at the surface. By contrast, trace elements like Sr and Ba

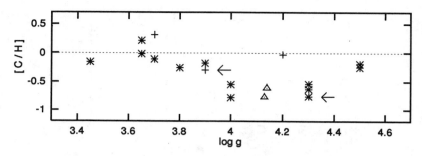

FIGURE 4. Carbon vs. surface gravity in normal A stars (asterisks and plus signs) and two blue stragglers in M67 (triangles). Plus signs denote stars whose Ba/Fe ratio is close to solar. All other A stars, the blue stragglers included, show large Ba excesses ([Ba/Fe]> 0.5) indicative of diffusion. Arrows: Sirius (right) and Vega (left). The ZAMS is near log g = 4.3.

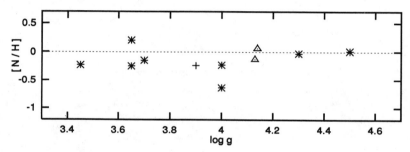

FIGURE 5. Nitrogen vs. surface gravity. The symbols are the same as in Fig. 4. Some of the program stars have not yet been analysed.

tend to float, as testified by Am and Ap stars. Carbon seems to become deficient when the star leaves the ZAMS and to recover later on; the same may happen with N (Figs. 4 and 5). The abundance pattern of the *blue stragglers*, discussed in the context of anomalous stellar evolution (6), strikingly resembles that of normal A stars affected by diffusion (Figs. 1-5). Their deficiency of C obviously is not due to CN processing. Moreover, N is not notably enhanced.

REFERENCES

1. Michaud, G., ApJ **160**, 641–658 (1970).
2. Schwarzschild, M., Spitzer, L.,Jr., Wildt, R., ApJ **114**, 398–406 (1961).
3. Venn, K.A., Lambert, D.L., ApJ **363**, 224–244 (1990).
4. Holweger, H., Stürenburg, S., in *Peculiar Versus Normal Phenomena in A-Type and Related Stars*, Dworetsky,M.M., Castelli,F., Faraggiana,R. (eds.), 356–366 (1993).
5. Stürenburg, S., A&A **277**, 139–154 (1993).
6. Mathys, G., A&A **245**, 467–484 (1991).
7. Michaud, G., Charland, Y., ApJ **311**, 326–334 (1986).

Triple-alpha burning products on the surface of peculiar post-AGB stars

Klaus Werner(1), Stefan Dreizler(2), Ulrich Heber(2), Thomas Rauch(1)

(1) Institut für Theoretische Physik und Sternwarte der Universität Kiel, Germany
(2) Sternwarte Bamberg, Universität Erlangen-Nürnberg, Germany

Abstract. The so-called PG 1159 stars form a new spectroscopic class of extremely hot hydrogen-deficient post-AGB stars. Our spectral analyses with model atmospheres show that their chemical surface composition is dominated by carbon, helium, and oxygen. We suggest that these peculiar stars have suffered a late helium-shell flash which has caused the removal of the hydrogen-rich envelope and even most of the helium-rich intershell matter. The idea that the former helium-burning region is now exposed at the surface of the PG 1159 stars is supported by the recent detection of a very high amount of neon in some objects.

The most extreme object is H 1504+65 which is one of the brightest X-ray sources in the sky and the hottest star ever analyzed with model atmosphere techniques (T_{eff}=170 000 K). The photosphere is devoid of hydrogen and helium and composed of oxygen and carbon by equal amounts! This means that we look at the naked core of the former Red Giant. This gives us the unique possibility to confine empirically the still uncertainly known $^{12}C(\alpha,\gamma)^{16}O$ nuclear reaction rate.

Key words: Stellar Atmospheres — $^{12}C(\alpha,\gamma)^{16}O$ reaction rate.

1. PG 1159 stars

The PG 1159 stars currently comprise 22 known objects. They are very hot hydrogen-deficient post-AGB stars characterized by their unusual spectral appearance. They display broad and shallow absorption lines of He II and C IV and occasionally O VI. Most prominent is an absorption trough region made up by He II 4686Å and a blend of several C IV lines. Eleven members of this group are associated with old extended planetary nebulae (PNe) and it is speculated that the other objects lack a visible PN because they have dispersed in the interstellar medium. Seven PG 1159 stars were shown to be non-radial g-mode pulsators. In particular the prototype star (PG 1159−035) has been studied photometrically with outstanding scrutiny by the Whole Earth Telescope project (Winget et al. 1991). These stars are primary examples demonstrating the power of the recently developed asteroseismological tools to study the interior structure of stars.

Analyses of high-S/N medium-resolution spectra obtained at telescopes of the 3-meter class in Chile (ESO) and Spain (Calar Alto) were performed with non-LTE model stellar atmospheres. The nine objects examined so far span a wide range of effective temperatures, from 65 000 K up to 170 000 K. The surface gravities are high, ranging from $\log g$ =5.9 (cgs units) for the luminous central stars to $\log g = 8.0$ for the most evolved object. The surface composition is generally dominated by carbon and helium in roughly equal amounts (by mass) and high admixtures (up to 17%) of oxygen. Comparison with stellar evolutionary tracks gives masses typical for white dwarfs (about 0.6 $M_\odot$), with the two extreme values 0.5 and 0.87 $M_\odot$. The luminosities range from about 10 to 10^4 $L_\odot$ and the distances are between 0.5 and 3 kiloparsec (for a recent review see Werner 1993).

The chemical surface abundances indicate that the PG 1159 stars have lost their hydrogen-rich envelope and even most of the helium-rich intershell region almost down to the carbon-oxygen-rich core. This is supported by pulsational models which predict that non-radial pulsations are driven by the kappa-effect involving carbon and oxygen. High amounts of C and O (roughly 80%) are required for this mechanism, a low amount of helium (at most about 20%) and no hydrogen at all may be present. Most recent results indicate indeed that the helium layer in PG 1159−035 is thinned down (from about $10^{-2} M_\odot$) by a factor of 5 (Kawaler & Bradley 1994). As far as accessible by both, spectroscopic and pulsational analysis methods, the derived stellar parameters (effective temperature, mass, surface composition) are in excellent agreement in the best studied case, PG 1159−035.

2. Evolutionary state

What is the evolutionary history of the PG 1159 stars? The answer to this question is still open. One possible interpretation of the peculiar chemical abundances is that the PG 1159 stars are the result of a late helium-shell flash in the precursor star. Such an event may happen to a post-AGB star that has almost contracted to white dwarf dimensions. As a consequence the star re-expands to giant dimensions ("born-again AGB star") and retraces the post-AGB evolution for the second time. During this course the star might strip off its envelope layers down to layers where triple-alpha burning has taken place. The recent detection of high amounts of neon (20 times solar) in some PG 1159 stars supports this idea (Werner & Rauch 1994). Neon is strongly enhanced in the helium burning shell by the reaction $^{14}N(\alpha,\gamma)^{18}F(\beta,\nu)^{18}O(\alpha,\gamma)^{22}Ne$. This chain starts with ^{14}N which was produced previously during CNO burning.

Detailed evolutionary calculations are lacking for this highly dynamic phenomenon of a late helium flash, therefore other explanations (like mixing and burning of the surface hydrogen) cannot be ruled out at the moment.

Immediate precursors of the PG 1159 stars may be identified with the Wolf-Rayet central stars of PNe, more exactly with the early type [WC] stars (i.e. spectral type [WCE]). This is based on the abundance analyses of the [WC 2] central star NGC 6751 (Hamann & Koesterke 1993) as well as two transition objects (classified as [WC]–PG 1159), the well known central stars of A 30 and A 78 (Werner & Koesterke 1992, Leuenhagen et al. 1993), which are PNe with strongly H-deficient knots expelled from the central stars. These stars show abundance patterns similar to the PG 1159 class. A very strong confirmation of the [WC]–PG 1159 evolutionary link represents the observation of a mass-loss event in the central star of Longmore 4 which turned its spectral type from PG 1159 to [WC] and back within a few weeks (Werner et al. 1992).

PG 1159 stars should evolve into hot helium-rich (DO) white dwarfs as soon as steadily increasing gravitational forces remove the heavy elements (C and O) from the atmospheres. A prerequisite is that the stars have not retained traces of hydrogen in the envelope which would eventually float atop and turn the star into a hydrogen-rich (DA) white dwarf. At this point we emphasize that the upper limits of the hydrogen abundance from the spectroscopic analyes are not very strict. A fraction of 10% would easily escape detection. Such an amount is certainly enough to turn the star into a DA. If the envelope "peeling" scenario for the PG 1159s is right, however, then we can in fact assume that their fate is to become a DO. But caution is necessary and this is underlined by the detection of the hybrid central stars (objects with high He and C abundances *and* detectable hydrogen).

In any case, the existence of DO white dwarfs makes at least a subgroup of the PG 1159 stars likely progenitors (although doubt is cast even on this statement by two known central stars with almost pure helium atmospheres, namely K 1-27 and LoTr 4 [Rauch et al. 1994a,b]). About a dozen hot DO white dwarfs (defined by the absence of He I lines) are known at present and detailed non-LTE model atmosphere analyses are now becoming available. KPD 0005+5106 was studied in the optical, UV and X-ray regions and found to be extremely hot ($T_{\rm eff}$=120 000 K).

3. H 1504+65 and the ^{12}C$(\alpha,\gamma)^{16}$O nuclear reaction rate

The most extreme PG 1159 star is H 1504+65. It is not only the hottest star ever analyzed with model atmosphere techniques ($T_{\rm eff}$=170 000 K) but it has a photosphere that is hydrogen *and* helium-deficient (Werner 1991). Thus the atmosphere is solely composed of carbon and oxygen. We have interpreted this composition as the exposition of the naked C-O stellar core of the former Red Giant. Thus we are in the advantageous situation to constrain empirically the ^{12}C$(\alpha,\gamma)^{16}$O nuclear reaction rate because we directly see the immediate burning product.

The knowledge of this reaction rate is still scarce but it is of high astrophysical interest for many reasons. Time scales of late stages of massive star evolution and the position in the HR diagram depend sensitively on the adopted value for this rate. Additionally the carbon and oxygen enrichment in the interstellar medium during the course of galactic chemical evolution depends on yields of stellar evolution calculations which in turn depend on the ^{12}C$(\alpha,\gamma)^{16}$O nuclear reaction rate. The internal composition of C-O white dwarfs and through this the cooling time scales are essentially depending on this rate, too. The rate proposed by Fowler et al. (1975) (F) has been increased by Harris et al. (1983) by a factor of three ($F \cdot 3$). Subsequently the rate was decreased again down to $F \cdot 2$ by Caughlan & Fowler (1988).

Constraining the rate from our C/O abundance ratio determination is not straightforward for several reasons. Besides the error bars of our analysis (O= 50 ± 20%, by mass) uncertainties in stellar evolution calculations, particularly concerning mixing processes, render a conclusive result difficult.

According to evolutionary calculations by Iben & Tutukov (1985) for a model with a mass appropriate for H 1504+65 and adopting a $F \cdot 2$ rate, we may expect an oxygen abundance of 40%. Following calculations of D'Antona & Mazzitelli (1991) we may estimate that a change of the $^{12}C(\alpha,\gamma)^{16}O$ rate to F or $F \cdot 3$ roughly results in O abundances of 30% and 50%, respectively. We therefore conclude that a rate as high as $F \cdot 3$ agrees best with our observations. This result agrees e.g. with the analysis of ejecta from the supernova 1987A (Thielemann et al. 1990).

Currently we work on the improvement of our photospheric analysis. Important tools are the ROSAT, Hubble Space Telescope (HST) and Extreme Ultraviolet Explorer (EUVE) satellites. X-ray, UV and EUV spectra of H 1504+65 were already obtained and are now being examined in detail (Barstow et al. 1994). The results are expected to further constrain our estimates for the $^{12}C(\alpha,\gamma)^{16}O$ nuclear reaction rate.

ACKNOWLEDGEMENTS

K.W., S.D., T.R. are supported by the DFG under grants We 1312/6-1 and He 1356/16-1 and by the BMFT under grant 50 OR 9409 1. K.W. thanks the organizers of the conference for generous financial support.

REFERENCES

Barstow M.A., Holberg J.B., Werner K., Nousek J.A. 1994, Proc. COSPAR Symp. Hamburg 1994, in press

Caughlan G.R., Fowler W.A. 1988, Atomic Data and Nuclear Data Tables 40, 283

D'Antona F., Mazzitelli I. 1991, IAU Symp. 145, p. 399

Fowler W.A., Caughlan G.R., Zimmerman B.A 1975, ARA&A 13, 69

Harris M.J., Fowler W.A., Caughlan G.R., Zimmerman B.A. 1983 ARA&A 21,165

Hamann W.-R., Koesterke L. 1993, in Planetary Nebulae, IAU Symp. 155, p. 87

Iben I. Jr., Tutukov A.V. 1985, ApJS 58, 661

Kawaler S.D., Bradley P.A. 1994, ApJ 427, 415

Leuenhagen U., Koesterke L., Hamann W.-R. 1993, Acta Astronomica 43, 329

Rauch T., Köppen J., Werner K. 1994a, A&A 286, 543

Rauch T., Köppen J., Werner K. 1994b, A&A submitted

Thielemann F.-K., Hashimoto M., Nomoto K. 1990, ApJ 349, 222

Werner K. 1991, A&A 251, 147

Werner K. 1993, in White Dwarfs: Advances in Observation and Theory, NATO ASI Series C, Vol. 403, Kluwer, Dordrecht, ed. M.A.Barstow, p. 67

Werner K., Koesterke L. 1992, in Atmospheres of Early-Type Stars, eds. U.Heber and C.S.Jeffery, Springer, Berlin, p. 288

Werner K., Rauch T. 1994, A&A 284, L5

Werner K. et al. 1992 A&A 259, L69

Winget D.E. et al. 1991, ApJ 378, 326

Abundances for Type I Planetary Nebulae: Evidence for Convective Envelope Burning?!

Robin L. Kingsburgh

Instituto de Astronomía–Ensenada, Universidad Nacional Autónoma de México

Abstract. Elemental abundances are derived for a sample of southern galactic planetary nebulae. We define Type I PN as those which have experienced envelope-burning conversion to nitrogen of dredged-up primary carbon. Such nebulae have nitrogen abundances which exceed the total (C+N) abundance of H II regions in the same galaxy. In the current galactic sample, 11 nebulae are classified as Type I, having N/O>0.8. For these PN, no evidence is found for oxygen depletion, compared to non-Type I PN, hence no evidence for the ON-cycle (predicted to operate during the 2nd dredge-up) significantly altering the surface abundances of the progenitor stars is found. A comparison between the N abundances in the PN and the (C+N) abundances in galactic H II regions indicates that ∼36% of the initial C is converted into N in the non-Type I PN, consistent with model predictions for the 1st dredge-up (Becker & Iben 1980). However, the high nitrogen abundances found in the Type I PN require envelope-burning of dredged-up carbon into nitrogen, during the 3rd dredge-up phase. Total C+N+O abundances are correlated with C/H for the entire sample; carbon has been enhanced by He-burnt material brought up by the 3rd dredge-up.

Key words: planetary nebulae: abundances — nucleosynthesis

1. INTRODUCTION

Elemental abundances have been derived for 68 southern galactic planetary nebulae (PN). Spectrophotometric observations were obtained at the 3.9-m Anglo Australian Telescope, with the RGO spectrograph and the IPCS as the detector. In addition, archive UV spectra of 25 PN, obtained with the IUE satellite, were also incorporated into this analysis, which is fully presented in Kingsburgh & Barlow (1994, KB94). Here, the abundances derived for the Type I PN are discussed in terms of constraining the extent of various dredge-up episodes in the progenitor star.

For each PN, the reddening, the electron temperatures and densities and the ionic abundances were derived using standard techniques. Total elemental abundances were dereived either by adding all the stages of ionization present (facilitated by the incorporation of UV data) or by using ionization correction factors based on detailed photoionization modelling of PN by Walton et al. (1994), which more accurately account for the unseen stages of ionization in the high excitation objects. Table 1 presents the average abundances[1] found for the sample, along with solar and galactic H II region abundances.

---

[1]Note that temperature fluctuations (e.g. Peimbert et al. 1993) have not been taken

Table 1: Average PN Abundances

	Type I PN	non-Type I PN	HII Reg[a]	Solar[b]	Orion[e]
He/H	0.129±.037(11)	0.112±.015(43)	0.100	0.098	0.100
O/H×10^4	4.4±1.4(11)	4.9±2.2(42)	5.0	8.5	4.0
N/O	1.20(11)	0.20(36)	0.074[c]	0.12	0.17
C/O	0.68(3)	1.31(15)	0.58[d]	0.47	0.85

Notes: All abundance ratios in Table 1 and throughout text are by number. Numbers in parentheses are the number of PN included in each average. [a] – HII region abundances from Dufour (1984, IAU Symp. No. 108, p.353), [b] – solar abundances from Grevesse & Anders (1989, AIP conf Proc 183, p.1) except for [c] – nitrogen (Grevesse et al. 1990, A&A, 232, 225) and [d] – carbon (Grevesse et al. 1991, A&A, 242, 488), [e] – Orion abundances from Rubin et al. (1991, ApJ, 374, 564).

2. TYPE I PN

2.1. General Properties

The 'Type I' PN were defined by Peimbert (1978) as being the most nitrogen- and helium-rich PN, having N/O≥1 and He/H≥0.14 by no. Type I PN generally have bipolar morphology, filamentary structure, a great range of observed ionization stages, and suffer from heavy extinction as they tend to be located close to the galactic plane. Recently Huggins(1993) has found that the degree of bipolarity of PN is correlated with the amount of molecular emission. Type I PN are thought to arise from the more massive end of the progenitor star distribution. Peimbert & Serrano (1980) estimated the mass of the progenitor star for the Type I PN NGC 2818, located in the open cluster of the same name, to be 2.1 $M_\odot$.

Peimbert & Torres Peimbert (1983) relaxed the original Type I abundance criterion of Peimbert (1978) to N/O≥0.5 or He/H≥0.125. Fig. 1 plots log(N/O) vs log(He/H) for the objects in our sample. Filled circles are the Type I PN, given by the definition below. Note that a continuous distribution of points is found; no clear separation between the Type I and non-Type I PN is immediately obvious. Let us examine the possible sources of nitrogen for these objects, to see if a physical criterion separating the Type I from the non-Type I PN can be established.

2.2. Revised Definition for Type I PN

Nitrogen enhancement of the surface of the PN progenitor star can occur at 3 stages: after the 1st dredge-up while the star is on the RGB and CN cycle products are dredged up; after the 2nd dredge-up phase, where ON cycle

---

into account here. If they were, the abundance number ratios relative to hydrogen would be somewhat increased, but the abundance ratios relative to oxygen would not be altered by more than 10% (Liu et al. 1994), and the conclusions drawn here would be unaffected.

products are brought to the surface; and throughout the 3rd dredge-up in the AGB thermal pulse phase, where, predicted for the most massive stars, the convective envelope reaches into the H-burning shell and dredges up the products of CN burning to the surface.

Following the 1st dredge-up, the maximum nitrogen abundance of the surface of the RGB star is the sum of the original C+N abundance, where the C would have been completely converted to N via the CN cycle. However, in some PN, the nitrogen abundance exceeds that of the initial C+N abundance, given by the C+N abundance of H II regions. Such PN are those which we define as Type I. In our galaxy, the criterion separating the Type I's from the non-Type I's is N/O>0.8. In the SMC and LMC, N/O> 0.2 and 0.35 respectively. In our sample we have identified 11 objects as Type I. For the non-Type I PN, we find that on average, 36% of the original C present was converted to N, in accord with the model predictions of Becker & Iben (1980).

We have not included helium in our definition for the Type I PN, as the He abundances are sensitive to collisional effects (Clegg 1987). Additionally, neutral helium is likely to be present in some objects, given that emission by neutral, singly ionized species, and molecules originates from some regions of some PN, hence some derived He/H ratios could be lower limits. The average He/H ratio for Type I PN is 1.2 times that of non-Type I PN.

2.3. Surface Nitrogen Enhancement of Type I Progenitor Stars

In order for the nitrogen abundance in Type I PN to exceed the initial C+N abundance, the progenitor's surface nitrogen abundance would be enhanced after the 2nd or 3rd dredge-ups, or both. Let us examine the signature of the 2nd dredge-up. Fig. 2 plots $\log(N/O)$ *vs* $\log(O/H)$, and if a significant amount of ON cycle products were brought to the surface by the 2nd dredge-up, an inverse trend would be expected, as N is produced at the expense of O in the ON cycle, however no trend is seen. Additionally, the mean O/H ratios for the Type I and non-Type I PN (Table 1) are the same within the errors, and also equivalent to O/H found for H II regions (Table 1). Past investigators have found an inverse trend in this diagram (e.g. Fig. 2 of Peimbert & Torres Peimbert 1983), where some objects with high nitrogen abundances also show oxygen depletion. However, as more fully discussed in KB94, we find that such trends are weighted by a few objects with low and somewhat uncertain O/H ratios, and more recent studies which incorporated UV spectra and hence more accurately derived the highest stages of ionization have found no oxygen depletion (e.g. the study of PB 1 by Kaler et al. 1991). Thus past studies relying only on optical data have underestimated the contribution from the highest stages of ionization, which can be important in the high excitation objects and thus underestimated the total oxygen abundance. We find no evidence for oxygen depletion and significant contamination of the surface abundance of the progenitor stars by products of the 2nd dredge-up, therefore the progenitior stars to Type I PN *have necessarily undergone convective envelope burning*, and dredged up nitrogen-rich material to their surfaces.

3. CARBON ABUNDANCES

We find a trend of increasing (C+N+O)/H with increasing C/H (see Fig. 7 of KB94), in agreement with the scenario of C produced via the 3α process being brought to the surface by the 3rd dredge-up. From the mean abundances of Table 1, the average increase in the total C+N+O abundance over the H II region value is 0.19 dex for both the non-Type I and Type I PN, while the increase in the total C+N abundance over the H II value is 0.38 dex for the non-Type I PN and 0.41 dex for the Type I PN. Almost all PN appear to have undergone the 3rd dredge-up phase, but only the relatively massive progenitors of the Type I PN had high enough temperatures at the base of their hydrogen convective envelopes for envelope burning to occur at a significant level.

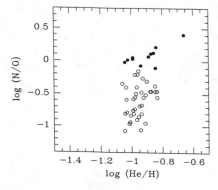 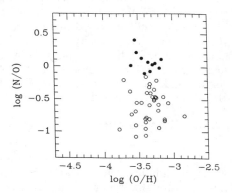

Fig. 1. — Log(N/O) vs. log(He/H). Type I PN are filled circles, non-Type I PN are open circles in all plots. A trend of increasing N/O with increasing He/H is found for only the Type I PN.

Fig. 2. — Log(N/O) vs. log(O/H). No evidence for ON cycle products is found, as predicted by Becker & Iben (1980) for the 2nd dredge-up phase, even for the Type I PN.

Acknowledgements: I thank M.J. Barlow for helpful discussions.

REFERENCES

Becker S. A., Iben I., 1980, ApJ, 237, 111
Clegg R. E. S., 1987, MNRAS, 221, 31p
Huggins. P.J., 1993, in: IAU Symp. No. 155, (Kluwer) p.147
Kaler J. B., Shaw R., Feibelman W. A., Imhoff C. L., 1991, PASP, 103, 67
Kingsburgh R. L., Barlow M. J., 1994, MNRAS, in press (KB94)
Liu, X.-W. et al., 1994, MNRAS, in press.
Peimbert, M., 1978, in: IAU Symp. No. 76, (Reidel) p. 215
Peimbert M., Serrano A., 1980, Rev. Mex. Astr. Astrophys., 5, 9
Peimbert M., Storey P.J., Torres-Peimbert S., 1993, ApJ, 414, 626
Peimbert M., Torres-Peimbert S., 1983, in: IAU Symp. No. 103, (Reidel) p. 233
Walton N.A., Barlow M.J., Clegg R.E.S. & Monk D.J., 1994, in prep.

EXPERIMENTAL RESULTS ON HIGH ENERGY COSMIC RAYS

G. Navarra

Istituto di Fisica dell'Università, Torino, Italy

Abstract. Some experimental data on the energy spectrum and composition of high energy cosmic rays are discussed.

Key Words: Cosmic rays, primary spectra, composition.

INTRODUCTION - THE DIRECT MEASUREMENTS

The energy spectrum and composition of cosmic rays provide informations on the acceleration mechanisms, the propagation processes and the chemical composition at their acceleration region. Due to the extremely wide energy range interested by the phenomenon, a significant information is expected to be carried by the energy dependence of such data. We want therefore to summarize here some experimental aspects concerning the high energy component.

The energy and nature of the c.r. particles is measured by direct observations up to $E_o \simeq 10^{13} - 10^{14}$ eV/particle by means of detectors operating on balloons and satellites. Update reviews can be found in refs. (1,2); original measurements are reported in refs. (3-9). Most interesting, at these energies, is the shape of the proton spectrum, that is steeper than the spectrum of the heavier nuclei, and possibly breaks at $\simeq 10^{13}$ eV. In fact, if interpreted with a unique power law spectrum, it has differential index $\gamma_p \simeq 2.8$, while the spectra of helium and iron primaries have respectively $\gamma_{He} \simeq 2.6 - 2.7$ and $\gamma_{Fe} \simeq 2.55 - 2.60$ (which implies that He is expected to dominate over protons at 10^{14} eV). The analysis in terms of a break leads to different results: $\gamma_{p1} = 2.62$, and $\gamma_{p2} = 3.14$ (1 and 2 meaning before and after the break at energy E_b) with $E_b = 3 \cdot 10^{12}$ eV (2), or $\gamma_{p1} = 2.74$, and $\gamma_{p2} = 3.19$ with $E_b = 4 \cdot 10^{13}$ eV (8). This shows that the uncertainties are still large, and further data from different approaches are necessary.

THE EAS MEASUREMENTS ($E_o > 10^{14}$ eV)

The 'all particle' intensity is $\simeq 5$ particles/(m^{2}d sr) above 10^{14} eV/nucleus and $\simeq 0.12$ particles/(m^{2}d sr) above 10^{15} eV/nucleus (3), that explains the difficulties of obtaining reasonable statistics by direct observations. Measurements have therefore to be performed from ground based detectors, by exploiting the cascades that the primaries induce in the atmosphere (Extensive Air Showers, EAS). In this way large collecting areas can be realized, but the measurements are constrained by the fixed atmospheric depth, i.e. target thickness (e.g. the influence of fluctuations is strongly enhanced). Moreover the interpretation of the data requires the knowledge of the hadron interaction cross sections, and the energies of interest are at the limits of the present accelerators. We have therefore to work on extrapolations of the collider measurements, and not only in energy, but also in kinematics (i.e. in the extreme fragmentation region) and projectiles (nuclei in spite of protons and antiprotons). Experiments have therefore both to study the cosmic ray primary energy and composition, and test the interaction models[1].

The primary **energy spectrum** is mainly obtained through the measurement of the electromagnetic component, i.e. through the spectrum of size (Ne = total number of charged particles at the observation level). The size spectra obtained from different detectors show a steepening ('knee'(13)) at the intensity $I_k \cong 3 \cdot 10^{-7}$ $m^{-2} s^{-1} sr^{-1}$. This scales in Ne with the atmospheric depth, as expected from its occurring at fixed primary energy. The size data (Ne) are converted to primary energy (Eo) following different techniques. The 'all particle spectrum', so obtained by different experiments, shows the change in slope at primary energy $Eo \cong (2-3) \, 10^{15}$ eV, from differential index of the power law spectrum $\gamma_1 = 2.7$, to $\gamma_2 = 3.0$ (14,15). The question is opened, whether such steepening is preceded at $\cong 1$ decade lower energies by a flattening ('bump'), that is not well defined having different features in different measurements.

Following a different approach, the size spectrum obtained by the quoted EAS-TOP experiment has been interpreted by means of the extrapolation of the direct measurements, and an hadronic interaction model, derived from an extrapolation of the collider data (16). Such procedure provides a good consistency between the data and the expectations (17).

At the highest energies experimental data have been reported by the Haverah Park, Yakutsk, Akeno (e.m. detectors), and Fly's Eye (atmospheric fluorescent light detector). The measured energy spectra are similar in shape, while the maximum difference in the absolute intensities at $Eo \cong 10^{19}$ eV is $\cong 50$ %. The spectrum has a constant slope from the knee up to $Eo > 10^{17}$ eV ($\gamma = 3.02$), steepens at $5 \, 10^{17} - 10^{18}$ eV ($\gamma = 3.20$), and becomes flatter above $5 \, 10^{18} - 10^{19}$ eV ($\gamma = 2.6$) (18), these features being significant at $\cong 4$ s.d. level. At these energies the overhall statistics is quite poor (7 events above 10^{20} eV). The highest energy event ($Eo \cong 3 \, 10^{20}$ eV) has been recorded by the Fly's Eye array on 15 October 1991 (19).

Concerning the **primary composition**, the relative abundances obtained at

---

[1] EAS arrays tend therefore to include detectors of all the EAS components. As an example the EAS-TOP array (10) operating at Campo Imperatore, 2000 m a.s.l., above the underground Gran Sasso Laboratories includes detectors of: the electromagnetic, GeV muon, hadron, atmospheric Cherenkov light, radio components of EAS, and has the additional possibility of measuring in coincidence with the muon detectors ($E_\mu > 1.4$ TeV) operating in the underground G.S. laboratory (11,12).

10^{14} eV/nucleus from the extrapolation of the direct measurements are shown in Tab. 1 (refs. 1,13,26), together with the low energy composition (LEC). In general the muon multiplicity distributions recorded deep underground (e.g.: Baksan, at E_μ > 220 GeV (21), and NUSEX, at E_μ > 3 TeV (22)), sensitive in this energy range, are compatible with the low energy composition, although are not sensitive to the differencies of the spectral indexes. A widely used method for studying the primary composition at energies Eo > 10^{14} eV is provided by the Nµ-Ne correlated data (for E_μ << Eo, the expected dependence of Nµ over Ne, as a function of the primary mass A, being: Nµ = k $A^{(1-\alpha)}Ne^\alpha$, with $\alpha \cong$ 0.8). The experiment has been performed at different muon energy thresholds from a few GeV (23) to 100 GeV (24), 220 GeV (25), and 1.4 TeV (26). These data below the 'knee' are also compatible with a mixed composition, as obtained from the extrapolation of the low energy spectra.

Above the knee the requirement of a mixed composition, including a 'significant' fraction of heavy (H) and very heavy (VH) nuclei, is supported by different data: the combined measurement of the e.m., muon, and hadron content of EAS (27), the rate of very large muon bundles (Nµ > 3000, for Eµ >220 GeV), recorded at the Baksan underground laboratory (28), the fluctuations of Nµ for fixed Ne (29). On other side, limitations to the fraction of H + VH primaries are required from measurement (25), and mostly from the muon multiplicity distribution recorded underground by MACRO (E_μ > 1.4 TeV), that restricts the fraction f(H+VH) to about 15 % of the total rate at Eo $\cong 10^{16}$ eV (30)[2].

An independent and different approach is provided by the study of the longitudinal development of the cascades, as can be deduced from the observation of the optical emission (Cherenkov or fluorescent light) in the atmosphere. The geometrical reconstruction can be obtained through time (31) or angular measurements (19,32). The measurement provides the height of the shower maximum Xm in the atmosphere. The relation between <Xm>, primary energy Eo, and atomic number A is: <Xm> = C + (1- B) Xo ln (Eo/A), where Xo is the radiation length and B and C are obtained from the interaction model (33). Data are generally consistent with the expectations from a mixed composition over the whole energy range 10^{15}-10^{19} eV (34).

TABLE 1. Extrapolated compositions at 10^{14} eV/nucleus.

A	1 (p)	4 (α)	14 (M)	24 (H)	56 (VH)	
LEC (20)	41	22	13	11	13	%
Ref. (1)	20	36	19	12	13	%
Ref. (17,26)	26	43	13	5	13	%

---

[2] Following the combined MACRO/EAS-TOP data (26), at these energies, extreme light and VH compositions can be excluded with good confidence levels. The presence of nuclei with A > 4 is required with c.l. $\cong$ 99.5 %.

By means of such technique the Fly's Eye group has observed a significant change, correlated with the 'flattening' of the spectrum at Eo $\cong 5 \cdot 10^{18}$ eV (19), both for <Xm> and d <Xm>/dEo. In fact the depth of <Xm> is observed to shift from 630 gr cm^{-2} at $3 \cdot 10^{17}$ eV (expected for heavy primaries) to 770 gr cm^{-2} at $2 \cdot 10^{19}$ eV (compatible with light primaries; the extragalactic component?).

While the problem of cosmic ray composition at high energies is far from being solved, it has to be noticed that the presence of all nuclear species, including H and VH primaries, seems to be required up to primary energies of at least 10^{18} eV to explain the main features of different observations. The new generation of experiments, from which we expect significant improvements, is based on new methods, and a better definition of the events through simultaneous and accurate measurements of the different EAS components.

REFERENCES

1. S. Swordy, Proceedings of the XXIII International Cosmic Ray Conference (I.C.R.C.), Rapporteur Paper, Ed. D A Leahy, RB Hicks & D Venkatesan, World Scientific, 243 (1993).
2. V.I. Zatsepin, ibid., Highlight Paper, 439.
3. N.L. Grigorov et al., Proc. 12th I.C.R.C., 5, 1746 (1971)
4. M.J. Ryan et al., Phys. Rev. Lett., 28, 15 (1972)
5. M. Simon et al., Ap. J., 239, 712 (1980)
6. J.J. Engelmann et al., Astron. Astrophys., 148, 12 (1985)
7. D. Muller et al., Ap. J., 374, 356 (1991)
8. T.H. Burnett et al., Ap. J., 349, L25 (1990); K. Akasimori et al., Proc. XXII I.C.R.C., 2, 57 (1991)
9. M. Ichimura et al., Phys. Rev. D, 48, 1949 (1993)
10. EAS-TOP Coll. (M. Aglietta et al.), Nuovo Cimento C, 9, 262 (1986)
11. MACRO and EAS-TOP Coll. (R. Bellotti et al.) Phys. Rev. D, 42, 1396 (1990)
12. EAS-TOP and LVD Coll. (M. Aglietta et al.), Nuovo Cimento A, 105, 1815 (1992)
13. G.B. Khristiansen and G.V. Kulikov, JETP 35, 635 (1958)
14. M. Nagano et al, J. Phys. G, 10, 1295 (1984)
15. A.A. Watson, Proc. 19th I.C.R.C., 9, 111, Rapp. Paper (1985)
16. C. Forti et al, Phys. Rev. D, 42, 3668 (1990)
17. EAS-TOP Coll., Proc. XXIII I.C.R.C., 4, 247 (1993); Nuclear Physics B, 35, 254 (1994)
18. M. Teshima, Proceedings of the XXIII I.C.R.C., Rapporteur Paper, 257 (1993).
19. D.J. Bird et al., Phys. Rev. Lett., 71, 3401 (1993)
20. S. Hayakawa, "Cosmic Ray Physics", 538 (Wiley Interscience, 1969)
21. E.V. Budko et al., Proc. XIX I.C.R.C., 8, 24 (1985)
22. M. Aglietta et al., Nuclear Physics B, 14 B, 193 (1990)
23. e.g. G.B. Khristiansen et al, "Cosmic Radiation of Very High Energy", Atomizdat, 13º (1975)
24. V.V. Vashkevich et al, Yadernaya Physica, 47, 1054 (1988)
25. B. S. Acharia et al, Proc. 18th I.C.R.C., 9, 191 (1983)
26. MACRO and EAS-TOP Coll., Nuclear Physics B, 35, 257 (1994); Physics Letters, in press
27. T.V. Danilova et al., J. Phys. G, 19, 429 (1993)
28. V.N. Bakatanov et al., Pis'ma Eksp. Teor. Fiz. 56, No 5, 237-241 (1992)
29. G.B. Khristiansen et al., Astroparticle Physics, 2, 127 (1994)
30. MACRO Coll. (S.P. Ahlen et al.,) Phys. Rev. D46, 895 (1992)
31. Yu. A. Fomin and G.B. Khristiansen, JETP Lett., 45, 56 (1978)
32. EAS-TOP Coll. (M. Aglietta et al.), Nuovo Cimento C, 16, 813 (1993)
33. J. Linsley, Ap. J., 235, L167 (1980)
34. see e.g. the review of B.A Khrenov, Nuclear Physics B, 33A,B, 18 (1993)

Recent Results of the GALLEX Solar Neutrino Detector and their Implications

Wolfgang Hampel

Max-Planck-Institut für Kernphysik
P.O. Box 103980, 69029 Heidelberg, Germany

for the GALLEX Collaboration[†]

Abstract. Data of the first 30 solar neutrino runs of the GALLEX detector, covering the time period from May 1991 to October 1993, are presented along with a discussion of their implications.

INTRODUCTION

The long-term averaged solar neutrino production rate measured with the radiochemical Homestake ^{37}Cl Detector is 2.55 ± 0.17 (stat.) ± 0.18 (syst.) SNU (1) whereas current Standard Solar Models (SSMs) predict values between 6.5 SNU (2) and 8.0 SNU (3). Besides a ∼ 15% contribution from ^{7}Be neutrinos the overwhelming part (∼ 80%) of the calculated rates for the Homestake experiment is due to ^{8}B neutrinos. The discrepancy between experiment and theory, called the "Solar Neutrino Problem" (SNP), is thus usually attributed to a deficiency of these ^{8}B neutrinos. This conclusion has been confirmed by the Kamiokande water Cerenkov detector (4) which observes only [0.51 ± 0.04 (stat.) ± 0.06 (syst.)] of the ^{8}B neutrino flux predicted by the SSM of (3) (the corresponding fraction is 0.65 if compared to the SSM of (2)).

Both detectors have energy thresholds much above the maximum energy (0.42 MeV) of the pp neutrinos which are produced in the primary fusion reaction in the sun. Detection of these neutrinos has been reported for the first time two years ago by the GALLEX Collaboration (5). Here we present updated results from the GALLEX detector and discuss their implications in view of the SNP.

RECENT RESULTS OF GALLEX

The radiochemical GALLEX detector, located at the Gran Sasso Underground Laboratory in Italy, is based upon the neutrino capture reaction ^{71}Ga(ν_e,e$^-$)^{71}Ge. The expectations from the above mentioned SSMs are 123

[†] See reference (6)

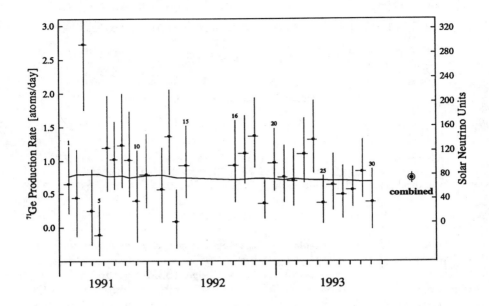

Fig. 1: Results of the first 30 solar runs of GALLEX (see text)

SNU (2) and 132 SNU (3), respectively. Relative contributions to these signals are 56–60% from pp and pep neutrinos, ~ 26% from ^{7}Be neutrinos and 8–10% from ^{8}B neutrinos. Up to July 8, 1994, GALLEX has carried out 39 solar neutrino runs, 27 blank runs and three ^{51}Cr neutrino source runs. Here we report the results of the first 30 solar runs, performed from May 1991 to October 1993. These data have been published recently (6). Details of the experimental procedures will not be repeated here, see for example (5,7,8).

The results of individual runs are plotted in Figure 1. Since in a single run on the average only 4.5 solar neutrino events are observed, there are rather large statistical fluctuations from run to run. If all 30 runs are combined, a ^{71}Ge production rate of 0.71 atoms per day or [79 ± 10 (stat.) ± 6 (syst.)] SNU is obtained (data point labelled "combined" in Figure 1).

Another gallium solar neutrino experiment is carried out by the Russian-American SAGE Collaboration (9). Initially, SAGE reported rather low ^{71}Ge production rates, see for instance Figure 3 in Ref. (6). On the other hand, the most recent SAGE average, based on 21 runs, is [74 +13/-12 (stat.) +5/-7 (syst.)] SNU (10) which is in good agreement with the above GALLEX result. It should be noted, however, that the overall efficieny factor resulting from the experimental conditions for these 21 SAGE runs is only 49 % of that for the 30 GALLEX runs.

IMPLICATIONS OF THE GALLEX RESULT

The GALLEX signal is only (60-64)% ± 10% of the SSM predictions (2,3), thus GALLEX confirms the SNP at a level of more than 3.5σ. In addition to the ^{8}B neutrino flux deficit, however, the GALLEX data suggest also a problem for ^{7}Be neutrinos. A conservative estimate based on the only assumption that the energy generation by hydrogen burning is in equilibrium with the solar luminosity yields an upper limit (2σ) for the ^{7}Be neutrino flux which is only 70 % of the SSM flux (3). In the same way the GALLEX result also allows to set an experimental upper limit (2σ) of 3.6 % on the CNO cycle contribution to the energy generation in the sun.

We have investigated the GALLEX data for time variations. Here we report on a search for an anticorrelation of the ^{71}Ge production rate with the sun spot activity. There have been indications that such an anticorrelation exists in the Homestake data (see for instance (11)). In this connection it is interesting to note that though the GALLEX data presented here have been accumulated only over a small fraction of the eleven-year solar cycle, the sun spot activity in that period covered almost the whole range of values between solar maximum and solar minimum.

In our maximum likelihood analysis we used the model which has been applied to the Homestake data by Filippone and Vogel (11). The ^{71}Ge production rate $P(t)$ is described as a function of the sun spot number $SN(t)$ by $P(t) = a + b\,[(SN(t) - \overline{SN})/\overline{SN}]$, where a and b are free parameters and $\overline{SN}$ is the sun spot number averaged over the GALLEX data taking period. The output from that model is $a = 82 \pm 11$ SNU and $b = 10 \pm 27$ SNU, it is also plotted in Figure 1 (solid line). Note that b is positive which implies that there is a (small) correlation with the solar activity rather than an anticorrelation. However, the amplitude b is in any case much smaller than its error. We therefore conclude that within present statistics there is no indication for a variation of the GALLEX signal with the sun spot activity.

COMBINED ANALYSIS OF ALL EXPERIMENTS

Finally we analyze the experimental data of all four existing solar neutrino detectors in order to obtain information about the fluxes of the different solar neutrino sources. For this we have performed a χ^2 minimization in which the expected signal for each of the solar neutrino experiments was calculated as a function of the fluxes of solar pp, ^{7}Be and ^{8}B neutrinos and then compared to the measured values. There were no constraints on the fluxes from solar models except that the resulting neutrino fluxes from all sources were required to be consistent with the observed solar luminosity.

The results are presented in Table 1. Errors were estimated by a Monte Carlo simulation. In the case of the ^{7}Be neutrino flux the output from the minimization was always zero, therefore an upper limit (95% c.l.) is given. It is clear from Table 1 that independent of the experiments selected for the

Table 1: Combined χ^2 fit for ^{7}Be and ^{8}B neutrino fluxes (see text).

Experiments included in the fit [a]	Best fit neutrino fluxes [b]	
	^{7}Be [c]	^{8}B
H K	< 0.61	0.42 ± 0.06
H G	< 0.70	0.36 ± 0.06
H G S	< 0.49	0.36 ± 0.06
K G	< 0.64	0.50 ± 0.10
K G S	< 0.46	0.49 ± 0.10
H K G	< 0.40	0.42 ± 0.06
H K G S	< 0.33	0.42 ± 0.05

[a] H = Homestake, K = Kamiokande, G = GALLEX, S = SAGE
[b] in units of the Standard Solar Model fluxes (3)
[c] upper limit (95% confidence level)

analysis, the outcome is always the same: ^{8}B neutrinos have been detected at a level of 40–50% of the SSM flux, whereas there is no room for a sizeable ^{7}Be neutrino flux contribution to the measured data. Since solar models have difficulties to account for this observation, neutrino mixing is the more attractive solution to the SNP (see for instance (12)).

REFERENCES

1. Lande, K., in Proceedings 16th Int. Conf. on Neutrino Physics and Astrophysics, Eilat, Israel, May 29 – June 3, 1994 (to be published).
2. Turck-Chièze, S., and Lopes, I., Ap. J. **408**, 347 (1993).
3. Bahcall, J.N., and Pinsonneault, M.H., Rev. Mod. Phys. **64**, 885–926 (1992).
4. Suzuki, Y., in Proceedings 16th Int. Conf. on Neutrino Physics and Astrophysics, Eilat, Israel, May 29 – June 3, 1994 (to be published).
5. Anselmann, P. et al., (GALLEX Collab.), Phys. Lett. B **285**, 376–389 (1992).
6. Anselmann, P., Hampel, W., Heusser, G., Kiko, J., Kirsten, T., Laubenstein, M., Pernicka, E., Pezzoni, S., Rönn, U., Sann, M., Schlosser, C., Wink, R., Wojcik, M., v. Ammon, R., Ebert, K.H., Fritsch, T., Hellriegel, K., Henrich, E., Stieglitz, L., Weirich, F., Balata, M., Ferrari, N., Lalla, H., Bellotti, E., Cattadori, C., Cremonesi, O., Fiorini, E., Zanotti, L., Altmann, M., v. Feilitzsch, F., Mößbauer, R., Schanda, U., Berthomieu, G., Schatzman, E., Carmi, I., Dostrovsky, I., Bacci, C., Belli, P., Bernabei, R., d'Angelo, S., Paoluzi, L., Bevilacqua, A., Charbit, S., Cribier, M., Gosset, L., Rich, J., Spiro, M., Stolarcyk, T., Tao, C., Vignaud, D., Hahn, R.L., Hartmann, F.X., Rowley, J.K., Stoenner, R.W., Weneser, J. (GALLEX Collaboration), Phys. Lett. B **327**, 377–385 (1994).
7. Anselmann, P. et al., (GALLEX Collab.), Phys. Lett. B **314**, 445–458 (1993).
8. Hampel, W., Phil. Trans. R. Soc. Lond. A **346**, 3–13 (1994).
9. Abdurashitov, J.N. et al., Phys. Lett. B **328**, 234–248 (1994).
10. Gavrin, V., in Proceedings 16th Int. Conf. on Neutrino Physics and Astrophysics, Eilat, Israel, May 29 – June 3, 1994 (to be published).
11. Filippone, B.W., and Vogel, P., Phys. Lett. B **246**, 546–550 (1990).
12. Hampel, W., J. Phys. G: Part. Phys. **19**, S209–S220 (1993).

Mid-infrared imaging of AGB Star Envelopes

Massimo Marengo(1), Giovanni Silvestro(1),
Paolo Persi(2), Livia Origlia(3)

(1) Istituto di Fisica Generale, Università di Torino, (2) Istituto di Astrofisica Spaziale, CNR, Frascati, (3) Osservatorio Astronomico di Torino

Abstract.
The recent development of bi-dimensional detectors operating in the mid-infrared has opened new possibilities for the observation and analysis of the high mass loss processes in the last stages of stellar evolution. Radio (CO) and maser (OH) observations show that an important rôle is played by intermediate and low mass stars (1-8 $M_\odot$) in the TP–AGB phase, and by their circumstellar envelopes of gas and dust. We analyse the problem of characterizing the chemical nature of the AGB envelopes (C-rich or O-rich) through mid-infrared observations by means of a suitable photometric system.

Key words: Stars: evolution — Stars: circumstellar shells — Infrared: sources

1. INTRODUCTION

Intermediate and low mass stars (1-8 $M_\odot$) in Asymptotic Giant Branch (AGB) stage are characterized by the formation of an optically opaque, circumstellar envelope of gas and dust, which will later evolve into a planetary nebula. Nucleosynthesis and mixing processes (third dredge-up) occuring in the star during the thermal pulsing phase (TP–AGB) involve also the external layers (Iben, 1981) and can cause the transition from oxygen- to carbon-rich envelope (Willems and deJong, 1988) as a consequence of the large variations in mass loss rates triggered by the thermal pulses themselves.

This transition is a consequence of the periodic C enrichment in the star convective envelope; observational and theoretical evidences (Kwok and Chan, 1989) show that stars in a mass range between 3 $M_\odot$ (or less) and 5 $M_\odot$ experience a sufficient number of dredge-up events to reach an abundance ratio [C]/[O]>1 before the expulsion of the convective envelope is completed. The stars in such a range of masses will lose their O-rich circumstellar envelope, developing a new one rich of carbon.

An oxygen excess in the envelope locks all carbon atoms in CO and allows the formation of silicate dust grains, which are globally responsible for a broad emission/absorption band at 9.7 μm, and an emission feature at 20 μm. On the other hand, an excess of carbon would form dust rich in graphite, amorphous carbon and a small but significant amount of SiC, which is detectable at 11.3 μm. The determination of the dust chemical abundances is thus favoured in the mid-infrared.

© 1995 American Institute of Physics

2. AGB ENVELOPES OBSERVATIONS

A collaboration between the Astronomical Observatory of Torino (OATo), the Department of Experimental Physics of the University of Torino, and the Institute for Space Astrophysics (IAS) of CNR, has made available a mid-infrared camera (TIRCAM, Tirgo InfraRed CAMera, Persi et al., 1994). The camera is equipped with a 10×64 Si:As array sensitive in the 5–25 μm wavelength range.

The photometric system of the camera is specifically designed for the observation of interstellar and circumstellar dust, providing the spectral resolution necessary to detect the known emission and/or absorption features of O-rich and C-rich dust grains.

Filters	Band (μm)	λ_{eff} (μm)	$\Delta\lambda$ (μm)	Chemical characteristics
F3	8.4–9.2	8.81	0.87	Dust continuum
F4	9.3–10.3	9.80	0.99	Silicates emission features (9.7 μm)
F5	9.8–10.8	10.27	1.01	Dust continuum
F6	11.2–12.3	11.69	1.11	SiC emission feature (11.3 μm)
F7	11.9–13.1	12.49	1.16	Dust continuum

Table 1: TIRCAM photometric system

The TIRCAM camera has been used in a few observing runs at the National Mexican Observatory of San Pedro Martir (Baja California) and at the Italian Infrared Telescope (TIRGO), in 1992–1994 years, for imaging of circumstellar envelopes around AGB stars.

The observed sources have been selected from a statistically complete sample of AGB stars extracted from the IRAS Point Source Catalogue (1985), in order to satisfy the observational constraints for an infrared camera like TIRCAM. The spatial extension of each source has been previously evaluated by a model developed on the basis of the radial brightness distribution computed by Martin and Rogers (1987) for the C-star CW Leonis.

A total number of 16 AGB stars with circumstellar envelopes (9 O-rich and 7 C-rich), and one post-AGB object have been observed. For each source the images were collected in the available photometric system and at a spatial resolution substantially limited by the telescope diffraction (2.4 and 3.4 arcsec at San Pedro Martir and at TIRGO, respectively). The complete results of the observations and the data analysis will be published in a forthcoming paper.

The obtained mid-infrared photometry was compared with the Low Resolution Spectra (LRS, 1986) measured by IRAS in the wavelength range 8–23 μm and a general good agreement was found. In Figure 1 the case of CW Leonis is shown. Besides, two of the sources (CW Leonis and R Leonis) after deconvolution with the instrumental Point Spread Function (using the iterative algorithm by Lucy, 1974), seems to be extended as a consequence of a possible spatially resolved envelope.

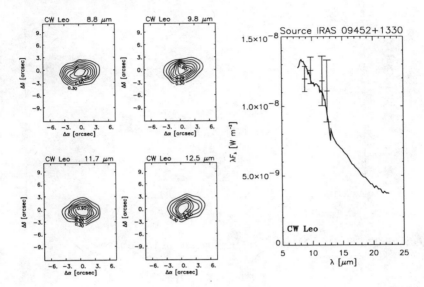

Fig. 1.— The C-rich source CW Leonis (IRC+10216) observed at 8.8, 9.8, 11.7 and 12.5 μm. The photometry of the source is compared with the IRAS LRS spectrum in the wavelength range 7–23 μm.

3. CHEMICAL CLASSIFICATION OF AGB ENVELOPES

Color-color diagrams in mid- and far-infrared allow to classify AGB and post-AGB sources on the basis of their evolutionary stage. Evolutionary sequences for O-rich, C-rich and transition objects can be drawn on the IRAS color-color diagram (van der Veen and Habing, 1988), based on the IRAS fluxes at 12, 25 and 60 μm. AGB stars in different evolutionary stages are located in different regions on the diagram, but there is not a complete separation between sources with C-rich or O-rich envelopes.

In order to obtain a chemical classification of the observed objects we have constructed a suitable color-color diagram in the TIRCAM photometric system The [8.8]−[12.5] color appears mainly related to the chemical nature of the dust in the envelope, being less than 0.6 for all O-rich envelopes, and greater than 0.6 for the C-rich sources. The only post-AGB object observed has the reddest color ($\sim$1.5) and is located in a separate region of the diagram. There are also some evidences that the [8.8]−[9.8] color can be associated with the evolutionary stage of the envelopes.

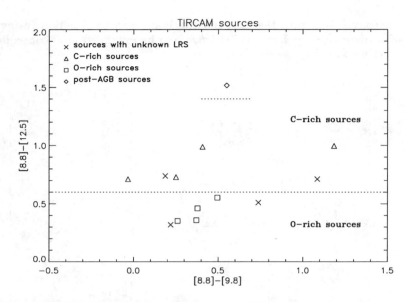

Fig. 2.— The TIRCAM color-color diagram of the observed sources

REFERENCES

Iben, I. Jr. 1981, in *Physical processes in Red Giants*, ed. I. Iben Jr., A. Renzini, p. 115, Dordrecht Reidel

IRAS Catalogue and Atlases. Atlas of low-resolution spectra, Iras Science Team 1986, A&AS, 65, 607

IRAS Catalogues and Atlases, Point Source Catalogue, 1985, US Government Pubblication Office

Kwok, S. and Chan, S. J. 1989, in *From Miras to Planetary Nebulæ: Which Path for Stellar Evolution ?*, ed. M. O. Mennessier and A. Omont, p. 297, Editions Frontieres, Gif sur Yvette Cedex - France

Lucy, L., B. 1974, AJ, 76, 754

Martin, P.,J., and Rogers, C. 1987, ApJ, 322, 374

Persi, P., Shivanandan K., Busso, M., Bonazzola, G., Corcione, L., Ferrari-Toniolo, M., Nicolini, G., Racioppi, F., Robberto, M., Tofani, G. 1994, Experimental Ap., submitted

van Der Veen, W., E., C., J. and Habing, H., J. 1988, A&A, 194, 125

Willems, F. J., and deJong, T. 1988, A&A, 196, 173

Mass and radius of the ellipsoidal variable barium star HD 121447

A. Jorissen

Institut d'Astronomie et d'Astrophysique, Université Libre de Bruxelles

Abstract. The parameters of the binary system HD 121447, the coolest known barium star with the second shortest orbital period (P = 185.7 d), have been derived from the spectroscopic orbit and the lightcurve indicating that the system is an ellipsoidal variable. A mass of about 1.6 $M_\odot$ and M_{bol} $\sim$ −2.2 follow from the assumption that the barium star is located on the upper giant branch, with the companion mass ($\sim$ 0.6 $M_\odot$) compatible with that of a white dwarf. This analysis provides the first direct determination of these physical parameters for a barium star.

Key words: Barium stars — Ellipsoidal variable — HR diagram

1. INTRODUCTION

The barium star HD 121447 is the coolest known barium star (K7IIIBa5), sometimes classified as a SC star. An effective temperature $T_{\rm eff}$ = 3800 K is derived from the $J - K$ index of 0.97 (1). With a metallicity [Fe/H] = +0.05 and overabundances of s-process elements as large as [Sr/Fe] = 1.2, HD 121447 exhibits the typical chemical peculiarities of barium stars (2).

In an effort to derive the orbital elements of all southern barium stars with strong chemical anomalies, the radial velocity of HD 121447 has been monitored since 1984 with the CORAVEL spectrovelocimeter (3), and it is included in the Long-Term Photometry of Variables (LTPV) program operating on the Danish 50 cm telescope at ESO since 1986 (4,5). HD 121447 deserves special attention because (i) it has the second shortest period among barium stars, and (ii) it is the barium star with the largest photometric variations among the sample monitored in the framework of the LTPV program.

2. LIGHTCURVE OF HD 121447

The LTPV observations are performed in the Strömgren $uvby$ system in a differential way, with HD 121699 (K0) and HD 117246 (K4III) as comparison stars (denoted by A and B in the following). The frequency of observations is about 1 per week for several months per year. The monitoring of HD 121447 has been started in February 1986, and 53 observing sequences have been collected up to August 1993. The internal accuracy on the differential magnitudes amounts to 10, 3.5, 2.6 and 1.9 millimagnitudes in the u, v, b and y bands, respectively, as derived from the rms deviation of all pairs of contiguous stars in a nightly sequence $APBPBPA$ [see (4) for details about the reduction method

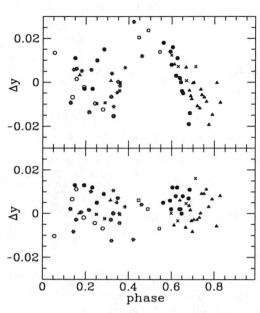

Fig. 1.— Upper panel: The y lightcurve of HD 121447 with respect to the average of the two comparison stars [i.e. $P - (A + B)/2$], as a function of the orbital phase. Phase 0 corresponds to the time of maximum radial velocity, and Δy corresponds to the residuals with respect to the average value over the whole monitoring. Different symbols refer to different orbital cycles. Measurements in the Geneva V band are represented by filled circles. The residuals in the Geneva V band were computed by adopting identical values for the average V and y magnitudes.
Lower panel: Differential $A - B$ lightcurve for the comparison stars in the y band. Symbols are as in the upper panel

and (6) for individual measurements]. Measurements have also been collected in the Geneva system during one observing campaign in April-August 1992.

Figure 1 presents the lightcurve of HD 121447 in the y band phased with the orbital period of 185.75 d. It shows a clear double-peak pattern with minima around phases 0.25 and 0.75. The noise in the lightcurve is larger than the internal accuracy and is due to the intrinsic jitter of the rather red comparison stars [see Fig. 5 in (4)]. There is thus some uncertainty on the peak-to-peak amplitude in the y band, which lies in the range $0.030 \leq \Delta y \leq 0.040$.

A period analysis of the lightcurve in the y band, based on phase dispersion minimization (7), clearly picks out a period $P_{\text{phot}} = 92.8$ d with a first-risk error of 0.2%. This period is exactly half the orbital period. This result, along with the fact that the system is the faintest when the companion is aligned with the line of sight and the absence of colour variations in $b - y$, is a strong indication that HD 121447 is in fact an ellipsoidal variable.

3. PARAMETERS OF THE BINARY SYSTEM DERIVED FROM THE RADIAL-VELOCITY AND LIGHT CURVES

The fact that HD 121447 is an ellipsoidal variable opens the way to the determination of the radius of the giant star and of the stellar masses, according to the method outlined in (8). The amplitude of the ellipsoidal light variations and the orbital elements available for a system with one observable spectrum

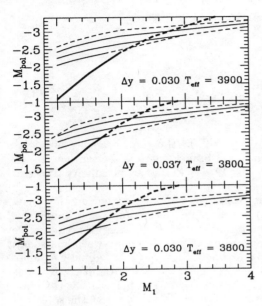

Fig. 2.— Absolute bolometric magnitude of the giant star as a function of mass for the dynamical solutions, for different choices of $T_{\rm eff}$ and Δy. The thin solid lines correspond to $M_2 = 0.5, 0.7$ and 1 $M_\odot$ (from bottom to top). Dashed lines delimit the area of possible solutions, as defined by $\sin i \leq 1$ (lower boundary) and $M_2 \leq 1.4$ $M_\odot$ in the case of a WD companion (upper boundary). In the case $\Delta y = 0.037$ mag and $T_{\rm eff} = 3800$ K, the upper left boundary is defined by the requirement that the giant does not fill its Roche lobe. The thick line represents the bolometric magnitude of a giant of given M_1 and $T_{\rm eff}$ according to evolutionary tracks (9) for solar-metallicity stars on the red giant branch (thick solid line) or on the early-asymptotic giant branch (thick dashed line)

are not sufficient to fully constrain the system parameters M_1, M_2, A, i and R_1 (respectively the masses of the giant and of its companion, the semi-major axis, the orbital inclination and the radius of the giant). Another constraint is provided by the requirement that the mass and radius of the giant derived from the dynamics of the system be consistent with the evolutionary tracks in the Hertzsprung-Russell diagram. For a given radius, the absolute bolometric magnitude of the giant can indeed be derived from the usual relation

$$M_{\rm bol} = 42.26 - 5\log(R_1/R_\odot) - 10\log T_{\rm eff},$$

with $T_{\rm eff} = 3800$ K for HD 121447 as quoted in Sect. 1. Acceptable dynamical solutions should thus have $(M_{\rm bol}, M_1)$ pairs consistent with those derived from the solar-metallicity evolutionary tracks of giants with $T_{\rm eff} = 3800$ K (Fig. 2).

Figure 2 shows that the dynamical solutions generally lie above the first giant branch, except for a narrow mass range if $\Delta y = 0.030$ mag and $T_{\rm eff} = 3800$ K. The system parameters consistent with the giant being located on the first giant branch as given by (9) are listed in Table 1. These acceptable solutions put the red giant close to the tip of the giant branch. Although the position of the giant branch depends quite sensitively upon parameters like α (the ratio of the convective mixing length to the pressure scale height), the solutions presented in Table 1 should not be affected by this uncertainty, since

Table 1: Parameters for the system HD 121447 consistent with the assumption that the giant is located on the RGB as given by (9). All symbols are defined in the text, and $T_{\text{eff}}= 3800$ K and $\Delta y = 0.030$ mag have been adopted

$1.5 \leq$	M_1 (M$_\odot$)	≤ 1.7
$0.5 \leq$	M_2 (M$_\odot$)	≤ 0.7
$68° \geq$	i	$\geq 48°$
$53 \leq$	R_1 (R$_\odot$)	≤ 61
$173 \leq$	A (R$_\odot$)	≤ 184
$0.65 \leq$	$R_1/R_{R,1}$	≤ 0.73
$-2.2 \geq$	M_{bol}	≥ -2.4

α in (9) has been chosen in such a way as to fit the giant branches of a wide range of clusters. The location of the giant branch is also quite sensitive to the metallicity. Since [Fe/H] = 0.05 for HD 121447 (2), solar-metallicity tracks as used in Fig. 2 are appropriate.

Figure 2 reveals that solutions where the giant is located on the early-asymptotic giant branch (E-AGB) are also possible in fact, and cannot be excluded on the ground of the present data. The $^{12}C/^{13}C$ ratio of 8 is suggestive of a low-mass for the giant (2), according to the trend between mass and $^{12}C/^{13}C$ ratio as derived for giants in open clusters (10). Giants in open clusters with turn-off masses above 2.2 M$_\odot$ have $^{12}C/^{13}C$ around 25, whereas for lower masses, the $^{12}C/^{13}C$ ratio decreases from 25 to below 10 as the mass decreases from 2.2 to 1 M$_\odot$. This argument would tend to reject the solutions with HD 121447 on the E-AGB as they imply $M_1 \geq 2$ M$_\odot$ and a $^{12}C/^{13}C$ ratio larger than observed. A detailed analysis will be presented in (11).

REFERENCES

1. Bessell M.S., Wood P.R., Lloyd Evans T., 1983, MNRAS 202, 59
2. Smith V.V., 1984, A&A 132, 326
3. Jorissen A., Mayor M., 1988, A&A 198, 187
4. Jorissen A., Sterken C., Manfroid J., 1992, A&A 253, 407
5. Sterken C., 1983, The Messenger 33, 10
6. Sterken C., Manfroid J., Anton K., Barzewski A., Bibo E., Bruch A., Burger M., Duerbeck H.W., Duemmler R., Heck A., Hensberge H., Hiesgen M., Inklaar F., Jorissen A., Juettner A., Kinkel U., Liu Zongli, Mekkaden M.V., Ng Y.K., Niarchos P., Püttmann M., Szeifert T., Spiller F., van Dijk R., Vogt N., Wanders I., 1993a. Second catalogue of stars measured in the long term photometry of variables program (1986 – 1990), ESO Scientific Report no. 11
7. Stellingwerf R.F., 1978, ApJ 224, 953
8. Hall D.S., 1990, AJ 100, 554
9. Schaller G., Schaerer G., Meynet G., Maeder A., 1992, A&AS 96, 269
10. Gilroy K.K., 1989, ApJ 347, 835
11. Jorissen A., Hennen O., Mayor M., Sterken C., Bruch A., 1994, A&A, submitted

Theoretical Modelling of Dust Shells around AGB Stars

E. Paravicini Bagliani, M. Marengo and G. Silvestro

Istituto di Fisica Generale, Universitá di Torino, 10125 Torino, Italy

Abstract.
A numerical code is developed, which integrates the equation of non-grey radiative transfer through a spherically symmetric dust shell around an evolved AGB star. The code computes, by an iterative method, a self-consistent thermal structure of the envelope, for multiple grain components of carbon- and oxygen-rich material. The emergent radial brightness distribution at different wavelengths is calculated taking into account the effect of non-isotropic scattering, absorption and thermal reemission by grains. Our model results, for different sets of stellar and nebular parameters, are compared with observational data on oxygen- and carbon-rich AGB stars with envelopes of different optical depths.

Key words: Stars: evolution — Stars: circumstellar shells — Radiative transfer

1. CIRCUMSTELLAR ENVELOPES OF AGB STARS

The last stages of evolution of stars with small and intermediate masses (1-8 $M_\odot$) are governed by mass loss. Thousands of stars on the Asymptotic Giant Branch (AGB) are observed to have infrared (IR) excess, due to circumstellar dust which forms in the outflow of the star. The radiation from the dust shell includes contributions from several layers having different physical parameters, and the interpretation of its complex spectrum demands appropriate theoretical modelling. Radiative transfer in circumstellar dust shells around evolved stars has been discussed by many authors (Rowan-Robinson, 1980; Griffin, 1990; Justtanont & Tielens, 1992). In the last few years, the advent of mid-IR cameras made it possible to spatially resolve some circumstellar dust shells and thus obtain information on their geometrical structure (Persi et al.,1992). New laboratory and astronomical data on the dust grains are now becoming available, that will allow more accurate chemical characterization of the sources. Our numerical code will make use of such new data for giving an interpretation of the photometric and imaging observations collected with the TIRCAM infrared camera.

2. ANALYTICAL DESCRIPTION OF THE MODEL

The model is based on the hypotheses: (1) spherical symmetry of the dust shell, with an $n(r) \propto r^{-2}$ density distribution, consistent with steady outflow at a constant velocity; (2) balance between absorption and emission by the dust grains, which ensures flux conservation in the envelope; (3) LTE dust radiation at the local temperature T(r).

Following Rowan-Robinson (1980), the radiative transfer equation is solved separately for the three components of the radiation intensity: $I_\nu^{(1)}$, the light from the central star; $I_\nu^{(2)}$, the radiation from grains; $I_\nu^{(3)}$, the scattered light. The temperature profile $T(r)$ is evaluated by an iterative method: the zero-order approximation accounts for the stellar radiation alone, then each new iteration leads to a new temperature distribution, subject to the condition of flux conservation, until $T(r)$ is found to be stabilized within the required accuracy. A procedure is employed which corrects the temperature profile in order to account for the whole radiative flux.

The code is used to calculate the temperature distribution at $\sim$60 radial points through the dust shell, and the emergent spectrum at up to 130 wavelength points from 0.4 μm to 250 μm.

The model parameters are:

- d_*, stellar distance from the Sun;
- T_*, stellar temperature;
- R_*, stellar radius;
- a , dust grain radius (around 0.1 μm);
- R_1, R_2, inner and outer radius of dust shell
- τ_{10}, optical depth of dust shell at $\lambda = 10$ μm

2.1. The model opacity profile

An excess of oxygen in the envelope favours the formation of silicate dust grains, which are globally responsible for a broad emission/absorption band at 9.7 μm, and a feature in emission at 20 μm. On the other hand, an excess of carbon would give rise to dust rich in graphite, amorphous carbon, and a small but significant amount of SiC, which is detectable through its emission feature at 11.3 μm. We use the "dirty silicate" model of Jones & Merril (1976) for the grain absorption and scattering efficiency of an oxygen-rich envelope; the graphite, amorphous carbon, and SiC opacity profiles of C-rich envelopes are from Draine & Lee (1984), Martin & Rogers (1987) and Chan & Kwok (1990) respectively.

3. MODEL RESULTS

Our model evaluates the temperature profile $T(r)$ for optically thin and thick, O- and C-rich envelopes. The spectral energy distribution λF_λ is estimated in a wide range of wavelengths for envelopes with different parameters and various dust grain opacity profiles. The spectral features are reported for optically thin and thick models of O-rich and C-rich envelopes of AGB stars.

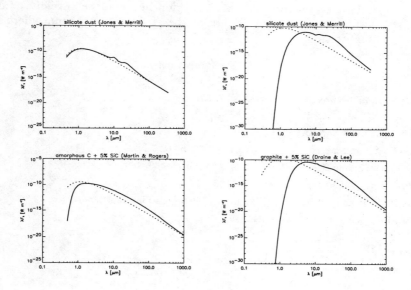

Fig. 1.— Simulations of AGB envelopes with different optical depth and composition: (A) $\tau_{10\mu m} = 0.1$, (B) $\tau_{10\mu m} = 9.0$, (C) $\tau_{10\mu m} = 0.1$ and (D) $\tau_{10\mu m} = 1.0$. The envelope's dust is O-rich in model (A) and (B), and C-rich in (C) and (D). The dotted line is the black-body spectra at the star's temperature (2500 K).

4. DIRECTIONS OF FUTURE WORK

We want to:

1. obtain a detailed fitting of the spectrum of individual sources; the dust opacity profiles will be modified in order to take into account recent astronomical and laboratory data for various materials. Comparison of our model results with observation will permit to improve estimates of mass loss rates along the AGB sequence;

2. start a program of coordinated IR camera observations and model building of a wide grid of spectra to account for the spectral profiles of cool stars with different characteristics;
3. model the variability of IR spectra caused by stellar (Mira-like) variability;
4. model the dynamics of mass loss, in order to follow the time development of the geometrical structure of circumstellar shells with different radial density distributions for the dust. This will require a detailed investigation of the region between the envelope's inner radius and the outer layers of the stellar atmosphere, where the dust forms.

REFERENCES

Chan, S. J., and Kwok, S. 1990, A&A, 237, 354

Draine, B. T., and Lee, H. M. 1984, ApJ, 285, 89

Griffin, I. P. 1990, MNRAS, 247, 591

Jones T. W., and Merrill, K. M. 1976, ApJ, 209, 509

Justtanont, K., and Tielens, A. G. G. M. 1992, ApJ, 389, 400

Martin, P. J., and Rogers, C. 1987, ApJ, 332, 374

Persi, P., Shivanandan K., Busso, M., et al. 1992, OATo internal report, No. 22/92

Rowan-Robinson, M. 1980, ApJS, 44, 403

Is The Cosmic Ray Spectrum Form By Split of Nuclei ?

W. Tkaczyk

*Department of Experimental Physics University of Łódź,
ul. Pomorska 149/153, 90236 Łódź*

Abstract. The paper proposes a model in which the cosmic ray spectrum is formed by propagation processes in magnetic and photon fields of Galaxy, for energy range from JACEE to Fly's Eye experiments. The turbulence effect of Galactic magnetic field on the propagation of cosmic ray nuclei has been investigated. We have found that the way of propagation of the protons (and nuclei respectively) is changed at the same energies where JACEE and Fly's Eye experiments observe alteration in the spectrum of protons. The diffusion equations for the nuclei in field of soft photons and turbulent magnetic field have been solved. We argue that the anisotropy of secondary protons should follow the heavy nuclei one. A good agreement with the whole features in total cosmic ray energy spectrum was obtained for parameters widely accepted for the cosmic environment.

INTRODUCTION

New results on energy spectrum above 10^{17}eV are reported by the Fly's Eye (1) group. The energy spectrum steepens above 10^{17}eV and again flattens above $10^{18.5}$eV. The data and Monte Carlo simulation show that at dip energy region the composition change to lighter with energy. The anisotropy above $2 \div 5 \times 10^{18}$eV does not contradict expectation, if we assumed heavy component (i.e. pure iron) of cosmic ray in turbulent Galactic magnetic field (2). The existence of super high energy cosmic ray above expected Greisen Zatsepin cutoff causing by interaction protons with 2.7K micro wave background photons indicates on difficulties their origin in extragalactic models.

JACEE group reported energy spectra for several (p, He, CNO, Ne-S, Fe) components (3,4). The average mass number of primary cosmic ray increases rapidly with energy at around 100TeV. The compilation of data from direct measurements shows the ratio of the flux of He/proton seems to increase with increasing energy (2). In fact, the protons seem to show some sign of a spectral "cut off" around 10^{14}eV. The paper proposes the Galactic model of cosmic rays origin in whole energy bands from GeV to end of the spectrum. The model explain the anisotropy and mass composition data

MODEL DESCRIPTION

We have considered the model of galactic cosmic ray sources with the following assumptions:

Assumptions

- The cosmic ray sources accelerate nuclei with mass number A=1÷56 (from hydrogen to iron) continuously with spectrum $Q(E) = k \cdot E^{-\gamma}$ where γ-is the spectral index of accelerate nuclei. We are following JACEE results using value γ=2.75 for protons and γ=2.55 for nuclei from helium to iron groups.
- The spectrum of soft photons filling cosmic space is a type of thermal bremsstrahlung with temperature kT=20eV. The concentration of plasma is equal to unity and photons residential time τ=R/c=1 sec.
- The relative mass composition of produced nuclei by sources we have took as observe JACEE group (3,4) et energy 10^{14}eV.
- The nuclei exiting source part through magnetic field composed with regular (concentric ring) and turbulent components $B = 7\%B_{regular} + 93\%B_{turbulent}$
- The cosmic ray sources displaced in disc are inside the sphere of irregular field with radius R_H

Propagation of nuclei through the radiation fields

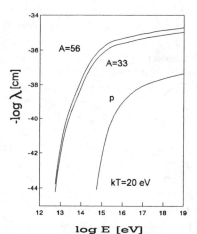

FIGURE 1 The reciprocity of the mean free path on fragmentation of nucleus with mass number A=56, A=33 and photoproduction process (curve p).

A useful summary of the cross-sections and other parameters for photons-nucleus reactions has been give by Karakuła and Tkaczyk (5). A mean free path λ for fragmentation of nuclei, from deuterium to iron, was calculated using that approximation. The results of nucleus fragmentation in the bremsstrahlung radiation bath with temperature kT=20 eV are shown in Figure 1, where λ^{-1} is plotted against energy for nuclei with mass A=56 and A=33.

Also shown in Figure 1 is the reciprocity of the proton mean free path for photo production process, (curve p). The solutions of the diffusion equation for particular nuclei and protons have been found and a computer code has been made using as input parameters specified in the assumptions to the proposed model. The groups of nuclei are collected as follows PP-primary protons, PS-secondary protons (from photo disintegration), He (A=4), CNO(A=14), Ne-S(A=24), Fe(A=56).

Propagation of nuclei in the turbulent magnetic field

The diameters of disk are the high z_0=0.2 kpc and the radius R_0=15 kpc. The disk

magnetic lines are concentric ring with constant strange inside disk $B_{regular}$ and change the sign up and below plain z=0. The strange of regular component outside the disk decreases as follow:
$B = B_{regular} \cdot \exp(\frac{z-z_o}{z_o}) \cdot \exp(\frac{r-R_o}{R_o})$.
The irregular component has the random direction and strange in the range $0 \div B_{regular}$ The numbers of N=1000 trajectories, for particular energy, of protons or nuclei, starting randomly from the disk, have been calculated. The calculations of trajectory of nuclei were stopped when exceed radius of halo R_H=50000 light years. The distances, real RW[ly] passed by nuclei and geometrical distance dg[ly] between starting and stopped points are calculated.

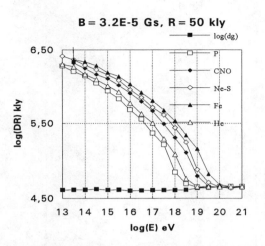

FIGURE 2. The average real distances RW[ly] passed by nuclei and geometrical distance dg[ly] between starting and stopped points.

RESULTS

The average values for RW and dg distances have been calculated as function of energy for particular group of nuclei. Figure 2 shows RW and dg distances. The leakage probability from the halo region is proportional to reciprocal of RW. The leakage effect has been taken in to account for the nuclei after its destructions in the field of photons. Figure 3 shows the cosmic ray energy spectrum data points and predictions by the model for the total (full line) and particular group of spectra (dashed lines).

DISCUSSION AND CONCLUSIONS

The total energy spectrum from the model indicates the same features as experimental data. The steepening in the spectrum of protons, observed by JACEE experiment at energy $\sim 10^{14}$eV is caused by theirs non diffusive leaked from acceleration region. The "knee" in the energy spectrum is due to photo disintegration process of nuclei, in a soft photon field type thermal bremsstrahlung with kT=20eV. A break among 2×10^{18} to 5×10^{18}eV observed in the energy spectrum of Fly's Eye is caused because above this energy the trajectories of protons are strike line in Galactic magnetic field, so the propagation length is simple geometric path of nucleus and does not depend from energy. The trajectories of iron are bending by magnetic field up to energy 10^{20}eV, so the secondary proton from destructions of heavy nuclei misplaces the disc

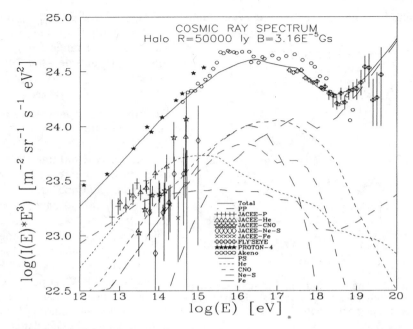

FIGURE 3. The cosmic ray energy spectrum data points and predictions by the model for the total (full line) and particular group of spectra (dashed lines).

displacements of sources. The argument against the galactic origin of cosmic ray supper high energy is enhancement diameter of sources by the giro-radious. The extended halo and the acceleration of nuclei in source act the above arguments weaker. The proposes strong turbulence of
magnetic field in extended halo do fluxes of heavy nuclei and subsequently secondary protons isotropic as is observed

ACKNOWLEDGEMENTS

This work was supported by the University of Łódź grant UŁ 505/468.

REFERENCES

1. Bird D. J., et. al., Phys. Rev Letters **71**, 3401-3404 (1993)
2. Teshima M., "Origin of Cosmic Ray above 10^{14}eV": in *Proceedings of 23rd ICRC*, invited, rapporteur & highlight papers, 1993, pp 257-278
3. Asakimori K., et. al., "Cosmic Ray Composition and Spectra:(I). Protons The JACEE Collaboration": in *Proceedings of 23rd ICRC* 1993, **2**, pp 21-23
4. Asakimori K., et. al., "Cosmic Ray Composition and Spectra:(II). Helium and Z>2 The JACEE Collaboration": in *Proceedings of 23rd ICRC* 1993, **2**, pp 25-27
5. Karakuła S. and Tkaczyk W., 1993 Astroparticle Phys., **1**, 229-237.

Cosmic Ray Tracking with Streamer Tubes

U. Brandt[2], K. Daumiller[1], P. Doll[1], R. Gumbsheimer[1],
H. Hucker[1], H.O. Klages[1], P. Kleinwächter[3], G. Kolb[2], J. Lang[2]

[1] *Nuclear Research Center Karlsruhe, 76021 Karlsruhe, Germany*
[2] *University of Karlsruhe, 76128 Karlsruhe, Germany*
[3] *Research Center Rossendorf, 01474 Dresden, Germany*

Abstract. The EAS experiment KASCADE (KArlsruhe Shower Core and Array DEtector) (1) which is presently under construction at the Karlsruhe Nuclear Research Center consists of an extended array of detectors for the electron / photon component and for muons far from the shower core and of a compact hadron calorimeter. In addition, close to the calorimeter a large area muon detector will be implemented employing streamer tubes and persuing the tracking concept.

MUON DETECTOR

The precise determination of the muon content in extensive air showers is of crucial importance for the study of the primary elemental composition of cosmic rays. A clean separation of the muon signals from background due to electromagnetic or hadronic processes is important. Therefore, in addition to a distributed array of heavily shielded scintillators and a central detector, a tracking detector system under a ~ 15 r.l. shielding is set up for the KASCADE experiment (1).

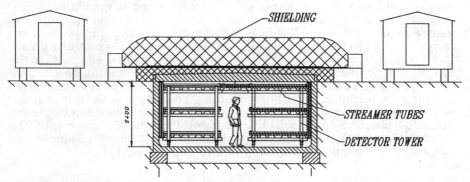

Figure 1. Muon detector tunnel located between two detector stations.

3) supported by BMFT, 211-4006-06DR 103

Figure 1 shows that the streamer tube detectors will be arranged in three horizontal planes and will also cover the walls of a tunnel. The streamer tube planes are mounted on a support structure (tower) which carries the tubes of 4m length. Nine pairs of towers will be housed by the tunnel making a total detector area of 144 m^2 for vertical particles. The total gas volume of 4,5 m^3 will be supplied by an argon-carbondioxid-isobutane gas mixture. The operation stability of the detector system under variations of pressure and temperature is presently investigated.

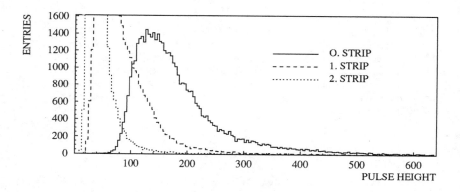

Figure 2. Induced charge spectra for adjacent strips.

Four modules are carried by a tower. Each module (2 x 4 m^2) will be read out on 96 pairs of anode wires and on 192 strips oriented perpendicularly to the wires and on 96 strips oriented under 60° to the wires. Figure 2 shows induced charge spectra on adjacent strips with the central strip (0. strip) exhibiting the largest values. The streamer tube cell geometry is 9 x 9 mm^2. Because of the wire grouping and pitch of the strips (~ 20 mm) the digital readout for both coordinates provides an angular resolution of about 1°. This resolution is sufficient to identify muons and to discriminate against 'punch-through' events and accompanying particles as indicated by Monte Carlo calculations for extensive air-showers initiated by 10^{15} eV protons. Therefore, with the construction of a large area muon detector close to the central calorimeter, muon detection for KASCADE will be considerably improved. Especially the comparison of different detection techniques will be enabled employing the tracking concept like e.g. the GRAND experiment (2).

REFERENCES

1. P.Doll, K. Daumiller, H.O. Klages, Nucl. Instr. and Meth. **A323** (1992) 327
   P.Doll, K. Daumiller, H.O. Klages, G. Mondry, H. Müller, Nucl. Instr. and Meth. **A342** (1994) 495
2. J.Gress, Y. Lu, A. Anagnostopoules, J. Kochocki, J. Poirier, S. Mikocki, A. Trzupek, Nucl. Instr. and Meth. **A302** (1991) 368

SECTION II

NUCLEAR ASTROPHYSICS

Charged-Particle Thermonuclear Reactions

H.-P. Trautvetter, M. Möhle and S. Schmidt

Institut für Experimentalphysik III, Ruhr Universität, 44780 Bochum, Germany

1. INTRODUCTION

In this presentation no details on specific problems will be given, they could be found in recent review articles [1],[2] or text books e.g. [3] and references therein. It is well known that the energy dependence of charged particle reaction cross sections for low energies are governed by the coulomb penetration factor and the De Broglie wave length. It is therefore common to transform the cross section into the so called astrophysical S-factor: $S(E) = \sigma(E)E\exp(2\pi\eta)$, where η is the Sommerfeld parameter ($\eta = Z_1 Z_2 e^2/\hbar v$).

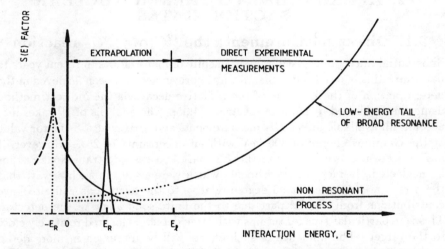

Figure 1. The steep drop of $\sigma(E)$ due to the Coulomb barrier results in a lower limit in energy, E_l, below which direct measurements are not feasible. The $S(E)$ data are extrapolated to zero energy with the guidance of theory and any other available knowledge.

A typical S-factor curve is shown in fig. 1,[1]. The low energy tail of broad resonances or non resonant processes can be studied experimentally down to a limiting energy E_l. However, the most interesting region for astrophysics is the vicinity of the Gamow peak, which arises at the overlap region of the Maxwell-Boltzmann distribution of the stellar plasma and the exponentially decreasing cross section [1],[3]. It is this region where the reaction rate is needed, but it is also the region which is hardly accessible by experiment. Therefore, the non resonant contributions have to be extrapolated (dotted line in figure 1) and the properties of the subthreshold resonances as well as other resonances located near the Gamow energy have to be investigated by other means. In chapter 2. three examples of experimental approaches will be presented: i) the direct measurement on $^{12}C(\alpha,\gamma)^{16}O$; and ii) the indirect method using direct capture or transfer reactions to determine the reaction rate will be discussed on the example of the $^{17}O(p,\alpha)^{14}N$ reaction (sect.2.2.1) and the $^{22}Na(p,\gamma)^{23}Mg$ reaction (sect. 2.2.2.).

2. APPROACHES TO THERMONUCLEAR REACTION RATES

2.1. Direct measurement: the $^{12}C(\alpha,\gamma)^{16}O$- reaction

Substantial uncertainties still remain in spite of many efforts in recent years to determine this reaction rate. Significant improvement has been achieved in the determination of the $E1$- part of the radiative decay via the indirect method using the β-delayed α- particle spectrum [4],[5]. The analysis of [4] takes into account all available direct $E1$- measurements and arrives at a S- factor value at the Gammov energy of 300 keV with an uncertainty of 26%. However, it has also been shown [6,7], that the $E2$- part and cascade transitions cannot be neglected. In fact, all experiments which were set up in such a way, that the γ-rays are observed at 90 degree relative to the beam axis will also have contributions from the $E2$ part due to the finite solid angle of the detector. This is in particular true for set ups with large detectors located relatively close to the target. The influence of such effects will be discussed in more detail in the contribution of Roters et al. in these proceedings, together with a new attempt to disentangle experimentally the contributions of the two radiation types $E1$ and $E2$.

2.2. Indirect method

It is possible to obtain in some cases information on the resonant thermonuclear reaction rate even if a resonance is located in the low energy region which is not accessible to direct experimental investigations (figure 1). Such a rate has the form $< \sigma v > \propto \omega\gamma \exp(E_r/kT)$, where $\omega\gamma \propto \Gamma_{in}\Gamma_{out}/\Gamma_{tot}$ and ω is a statistical factor. The energy of excited states can often be determined precisely by γ-spectroscopy using Ge-detectors after populating the state of interest via various reaction mechanisms and hence, E_r can be deduced with the known Q-value. The partial width of the incoming channel Γ_{in} is usually very small for states of astrophysical interest compared to the partial width of the outgoing channel Γ_{out} such that $\Gamma_{tot} \approx \Gamma_{out}$ and hence $\omega\gamma \propto \Gamma_{in}$. The partial width is given by $\Gamma \propto P_l\,\theta^2$, where θ^2 is the reduced particle width and P_l is the penetration factor. The later quantity can be extracted from either direct capture measurements into the states of interest (see 2.2.1.) or transfer reactions such as (d,p), (d,n), $(^3He,t)$, $(^3He,d)$, $(^6Li,d)$ or $(^7Li,t)$ (see 2.2.2.) and thus the resonance strength $\omega\gamma$ can be composed.

2.2.1. The $^{17}O(p,\gamma)^{18}F$ and $^{17}O(p,\alpha)^{14}N$- reaction

Hydrogen burning in massive stars proceeds via the CNO-tri-cycle. The $^{17}O(p,\alpha)^{14}N$ - reaction is a member of these cycles and its rate at relevant stellar temperatures is dominated by a resonance at 70 keV. There are three experiments reported [8,9,10] in an effort to determine the resonance strength $\omega\gamma$ for this 70 keV- resonance: In [8] an upper limit of $\omega\gamma < 3\cdot10^{-10}$ eV was reported, in [9] a value of $(1.75^{+1.0}_{-1.4})\,10^{-8}$ eV is given and in [10] $\omega\gamma < 1.6\cdot10^{-9}$ eV (2 s.d.) was quoted. The only direct determination of the strength $\omega\gamma$ is that of ref. [10]. This upper limit is a factor 6 higher than that of [8] and a factor 2 lower than the lower limit of [9]. In view of these discrepancies it was desirable to reinvestigate this resonance strength. The work from [9] is an indirect determination using the $^{17}O(^3He,d)^{18}F$- reaction much in the same way as described below for the $^{22}Na(^3He,d)^{23}Mg$- reaction (sect.2.2.2) and the work of [8] is also an indirect method using the direct capture approach (chapt. 2.2.). For the latter case it was necessary to search for a direct capture γ-transition into the 5668 keV- state of ^{18}F. However, the energy of this state was reinvestigated recently [11] and found to be 5 keV higher then assumed in [8]. Therefore, the result of [8] is not valid any more.

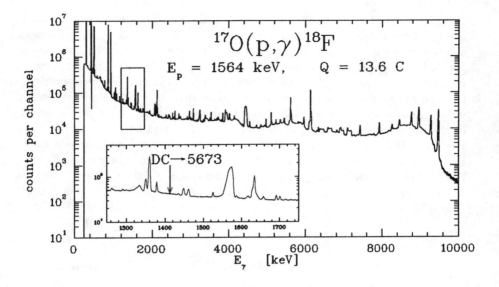

Figure 2. Relevant $^{17}O(p,\gamma)^{18}F$ -spectrum. The location of the DC-5673 keV transition is indicated in the inset of this figure.

We have made new investigations along the same line as described in [8] and a spectrum we obtained is shown in figure 2. The region were the direct capture transition is expected, according to the new excitation energy, is shown in the blow up of fig. 2. Preliminary analysis resulted in an upper limit for the cross section of $\sigma(DC - 5673 keV) < 4 \cdot 10^{-3} \mu b$ which corresponds to an upper limit of the $\omega\gamma < 8 \cdot 10^{-9}$ eV, which is within the limits of [9] and a factor 2 higher then the lower limit of [9]. From fig. 2 it is evident that the spectrum is dominated by backgrounds of various sources. New investigations have shown that this background could be reduced by a factor 9 when using implanted targets in W-backings. Such targets are now being prepared and it is hoped that the statistics can then be improved by a factor 3. With this new target it will therefore be possible to decide between both references [9] and [10].

2.2.2. The $^{22}Na(^{3}He, d)^{23}Mg$- reaction

The $^{22}Na(^{3}He, d)^{23}Mg$ reaction plays a major role in the understanding of the so called Ne-E problem. The use of indirect cross section measurements via the $^{22}Na(^3He,d)^{23}Mg$ reaction is a good tool to approach the necessary stellar reaction rates at temperatures below $2 \cdot 10^8$ Kelvin which is impossible to obtain in direct measurements due to the low cross section near the threshold [12],[13]. For such an experiment a "thin" ^{22}Na target is required for the observation of the deuteron reaction particles which can be produced by implanting ^{22}Na into a carbon foil. The best foil thickness for either stability under implantation conditions as well as for minimum energy loss for ^3He particles was found to be $75\mu g/cm^2$. The implantation was carried out at the ISOLDE separator at CERN which fulfills the requirements of high mass resolution and beam transport efficiency. In order to produce a ^{22}Na beam a surface ionization source was loaded with activated aluminum pieces (irradiated by a 70 MeV proton beam at the PSI in Zürich) that contained a total ^{22}Na activity of 6 mCi. This beam (beam current: 10 - 80 nA on the targets) was focused into a special set-up that contained a foil "ladder" and a second target holder for thick Ta-Ni backings that can be used for direct measurement of the $^{22}Na(p,\gamma)^{23}Mg$ reaction. The beam transport efficiency was found to be 95% under best conditions. Two carbon foils and one Ta-Ni backing were implanted at an energy of 60 keV. These samples contained ^{22}Na activities of 0.7, 0.3 and 0.5 mCi, respectively, after the implantation. An activity of about 0.5 mCi was found at apertures inside the implantation set up that limited the implanted area to a size of 1.5 x 3 mm^2. Deuteron groups of the $^{22}Na(^3He,d)^{23}Mg$ reaction were detected using the Q3D spectrometer at Munich which is especially suitable for these measurements due to its high energy resolution (E/ΔE = 5000) and large solid angle (11.3 msr). The ^3He beam was provided by a 15 MeV Tandem accelerator. Beam currents of about 100 nA could be achieved at a ^3He energy of 30 MeV. This energy was chosen to minimize the number of background peaks (due to the foil material and target contaminations) inside the relevant energy region. Deuteron groups corresponding to the ^{23}Mg levels at E_x = 7621, 7641, 7780, 7795, 7852, 8016, 8058, 8076 keV and a large number of levels below the proton threshold (at E = 7578 keV) could be observed at angles of 7.5, 10, 12.5, 15, 18, 21, and 31 degree. Because of the small ^{22}Na-implanted area it was possible to separate these deuteron groups clearly from background peaks (arising from ^{12}C, ^{13}C, ^{16}O, ^{14}N, ^{28}Si and small amounts of ^{35}Cl and ^{37}Cl), taking background spectra by steering the ^3He beam besides the implanted area on the carbon foil. The analysis of the angular distributions via the DWBA method allowed a determination of proton partial widths Γ_p for the ^{23}Mg-states near the proton threshold. In the case of small proton widths the required resonance strength $\omega\gamma$ can be deduced without knowledge about γ-widths:

$$\omega\gamma = \omega\Gamma_p\Gamma_\gamma/(\Gamma_p+\Gamma_\gamma); \quad if \quad \Gamma_p \ll \Gamma_\gamma \quad then \quad \omega\gamma = \omega\Gamma_p$$

The most probable value for the $\omega\gamma$- value was computed for two cases: i) an $l = 0$ component was allowed for in the fit for the angular distributions of the $(^3He, d)$- reaction and the value obtained at the minimum of the χ^2 was used to determine the contribution of this $l = 0$ part via the penetrability for s-waves; ii) the s-wave contribution was set to zero, when this was allowed by the χ^2-values. The lower and upper limits for the $\omega\gamma$ values are then obtained as follows. First the spectroscopic factors were deduced from the angular distribution data using the DWBA fit procedure for $l = 0, 2$ and $l = 1, 3$ and the errors for these values were obtained by one σ deviation through standard χ^2 analysis. The lower (upper) limits were then multiplied by 0.7 (1.3) to allow for a general 30% uncertainty in the DWBA-calculation. As a second step ranges for the $\omega\gamma$ were calculated using the C^2S- values with their respective ranges in combination with the corresponding penetrability functions. It should be emphasized that no uncertainty of the excitation energy of the corresponding levels in ^{23}Mg are included in above determinations. However, this has been done in determining the reaction rate. In the determination of the reaction rate the $\omega\gamma$- values as discussed above were used unless a value of the direct (p, γ) measurement existed or the upper limit was lower then from the $(^3He, d)$ work. In these cases the values or upper limits given by the (p, γ) work were used. With this information the most probable reaction rate (fig. 3) was calculated. In addition the upper and lower limits of the reaction rate have to include the uncertainty in the excitation energy of the compound states in ^{23}Mg as quoted in ref.[14]. Since the reaction rate is directly proportional to $\omega\gamma$ and hence to the proton width a change in the resonance energy results in a different penetrability for the incoming protons. Such changes are in part counterbalanced by the Boltzman factor $exp(-E_r/kT)$ in the reaction rate or can even change the upper limit into a lower limit (or vice versa), depending on the specific value of E_r. Thus, a variation of the reaction rate was calculated using the limits of the $\omega\gamma$ values as described above and varying at the same time the excitation energies within the quoted [14] limits. Additionally a variation of $\pm 0.05 fm$ of the nuclear radius $r_0 = 1.25 fm$ was allowed for. The resulting highest rate is then the upper limit and correspondingly the lowest rate the lower limit in fig. 3. The new results lead to a considerable reduction of the uncertainties in the stellar $^{22}Na(p, \gamma)^{23}Mg$ reaction rate (fig. 3) for the astrophysical important temperature region $T_8 > 1.0$. Remaining uncertainties are mainly due to uncertainties in small admixtures of s-components in the angular momentum transfer in the $(^3He, d)$-reaction. One obtains the dashed line in fig. 3 if this s-component is set to zero.

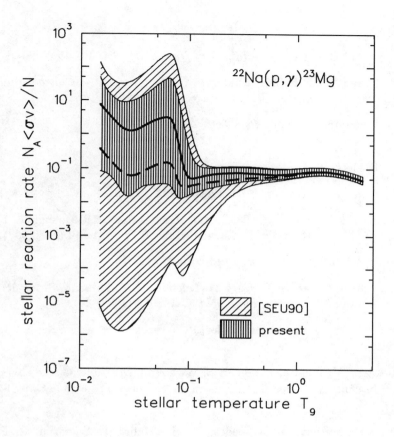

Figure 3. Uncertainties in the stellar $^{22}Na(p,\gamma)^{23}Mg$ reaction rate, normalized to an approximation by Wiescher and Langanke [15]. The vertically hatched area represents the new allowed region as compared to previous results [12] (cross hatched region). The solid line represents the most probable rate including s-components in the DWBA analysis and the dashed line is the most probable rate were s-components are set to zero. For more details see text.

References

[1] C. Rolfs, H.P. Trautvetter and W.S. Rodney, *Rep. Prog. Phys.* **50** 233

(1987).

[2] C. Rolfs and C.A. Barnes, *Annu. Rev. Nucl. Part. Sci.* **40** 45 (1990).

[3] C. Rolfs and W.S. Rodney, *Cauldrons in the Cosmos - An introduction to Nuclear Astrophysics*, Univ. Chicago Press (1988)

[4] L. Buchmann et al., *Phys. Rev. Lett.* **70** 762 (1993).

[5] Z. Zhao, R. H. Frances III, K. S. Lai, S. L. Rugari, M. Gai, *Phys. Rev. Lett.* **70** 2066 (1993).

[6] P. Dyer and C. A. Barnes, *Nucl. Phys.* **A233** 49 (1974).

[7] A. Redder, H. W. Becker, C. Rolfs, H. P. Trautvetter, T. R. Donoghue, T. C. Rinckel, J. W. Hammer, K. Langanke, *Nucl. Phys.* **A462** 385 (1987).

[8] C. Rolfs, W. S. Rodney, *Nucl. Phys.* **250** 295 (1775).

[9] V. Landre, P. Aguer, G. Bogaert, A. Lefebvre, J. P. Thibaud, S. Fortier, J. M. Maison, J. Vernotte, *Phys. Rev.* **C40** 1972 (1989).

[10] M. Berheide, C. Rolfs, U. Schroeder, H.P. Trautvetter, *Z. Phys.* A **343** 483 (1992) .

[11] G. Bogaert, V. Landre, P. Aguer, S. Barhoumi, M. Kious, A. Lefebvre, J. P. Thibaud, S. Fortier, J. M. Maison, J. Vernotte, *Phys. Rev.* **C39** 265 (1989).

[12] S. Seuthe, C. Rolfs, U. Schroeder, W. H. Schulte, E. Somorjai, H. P. Trautvetter, F. B. Waanders, R. W. Kavanagh, H. Ravn, M. Arnould, G. Paulus *Nucl. Phys.* **A514** 471 (1990).

[13] J. Goerres, M. Wiescher, K. L. Kratz, B. Leist, K. H. Chang, B. W. Filippone, L. W. Mitchell, M. J. Savage, R. B. Vogelaar, *Nucl. Instr. Meth.* **A267** 242 (1988).

[14] P. M. Endt, *Nucl. Phys.* **A521** 1 (1990).

[15] M. Wiescher, K. Langanke *Z. Phys.* **A325** 309 (1986).

Experiments with Radioactive Beams

W. Galster

*Institut de Physique Nucléaire, Université Catholique de Louvain,
B-1348 Louvain-la-Neuve, Belgium*

Abstract. In explosive astrophysical sites reactions involving shortlived unstable nuclei determine the timescale, energy production and nucleosynthesis. Radioactive beam (RB) facilities have opened this area of considerable astrophysical interest to direct measurements of reaction rates. Experimental results obtained with RB are presented and the techniques used are discussed.

MOTIVATION

Originally based upon astronomy, astrophysics has expanded to virtually all fields of science in the last 50 years. While astronomical observation continues to play an essential role, nuclear physics has been providing an input of increasing importance. The time scale, energy production and nucleosynthesis in the stellar environment are governed by sequences of nuclear reactions. Initially these reactions rates were mostly estimated by employing models that appeared to reproduce trends (1). In recent years some n, p, α capture rates on stable nuclei have been determined experimentally (2). These reactions dominate the quiescent phases of stellar burning and although we have a fair understanding of these processes, there is a definitive need for new nuclear physics data (3). In most cases this calls for accelerator based experiments and many aspects have been discussed in detail (2,4,5,6).

The lightest elements ($Z \leq 3$) were initially created in the big bang nucleosynthesis, which is essentially terminated by the mass = 5 and = 8 gaps ; the light elements with $3 \leq Z \leq 5$ are being produced in cosmic ray induced spallation in the interstellar medium (7). The thermo-nuclear pp chains power the normal stars (e.g. sun), gradually being superseded by the CNO cycle with increasing temperature. The quiescent phase of CNO operates as a closed cycle but evolves to high complexity on short timescales of burning. Proton capture on β-unstable nuclei plays an important role in the hot CNO cycles occuring in novae. Eventually a breakout or leakage from the hot CNO cycles may occur e.g. through $^{15}O(\alpha,\gamma)^{19}Ne(p,\gamma)^{20}Na$ creating the Ne Na and Mg Al chains. In a rapid sequence of p- and α-capture heavy elements beyond Fe can be synthesized in the rp- and αp-process (8), thought to occur in X-ray bursts or in type I supernovae. These heavy nuclei might be the seeds for the slow s-process that are necessary to synthesize the heavier elements ($A \leq 200$) in the presence of a neutron source (6). The fast r- and p-processes also require heavy seed nuclei. The r-process is thought to be initiated by a rapid sequence of α-capture followed by n-capture into the very neutron-rich side of the mass table (9). The p-process builds on the p-nuclei that may have been produced by photodisintegration (γ,n) of s-nuclei (10). The astrophysical sites associated with both processes are type II supernovae.

There is an urgent need for experimental data as well as for theoretical models in order to solve the extensive reaction networks of explosive burning that describe spectacular cosmic events such as novae, X-ray bursts and supernovae. Unlike the quiescent phases, where the reaction path lies in the valley of stability, the explosive phases occur in the vicinity of the proton and neutron drip lines.

The statistical model often describes the quiescent scenarios adequately, where the nuclear reactions proceed in general at high level densities in the continuum. Explosive burning on the other hand proceeds through isolated resonances and a good knowledge of reaction rates and nuclear structure is required. This calls for extensive measurements in the laboratory using radioactive beams. The general aspects have been described previously (11, 12, 13). The availibility of RB constitutes a major breakthrough in experimental nuclear astrophysics. First experiments with RB that are of astrophysical interest are described in the following.

NUCLEAR ASTROPHYSICS WITH RADIOACTIVE BEAMS

The first operational RB facility dedicated mainly to nuclear astrophysics is the one at Louvain-la-Neuve. RB's of ^{6}He, ^{13}N, ^{19}Ne with intensities up to 10^9 pps are available, other beams such as ^{11}C, ^{15}O, ^{18}F, ^{18}Ne, ^{35}Ar are currently being developed. Details of RB facilities have been discussed in three conferences on Radioactive Nuclear Beams (14) and in recent workshops (15). Figure 1 shows the layout of the two cyclotron facility at Louvain-la-Neuve.

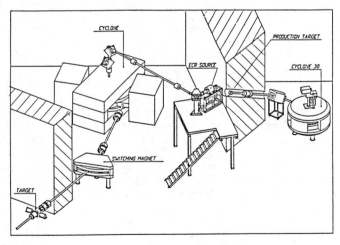

FIGURE 1. Layout of the Louvain-la-Neuve facility

The radioactive species is produced by the first cyclotron, extracted from the production target, transported in gaseous form to the ion source (ECR) and injected into the second cyclotron for acceleration. It is possible to change between isobaric beams (e.g. ^{13}N/^{13}C or ^{19}Ne/^{19}F) within seconds. Other RB-facilities that are limited operational for astrophysics purpose either lack a postacceleration (e.g. Isolde-CERN, Triumf-TISOL) or are of the fragmentation method type (RIKEN, MSU, GANIL) that delivers high energy low intensity beams (useful for β-decay and Coulomb breakup studies).

As the cross sections of astrophysically relevant reactions are very small and because a RB facility constitutes a high background environment, most experiments are extremely difficult. Every experiment has to be evaluated carefully in order to find a suitable detection technique. Radiative capture (p,γ), (α,γ) and (p,α) reactions can be measured directly on line by detecting deexcitation γ-rays, charged particles or heavy recoils. Activation methods are a promising alternative: the radioactive nuclei are implanted in a catcher foil, moved to a detection system and counted off line following their decay. High energy β particles can be focused into a stack of scintillators by means of a solenoid. The alternative detection methods are summarized in fig.2. It is highly desirable to measure the same quantities with different methods if possible to enhance the reliability of the obtained results.

The radiative capture of a proton on ^{13}N characterizes the transition from the CNO to the hot CNO cycles. This ^{13}N(p,γ)^{14}O reaction rate has been measured directly using RB recently (16). The resonant capture is dominated by a $J^\pi = 1^-$ resonance in ^{14}O (E* ~ 5.17 MeV) with a partial width $\Gamma_\gamma = 3.3 \pm 1.3$ eV.

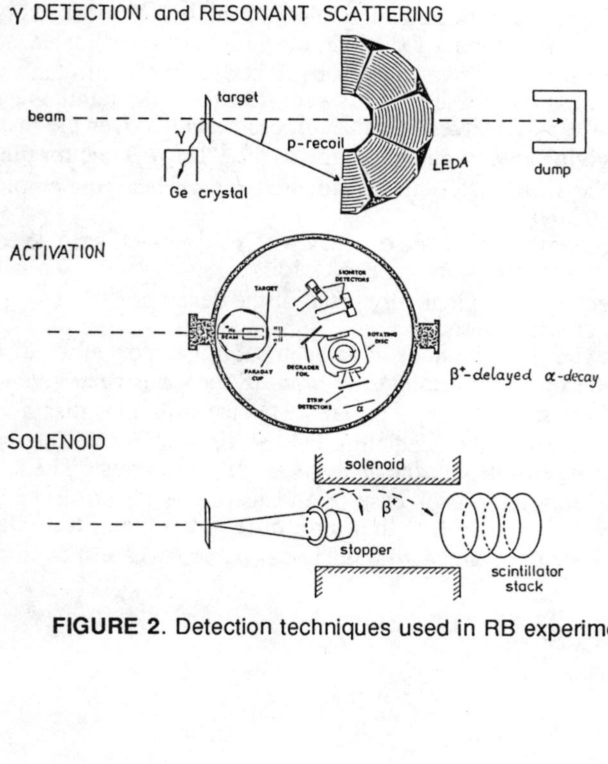

FIGURE 2. Detection techniques used in RB experiments

The exact energy and total width of this resonance was deduced from a Breit Wigner analysis of the resonant ^{13}N(p,p) scattering (14,16) to be $E_{res}^{(cm)} = 526 \pm 1$ keV and $\Gamma = 37.0 + 1.1$ keV. These experiments were carried out with intense (3.10^8) beams of ^{13}N on thick (200 µg/cm^2) polyethylene targets ; the γ-rays were detected in two large volume Ge as indicated in fig.2. The direct capture was inferred from the ^{13}N(d,n)^{14}O reaction studied at the same time (17). Although the direct capture rate is much smaller, interference between resonant and direct amplitudes leads to considerable deviations off resonance as compared to the dominant resonant process alone. Here the total capture rate is enhanced below resonance. These measurements constitute the first complete set of data obtained for an important reaction in the explosive burning phase in novae.

The cross section for the resonant part of ^{13}N(p,γ) radiative capture was also deduced from indirect measurements using the coulomb breakup of high energy (~ 100 MeV/u) ^{14}O beams. Two groups obtained $\Gamma_\gamma = 3.1 \pm 0.6$ eV (RIKEN) and $\Gamma_\gamma = 2.4 \pm 0.9$ eV (GANIL), respectively (18). Here, it is assumed that the coulomb breakup at very high energies is the inverse process of the radiative capture at very low energies. However, higher order processes and contributions from nuclear breakup cannot simply be ruled out (19), although there may be cases where the simple coulomb dissociation dominates. Coulomb dissociation can at best in a few cases be regarded as an alternative to direct measurements of (p,γ) rates.

At sufficiently high temperatures ($T_9 \sim 0.4$) a leakage may occur from the hot CNO into the Ne Na-Mg Al cycles, feeding the rp-process. The most likely sequence is via ^{15}O(α,γ)^{19}Ne(p,γ)^{20}Na ... and both reactions can be considered to be bottlenecks for the rp-process. The slower of the two is most likely ^{15}O(α,γ) due to the high coulomb barrier for α-particles, however, the nuclear structure of the relevant resonances in ^{19}Ne and ^{20}Na enters decisively into the estimates. A precise measurement of these rates is required to understand the rp-process nucleosynthesis (19). We have carried out measurements over the last two years with the aim of giving at least an upper limit to the ^{19}Ne(p,γ) rate for the 2.65 MeV resonance in ^{20}Na. Three different experimental approaches were employed based on the activation method :

(i) measurement of the β-delayed α-decay with silicon microstrip detectors ;
(ii) the same but using polycarbonate track foils ;
(iii) measurement of the high energy β-tail of the decay of ^{20}Na using a solenoid and a stack of scintillators.

Results of the first two methods have been submitted for publication (20). An upper limit for the resonance strength of $\omega\gamma \leq 18$ meV has been given with 90 % C.L., consistent with $J^\pi = 1^+$, $\omega\gamma \sim 6$ meV (21) but in obvious disagreement with a more recent calculation (22) predicting $J^\pi = 3^+$, $\omega\gamma \sim 80$ meV. The third method also gave promising results, but the analysis is still in progress. The experimental apparatus was calibrated in each case by means of the ^{19}Ne(d,n)^{20}Na reaction, which exhibits cross sections that are orders of magnitude larger than ^{19}Ne(p,γ)^{20}Na. Spectra obtained with methods (i) and (iii) are shown in fig.3.

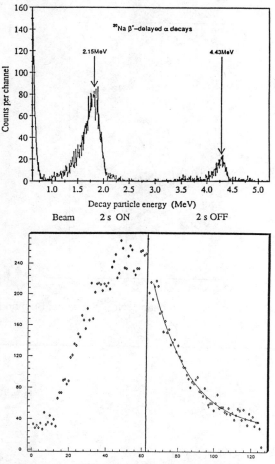

FIGURE 3. Calibration of ^{19}Ne(p,γ) detectors with (d,n) reaction : upper, β-delayed α-decay of ^{20}Na observed in microstrip detector ; lower, β-solenoid yield beam on/off = 2s/2s, fit τ(^{20}Na) = 460 ± 34 ms.

At the same time, we studied the resonant scattering of isobaric ^{19}Ne/^{19}F beams on thick (200 - 600 μg/cm^2) CH$_2$ targets. The spin, parity, resonance energy and total width of two broad resonances in ^{20}Na could be assigned unambiguously. Fig. 4 shows data obtained with a large Si-strip detector. A very careful analysis was carried out employing Breit-Wigner (BW), R- and K-matrix fits. The results are summarized in table I (23). The agreement between the different methods is excellent. Using wellknown resonances in the isobaric beam scattering is an extremely precise method. We find deviations of the order of 50 keV compared to (^{3}He,t) spectrograph works as can be seen from fig.5.

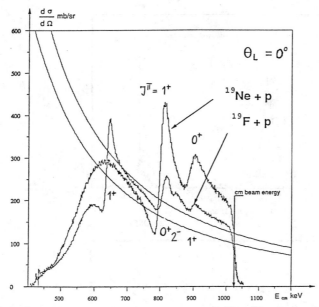

FIGURE 4. Resonant scattering on CH_2 of isobaric $^{19}Ne/^{19}F$ beams. Spin and parity of resonances in $^{20}Na/^{20}Ne$ are indicated.

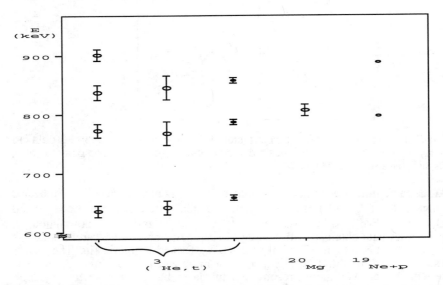

FIGURE 5. Comparison of resonance energies obtained with different techniques : $^{20}Ne(^3He,t)$ spectrograph, β-delayed p-decay of ^{20}Mg, direct resonant scattering using RB.

The energy of the $J^\pi = 1^+$ resonance in ^{20}Na ($E^* \sim 2996$ keV) is, however, in excellent agreement with recent studies of the β-decay of ^{20}Mg (24), in particular with the work of Görres et al from Notre Dame/MSU. The β-decay of ^{20}Mg populates $J^\pi = 1^+$ states in ^{20}Na preferentially. The main aim of these experiments was to search for the lowest level above the p-threshold ($E^* \sim 2646$ keV). The fact that this level has not be seen in ^{20}Mg β-decay to ^{20}Na supports the assignment as a $J^\pi = 1^+$ intruder state (21).

TABLE I. Energies and widths of the ^{20}Na resonances ; γ and g are the R-matrix and K-matrix reduced width amplitudes

J^π	Formalism	E^* (^{20}Na) (KeV)	γ or g (MeV$^{1/2}$)	Γ (keV)
1^+	BW	2996 ± 7		19.8 ± 2
	R-matrix	2996	0.92	19.8
	K-matrix	2996	15.6	19.8
0^+	BW	3086 ± 7		35.9 ± 2
	R-matrix	3086	1.00	35.9
	K-matrix	3086	15.8	35.4

The ^{12}C$(\alpha,\gamma)^{16}$O reaction in red giant stars determines the equilibrium of ^{12}C/^{16}O after He-burning. As we all realize the importance of this two elements, a precise measurement of the reaction rate is of utmost importance. The rate is dominated by the tails of three resonances, two of which are below the α threshold. In early experiments (2) γ-rays were detected directly and improved new measurements are being carried out at Bochum (25). Recently, a group at TRIUMF has used the β–decay of ^{16}N to feed the resonances in the region of astrophysical interest. Low energy (12 keV) ^{16}N beams were implanted in thin catcher foils for a few seconds, then rotated in front of Si detector pairs. The procedure was repeated continuously. A very careful and precise analysis yielded much improved results for the E1 part of the reaction rate (26).

CONCLUSIONS

With the advent of RB, experimental nuclear astrophysics is entering into the domain of explosive burning. The experimental difficulties are manifold : the available RB intensities are low ($\sim 10^9$ pps) as compared to stable beams ($> 10^{13}$ pps), the background rates (γ,β,n) are high and the cross sections are very small ($\leq$ μbarn). For short, this is a real challenge for the experimentalist, who has to take a fresh look at novel techniques. The theorist is not to be envied either, as extensive reaction networks have to be modeled and calculated. One cannot expect that the enormous amount of reaction rates ($\sim 10^3$) needed could all be determined experimentally in a reasonable span of time. A concerted effort is required from experimentalists and theorists to lay the foundations for reliable nuclear models based upon and constantly being crosschecked by experimental data. As our knowledge of nuclear physics is biased towards stable nuclei, we have to learn more about nuclei close to the limits of stability, the neutron/proton drip lines.

These are the nuclei that play a vital role in explosive burning. Reaction rates around certain key points in the reaction sequences should be determined with the highest priority. These are bottlenecks (where the rate is extremely low), waiting points (where the sequence has to await the β-decay) and the onset of a new reaction cycle/sequence. A better understanding of the nuclear structure of exotic nuclei is needed, in particular of low lying levels in light and medium heavy nuclei. As far as RB facilities are concerned, it is hoped that some of these projects will be realized soon and that a significant part of operation time will be devoted to nuclear astrophysics studies.

ACKNOWLEDGMENTS

The work described here is part of the Radioactive Ion Beam Programme in Louvain-la-Neuve supported by a special grant (PAI) from the Belgian government. The experimental work referred was carried out in collaboration with R. Coszach, Th. Delbar, P. Leleux, I. Licot, E. Liénard, P. Lipnik, C. Michotte, A. Ninane and J. Vervier (Université Catholique de Louvain) ; F. Binon, P. Duhamel and J. Vanhorenbeeck (Université Libre de Bruxelles) ; C. Bain, T. Davinson, R. Neal, R. Page, A. Shotter and P. Woods (University of Edinburgh) ; P. Decrock, M. Gaelens, M. Huyse, G. Vancraeynest, P. Van Duppen and J. Wauters (K.U.Leuven) ; M. Wiescher (University of Notre Dame) ; C. Sükösd (Technical University of Budapest).

I would like to thank P. Leleux for careful reading of the manuscript.

REFERENCES

1. E.M. Burbridge, G.R. Burbridge, W.A. Fowler, F. Hoyle, *Rev. Mod. Phys.* **29** (1957) 547 ; R.V. Wagoner, W.A. Fowler, F. Hoyle, *Ap. J.* **148** (1967) 3 ; W.A. Fowler, G.R. Caughlan, B.A. Zimmermann, *Ann. Rev. Astr. Ap.* **5** (1967) 525, **13** (1975) 69.
2. C. Rolfs, W.S. Rodney, *Cauldrons in the Cosmos*, University of Chicago Press (1988).
3. NuPECC report on "Impact and Applications of Nuclear Science", Nuclear Astrophysics 1994.
4. M. Wiescher in "Astrophysics and Neutrino Physics", World Scientific 1993, ed. Da Hsuan Feng, p. 184.
5. H.P. Trautvetter in ref. (27) p. 139.
6. F. Käppeler, H. Beer, K. Wisshak, *Rep. Prog. Phys.* **52** (1989) 945.
7. H. Reeves, *Rev. Mod. Phys.* **66** (1994) 193.
8. A.E. Champagne, M. Wiescher, *Ann. Rev. Nucl.* **42** (1992) 39 ; R.K. Wallace, S.E. Woosley, *Ap. J. Suppl.* **45** (1981) 389.
9. K.L. Kratz et al., in ref. (27), p. 349.
10. M. Rayet, N. Prantzos and M. Arnould, *Astron. Astrophys.* **227** (1990) 221 ; W.H. Howard, B.S. Meyer and S.E. Woosley, *Ap. J. Letters* **373** (1991) L5.
11. P. Leleux in ref. (27) p. 267.

12. W. Galster in "The future of nuclear spectroscopy" ed. W. Gelletly, INP-NCSR Demokritos Publishing 1994, p. 337.
13. B. Sherrill ibid. p. 323.
14. Conferences on "Radioactive Nuclear Beams", I, II, III,
    I RNB, Ed. W.D. Myers, World Scientific 1991
    II RNB, Ed. Th. Delbar, Adam Higer 1992
    III RNB, Ed. D.J. Morrissey, Ed. Front. (Gif-sur-Yvette) 1993.
15. RB Workshops at Oak Ridge, USA 1993 and Lake Louise, Canada 1994
16. P. Decrock et al., *PRL* **67** (1991) 808 ; *PLB* **304** (1993) 50.
17. P. Decrock et al., *PRC* **48** (1993) 2057.
18. T. Motobayashi et al., *PLB* **264** (1991) 259 ; J. Kiener et al., in ref. (27) p. 317.
19. A.C. Shotter in ref (27) p. 303 ; H. Rebel in ref (27) p. 309.
20. R. Page et al., submitted to *PRL*.
21. L.O. Lamm et al., *NPA* **510** (1990) 503.
22. B.A. Brown et al., *PRC* **48** (1993) 1456.
23. R. Coszach et al., accepted for publication by *PRC*.
24. J. Görres et al., *PRC* **46** (1992) R833 ; S. Kubono et al., *PRC* **46** (1992) 361 ; A. Piechaczek et al., in ref. (14) RNB III.
25. G. Roters et al., contr. to this conference.
26. L. Buchmann et al., *PRL* **70** (1993) 726 ; R. Azuma et al., accepted for publication by *PRC*.
27. Nuclei in the Cosmos II edited by F. Käppeler and K. Wisshak, Institute of Physics Publishing 1993.

New Experimental Data for the s–Process: Problems and Consequences

F. Käppeler

Kernforschungszentrum Karlsruhe, IK-III, Postfach 3640, D-76021 Karlsruhe, Germany

Abstract.
The interpretation of the s–process abundances in the solar system, in stellar atmospheres, and in the form of isotopic anomalies in meteoritic material represents an important test for various He–burning scenarios. The information obtained from such analyses depends to a large extent on the quality of the nuclear physics that is available. This contribution deals with (i) the status of the main neutron producing reactions for the s–process, and (ii) the progress in the determination of stellar (n,γ) rates, in particular for the unstable branch point isotopes, and the problems related to p–process corrections and neutron captures from excited states. The analysis of s–process branchings in the mass range $90 < A < 200$ is briefly discussed including an update of the most important cases.

Key words: nuclear astrophysics — neutron capture — s–process

1. INTRODUCTION

In the basic paper of Burbidge, Burbidge, Fowler, and Hoyle (1) the section, which lists the modes of nucleosynthesis, ends with: *An auxiliary but indispensible process which is also demanded in our description of element synthesis is a nuclear process which will provide a source of free neutrons for both the s–process and the r–process.* The problem of the neutron source(s) for the s–process is repeatedly addressed in this paper but was not pursued with much emphasis lateron. Only recently the quest for the various (α,n) reactions during stellar helium burning was revived by a series of investigations (Section 2).

For all s–process studies, reliable cross sections are the most important nuclear physics data, since these define the emerging abundances as well as the location of the reaction path. Considerable progress has been achieved in this field during the last decade as it is illustrated by a number of contributions to this conference. (Section 3).

Among the various modes of nucleosynthesis, the s–process is distinguished by the fact that the s–abundances can be determined quantitatively. Therefore, it represents an important possibility for testing and characterizing He–burning scenarios. The overall abundance distribution produced in the s–process is to good approximation independent of temperature and only determined by the integrated neutron exposure. The reason for this feature is that the s–abundances are inversely proportional to the respective (n,γ) cross sections. Since these cross sections exhibit roughly a $1/v_n$ trend, where v_n is the neutron velocity, one finds that the neutron capture rate,

$$\lambda_n = n_n \, v_n \, <\sigma>$$

is practically constant, so that the abundances are not sensitive to the temperatures during He–burning between $1 < T_8 < 3$. Therefore, the overall shape of the s–process abundances reflects the total neutron exposure during helium burning rather than the physical conditions of the respective scenarios. This means that almost any He–burning scenario will reproduce this gross structure provided that it predicts the proper neutron exposure.

Different s–process scenarios can be tested in greater detail via the fine structure in the s-process abundances. This fine structure is due to the effect of s–process branchings, which result from the competition between β–decay and neutron capture whenever the neutron capture chain encounters an unstable isotope with a half–life comparable to the neutron capture time. The relative abundances in the two parts of such a branching are determined by the ratio of the β–decay rate and the neutron capture rate. This feature allows to deduce the effective neutron capture rate and, hence, the s–process neutron density. In a number of cases, the stellar β–decay rates can depend on temperature and/or on electron density, so that these parameters can be checked as well. Therefore, the s–process branchings are a more stringent test for s–process models (Section 4).

Other aspects of the s–process – which can not be discussed in detail here – are related to the fact that the s–abundances are well characterized by the respective (n,γ) cross sections. This is quite unique compared to other nucleosynthesis mechanisms where the abundance yields are much more difficult to determine, either due to uncertain nuclear physics data (r–process, p–process) or due to uncertainties of the related models (explosive scenarios). Therefore, subtraction of the well defined s–abundances from the solar distribution yields the abundances from these other processes often more reliably than by direct model calculations. The best established example is the r–process distribution which is confirmed by the good agreement with the ensemble of r–only isotopes. This procedure is less reliable for the p–process because the p–abundances are, in general, very small. Similar problems occur in the mass region below iron, but there the s–abundances are usually smaller than the abundances produced during the advanced burning stages. Nevertheless, there are a number of relatively rare neutron rich isotopes, where the s–contributions are significant. In this domain, (n,α) and (n,p) reactions are often the dominant reaction channels (see contributions by Wagemans *et al.* (2) and Gledenov *et al.* (3)).

In this context, isotopic anomalies are another fascinating subject. The interpretation of these abundance patterns open new and exciting views to nucleosynthesis in AGB stars as discussed in several contributions to this conference.

2. STELLAR NEUTRON SOURCES

The most efficient neutron sources in stellar He–burning are the reactions $^{13}C(\alpha,n)^{16}O$ and $^{22}Ne(\alpha,n)^{25}Mg$. As already noted by Burbidge *et al.* (1), there is a principal difference between these reactions. During He–burning ^{13}C is produced by proton captures on freshly synthesized ^{12}C, independent of any preexisting seed. Hence ^{13}C is a primary isotope and its production does not depend on metallicity. This is different for ^{22}Ne which is produced by the reaction sequence
$$^{14}N(\alpha,\gamma)^{18}F(\beta^+\nu)^{18}O(\alpha,\gamma)^{22}Ne(\alpha,n)^{25}Mg.$$
The seed abundance corresponds to the initial carbon, nitrogen, and oxygen abundances which are converted to ^{14}N in the CNO–cycle. The ^{22}Ne available for neutron production depends, therefore, on metallicity, it is a secondary isotope. While this represents a restriction for ^{22}Ne, the problem of ^{13}C is its production which requires a certain amount of mixing from the envelope into the He–burning zone. This mixing is strongly model–dependent and subject for ongoing investigations.

The absence of ^{25}Mg in the atmospheres of thermally pusing AGB stars, which are enriched in s–processed matter, suggests that the $^{13}C(\alpha,n)^{16}O$ reaction is the main neutron source in these stars. This scenario is expected to yield the *main* s–process component, which accounts for the s–abundances in the mass range from Zr to Pb (4). In a recent experiment, the cross section of that reaction has been studied close to the astrophysically relevant energy range (5;6). In this measurement a remarkable sensitivity limit of 60 pbarn could be reached which was set by the cosmic ray background. In terms of the S–factor these results indicate an increase towards lower energies, possibly due to a subthreshold resonance at $E_{CM} = -2$ keV. Compared to the compilation of Caughlan and Fowler (CF88, Ref.7) this effect would increase the reaction rate by a factor 10 at temperatures of $T_8 = 0.1$, but at He–burning temperatures there is agreement within the experimental uncertainties. With these improved experimental data, the status of the $^{13}C(\alpha,n)^{16}O$ reaction appears satisfactory.

This does not hold for the $^{22}Ne(\alpha,n)^{25}Mg$ reaction, which is considered to be the main neutron source for the s–process during core He–burning in massive stars. At this site, the resulting yields are concentrated in the mass range $70<A<90$, giving rise to the so–called *weak* s–process component (8; 9). This reaction also contributes noticeably to the s–process during He–shell burning in low mass stars (4).

Considerable efforts have been made in recent years to determine the $^{18}O(\alpha,\gamma)^{22}Ne$ rate (10;11;12), which is important for the reaction sequence to ^{22}Ne. While the rate for the competing $^{18}O(\alpha,n)^{21}Ne$ reaction (13) was found to be about 10 times smaller than previously assumed (7), the (α,γ) channel turned out to be enhanced by two orders of magnitude due to a resonance at 470 keV. This means that the rate ratio $(\alpha,n)/(\alpha,\gamma)$ at helium burning temperatures is definitely $<10^{-3}$. Consequently, neutron production via (α,n) reactions on ^{18}O and ^{21}Ne is negligible in the s–process.

New results for the ^{22}Ne$(\alpha,\gamma)^{26}$Mg reaction by Giesen et al. (14) have improved previous data (15) for the 828 keV resonance and include a new resonance at 400 keV which dominates the rate below $T_8 = 2$. This implies fair agreement with the CF88 rate at s–process temperatures.

For the ^{22}Ne$(\alpha,n)^{25}$Mg reaction the situation is more difficult. In this case the astrophysically important energy range could not be reached in direct experiments (5;16;17). Therefore, the influence of possible low energy resonances has only been estimated because the anticipated resonance strengths (7) are too weak to be measured with current γ– and n–detector systems. Though ^{26}Mg has a high level density above the α threshold at 10.611 MeV (18) few resonances are expected in the α cross sections of ^{22}Ne since only natural parity states are populated in these reactions. In a recent study of the natural parity states in ^{26}Mg by an α-transfer experiment via the reaction ^{22}Ne$(^{6}$Li,d), Giesen et al. (14) confirmed the energy and strength of the 828 keV resonance observed in the direct experiment of Drotleff et al. (5). A second resonance at 633 keV would be compatible with corresponding resonances in the ^{26}Mg(γ,n) and the ^{25}Mg(n,γ) cross sections but could not be identified unambiguously. However, it was found that this resonance – if it exists – would cause a strong enhancement of the ^{22}Ne(α,n) rate.

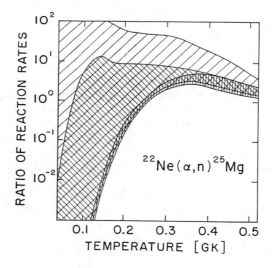

Fig. 1.— Ratio of the ^{22}Ne$(\alpha,n)^{25}$Mg rate and the tabulated rate of Caughlan and Fowler (7). The hatched area indicates the enhancement if all possible low lying resonances are considered, the cross–hatched are shows the situation after most of these resonances were excluded because of unnatural parity, and the vertically hatched area corresponds to the effect of the well established resonance at 828 keV. For more details see text.

This is illustrated in Fig. 1, which shows the ratio of the resulting reaction rate to the CF88 rate (7). The hatched area indicates the enhancement if all possible low lying resonances are considered. With the results of Giesen *et al.* (14), this region can be reduced to the cross-hatched band which considers only the upper limit for the 633 keV resonance but excludes the resonances with unnatural parity. What is known for sure so far is the vertically hatched area, which corresponds to the effect of the well established resonance at 828 keV.

Since the possible enhancement of the ^{22}Ne(α,n) rate would have far reaching consequences for the s-process during core He-burning in massive stars (19;20), an experimental solution for this problem of the 633 keV resonance is very important.

3. NEUTRON CAPTURE RATES

3.1. Small (n,γ) Cross Sections

The reaction flow in the s-process can be characterized reliably by the $<\sigma>N_s(A)$ curve. Hence, the s-process abundances can be given with typical uncertainties of a few percent, depending on the quality of the respective cross sections. The resulting abundance distribution and the role of the small (n,γ) cross sections at magic neutron numbers are discussed in contributions to this conference by Beer *et al.* (21), Corvi *et al.* (22), and Rauscher *et al.* (23). Another important problem is related to the abundant nuclei present in the He-burning plasma such as ^{12}C, ^{16}O, and ^{22}Ne. Despite of their small neutron reaction cross sections these isotopes may have a significant influence on the neutron economy of the s-process (Nagai *et al.* 24).

Suited experimental methods for such experiments are high resolution time of flight (TOF) studies in search of complete resonance schemes (22;25) or TOF measurements in a very high neutron flux as presented in the following contribution (26). In some cases, the activation method represents an attractive complement to TOF experiments, especially for the investigation of very small amounts of sample material (Section 3.3).

3.2. Cross Sections of s-Only Isotopes

The isotopes which can be directly assigned to the s-process are particularly important for normalizing the $<\sigma>N_s$ curve as well as for defining the s-process branchings. This set of data is currently being measured with improved accuracy using the Karlsruhe 4π BaF$_2$ detector as described in the contribution by Wisshak *et al.* (27).

3.3. Cross Sections of Unstable Isotopes

Apart from the involved s-only nuclei, branching analyses require also the neutron capture rates of the branch point nuclei with adequate accuracy. Until recently, there was no experimental information on the (n,γ) cross sections of these unstable isotopes. Because of the sample activity, only small amounts of material can be used in such measurements in order to avoid overloading of the detectors. Therefore, the most sensitive experimental techniques have to be used in these cases (28).

The activation technique combines superior sensitivity with good selectivity (which means that isotope mixtures can be studied via the characteristic γ–ray energies of the respective decay products). This technique is based on the fact that quasi–stellar neutron spectra can be produced in the laboratory via the ^{7}Li(p,n)^{7}Be (29;30) and the ^{3}H(p,n)^{3}He reactions (31). With respect to s–process branchings, the activation technique allows to investigate a number of unstable nuclei. This is important, since experimental (n,γ) cross sections are required to replace the values obtained from statistical model calculations which exhibit uncertainties of 20 to 30% even in the most favorable cases.

First successful measurements are reported at this conference for the branch point isotopes ^{155}Eu and ^{163}Ho (32). Another example relates to the branchings at ^{134}Cs and ^{135}Cs. Of these nuclei, only ^{135}Cs appears to be accessible to a direct experiment, since the activity of a ^{134}Cs sample would be prohibitive for any technique. For the relatively long–lived ^{135}Cs ($t_{1/2} = 2 \times 10^6$ yr) there is the problem of ^{134}Cs and ^{137}Cs impurities. Therefore, the sample was prepared using a mass separator for implanting ^{135}Cs into a thin carbon disk. Starting from a batch that was practically free of ^{134}Cs, the contamination with ^{137}Cs could be reduced to less than 0.2%. This sample with a ^{135}Cs mass of about 10^{15} atoms was sandwiched between two gold foils and irradiated for 14 days in a mean neutron flux of about 3×10^9 s^{-1}.

The γ–ray spectrum of the irradiated sample is given in Fig.2. The dominating line at 662 keV and is due to the small remaining ^{137}Cs contamination. The ^{135}Cs cross section can be determined by the γ–ray lines at 809 and 1048 keV from the decay of ^{136}Cs, which both exhibit a very good signal/background ratio. The last step in the experiment requires the determination of the ^{135}Cs/^{137}Cs ratio by mass spectroscopy in order to define the sample mass.

3.4. Corrections for s–Only Isotopes

The refined experimental techniques have now reached an accuracy where even small corrections become important. In the first place these corrections refer to the abundances and cross sections of the s–only isotopes. While these isotopes are shielded against the r–process, a minor but significant fraction may have been produced in the p–process. This possibility must always be kept in mind when talking about s–only nuclei. At s–process temperatures, the stellar (n,γ) cross sections can differ from the laboratory values if low–lying excited states are efficiently populated.

The p–process contributions have long been considered in an empirical fashion assuming a smooth abundance distribution that was determined by means of nearby p–only nuclei (33). Recently, more realistic p–process yields have been calculated for the explosive burning of Ne/O layers (34;35;36). The

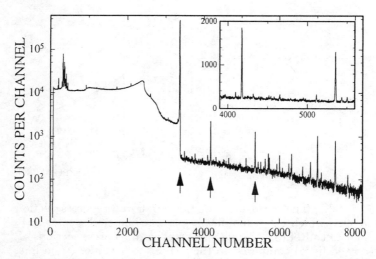

Fig. 2.— The γ-ray spectrum of an activated ^{135}Cs sample. The spectrum is dominated by the 662 keV line of a small contamination with ^{137}Cs (left arrow). The induced ^{136}Cs activity shows up via the lines at 809 and 1048 keV, which are given with better resolution in the inset.

p–only abundances are reproduced by these calculations within a factor of 3. Despite of their relatively large uncertainties, these data have to be considered in determining the p–process corrections of the s–abundances. In many cases, the empirical and calculated p–process corrections are similar, but there are also strikingly large differences. An extreme example is ^{152}Gd which determines the temperature–sensitive branching at A = 151, 152. The empirical p–correction for this isotope was estimated to 33%, whereas the calculations predict values of less than 6%! Obviously, such differences have a significant impact on the interpretation of the observed s–only abundances with respect to both, the $<\sigma>N_s$ curve and the s–process branchings.

Thermal effects on stellar cross sections were intensively discussed in the context of the ^{187}Re/^{187}Os clock (Ref.37 and references therein). Usually, the differences between laboratory and stellar cross sections are small, but for low-lying states with large spins the population probability in the stellar photon bath as well as the (n,γ) cross sections are enhanced, and can result in an enhancement of the effective stellar cross section. Those cases are included in the left part of Fig. 3, which compares the ratios of stellar and laboratory cross sections obtained in two statistical model calculations (38;39). One finds that this correction is important in the mass range above A = 150 where all s–only isotopes between ^{154}Gd and ^{187}Os seem to be affected. However, the uncertainties implied by the relatively large differences between the calculated values require further studies in order to put these corrections on safer grounds.

The right part of Fig. 3 shows the temperature–dependence of the cross

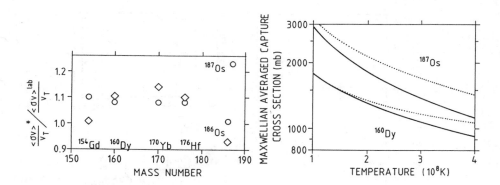

Fig. 3.— *Left:* The ratio of stellar over laboratory cross section for those s–only nuclei, which show significant effects. The circles refer to Ref(38) and the diamonds to Ref.(39). *Right:* The stellar (n,γ) cross sections of ^{160}Dy and ^{187}Os (dotted lines) compared to their laboratory values (solid lines).

sections of ^{160}Dy and ^{187}Os with and without considering the excited states. The effect is larger for ^{187}Os because the stellar cross section of that isotope is dominated by the first excited state $(3/2^+)$ at 9.75 keV. Already at kT = 20 keV ($T_8 = 2.3$) this state is populated to 52%, whereas the ground state population of 42% is due to the lower spin of $1/2^-$. In ^{160}Dy, the first excited state at 86.8 keV (2^+) is populated only at higher temperatures (6% and 21% at kT = 20 and 30 keV, respectively), but the difference between laboratory and stellar cross section grows faster because of the larger spin difference. Note that this effect can also be used to determine the stellar temperature if the effective cross sections of these isotopes can be determined from the $<\sigma>N_s$ systematics.

4. BRANCHINGS IN THE s–PROCESS PATH

Analyses of the abundance patterns in the various branchings are expected to yield detailed information on the stellar s–process site. In particular, it should be possible to distinguish between the different s–process scenarios such as the steady situation assumed by the classical approach (8) and the dynamic s–process suggested by stellar models for low mass stars (4;40;41). This distinction, however, can only be worked out if reliable cross sections are available for describing the abundance patterns of the various branchings.

As an example, Fig.4 shows the s–process flow in the mass region between xenon and barium, with branchings at ^{134}Cs and ^{1135}Cs. Note that ^{134}Ba and ^{136}Ba are shielded against the r–process by their xenon isobars. If these branchings are significant, the $<\sigma>N_s$ value of ^{136}Ba should be larger than that of the partially bypassed ^{134}Ba. That this holds true is shown in the contribution of Wisshak et al. (27), who measured the cross section ratio of these isotopes for the first time with sufficient accuracy.

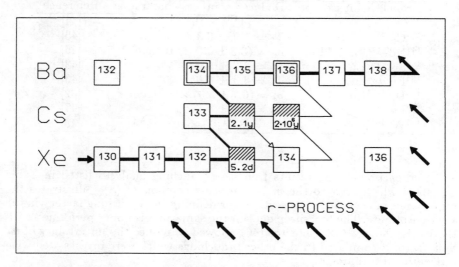

Fig. 4.— The s-process flow between xenon and barium.

The strength of a branching can be expressed in terms of the rates for β-decay and neutron capture at the branch point nuclei as well as by the $<\sigma>N_s$ values of the involved s-only isotopes,

$$f_\beta = \frac{\lambda_\beta}{\lambda_\beta + \lambda_n} \approx \frac{(<\sigma>N_s)_{134\text{Ba}}}{(<\sigma>N_s)_{136\text{Ba}}}.$$

By measuring the cross section ratio, the branching factor f_β can be determined. Branchings, where the stellar β-decay rate equals the terrestrial rate are suited for determining the effective neutron capture rate, λ_n, at the branching point and, hence, the neutron density during the s-process. Such examples are the branchings at A = 95, 147/148, and 185/186 (Table 1). The branching at A = 134/135, however, belongs to those cases which are sensitive to temperature because the decay of ^{134}Cs is significantly accelerated during the s-process (42).

In total, there are about 15 to 20 significant branchings on the s-process path. For a systematic investigation of the physical conditions at the stellar site, the neutron density must be obtained from those branchings which are not affected by temperature. With this information, the branching factors of the remaining examples can be derived and the effective stellar decay rates be determined. Finally, the dependence of these rates on temperature and/or electron density can be used to estimate the mean s-process temperature and mass density (43). The present status of the branching analyses with the classical approach is summarized in Table 1.

Table 1: Results from various branching analyses of relevance for the *main* component

Branch point isotope	Deduced s-process parameter	Reference
^{95}Zr	$n_n = (4^{+3}_{-2}) \times 10^8$ cm^{-3}	[44]
^{134}Cs	$T_8 = 1.9 \pm 0.3$	[45]
^{147}Nd–^{147}Pm–^{148}Pm	$n_n = (3.8 \pm 0.6) \times 10^8$ cm^{-3}	[46]
^{151}Sm–^{152}Eu	$T_8 = 3.3 \pm 0.6$	[43]
^{154}Eu–^{155}Eu	$T_8 = 3.3 \pm 0.6$	[43]
^{163}Dy/^{163}Ho	$\rho_s = (6.7 \pm 3.3) \times 10^3$ gcm^{-3}	[32]
^{176}Lu	$T_8 = 3.1 \pm 0.6$	[47]
^{185}W–^{186}Re	$n_n = (3.5^{+1.7}_{-1.1}) \times 10^8$ cm^{-3}	[48]

Comparison of the results from the branching analyses listed in Table 1 – all of which refer to the *main* s-process component (8) – illustrates the importance of accurate cross sections for defining the branching factor via the s–only isotopes: The measurements on the samarium isotopes performed with the Karlsruhe 4π BaF$_2$ detector (46) allowed to reduce the uncertainty of n_n by a factor 2 compared to the other branchings which were investigated with conventional techniques.

As far as the neutron density is concerned, all values are consistent within uncertainties. For the mean temperature, however, the new analysis of the branching at ^{134}Cs seems to be in conflict with the previous value of $T_8 = 3.3 \pm 0.6$. Whether this reflects simply the uncertainty of the stellar decay rate of ^{134}Cs (see comments in Ref.42), or whether there is an astrophysical reason can only be decided after more branchings have been studied with an improved set of cross sections.

A first attempt to determine the mass density in the s–process was carried out by Beer *et al.* (49) by analyzing the branching to ^{163}Ho and ^{164}Er. Their value, $\rho_s = (8 \pm 5) \times 10^3$ gcm^{-3}, is in reasonable agreement with the new result of Ref.(32) and with stellar models for helium shell burning (40;41).

So far, the results from the different branchings are still consistent with each other. Therefore, the assumption of the classical approach, that neutron density and temperature are constant during the s–process, is not yet ruled out, although the problem with the ^{134}Cs branching may be a first hint towards the scenario depicted by the stellar models which suggest these parameters to vary strongly with time. Consequently, the freeze-out behavior at the end of each neutron exposure may cause a distortion of the abundance pattern predicted by the classical approach. Further efforts have to concentrate on improving the relevant cross sections for as many branchings as possible, in order to achieve a comprehensive s–process systematics.

REFERENCES

1. Burbidge, E.M., Burbidge, G.R., Fowler, W.A., and Hoyle, F., *Rev. Mod. Phys.* **29**, 547 (1957).
2. Wagemans, C., Druyts, S. and Barthelemy, R., *contribution to these proceedings*.
3. Gledenov, Yu.M., Salatski, V.I., Sedychev, P.V., and Koehler, P.E., *contribution to these proceedings*.
4. Gallino, R., Busso, M., Picchio, G., Raiteri, C.M., and Renzini, A., *Ap. J.* **334**, L45 (1988).
5. Drotleff, H.W.,Denker, A., Knee, H., Soiné, M., Wolf, G., Hammer, J.W., Greife, U., Rolfs, C., and Trautvetter, H.-P., *Ap. J.* **414**, 735 (1993).
6. Denker, A., and Hammer, J.W., *contribution to these proceedings*.
7. Caughlan, G.R., and Fowler, W.A., *Atomic Data Nucl. Data Tables* **40**, 291 (1988).
8. Käppeler, F., Beer, H., and Wisshak, K., *Rep. Prog. Phys.* **52**, 945 (1989).
9. Raiteri, C.M., Busso, M., Gallino, R., Picchio, G., and Pulone, L., *Ap. J.* **367**, 228 (1991).
10. Trautvetter, H.-P., Wiescher, M., Kettner, K.U., Rolfs, C., and Hammer, J.W., *Nucl. Phys. A* **297**, 489 (1978).
11. Vogelaar, R.B., Wang, T.R., Kellogg, S.E., and Kavanagh, R.W., *Phys. Rev. C* **42**, 753 (1990).
12. Giesen, U., Browne, C.P., Görres, J., Ross, J.G., Wiescher, M., Azuma, R.E., King, J.D., and Buckby, M., *Nucl. Phys. A* **567**, 146 (1994).
13. Denker, A., Drotleff, H.W., Grosse, M., Hammer, J.W., Knee, H., Kunz, R., Mayer, A., Seidel, R., and Wolf, G., in *Heavy Element Nucleosynthesis*, eds. E. Somorjai and Zs. Fülöp, Debrecen: Institute of Nuclear Research of the Hung. Acad. of Sci., 1994, p. 145.
14. Giesen, U., Browne, C.P., Görres, J., Graff, S., Iliadis, C., Trautvetter, H.-P., Wiescher, M., Harms, V., Kratz, K.-L., Pfeiffer, B., Azuma, R.E., Buckby, M., and King, J.D., *Nucl. Phys. A* **561**, 95 (1993).
15. Wolke, K., Harms, V., Becker, H.W., Hammer, J.W., Kratz, K.-L., Rolfs, C., Schröder, U., Trautvetter, H.-P., Wiescher, M., and Wöhr, A., *Z. Phys. A* **334**, 491 (1989).
16. Harms, V., Kratz, K.-L., and Wiescher, M., *Phys. Rev. C* **43**, 2849 (1991).
17. Drotleff, H.W., Denker, A., Hammer, J.W., Knee, H., Küchler, S., Streit, D., Rolfs, C., and Trautvetter, H.P., *Z. Phys. A* **338**, 367 (1991).
18. Endt, P.M., *Nucl. Phys. A* **521**, 1 (1990).
19. Meynet, G., and Arnould, M., in *Nuclei in the Cosmos*, eds. F. Käppeler and K. Wisshak, Bristol: IOP, 1993, p. 487.
20. Käppeler, F., Wiescher, M., Giesen, U., Görres, J., Baraffe, I., El Eid, M., Raiteri, C.M., Busso, M., Gallino, R., Limongi, M., and Chieffi, A., *Ap. J.*, in print.
21. Beer, H., Spettel, B., and Palme, H., *contribution to these proceedings*.
22. Corvi, F., Mutti, P., Athanassopulos, K., and Beer, H., *contribution to these proceedings*.
23. Rauscher, T., Balogh, W., Bieber, R., Kratz, K.-L., Mohr, P., Oberhummer, H., Staudt, G., and Thielemann, F.-K., *contribution to these proceedings*.
24. Nagai, Y., Igashira, M., Shima, T., Ohsaki, T., and Masuda, K., *contribution to these proceedings*.

25. Beer, H., Corvi, F., and Athanassopulos, K., in *Capture Gamma-Ray Spectroscopy and Related Topics*, ed. J. Kern, Singapore: World Scientific, 1994, p. 698.
26. Koehler, P.E., and Käppeler, F., *contribution to these proceedings*.
27. Wisshak, K., Voss, F., and Käppeler, F., *contribution to these proceedings*.
28. Käppeler, F., in *Radioactive Nuclear Beams*, ed.Th. Delbar, Bristol: Adam Hilger, 1991, p. 305.
29. Beer H., and Käppeler, F., *Phys. Rev. C* **21**, 534 (1980).
30. Ratynski, W., and Käppeler, F., *Phys. Rev. C* **37**, 595 (1988).
31. Käppeler, F., Naqvi, A.A., and Al-Ohali, M., *Phys. Rev. C* **35**, 936 (1987).
32. Jaag, S., and Käppeler, F., *contribution to these proceedings*.
33. Beer, H., 1985, Kernforschungszentrum Karlsruhe internal report.
34. Rayet, M., Prantzos, N., and Arnould, M., *Astron. Astrophys.* **227**, 221 (1990).
35. Prantzos, N., Hashimoto, M., Rayet, M., and Arnould, M., *Astron. Astrophys.* **238**, 445 (1990).
36. Howard, W.M., Meyer, B.S., and Woosley, S.E., *Ap. J. Letters* **373**, L5 (1991).
37. Winters, R.R., and Macklin, R.L., *Phys. Rev. C* **25**, 208 (1982).
38. Holmes, J.A., Woosley, S.E., Fowler, W.A., and Zimmerman, B.A., *Atomic Data Nucl. Data Tables* **18**, 305 (1976).
39. Harris, M.J., *Ap. Space Sci.* **77**, 357 (1981).
40. Iben Jr., I., and Renzini, A., *Ann. Rev. Astron. Astrophys.* **20**, 271 (1983).
41. Hollowell, D.E., and Iben Jr., I., *Ap. J.* **340**, 966 (1989).
42. Takahashi, K., and Yokoi, K., *Atomic Data Nucl. Data Tables* **36**, 375 (1987).
43. Käppeler, F., Gallino, R., Busso, M., Picchio, G., and Raiteri, C.M., *Ap. J.* **354** 630 (1990).
44. Toukan, K.A., and Käppeler,F., *Astrophys. J.* **348**, 357 (1990).
45. Voss, F., Wisshak, K., Guber, K., Käppeler, F., and Reffo, G., *Phys. Rev. C*, submitted.
46. Wisshak, K., Guber, K., Voss, F., Käppeler, F., and Reffo, G., *Phys. Rev. C* **48**, 1401 (1993).
47. Klay, N., Käppeler, F., Beer, H., and Schatz, G., *Phys. Rev. C* **44**, 2839 (1991).
48. Käppeler, F., Jaag, S., Bao, Z.Y., and Reffo, G., *Ap. J.* **366**, 605 (1991).
49. Beer, H., Walter, G., and Macklin, R.L., in *Capture Gamma-Ray Spectroscopy and Related Topics*, ed. S. Raman, New York: American Institute of Physics, 1985, p. 778.

NUCLEAR PROPERTIES FAR FROM STABILITY AND THE R-PROCESS

Karl-Ludwig Kratz

*Institut für Kernchemie, Universität Mainz,
D-55099 Mainz, Germany*

Abstract: Within the 'waiting-point' assumption, we have investigated the influence of nuclear properties far from stability on the prediction of solar-system r-process abundance ($N_{r,\odot}$) distributions. When using data sets for masses and β-decay quantities from the Finite-Range Droplet Model (FRDM) and the Extended Thomas-Fermi plus Strutinski Integral (ETFSI) method, local deficiencies in the $N_{r,\odot}$ fits around $A\simeq115$ and 175 cannot be avoided. As pointed out in previous papers, they are due to model-inherent weaknesses in the shell-structure description very far from stability. New experiments at CERN/ISOLDE together with recent HFB calculations using the Skyrme P interaction, confirm our earlier qualitative predictions, in particular shell-quenching and neutron-skin effects in drip-line nuclei. These new nuclear-structure properties are shown to result in a considerable improvement of the global $N_{r,\odot}$ fits.

INTRODUCTION

There exist two possible ways to learn details of an astrophysical event: (a) In the first approach, a straightforward application of nuclear-physics and astrophysical models probes the detailed behaviour and results in a number of features to be compared with observations. (b) The second approach takes the observables as a *constraint* in order to conclude the necessary conditions required to reproduce these features. Given the history and long list of r-process scenarios [1], the fact that the stellar site for the rapid neutron-capture process is still not completely understood, and also that the recently suggested high-entropy bubble in type II supernova (SN II) explosions does not yet produce fully satisfying results [2-4], we have given priority to the second, more deductive approach [5,6]. There is, however, the promising hope [7] that both above approaches could merge; but it is still left to analyze the whole mass range of SN II progenitors.

In our attempt of determining stellar conditions from the observed solar-system r-abundance ($N_{r,\odot} \simeq N_\odot - N_s$) pattern [8] and a sophisticated nuclear-physics basis, we have shown that the $N_{r,\odot}$ distribution can already be described by a superposition of three time-dependent r-components, which are approximately in a steady-flow equilibrium of β-decays. Each of the components proceeds up to one of the abundance peaks, which are related to the neutron shell closures at N=50, 82 and 126. Still occurring local deviations from the $N_{r,\odot}$ pattern were interpreted as deficiencies in the predictive power of present nuclear models very far from stability, in particular overly strong shell corrections, effects from the neglect of the proton-neutron (p.n.) residual interaction, and correlated problems with describing shape transitions in the neutron mid-shell regions (for details, see Refs. [5,6]).

Concerns have been raised since then, that this might be an overinterpretation [3,9], and that an almost continuous superposition of a multitude of components would automatically prevent these deviations. Goriely and Bouquelle [10] used, e.g., a total of 1,335 components with stellar temperature (T_9), neutron density (n_n) and process duration (τ) being 'free' parameters within certain limits. In the present paper, we will show that – when using selfconsistent nuclear-data sets [5,11] – also a multicomponent fit is not able to prevent the above deviations. One obtains almost identical features, provided that the stellar conditions are not randomly chosen in the sense of a *Fourier analysis fit* as done in Ref. [10], but rather form a natural continuum among the previously discussed minimum set of components.

Since the deviations of the r-abundance fits obviously are not related to astrophysical models, it seems possible that nuclear physics near the neutron drip-line may obtain constraints from astrophysical observations. In the present paper, we will check our interpretation by applying – for the first time – masses and β-decay properties from a very recent microscopic mean-field model [12].

NUCLEAR-DATA REQUIREMENTS FOR THE CLASSICAL R-PROCESS

'Complete' r-process network calculations require a large number of astrophysical and nuclear-physics input parameters. In order to facilitate these complicated calculations, during the past three decades many attemps to predict the $N_{r,\odot}$ distribution were based on the simplified assumptions of the $(n,\gamma) \rightleftharpoons (\gamma,n)$ equilibrium concept. When assuming in addition a steady-flow equilibrium of β-decays, the prediction of r-abundances requires only the input of nuclear masses (respectively neutron separation energies, S_n), β-decay half-lives ($T_{1/2}$) and β-delayed neutron emission probabilities (P_n), as well as the stellar parameters T_9, n_n and τ. Whereas for a given n_n, the S_n determine the r-process path, the $T_{1/2}$ of the isotopes along this flow path, in principle, define the progenitor abundances and – when taking into account P_n branching during freeze-out – also the final r-abundances. Only in recent years, the validity of this 'waiting-point' approximation in combination with a steady β-decay flow could be confirmed locally for the $A \simeq 80$ and 130 $N_{r,\odot}$ peaks on the basis of first experimental information in the r-process path (see, e.g. Ref. [13]). With this, the *long* $T_{1/2}$ of the classical $N=82$ 'waiting-point' nucleus ^{130}Cd is, for example, directly correlated with the *large* $N_{r,\odot}$ value of its isobar ^{130}Te in the $A \simeq 130$ r-abundance peak.

Since the vast majority of nuclei in or close to the r-process path will probably never become accessible in terrestrial laboratories, a general understanding of their nuclear properties can only be obtained through theoretical means. In order to avoid a vanishing of real signatures or the creation of artificial r-abundance effects from the use of mass and half-life models of largely different sophistication [4,10], we have tried to perform the calculations in a *unified* approach [5,6] within which all relevant nuclear properties can be studied in an internally consistent way. The combination of nuclear masses from the Finite-Range Droplet Model (FRDM) [14] and β-decay properties from the QRPA approach of Möller and Randrup [15] is discussed in detail in Refs. [5,6]. Analogously, when adopting the masses from the Extended Thomas-Fermi plus Strutinski Integral (ETFSI) model [16], we use β-decay properties deduced from QRPA calculations with the Q_β values and deformation parameters given by this approach. For comparison with the above two macroscopic-microscopic theories, in special cases we also use the macroscopic formula of Hilf et al. [17], which is the only droplet-type model with a steep mass parabola.

In principle, with the above two global approaches – FRDM+QRPA and ETFSI +QRPA – we have two rather sophisticated, and internally consistent nuclear-data sets for astrophysical calculations which are expected to yield more reliable predictions of nuclear-physics parameters than earlier models. Nevertheless, being aware that even these approaches must have their deficiencies, we have tried to improve the data sets by taking into account all recent experiments on Q_β, S_n, $T_{1/2}$ and P_n. Furthermore, for localized extrapolations known nuclear-structure properties, either model-inherently not contained in or not properly described by the above global methods, were taken into account.

MULTICOMPONENT FITS TO SOLAR ABUNDANCES

In Ref. [5] we have shown that a superposition of three time-dependent components, where each had attained a steady-flow equilibrium between magic numbers, and which fitted each a $N_{r,\odot}$ peak and the adjacent interpeak region for lower mass numbers, gave already a quite impressive fit to the solar r-process abundances. The conditions were taken from a 'best fit' with static steady-flow assumptions to each of the above three mass regions. We will not yet discuss possible deviations from a perfect fit. In fact, in Ref. [5] we had combined the global FRDM masses [14] with the masses of Hilf et al. [17] for the $Z=39$ to 53 region, in order to avoid to some degree the otherwise huge abundance trough at $A \simeq 115$ before the $N=82$ shell closure.

In reality, an explosive astrophysical event will, of course, not only consist of three mass zones but a more continuous distribution. To simulate this, we have essentially just to transform the coarse grid of $n_n - T_9 - \tau$ conditions used in Ref. [5] into a finer one, which corresponds to a finer resolution of mass zones. As, in addition, each of the three $n_n - T_9$ curves in the 'band' shown in Fig. 1 of Ref. [6] give identical results for the respective abundance peaks, we can eliminate T_9 as free parameter. Figure 1 shows the three n_n-values resulting for a temperature of $T_9=1.35$ as open circles along the line marked by $\rho(n_n)$. Our three-component fit required a superposition with statistical weights of about 10:3:1. The calculations also required different time scales, covering a range from 1.2 s to 2.2 s, which are of the same order but slightly longer for the components which build the heavier nuclei. Higher neutron number densities lead to a more neutron-rich r-process path with shorter β-decay half-lives. This, however, does not compensate fully for the increasing number of β-decays

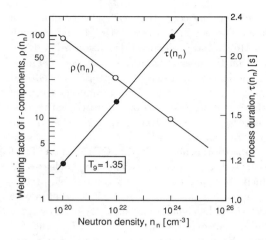

Figure 1: Weighting factors $\rho(n_n)$ and time scales $\tau(n_n)$ for r-process components. The conditions obtained for a 'best-fit' three-component superposition [5] are indicated by open circles and full dots, respectively.

necessary to populate all three r-abundance peaks. The sequence of three time scales is also shown in Fig. 1 as full dots along the line indicated with $\tau(n_n)$.

Among other things, stimulated by the *'words of caution'* by the Brussels group [9,10,3], we extended our three-component approach to a ten-component superposition by segmenting the neutron-density and time-scale ranges in ten equal logarithmic spacings. Figure 2 shows the resulting r-abundance curves for three [5] and ten components in comparison with the $N_{r,\odot}$ distribution. The nuclear-physics input is identical for both calculations. We do observe a slight improvement from the three-component to the ten-component superposition, but the major deviations persist: a trough around $A \simeq 115\text{-}125$ before the $A \simeq 130$ peak and two spikes at $A \simeq 112$ and $A \simeq 176$. The first deviation is essentially explained by a mass-formula effect (i.e. a too strong shell-effect when approaching $N=82$); and the two spikes are caused by deformation-dependent too long β-decay half-lives in transitional regions. Hence, the above result is a clear proof that our three-component fit [5] is a quite valid first-order approach, and that for reasonable superpositions which follow a natural continuum of stellar conditions, a quite good but not perfect fit is obtainable. The remaining deviations can, in fact, be utilized to set constraints on necessary nuclear-model features far from stability. This will be discussed in more detail later.

Dependence on Nuclear Masses

In the right part of Fig. 2 and in Fig. 3, we show global r-abundance curves from a superposition of ten $n_n - \tau$ components for three different nuclear mass sets with the corresponding QRPA calculations for $T_{1/2}$ and P_n values. In all three cases, identical conditions for the stellar parameters were used, as given in Fig. 1.

As already discussed above, the FRDM+Hilf 'hybrid' masses used in Fig. 2 yield the best possible agreement with the $N_{r,\odot}$ distribution. In the left part of Fig. 3, exclusively FRDM mass predictions are used. Apart from a more pronounced $A \simeq 115$ abundance trough, under the given $n_n - \tau$ conditions too little r-material is observed beyond the $A \simeq 130$ peak. This is due to a number of obviously too low S_n values in

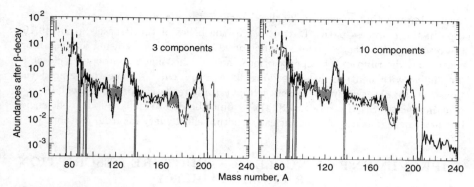

Figure 2: Global r-abundance fits with superpositions of three (left part) and ten time-dependent r-components (right part) calculated with weighting factors for neutron densities and time scales according to the exponential fits shown in Fig. 1. In order to test whether in a more continous distribution of $T_9-n_n-\tau$ components the (already 'minimized') deficiencies in the original fits [5] can be avoided, we again have used the FRDM masses below $Z=39$ and beyond $Z=53$. In the intermediate region corresponding to the $A\simeq 115$ abundance trough, the S_n values from the Hilf et al. mass formula [17] were chosen.

the r-process path just beyond the $N=82$ shell, which act as a bottle-neck in the r-process flow. In the right part of Fig. 3, the results from the ETFSI masses plus β-decay properties are shown. Here, pronounced abundance troughs at $A\simeq 115$ and 175 occur due to overly strong $N=82$ and $N=126$ shell corrections. As has already been discussed in Refs. [5,6], a consequence of the shell strengths is a wrong trend in the S_n values beyond neutron mid-shells. This behaviour of the FRDM and ETFSI

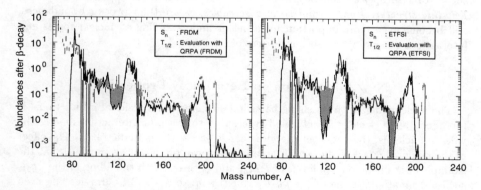

Figure 3: Global r-abundance fits with superpositions of ten r-components, calculated with the $n_n - T_9 - \tau$ conditions given in Fig. 1. Left part: S_n values and input for the QRPA calculations from FRDM [14]; right part: S_n-values and input for QRPA from ETFSI [16]. For discussion, see text.

masses implies that in certain localized regions there would exist *not a single* isotope in the r-process path. In contrast to the belief of the Brussels group [9,10], the resulting abundance troughs cannot be avoided, neither by using an even more continuous distribution of superpositions, nor by choosing r-components closer to β-stability with neutron densities down to $n_n \simeq 10^{17}$ cm^{-3}. Since the same deviations are also observed in the recent *astrophysically* realistic r-process calculations of neutrino-heated SN II ejecta [2,4,8], they cannot be due to stellar-model deficiencies but must be signatures of new nuclear-structure effects very far from β-stability.

IMPROVEMENT OF NUCLEAR-STRUCTURE DESCRIPTION FAR FROM STABILITY

Nuclear shell-structure between β-stability and unstable regions half-way to, and in limited cases *in* the r-process path are reasonably well studied by now, both experimentally and theoretically. Partly due to the recent developments in radioactive ion-beam (RIB) physics, the yet unknown structure of nuclei near and at the particle drip-lines has become one of the most exciting challenges today. Although the parameters of interactions used so far in mean-field theories were mainly determined as to reproduce known properties, these models have a surprisingly high global predictive power for unknown exotic nuclei (for nuclear-masses and β-decay properties see, e.g., Refs. [11] and [15], respectively). However, for the sometimes dramatic extrapolations to the limits of particle binding, the above parameters may not always be proper to be used at the drip-lines, in particular near classical shell closures. Therefore, new spectroscopic results from ISOL or RIB experiments far off stability will be essential to test the model predictions. Having in mind, however, that such exotic isotopes are only accessible in exceptional cases, our approach of *learning drip-line structure from astrophysical observables*, as are the r-process abundances, will become even more important for a better understanding of nuclear forces and interactions depending on isospin degrees of freedom.

New Information on Beta-Decay Properties

The shell model, with its single-particle (s.p.) basis and residual interactions, is fundamental to nuclear structure. Largely due to the p.n. interaction among the valence nucleons, the s.p. level energies change across the periodic table. The effect of the monopole part of the p.n. interaction has been discussed by Heyde [18] at the example of the neutron s.p. levels of $^{91}_{40}$Zr$_{51}$ and $^{131}_{50}$Sn$_{81}$. With the filling of the $\pi g_{9/2}$ shell a dramatic lowering of the $\nu g_{7/2}$ orbit (by up to 3 MeV) occurs, leading even to an inversion of the $\nu d_{5/2}$ and $\nu g_{7/2}$ states in the ^{132}Sn region. Consequences of this effect on the β-decay properties (in particular on the $T_{1/2}$), and on the validity of the 'waiting-point' concept for the $A \simeq 130$ r-abundance peak have been discussed in Refs. [5,6].

In the context of studying the influence of the p.n. interaction on $T_{1/2}$ values, very recently an experiment to identify neutron-rich Ag isotopes has been carried out at

Table 1: Comparison of experimental half-lives ($T_{1/2}$) of neutron-rich Ag isotopes with literature values and QRPA predictions (from Ref. [19]).

A	Beta-decay half-life, $T_{1/2}$ [ms]			
	Experiment		Theory	
	This work	Literature	QRPA-Nilsson	QRPA-F.Y.
121	1045 (40)	910 (60)	216	405
122	528 (11)	480 (80)	102	261
123	297 (6)	309 (15)	60	117
124	171 (10)	590 (80)	51	100
125	156 (7)	—	49	117
126	97 (8)	—	62	153
127	109 (15)	—	36	80

CERN/ISOLDE [19]. A newly developed chemically selective laser ion source based on resonance ionization of Ag atoms in a hot cavity was used in order to minimize disturbing isobar and molecular-ion contaminations. Although the astrophysically important 'waiting-point' nucleus ^{129}Ag$_{82}$ could not yet be observed unambiguously, the previously known $T_{1/2}$ of $^{121-124}$Ag were improved and the $T_{1/2}$ of $^{125-127}$Ag were measured for the first time. For all these isotopes, *longer* $T_{1/2}$ were obtained than predicted by our early straightforward QRPA calculations neglecting the p.n. interaction. In Table 1, the measured half-lives are compared with literature values and with two sets of QRPA shell-model predictions. The first data set represents results from straightforward calculations using Nilsson-model wave functions and neglecting effects from the p.n. residual interaction. In the second set, the more recent Folded-Yukawa s.p. model with an isospin-dependent energy term is used which takes into account the above p.n. interaction effect in an empirical way. With the observed $T_{1/2}$ trend, a rather reliable prediction of $T_{1/2} \simeq (120-140)$ ms is now possible for the $N=82$ 'waiting-point' isotope ^{129}Ag, in good agreement with our old *astrophysical request* of roughly 160 ms [13] deduced from the $A \simeq 130$ $N_{r,\odot}$ peak shape. As an example of possible effects on r-abundance fits due to improved nuclear-physics input, Fig. 4 shows the results of a *static* calculation with $N_{r,\odot}(Z)\lambda_\beta(Z)$=const. for the $A \simeq 130$ peak. In the left part, the fit obtained when using the straightforward $T_{1/2}$ predictions from our *old* QRP approach with Nilsson-model s.p. energies is displayed. The right part shows the calculated r-abundances derived from recently measured $T_{1/2}$ and P_n values of Ag, Cd and In isotopes, together with improved QRPA predictions using Folded-Yukawa or *experimental* s.p. levels. It is clearly evident from this figure, that a considerable improvement of the $A \simeq 130$ r-abundance fit can be achieved with the updated nuclear-physics input.

As has already been discussed qualitatively in Ref. [5], other effects due to the neglect of the p.n. residual interaction in both the mass models and the QRPA formalism to calculate GT strength functions (from which the theoretical $T_{1/2}$ and P_n values are derived) are the deficiencies in the development of quadrupole deformation

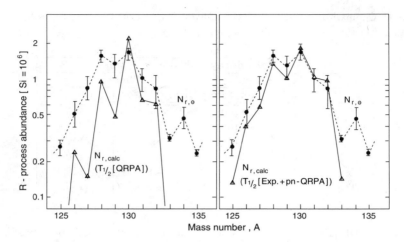

Figure 4: Static steady-flow fits $[N_{r,\odot}(Z)\lambda_\beta(Z)=\text{const.}]$ for the $A\simeq 130$ r-abundance peak. In the left part, $T_{1/2}$ and P_n values from the *old* straightforward QRPA calculations, neglecting effects from the p.n. interaction, are used. The right part shows a considerably improved fit, resulting from the application of *measured* β-decay properties together with *new* QRPA predictions with an empirical s.p. energy correction. For further discussion, see Refs. [5,6].

before and beyond neutron mid-shells ($N=66$ and 104). A first, more quantitative interpretation for the $A \simeq 120$ mass region in terms of a weakening of the $\pi f_{5/2} - \nu h_{11/2}$ residual interaction, simultaneously affecting S_n and $T_{1/2}$ values, was given by Walters [20]. More recently, similar consequences for drip-line nuclei have been discussed in terms of *neutron-skin* effects within state-of-the-art mean-field theories [12,21]. With the p.n. residual interaction being stronger than the p.p. and n.n. interactions, the proton potential becomes deeper when going away from β-stability, whereas the potential for neutrons becomes shallower and more diffused. As in the picture of Walters [20], this leads to the formation of a loosely-bound $\nu h_{11/2}$ neutron skin in the $A\simeq 120$ mass region, accompanied by a vanishing of the shell gap at $N=82$. It has already been discussed in Ref. [5], that another consequence for the exotic $40 \leq Z \leq 44$, $N \geq 72$ nuclides will be a less deformed shape as the *skin* neutrons feel smaller restoring forces against quadrupole deformation than the inner *core* neutrons. It was also speculated, that this could lead to longer half-lives as well as changes in the P_n values. Following the ideas given in Ref. [12], that the neutron s.p. spectrum for systems with large neutron excess can be approximated with that of a Nilsson model without the l^2 term in the one-body Hamiltonian, we have recalculated the $T_{1/2}$ and P_n values of 15 potential r-process-path isotopes (between $^{112}_{40}\text{Zr}_{72}$ and $^{126}_{44}\text{Rh}_{82}$) within the QRPA using the Nilsson model with the l^2 term reduced by 90%. As already discussed in Ref. [5] as a possible consequence of shape coexistence, we indeed obtained *longer* $T_{1/2}$ values, on the average by 50% compared to our previous calculations. Since within the 'waiting-point' concept, the $T_{1/2}$ of isotopes in the r-process path define the progenitor abundances, our *new* $T_{1/2}$ will help filling up the present $A \simeq 115$ r-abundance trough (see Figs. 2 and 3). It is worth to be mentioned in this context, that in all other mass regions – where the r-process path does not come close to the neutron drip-line – the $T_{1/2}$ will not be affected and, therefore, will not modify the present r-abundance fits.

New Information on Neutron Separation Energies

As is discussed in some detail in Refs. [5,6], a reproduction of the $N_{r,\odot}$ pattern in the vicinity of neutron shell closures requires a quenching of shell-effects close to the neutron drip-line which is obviously *not* contained in the global mass models [14,16] used so far. That such a request is not purely speculative, has already been shown for the $N=28$ and 50 shells in Refs. [22] and [23,5], respectively. Since meanwhile, the shape coexistence around $^{44}_{16}S_{28}$ has been confirmed by Skyrme-HF calculations [24], it seems worthwhile to check the results from such state-of-the-art mean-field models also for the $N=50$, 82 and 126 shells.

In a self-consistent, microscopic way nuclei can be described by the density-dependent Skyrme mean-field or by the relativistic mean-field (RMF) theory. In two recent papers [12,25], the behaviour of shell-effects in nuclei near the neutron drip-line have been examined within these approaches, leading to considerably different results. Whereas in the RMF [25], strong shell-effects for $N=82$ are obtained supporting the predictions of the FRDM and ETFSI mass models, the Skyrme P ansatz [12] results in a quenching of shell-effects as requested from our astrophysical considerations [5]. Although both microscopic theories still have their limitations, the second one seems to be more realistic. The contradictory results obviously originate from a different treatment of pairing correlations in drip-line nuclei. With the neutron Fermi energy being close to the top of the neutron potential well, the pairing force scatters part of the quasiparticles into continuum states. As the centrifugal barrier pushes up states of high angular momentum j, the energies of low-j states decrease relative to all other levels. In light and medium-heavy nuclei, low-j continuum states, which are located right above the shell gaps, enter these gaps and effectively lead to quenched shell-effects. For the $N=82$ region, this is shown in Fig. 4 at the example of the S_n values of the $N=81$ and 83 isotones below $^{132}_{50}Sn_{82}$. From the energy distance $[S_n(N=81)-S_n(N=83)]$, it is evident that the FRDM, the ETFSI and even the HFB model with the Skyrme III interaction [26], which all three use BCS or BCS-like pairing methods, show a very strong $N=82$ shell effect. In these approaches, $^{122}_{40}Zr_{82}$ would be (nearly) as neutron-magic as $^{132}_{50}Sn_{82}$. Only the Skyrme P ansatz with the *new* treatment of the pairing force results in a quenching of shell-effects.

With the promising features of the latter mean-field model [12], just prior to this conference we have started to perform static steady-flow r-abundance fits for the 'pathological' $A \simeq 115$ and 175 mass regions. Since so far the Skyrme P variant is limited to the calculation of the spherical shape, we only have replaced the mass predictions from the global FRDM [14] and ETFSI [16] models locally by the *new* S_n's around the magic $N=82$ and 126 shells. Under exactly the same $n_n - T_9$ conditions as requested for our 'best fits' in Ref. [5], we indeed observe a filling up of the abundance

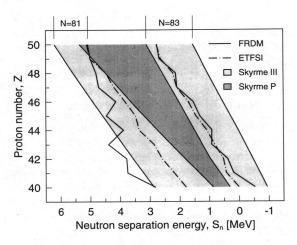

Figure 5: Comparison of S_n's for the isotones $N=81$ and 83 as obtained from different mass models (FRDM [14]: –; ETFSI [16]: –·–; Skyrme III [26]: light shaded area; Skyrme P [12]: dark shaded area). The energy distance $[S_n(N=81)-S_n(N=83)]$ between the isotones is a measure for the $N=82$ shell strength. Only the Skyrme P ansatz with the correct treatment of the pairing force leads to the required quenching of the $N=82$ shell-effect.

troughs to a large extent. A preliminary result for the $90 \leq A \leq 130$ mass region (the 2nd component with $n_n = 9.5 \times 10^{20} \text{cm}^{-3}$ and $T_9 = 1.20$) is shown in Fig. 6.

As has been discussed in Refs. [5,6], due to the too slow decrease of the S_n's of neutron-rich $Z>40$ isotopes in the FRDM and ETFSI models, *not a single* nuclide with $S_n \simeq 2$ MeV exists in the r-process path just below $N=82$. With the steeper S_n

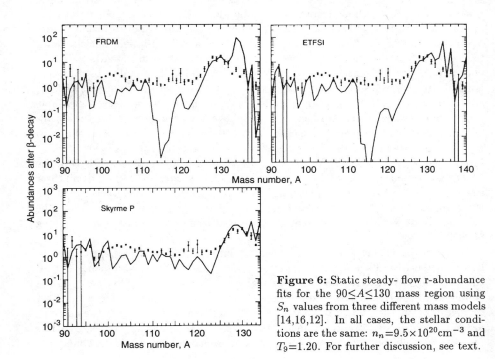

Figure 6: Static steady-flow r-abundance fits for the $90 \leq A \leq 130$ mass region using S_n values from three different mass models [14,16,12]. In all cases, the stellar conditions are the same: $n_n = 9.5 \times 10^{20} \text{cm}^{-3}$ and $T_9 = 1.20$. For further discussion, see text.

trend of the Skyrme P model, however, altogether seven isotopes (with $Z=40$ to 43, and $N=74$ to 80) become potential progenitors, their initial abundances being now sufficient to nearly reproduce the $N_{r,\odot}$ pattern in the $A \simeq 115$-125 mass region (see lower left part of Fig. 6). If, in addition to the *new* S_n values, also the *longer* $T_{1/2}$ discussed above will be used, a very good fit may be expected. A detailed analysis of multicomponent, time-dependent r-abundance calculations with the improved nuclear-physics input is in progress [27].

SUMMARY

We have shown that most of the recent concerns [3,9,10] about our deductive approach and the interpretation of the local deficiencies in the abundance fits are not justified. Obviously, when using selfconsistent nuclear-data sets based on the global FRDM and ETFSI models, also a continuous superposition of a multitude of r-components will not be able to prevent local deviations from the observed $N_{r,\odot}$ distribution. With this result, also the hope that the 'microscopic' ETFSI masses would be more reliable in regions of the r-process path than those from earlier macroscopic or macroscopic-microscopic models could not be realized. However, new experimental results together with recent mean-field calculations confirm our earlier nuclear-physics predictions [5,6], and indicate partly new nuclear-structure signatures of neutron drip-line nuclei which, indeed, seem to be preserved in the $N_{r,\odot}$ pattern.

In a forthcoming paper [28], we will present results on a detailed testing of effects from different nuclear-mass and half-life models on theoretical r-abundance distributions. Our main conclusion from this parameter study is, that with the availability of todays microscopic nuclear-structure models, r-process calculations using gross theories from the mid-seventies or applying *'mixed'* physics parameter sets of different sophistication, even when combined with the best SN models, should no longer be persued. It is mainly such inconsistent approaches that led to recent statements about the limited reliability of r-process calculations due to *'numerous uncertainties'* in nuclear and astrophysical parameters. When considered in a more optimistic way, on the one hand our deductive approach has given valuable constraints on stellar parameters supporting the hot-entropy scenario of SN II explosions, and on the other hand has helped to initiate recent improvements of nuclear-structure models. A deeper understanding of the coupling between nuclear physics at the drip-lines and explosive nucleosyntheses will remain an exciting challenge for future work.

ACKNOWLEDGEMENTS

This work was performed in collaboration with colleagues from many institutions: Academy of Sciences Troitzk, CalTech, CERN/ISOLDE, LANL, and Universities of Basel, Mainz, Maryland, Montréal, Warsaw and Wien. My special thanks are due to Friedel Thielemann; within our invariably stimulating collaboration over the last ten years many problems could be solved ... *from Heidelberg to Ostpreußen, around five o'clock in the morning.*

Financial support from BMFT (06 MZ 465), and DFG (Kr 806/3 and 436 RUS 17/26/93) is gratefully acknowledged.

REFERENCES

1. Cowan, J.J., et al., Phys. Rep. **208**, 267 (1991).
2. Meyer, B.S., et al., Ap. J. **399**, 656 (1992).
3. Howard, W.M., et al., Ap. J. **417**, 713 (1993).
4. Takahashi, K., et al., Astron. Astrophys. **286**, 857 (1994).
5. Kratz, K.-L., et al., Ap. J. **403**, 216 (1993).
6. Kratz, K.-L., et al., Proc. 2nd Int. Symp. on Nuclear Astrophysics, *'Nuclei in the Cosmos'*, IOP Publ. Ltd, p. 349 (1992).
7. Woosley, S.E., et al., Ap. J. (1994), in print.
8. Käppler, F., et al., Ap. J. **417**, 713 (1993).
9. Arnould, M., and Takahashi, K., *'Origin and Evolution of the Elements'*, Cambridge Univ. Press, p. 396 (1992).
10. Goriely, S., and Bouquelle, V., Proc. 2nd Int. Symp. on Nuclear Astrophysics, *'Nuclei in the Cosmos'*, IOP Publ. Ltd, p. 595 (1992).
11. Möller, P., and Nix, J.R., Proc. Int. Conf. NFFS-6 and AMCO-9, IOP Conf. Ser. **132**, 43 (1993).
12. Dobaczewski, J., et al., Phys. Rev. Lett. **72**, 981 (1994); and priv. communication.
13. Kratz, K.-L., Rev. Mod. Astron **1**, 184 (1988).
14. Möller, P., et al., At. Data Nucl. Data Tables (1994), in print.
15. Möller, P., and Randrup, J. Nucl. Phys. **A514**, 1 (1990).
16. Aboussir, Y., et al., Nucl. Phys. **A549**, 155 (1992).
17. Hilf, E.R., et al., Proc. Int. Conf. NFFS-3, CERN **76-13**, 142 (1976).
18. Heyde, K., Res. Rep. in Physics, *'Nuclear Structure of the Zirconium Region'*, p. 3 (1988).
19. Fedoseyev, V.N., et al., Proc. 7th Int. Symp. on *'Resonance Ionisation Spectroscopy and its Applications'*, AIP Press (1994), in print.
20. Walters, B.W., Proc. Int. Symp. on *'Nuclear Physics of our Times'*, World Scientific, p. 457 (1993).
21. Fukunishi, N., et al., Phys. Rev. **C48**, 1648 (1993).
22. Sorlin, O., et al., Phys. Rev. **C47**, 2941 (1993); and contrib. to this conference.
23. Kratz, K.-L., et al., Phys. Rev. **C38**, 278 (1988).
24. Werner, T.R., et al., Phys. Lett. **B** (1994), in print.
25. Sharma, M.M., et al., Phys. Rev. Lett. **72**, 1431 (1994).
26. Langanke, K., and Vogel, P., priv. communication (1994).
27. Chen, B., et al., submitted to Phys. Lett. **B** (1994).
28. Thielemann, F.-K., and Kratz, K.-L., to be submitted to Astron. Astrophys. (1994).

The rp-process, rates from direct and indirect reaction studies

Michael Wiescher

University of Notre Dame, Dept. of Physics, Notre Dame, IN 46556

The reaction sequences and the reaction flow in the CNO cycles and the rp-process are important for the understanding of the energy generation and nucleosynthesis of heavy elements in hot and explosive stellar hydrogen burning. Many of the involved reaction rates in particular for the proton capture on unstable short lived nuclei are only approximated and have not been verified experimentally. New experimental approaches are therefore neccessary to determine these rates more reliable. This requires either the use of intense radioactive beams to measure directly the reaction cross sections over a wide energy range, or in a more classical approach the study of the level structure of the involved compound nuclei in the corresponding excitation range. On particular examples of reactions in the hot CNO cycles and the rp- process, indirect experimental approaches will be presented and discussed.

I. INTRODUCTION

The nuclear processes and reaction sequences in stellar and explosive hydrogen burning scenarios determine the time scales, energy generation and nucleosynthesis at different temperature and density conditions (1). Large reaction network codes have been developed in recent years to predict, to understand, and to simulate these burning processes. The reliability of these calculations depends strongly on the 'nuclear input' parameters, the reaction rates of the potentially involved nuclear reactions. In the following sections I will discuss on two examples current experimental methods to determine the reaction rates of weak capture processes. Hydrogen burning takes place via two kinds of possible reaction sequences. Chains of (p,γ) reactions and β-decays fusion the initial material towards heavier masses and characterize the pp-chains and the rp- process. Quite often the reaction flow is not serial but cyclic. This occurs if the α threshold is lower than the proton threshold in a compound nucleus of a proton capture reaction. This is mainly the case for nuclei close to the line of stability and therefore occurs predominantly at lower stellar temperature conditions. Cyclic reaction sequences of (p,γ) and β-decays are closed by a (p,α) reaction, like the CNO cycles, and the initial catalytic material is stored in the cycle for a certain amount of time τ_{cycl}. This will cause a temporary enrichment of the abundances of the isotopes within the cycle which can only be depleted by breakout or leakage reactions. The dominant leakage reaction at lower temperature conditions is the (p,γ) competing with the (p,α)-reaction. These reaction branchings determine the time period τ_{cycl},

$$\tau_{cycl} = \left(\frac{<\sigma v>_{(p,\gamma)} + <\sigma v>_{(p,\alpha)}}{<\sigma v>_{(p,\gamma)}} \right) \cdot \left[\sum_j \tau_j^c \right], \qquad (1)$$

where τ_j^c are the lifetimes of the isotopes in the reaction cycle and $<\sigma v>_{(p,\gamma),(p,\alpha)}$ are the reaction rates for the (p,γ) and (p,α) reactions, respectively. Large break-out rates limit the storage time to the sum of the lifetimes of the isotopes in the cycle. For small leakage rates the storage time becomes large, eventually larger than the macroscopic time scale of the scenario. In this case the cycle is the endpoint of the rp-process nucleosynthesis and the equilibrium abundance distribution in the cycle will reflect the final nucleosynthesis abundance distribution (1). Figure 1 shows the various reaction cycles (CNO, NeNa, SiP, and SCl) at conditions of hot hydrogen burning. To understand the influence of reaction cycles it is important to

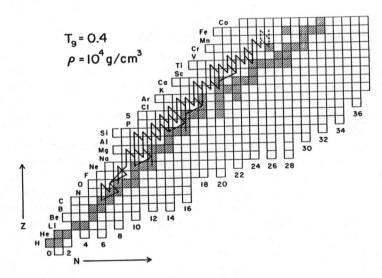

FIG. 1. Sequential and cyclic reaction sequences in hot hydrogen burning. Indicated are the reaction flows integrated over a period of t=100 s for a constant temperature of T=4·10^8 K and a density of ρ=10^4 g/cm^3. Solid lines indicate the dominant nuclear flow, the dashed lines show the flow which is about an order of magnitude weaker.

study the (p,α) reactions like ^{15}N(p,α), ^{17}O(p,α) and ^{18}F(p,α) closing the CNO-cycles as well as other (p,α) reactions on T=1/2 nuclei like ^{23}Na, ^{31}P and ^{35}Cl, which close the cycles in the higher mass ranges, like the NeNa-, SiP-, and SCl-cycle shown in figure 1.

Because of the typically high reaction Q-values the (p,γ) reactions are characterized by many resonances and the reaction rates can be very well approximated by Hauser-Feshbach calculations (2).

The competing (p,α) reaction rates are, however, in many cases determined by only a few single resonances depending on the selection rules for the α-decay of the populated compound levels. The reaction rates [cm^{3}s^{-1}mole^{-1}], given as a function of temperature T_9 (in 10^9 K) and the reduced mass A,

$$<\sigma v> = 2.56 \cdot 10^{-13} (A \cdot T_9)^{-3/2} \sum_i^n \omega\gamma_i \cdot exp\left(\frac{-11.605 E_{R_i}}{T_9}\right), \qquad (2)$$

depend mainly on the number of resonance levels n, the resonance energies E_{R_i} [MeV] and the resonance strengths $\omega\gamma_i$ [MeV],

$$\omega\gamma = \frac{2J+1}{2(2J_t+1)} \cdot \frac{\Gamma_p \cdot \Gamma_\alpha}{\Gamma_{tot}}; \qquad (3)$$

J and J_t is the spin of the resonance level and target nucleus, Γ_p, Γ_α, and Γ_{tot} is the proton and α partial width and the total width of the resonance level, respectively.

The cycles are only closed for temperature conditions at which the (p,α) rates are larger, or at least in the same order of magnitude as the competing (p,γ) rates, this will only occur in the case of a strong (p,α) resonance in the temperature range of the event, as e.g. in the first CNO-cycle and the NeNa-cycle (3).

In the following I will discuss on two examples, ^{18}F(p,α)^{15}O and ^{35}Cl(p,α)^{32}S, different experimental techniques to determine the stellar reaction rate of (p,α) reactions.

II. ^{18}F(p,α)^{15}O

The reaction ^{18}F(p,α)^{15}O competes with the β-decay of ^{18}F and becomes important in dense stellar environments of temperatures T$\geq$2·10^8 K. The reaction closes the second hot CNO-cycle, ^{16}O(p,γ)^{17}F(p,γ)^{18}Ne($\beta^+\nu$) ^{18}F(p,α)^{15}O (see figure 1) (4).

A direct measurement of ^{18}F(p,α)^{15}O is difficult because of the relatively short life time (τ=158.3 m) of the target nucleus ^{18}F. Therefore mesurements have been performed to study the nuclear structure of the compound nucleus ^{19}Ne above the proton threshold, to determine the level density, as well as the excitation energies and the decay partial widths of the proton unbound states. The measurement of these parameters allows a better estimate for the reaction rate.

To determine the energies of the excited states in ^{19}Ne, the excitation range above the proton threshold of S_p=6.411 MeV has been studied via the ^{19}F(^{3}He,t)^{19}Ne reaction at the FN-tandem accelerator of the University of Notre Dame. In addition the mirror reactions ^{16}O(^{6}Li,^{3}He)^{19}F and ^{16}O(^{6}Li,t)^{19}Ne have been investigated to determine the analog levels in ^{19}F and ^{19}Ne, respectively. Figure 2 shows the spectra for ^{16}O(^{6}Li,^{3}He)^{19}F and ^{16}O(^{6}Li,t)^{19}Ne measured at 32 MeV beam energy using the Notre Dame spectrograph at an angle of 15°. Also shown for comparison is the spectrum of

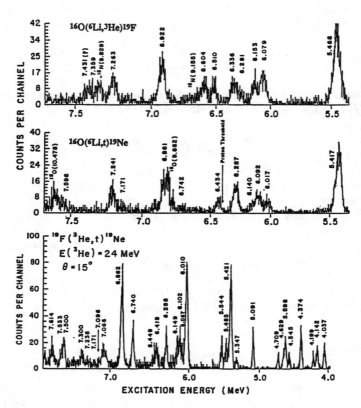

FIG. 2. ^{19}Ne and ^{19}F spectra obtained by ^{19}F(^{3}He,t)^{19}Ne, ^{16}O(^{6}Li,t)^{19}Ne and ^{16}O(^{6}Li,^{3}He)^{19}F, respectively.

^{19}F(^{3}He,t)^{19}Ne measured at an beam energy of 24 MeV and a spectrograph angle of 15°. The analog structure in the first two spectra can be clearly observed, but the density of populated states is considerably smaller than in the ^{19}F(^{3}He,t) measurement. The resonance energies in the ^{18}F(p,α) reaction can be directly calculated from the measured excitation energies.

To determine the resonance strengths additional information about spins and the partial widths of the resonance levels are required. The relative particle widths Γ_p and Γ_α have been obtained via ^{19}F(^{3}He,t-p,α) coincident measurements. These experiments were performed at the Princeton QDDD-spectrograph to take advantage of the significantly larger solid angle. The spectrograph was positioned at an angle of 0°, three large-area Si-detectors were positioned at 90°, 110° and 145° to measure the protons and α particles in coincidence with the tritons populating the unbound states in ^{19}Ne. The relative partial widths are directly proportional to the number of coincident events $N_{coinc}(t-p,\alpha)$,

$$\frac{\Gamma_{p,\alpha}}{\Gamma_{tot}} = \frac{N_{coinc}(t-p,\alpha)}{N(t)} \cdot (\Omega_{p,\alpha} W(\theta_{p,\alpha}))^{-1}, \qquad (4)$$

N(t) is the number of detected tritons populating the level, $W(\theta_{p,\alpha})$ is angular distribution, which can be dirctly obtained from the experiment, and $\Omega_{p,\alpha}$ is the solid angle of the Si-detectors, directly determined from coincidence measurements for states with $\Gamma_\alpha = \Gamma_{tot}$. Figure 3 shows the particle spectra in coincidence with the triton groups corresponding to the excited states above 6.74 MeV in ^{19}Ne. While the states at 6.740 and 6.862 MeV decay mainly into

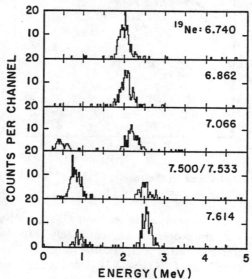

FIG. 3. Proton and α particle spectra obtained in coincidence with triton groups from ^{19}F(^{3}He,t)^{19}Ne.

the α channel, the higher excited states show also the decay into the proton channel.

The obtained experimental information is not sufficient to calculate the resonance strengths in ^{18}F(p,α),

$$\omega\gamma = \frac{2J+1}{6} \cdot \Gamma_p \left(\frac{\Gamma_\alpha}{\Gamma_{tot}}\right) \qquad (5)$$

and the remaining uncertainty is the proton widths Γ_p. This has been estimated by adopting the level parameters of the mirror states in ^{19}F. Figure 4 shows the resulting reaction rate for ^{18}F(p,α)^{15}O as a function of temperature. The reaction rate is dominated by the contributions of the broad low energy state at 6.442 MeV (E_R=0.038 MeV) and at higher temperatures T$\geq$ 2·10^{8}K by the resonance level at 6.741 MeV (E_R=0.329 MeV). Comparing

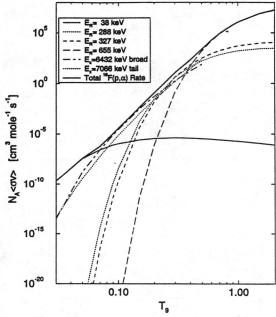

FIG. 4. Resonant contributions to the reaction rate of ^{19}F(p,α)^{19}Ne as a function of temperature

the rate with the rate of the competing reaction ^{18}F(p,γ)^{19}Ne indicates that the latter radiative capture process is approximately three to four orders of magnitude weaker over the entire temperature range 1·10^7 - 1·10^9 K. This indicates that the second hot CNO cycle is closed and that the initial ^{16}O material is rapidly processed towards ^{15}O.

III. ^{35}Cl(p,α)^{32}S

Another example is the reaction ^{35}Cl(p,α)^{32}S, which would close the SCl-cycle. Because of the high proton threshold, S_p=8.51 MeV, the level density in the compound nucleus is high and many resonances determine the reaction rate of ^{35}Cl(p,γ)^{36}Ar. Figure 5 shows the excitation curve for ^{35}Cl(p,γ)^{36}Ar. Several new resonances have been observed, but none of these resonance levels could be identified in the ^{35}Cl(p,α)^{32}S reaction. Large background from elastic scattering prevents a direct measurement of low energy (p,α) resonances. Therefore only upper limits have been obtained for the ^{35}Cl(p,α) resonance strengths (5). Because only natural parity states can decay into the α-channel and contribute to the reaction rate of ^{35}Cl(p,α)^{32}S, a considerably smaller resonance density is expected. To investigate the single particle as well as α-particle structure of the proton unbound states in ^{36}Ar single particle transfer reactions, ^{35}Cl(^{3}He,d)^{36}Ar, as well as α-transfer reaction, ^{32}S(^{6}Li,d)^{36}Ar, have been studied. These measurements were performed at the Notre Dame spectrograph. Figure 6 shows the respective spectra. Despite

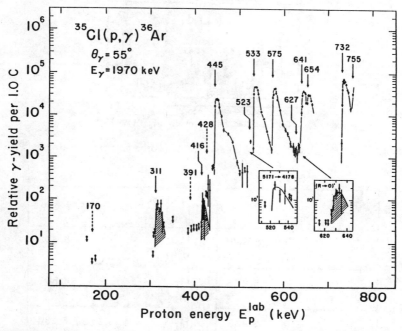

FIG. 5. The excitation function of ^{35}Cl(p,γ)^{36}Ar. Indicated are the energies of the observed resonances. The resonances marked by the dashed regions have been first reported by Iliadis et al. 1994.

the limited resolution in the (^{6}Li,d)-study, the (^{6}Li,d)-spectrum shows that only a few of the many levels seen in the (^{3}He,d)-spectrum are populated in the α-transfer, indicating natural spin for these states.

To determine the actual α- and γ-decay width of these states, ^{35}Cl(^{3}He,d-α) and ^{35}Cl(^{3}He,d-γ) coincidence measurements have been performed at the QDDD-spectrograph at Princeton using a set-up similar to the one decribed before. In addition a 12.7 cm x 10.2 cm NaI-detector was positioned in close proximity to the target. Figure 7 shows the particle- and gamma-spectra in coincidence with the different deuteron groups populating the proton unbound states in ^{36}Ar.

132   The rp-Process

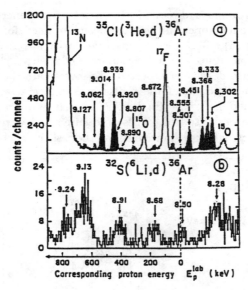

FIG. 6. ^{35}Cl(^{3}He,d) and ^{32}S(^{6}Li,d) spectra of proton unbound states in ^{36}Ar.

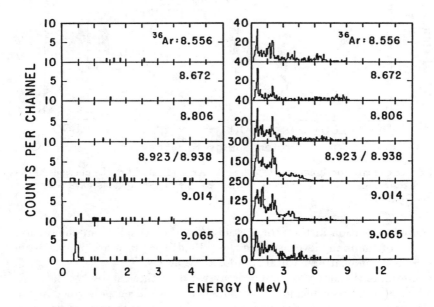

FIG. 7. Particle and γ spectra taken in coincidence with deuteron groups to proton unbound states from ^{35}Cl(^{3}He,d)^{36}Ar.

Except for the proton decay of the 9.065 MeV state ($\Gamma_p/\Gamma_{tot}=0.92\pm0.11$) no particle decay has been observed. This indicates an approximate upper limit for the relative α-width of $\Gamma_\alpha/\Gamma_{tot} \leq 0.1$. Analysis of the γ-d-coincidence spectra yields for all shown states a relative γ-width of $\Gamma_\gamma/\Gamma_{tot} \approx 1.0$ and for the 9.065 MeV state, $\Gamma_\gamma/\Gamma_{tot}=0.17\pm0.04$. These results show that for all low energy proton unbound states the α-partial width is at least one order of magnitude smaller than the γ-partial width. Therefore the reaction rate for ^{35}Cl(p,α) is considerably weaker than the reaction rate of ^{35}Cl(p,γ) at temperatures T$\leq 10^9$ K.

Figure 8 shows the ratio of the ^{35}Cl(p,α) and ^{35}Cl(p,γ) reaction rate as

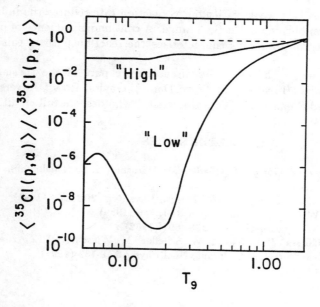

FIG. 8. The ratio of the reaction rate of ^{35}Cl(p,α)^{32}S and ^{35}Cl(p,γ)^{36}Ar as a function of temperature. The solid lines indicate the range of uncertainty in the ^{35}Cl(p,α)^{32}S reaction. The upper limit corresponds to the experimental upper limit for the relative α width, the lower limit neglects all low energy resonance contributions in the (p,α) channel.

a function of temperature. The results suggest that the ^{35}Cl(p,γ) break out reaction from the SCl-cycle dominates at low temperatures. Therefore only a weak cycle pattern characterizes the reaction flow in this mass range. This result also indicates that in hot hydrogen burning only less than ten percent of the processed S,Cl-material remains stored in the SCl-cycle, while the bulk of the material is being processed towards Ca.

IV. CONCLUSION

A direct measurement of both reactions, $^{18}F(p,\alpha)^{15}O$ and $^{35}Cl(p,\alpha)^{32}S$ is difficult. Indirect methods, such as discussed here, can give important additional information about the level structure of the compound nucleus. This allows a reliable estimate of the reaction rates usually sufficient for nucleosynthesis network calculations.

Based on these rates it can be concluded that the $^{18}F(p,\alpha)^{15}O$ reaction is four orders of magnitude faster than the competing $^{18}F(p,\gamma)^{19}Ne$ reaction. This causes a rapid conversion of initial ^{16}O into ^{15}O during hot hydrogen burning in the CNO cycles.

In the other case the weak $^{35}Cl(p,\alpha)^{32}S$ reaction rate reduces the influence of cyclic hydrogen burning in the S,Cl range. A continuous sequence of proton capture reactions and β-decays characterizes the reaction path in this mass range.

Acknowledgement: The examples discussed are part of the thesis work of Christian Iliadis (University of Notre Dame), Gaylon Ross (University of Notre Dame) and Sinan Utku (Yale University) who deserve full credit.

REFERENCES

1. L. Van Wormer, J. Görres, C. Iliadis, M. Wiescher, F.K. Thielemann, Ap.J. (1994) in press
2. J.J. Cowan, F.K. Thielemann, J.W. Truran, Phys.Rep. **208**, 268 (1991)
3. J. Görres, M. Wiescher, C. Rolfs Ap.J. **343**, 356 (1989)
4. M. Wiescher, K.U. Kettner, Ap.J. **263**, 891 (1982)
5. C. Iliadis, J. Görres, J.G. Ross, K.W. Scheller, M. Wiescher, R.E. Azuma, G. Roters, H.P. Trautvetter, H.C. Evans, Nucl.Phys.A**571** 132 (1994)

Weak Interaction Rates

E. Sugarbaker

Department of Physics, The Ohio State University, Columbus, Ohio 43210

> I review available techniques for extraction of weak interaction rates in nuclei. The case for using hadron charge exchange reactions to estimate such rates is presented and contrasted with alternate methods. Limitations of the (p,n) reaction as a probe of Gamow-Teller strength are considered. Review of recent comparisons between beta-decay studies and (p,n) is made, leading to cautious optimism regarding the final usefulness of (p,n)- derived GT strengths to the field of astrophysics.

INTRODUCTION

The rate at which a weak interaction process proceeds appears frequently as a critical component of astrophysical studies. Calculation of these rates requires knowledge of the relevant nuclear matrix elements. This paper shall concentrate on the experimental attempts to determine these needed matrix elements and shall leave presentation of astrophysical motivations to other papers at this conference.

Often the dominant weak interaction process of interest is associated with the "allowed" Fermi (F) and Gamow-Teller (GT) operators. Due to limitations associated with using weak interaction probes to directly determine GT strength distributions, charge-exchange reactions, primarily the (p,n) reaction, have been used to estimate the magnitude of these nuclear matrix elements. My aim here is to convey to potential users of such information, the strengths and weaknesses associated with using this strong interaction probe in place of better understood beta decay techniques.

WEAK INTERACTION COUPLING

Allowed beta decay proceeds via a coupling to the weak interaction field through two isovector operators: a spin-independent (Fermi) and spin-dependent (Gamow-Teller) operator. The reduced nuclear matrix elements B(F) and B(GT), respectively, of these operators between the initial and final nuclear wave functions are best determined from half-life and angular correlation measurements in beta decay. The coupling to the weak field is well understood, making extraction of the "weak interaction" matrix elements (the topic of this paper) very simple via the relation

$$\frac{K'}{ft} = g_V^2 B(F) + g_A^2 B(GT). \tag{1}$$

I shall usually employ units such that for free neutron decay, $B(F) = 1$ and $B(GT) = 3$. Alternate units for B(GT) for which neutron decay is $(g_A/g_V)^2$ times larger can also be found in the literature, indicative of the difficulty associated with actually performing the separation of interaction from nuclear structure implied by Eq. (1).

Determination of these nuclear matrix elements in more complex nuclei is often required to calculate the impact of weak-interaction processes in astrophysical problems. This raises complications, since while the Fermi strength remains concentrated in a narrow resonance, the GT strength is highly fragmented over at least the entire nuclear-excitation region. This leaves much of the GT strength distribution inaccessible to beta decay, forcing one to rely on alternate probes to determine most of these matrix elements.

One alternative is to use a calibrated neutrino flux of adequate energy to populate GT strength associated with numerous excited states. For instance, a very intense ^{51}Cr ν source is being used to determine the integrated capture cross section associated with transitions to the ground and first few excited states of ^{71}Ge in the GALLEX solar neutrino detector (1). Higher-lying GT strength can be probed with muon-decay neutrinos, as in the attempt at LAMPF to determine the total B(GT) below the neutron-emission threshold for a ^{127}I detector (2). Such techniques can only provide an integrated value of GT strength over all transitions accessible to the maximum energy neutrino. This may still prove to be useful information, given the uncertainties discussed below associated with the overall absolute normalization of GT strength functions from the (p,n) studies (3). However, when the maximum ν energy is large, significant corrections may need to be applied to the measured integrated capture cross section to account for forbidden strength contributions.

STRONG INTERACTION COUPLING

Nucleon charge-exchange reactions are also purely isovector in character, but proceed via a strong effective interaction derived from a potential having the form

$$\begin{aligned} v = &v_o(r) + v_\sigma(r)\vec{\sigma}\cdot\vec{\sigma} + v_\tau(r)\vec{\tau}\cdot\vec{\tau} + v_{\sigma\tau}(r)\vec{\tau}\cdot\vec{\tau}\vec{\sigma}\cdot\vec{\sigma} \\ &+ v_{LS}(r)\vec{L}\cdot\vec{S} + v_{LS\tau}\vec{L}\cdot\vec{S}\vec{\tau}\cdot\vec{\tau} + v_T(r)S + v_{T\tau}(r)S\vec{\tau}\cdot\vec{\tau}. \end{aligned} \tag{2}$$

The central third and fourth terms of this potential can connect via isospin-flip and/or spin-flip operators the same initial and final states as are of interest in the above weak interaction cases. While the possibility to utilize nucleon charge-exchange reactions to extract the desired GT matrix elements was recognized long ago (4), reasonable selectivity in the reaction mechanism to this component (fourth term in Eq. (2)) could only be achieved with the advent of the (p,n) studies at IUCF at beam energies above 100 MeV. An empirical proportionality between very forward-angle differential cross sections and the reduced nuclear matrix elements, B(F) and B(GT), has been suggested (5). A full discussion of this proportionality is presented in detail by Taddeucci et al. (6). The underlying "model" for such a proportionality is that at zero momentum transfer and energy loss, $(q,\omega) = 0$, the cross section for an allowed transition α in a simplified DWIA calculation should be represented approximately by

$$\sigma_\alpha(q,\omega = 0) \simeq K_\alpha N_\alpha |J_\alpha|^2 B(\alpha), \quad (3)$$

where K_α, N_α, and J_α are a kinematic factor, distortion factor, and the volume integral of the relevant interaction component, respectively. In reality, even at 0°, the above cross section cannot be observed. A correction $F(q,\omega)$, usually of order few % at lower excitation energies, is required to transform the measured $\sigma(0°)$ back to the value appropriate for q and ω equal to zero, at which Eq. (3) is believed to be most valid. It has therefore been convenient to define the "unit cross section" $\hat{\sigma}_\alpha$ as the cross section defined in Eq. (3) per unit $B(\alpha)$. Observation of large (40%) variation in unit cross sections among neighboring nuclei has thwarted all efforts at establishing a global proportionality based on unit cross sections (6). However, similar proportionality for all GT states in a given nucleus may still be possible, if numerous uncertainties in the reaction mechanism can be controlled. Using a $\hat{\sigma}_{GT}$ determined from one strong transition for which the B(GT) is known from beta decay, one may estimate the GT strength of other non-Fermi states "i" by

$$B_i(GT) = \frac{\sigma_i(0°)}{\hat{\sigma}_{GT} F(q,\omega)}. \quad (4)$$

If *no* state of known B(GT) is available for a given nucleus, one must normalize the full GT strength function with a $\hat{\sigma}_{GT}$ estimated from theory (via Eq. (3)), from neutrino capture measurements, or from yet another empirical (p,n) relationship between the $\hat{\sigma}$ for F and GT transitions in a specific nucleus. The latter approach has often been employed. It is based on the observation that in many cases studied, the ratio of GT to F unit cross sections,

$$R^2 = \frac{\hat{\sigma}_{GT}}{\hat{\sigma}_F} \propto \frac{|J_{\sigma\tau}|^2}{|J_\tau|^2}, \quad (5)$$

is reasonably A-independent. The ^{14}C$(p,n)^{14}$N(2.31 MeV) reaction provides the most isolated F transition with which to investigate the full energy dependence of R^2 (7). Over the energy range from 50 to 200 MeV, it was determined that $R = E_p/E_o$, where $E_o = 55.0 \pm 1.7$ MeV (6). The assigned error is based on the standard deviation of E_o values observed for other even-A nuclei. Unfortunately, it appears that some odd-A nuclei agree with this value and some do not (8). An example of each case will be discussed below. Until the source of this fluctuation is determined, absolute calibration of odd-A GT strength functions, *in cases where a moderate-sized B(GT) value from beta decay is not available for at least one isolated transition*, will be uncertain by at least ±20%.

EXAMPLES

A review of the results obtained from many intermediate energy (p,n) studies has recently been completed (9). Three specific cases shall be discussed below to provide examples of the success and present difficulties associated with this probe of GT matrix elements.

The ^{71}Ga case. Due to its low threshold energy for neutrino capture, the GT strength function in ^{71}Ge has been of significant interest. In this case, the B(GT) of the ground state transition is known from beta decay to be 0.091 (10). The B(GT) distribution obtained from the (p,n) reaction at $E_p = 120$ MeV (11) is presented in Fig. (1). If the nominal value of E_o is used to calibrate the GT strengths from the observed Fermi cross section (assumed to completely define the large narrow peak at an excitation energy of 8.9 MeV), a value of 0.089 ± 0.007 is obtained (11). These two values for the ground state B(GT) are in excellent agreement, even though this odd-A case might have been expected to be among those having an anomalous value of E_o.

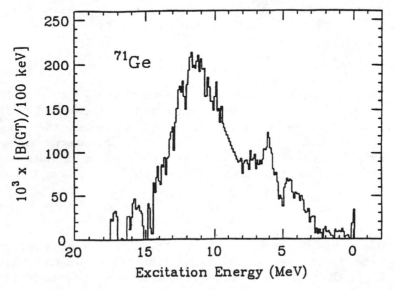

FIG. 1. The B(GT) distribution in ^{71}Ge from the ^{71}Ga$(p,n)^{71}$Ge reaction at $E_p = 120$ MeV.

TABLE 1. Comparison of the B(GT) values in ^{35}Ar from (p,n) and beta decay.

E_x (MeV)	B(GT)$_{\beta decay}$	B(GT)$_{(p,n)}$
1.18	0.0311±0.001	0.032±0.005
1.75	0.014±0.001	0.018±0.003
2.6	0.0385±0.0016	0.030±0.005
2.98	0.041±0.002	0.032±0.005
3.88	0.0585±0.007	0.065±0.008

A comparison of these results with those obtained at a much lower bombarding energy of 35 MeV (12) for the first excited state at 175 keV provides an excellent example of the intrusion of $\Delta L > 0$ reaction strength at 0° (13). Even at the higher bombarding energies, above an excitation energy of about 12 MeV, the dominant contribution to the 0° spectrum is no longer GT strength, but is "forbidden" $\Delta L > 0$ strength. Multipole decomposition of the forward angular distribution of the differential cross sections has often been used to estimate the degree to which this higher-multipolarity transition strength might "contaminate" the determination of B(GT). However, it has recently been pointed out that tensor components of the strong force could mix the amplitudes in such a way that multipole decompositions could lead to incorrect assignment of GT values (14). Indeed, experimental evidence of non-central interaction terms in polarization transfer observables has recently been reported (15). One must remain aware of the limitations in the validity of Eqs. (3-5).

The ^{35}Cl case. In the ^{35}Cl$(p,n)^{35}$Ar reaction the ground state has both F and GT components, and numerous weakly populated excited states below 5 MeV have B(GT) known from beta decay. This region of the (p,n) spectrum is displayed in Fig. (2). This case therefore provides an excellent test of the proportionality of B(GT) to cross section for many transitions within a given nucleus. Application of the observed average $\hat{\sigma}_{GT}$ of 11.92±1.38 mb/sr/(unit B(GT)) leads to the values given in Table 1. Under this assumption of a uniform value of $\hat{\sigma}_{GT}$, one obtains only fair (about ±20%) agreement with beta decay. As mentioned above, for such weak transitions, it is quite likely to have some non-GT operators contributing to the cross section. Polarization transfer measurements using the $(\vec{p}, \vec{n})$ reaction can be used to determine the fraction of the mixed F and GT ground state σ that is characteristic of a pure GT transition. The transverse spin-transfer coefficient $D_{NN}(0°)$ is the ratio of outgoing neutron polarization to incoming proton polarization. Using nominal values of $D_{NN}(0°)$ for the F and GT components of this mixed transition, one can find the $\sigma_{GT}(0°)$ (16). The above $\hat{\sigma}_{GT}$ can then be used to obtain a value of 0.054±0.009 for the B(GT) component of the ^{35}Ar ground state transition. This compares well with the beta decay value of 0.0496±0.0027. Interestingly, the implied $\hat{\sigma}_F$ results in a value of E_o of only about 32 MeV (16).

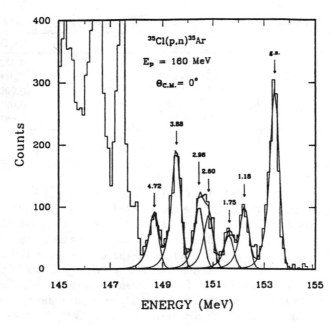

FIG. 2. Low-lying differential cross section in the ^{35}Cl$(p,n)^{35}$Ar reaction.

The ^{37}Cl case. Perhaps the most interesting comparison between beta decay and (p, n) can be made for the case of ^{37}Cl. The recent measurement of the β-delayed proton spectrum from excited ^{37}K states (17) has generated a great deal of interest in comparing these two probes of GT strength (18–21). A preliminary B(GT) distribution from (p, n) at $E_p = 160$ MeV of a recent remeasurement (22) of the earlier 120 MeV (p, n) study (23) is shown in Fig. (3). The results of all (p, n) studies are reasonably consistent and have been recently reviewed by Aufderheide et al. (24). Also shown in Fig. (3) are results from the beta decay study of Garcia et al. (17), but exhibited with resolution comparable to that in (p, n). Two major inconsistencies were noted (17,18) to exist between the results of these two probes. The first inconsistency is apparent in Fig. (3) and was based on the observation that the B(GT) values from (p, n) significantly varied from those from beta decay for the states at an excitation energy of 1.4 MeV and 3.2 MeV. This raised strong doubts regarding the validity of *any* proportionality between B(GT) and zero-degree (p, n) cross sections. The assumption of Garcia et al. (17) that gamma decay would not compete significantly with proton decay of the 3.2 MeV state was questioned by Goodman et al. (20) and disproved by Iliadis et al. (25). The beta decay result for the 1.4 MeV state was modified by this change, since its B(GT) had been inferred from the total strength observed at higher energies. Iliadis et al. (25) revised the B(GT) from beta decay in light of these new decay results. These revisions to the beta decay results significantly improve the agreement between beta decay and (p,n) B(GT) values for these states, as can be seen in Table 2.

The other discrepancy arose in a comparison of the B(GT) integrated up to about 8.5 MeV using these two probes. Adelberger et al. (18) observed that a large disagreement appears in the integrated strength, primarily above 5 MeV. It has, however, been noted (19) that 5 MeV is the very energy at which the Fermi IAS peak is located and that it appears likely that in the original (p,n) study (23) significant GT strength could have been misidentified as part of this poorly-resolved Fermi peak. The GT spectrum from (p,n) shown in Fig. (3) has only estimated the amount of cross section in the large peak at 5 MeV which should be assigned to the Fermi component. A meaningful determination of the GT strength in this region of excitation awaits analysis of (p,n) polarization transfer measurements. In the region of 8 MeV, further refinement of the calculation of the large (about 11 %) $F(q,\omega)$ correction is still required (22), prior to making a firm comparison between beta decay and (p,n) at high excitations.

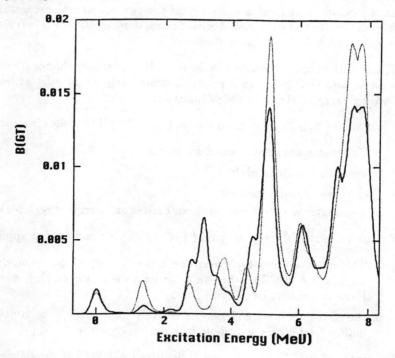

FIG. 3. Comparison of the B(GT) strength function of ^{37}Ar or ^{37}K based on β-delayed proton decay of ^{37}K (plotted with experimental width of (p,n) studies—dotted curve) (17) and on ^{37}Cl$(p,n)^{37}$Ar reaction at $E_p = 160$ MeV (solid line) (22)

TABLE 2. Comparison of the B(GT) values assigned to states in ^{37}Ar or ^{37}K based on recent ^{37}Cl(p,n) cross sections and on analog transitions in ^{37}K reported in a ^{37}Ca beta decay study.

Probe	B(GT;1.4 MeV)	B(GT;2.8 MeV)	B(GT;3.2 MeV)
"new" beta decay[a]	0.074±0.010	0.067±0.004	0.0039±0.0009
"new" (p,n)[b]	0.0144±0.0035	0.0774±0.0102	0.136±0.015
"revised" beta decay[c]	0.019±0.021	(same as "old")	0.16±0.05

[a] from Garcia et al. (17).
[b] as quoted in Aufderheide et al. (24).
[c] from Iliadis et al. (25).

OVERVIEW

The preceding discussion and examples indicate that care must be exercised when either beta decay, ν capture, or charge-exchange reaction studies purport to determine isovector reduced nuclear matrix elements. I attempt (with trepidation) below to suggest a summary of the current situation regarding the uncertainties that are associated with "extraction" of B(GT) values from intermediate energy (p,n) reaction studies.

- The empirical proportionality between hadron charge-exchange 0° cross sections and B(GT) (as in Eq. (4)) is more likely to be valid at bombarding energies above 100 MeV/nucleon.

- Estimates of B(GT) are probably good (±5 – 10%) **if** the following apply:
    1. The states are in an even-A nucleus **and**
    2. are constrained to **either**
        - strong transitions **or**
        - integrated strength below an excitation energy of ∼15 MeV.

- Estimates of B(GT) may be good (±10 – 20%) if the following applies:
    1. The states are moderately-strong transitions in an odd-A nucleus for which a B(GT) is known from beta decay for at least one of these transitions **or**
    2. the states are weak and thus more likely to be subject to "contamination" by higher multipole reaction components.

- Estimates of B(GT) have known difficulties ($\geq \pm 20\%$) if the following applies:
    1. The states are in an odd-A nucleus for which a direct GT calibration with respect to beta decay is not available for any state in the spectrum **or**

2. the states are at higher excitation energy where
   - the GT strength is no longer the dominant component such that the "forbidden" contributions dominate and/or
   - where the $F(q,\omega)$ correction becomes large and increasingly model dependent.

ACKNOWLEDGEMENTS

Many people have contributed over more than the past decade to our current understanding of intermediate energy charge-exchange reactions as a probe of nuclear isovector matrix elements. Too numerous to mention explicitly here, they are partially represented in the references below and more completely in those of Ref. 9.

This work was supported by the U.S. National Science Foundation under Grant No. PHY-9108242.

REFERENCES

1. W. Hampel, these conference prodeedings.
2. B. T. Cleveland et al., Proceedings of the 23rd International Cosmic Ray Conference (University of Calgary, Alberta, Canada , 1993) Vol. 3, p. 865.
3. J. Engel, S. Pittel and P. Vogel, Phys. Rev. Lett. **67**, 426 (1991).
4. J. I. Fujita, S. Fujii and K. Ikeda, Phys. Rev. **133** 546 (1964).
5. C. D. Goodman et al., Phys. Rev. Lett. **44**, 1755 (1980).
6. T. N. Taddeucci et al., Nucl. Phys. **A469**, 125 (1987).
7. E. Sugarbaker et al., Phys. Rev. Lett. **65**, 551 (1990).
8. Y. Wang et al., IUCF Scientific and Technical Report– 1991, p. 37.
9. J. Rapaport and E. Sugarbaker, to be published in Annu. Rev. Nucl. Part. Sci. **44** (1994).
10. W. Hampel and L. P. Remsberg, Phys. Rev. C **31**, 666 (1985).
11. D. Krofcheck, Ph.D. thesis, The Ohio State University, 1987.
12. H. Orihara et al., Phys. Rev. Lett. **51**, 1328 (1983).
13. D. Krofcheck et al., Phys. Rev. Lett. **55**, 1051 (1985).
14. S. Austin et al., Phys. Rev. Lett. **73**, 30 (1994).
15. D. J. Mercer et al., Phys. Rev. Lett. **71**, 684 (1993).
16. A. J. Wagner, Ph.D. thesis, The Ohio State University, 1988.
17. A. Garcia et al., Phys. Rev. Lett. **67**, 3654 (1991).
18. E. Adelberger et al., Phys. Rev. Lett. **67**, 3658 (1991).
19. J. Rapaport and E. Sugarbaker, Phys. Rev. Lett. **69**, 2444 (1992).
20. C. D. Goodman et al., Phys. Rev. Lett. **69**, 2445 (1992).
21. D. P. Wells et al., Phys. Rev. Lett. **69**, 2446 (1992).
22. C. D. Goodman and J. Cartwright, private communication.
23. J. Rapaport et al., Phys. Rev. Lett. **47**, 1518 (1981).
24. M. B. Aufderheide et al., Phys. Rev. C **49**, 678 (1994).
25. C. Iliadis et al., Phys. Rev. C **48**, R1479 (1993).

First Direct Measurement of a p(n,γ)d Reaction Cross Section at Stellar Energy

T.S.Suzuki*, Y.Nagai*, T.Shima*, M.Igashira+, T.Kikuchi*, H. Sato* and T.Kii*

Department of Applied Physics and +Research Laboratory for Nuclear Reactors, Tokyo Institute of Technology, O-okayama, Meguro, Tokyo 152, Japan

Abstract. The cross section of the p(n,γ)d reaction was measured for the first time in the neutron energy between 10 and 280 keV. The result is in good agreement with the theoretical values calculated recently by Hale et al. and Ohtsubo et al.. The present result, however, is not consistent with the estimated value by using the experimental value of the deuteron photo-disintegration.

The primordial abundances deduced from observed light elements D, ^{3}He, ^{4}He, and ^{7}Li have been compared with the abundances predicted by the standard big-bang nucleosynthesis theory[1]. From the comparison, the baryon to photon ratio η has been determined in a limited narrow range, which supports the standard theory, and the baryon density parameter of $0.01 < \Omega_b < 0.09$ has been also obtained[2]. Although the larger uncertainties in the present constraint on Ω_b are those in the Hubble constant and in the primordial abundances, respectively, it is important to obtain more accurate nuclear reaction rates to make further severe constraint on Ω_b. Several important nuclear reactions in the primordial nucleosynthesis have been discussed by Smith et al.[2]. A neutron capture on a proton $p(n,\gamma)d$ is one of the important reaction. All deuterium is produced by the reaction, and most ^{4}He, which is the most crucial element to determine the ratio η, is produced by the deuterium induced reactions. Namely, the cross section of the p(n,γ)d reaction, strongly affects the primordial abundances of D, ^{3}He, ^{4}He and ^{7}Li. It is therefore quite important to obtain the precise reaction cross section. However it has not ever been measured directly at the stellar energy between 10 and 300 keV. It was estimated by using the calculated cross section of the photodisintegration of the deuteron, and the measured neutron capture cross section on a proton at thermal energy[3]. Recently, the cross section of the p(n,γ)d reaction has been calculated by Hale et al.[4] and Ohtsubo et al.[5] in the neutron energy between 0 and 100 MeV. Ohstubo et al. took into account of the meson exchange effects for the electric dipole γ-ray transition. The calculted value is different from the value estimated by Fowler et al. and it gives the deuteron photo-disintegration cross section 15% smaller than the experimental value, measured at the γ-ray energies of 2.51, 2.615 and 2.754 MeV, which correspond to the neutron laboratory energy of 570, 780 and 1060 keV. Therefore it is very important to measure directly the cross section of the p(n,γ)d reaction at stellar neutron energy. In this study the cross section of the p(n,γ)d reaction was measured for the first time in the neutron energy between 10 and 280 keV. In the experiment a prompt γ-ray detection method was employed together with pulsed neutrons. The pulsed neutrons were produced by the ^{7}Li(p,n)^{7}Be reaction by using the 1.5 ns bunched proton beam, provided from the 3.2 MV Pelletron Accelerator at the Tokyo Institute of Technology[6]. Polyetylene samples with various thickness of 1,2, and 4mm were used to determine the capture cross section accurately after correcting for neutron multiple scattering effects in the samples propely as discussed later. A gold (Au) sample with thickness of 1.7mm and the diameter of 90mm was also used, as the capture cross section of Au has been well known. These samples were placed 12cm away from the neutron source at an angle of 0° with respect to the proton beam direction. Prompt γ-rays from a captured state were detected by an anti-Comton NaI(Tl) spectrometer[6]. The spectrometer was set at 125° with respect to the proton beam direction. Captured events were stored in a minicomputer in a list mode. The γ-ray spectra were obtained by putting the gates on the TOF spectrum measured by the NaI(Tl) spectrometer. Here it should be noted that 2.2 MeV γ-ray peak from the p(n,γ)d reaction is also observed in the background region of the TOF spectrum. This is due to the thermal and /or scattered neutron capture on hydrogen. A background subtracted γ-ray spectrum is shown in fig.1, where only the γ-ray from the p(n,γ)d reaction is clearly observed. The γ-ray intensity was analyzed by using the response function of the NaI(Tl) spectrometer. The absolute cross section σ_p was obtained by comparing the γ-ray intensity with that of Au as

$$\sigma_p = \frac{C_{Au}}{C_p} \frac{n_{Au}}{n_p} \frac{\phi_{Au}}{\phi_p} \frac{Y_p}{Y_{Au}} \frac{\Omega_{Au}}{\Omega_p} \sigma_{Au} \quad (1)$$

Where Ω, n and ϕ indicate the solid angle of the sample to the neutron source, the thickness of the sample and the neutron yield measured by a ^{6}Li-glass scintillation detector, respectively. Y_{Au} and σ_{Au} denote the γ-ray yield and the absolute capture cross section of Au, respectively. C is the correction factor for the multiple scattering effect and the shielding of the incident neutron in a sample, respectively. They were calculated by using the Monte Carlo code TIME-MULTI[7]. In order to measure the cross section for the neutron energies above 80 keV the γ-ray spectrum was obtained at the proton energy of 2.03 MeV. Here it should be noted that as the neutron energy becomes higher, two γ-ray lines of 6.8 and 7.3 MeV become prominent in the foreground spectrum,

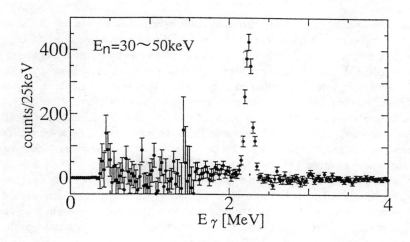

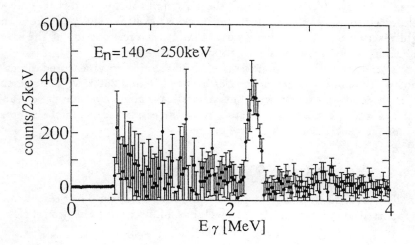

FIGURE 1. Background subtracted γ-ray spectra measured fot the neutron energies between 30 and 50 keV (upper), and between 140 and 250 keV (lower), respectively.

but not in the background spectrum and they were not observed for the lower neutron energies below 80 keV. From the facts these-rays were considered to be due to the direct capture reactions of the incident neutrons on the materials around the NaI(Tl) spectrometer. Here it should be mentioned that when the neutron energy is low, the width of the cone of the emitted neutron is narrow. Therefore, the neutron is not captured directly by the materials around the spectrometer. However, as the neutron energy becomes higher, the width becomes wide and therefore the neutrons can be captured directly by these materials. In order to confirm this point and identify the materials the γ-ray spectrum was measured without the paraffin sample. Two γ-rays of 6.8 and 7.3 MeV were clearly seen. Because of these facts and the γ-ray energies, these γ-rays can be identified to be due to the ^{6}Li(n,γ)^{7}Li reaction. ^{6}Li is used as ^{6}LiH to prevent the neutrons scattered by the sample from entering into the NaI(Tl) detector. The reason why these γ-rays become prominent at higher neutron energies is due to the 5/2$^-$ resonance at 7459.5 keV in ^{7}Li, 244.5 keV above the neutron threshold. As the γ-rays from the ^{6}Li(n,γ)^{7}Li reaction were observed in the foreground spectrum, the γ-ray from the p(n,γ)d reaction was also expected to be observed. However, it was shown that the γ-ray from the p(n,γ)d reaction is not observed clearly. Finally,the cross section at the average neutron energy of 185 keV was obtained accurately.

In this study the cross section of the p(n,γ)d reaction was measured directly for the first time in the neutron energies between 10 and 280 keV. The present result is in good agreement with the theoretical values by Hale et al. and Ohtsubo et al., but it is not consistent with the estimated value of the cross section of the deuteron photo disintegration.

ACKNOWLEDGEMENTS

Finally, we thank Prof. H.Ohtsubo for useful discussions. The present work was supported by a Grant-in-Aid of the Japan Ministry of Education, Science and Culture.

REFERENCES

1. Boesguard,a.M., and Steigman,G.,*Ann. Rev. Astr. Ap.***23**,319(1985).
2. Smith,M.S., Kawano,L.H., and Malaney,R.A.,*Ap.J.(supplement)* **85**,219(1993)
3. Fowler,W.A., Caughman,G.R., and Zimmerman,B.A.,*Annu.Rev.Astron.Astrophys.***5**,525(1967)
4. Hale,G.M., Dodder,D.C., Siciliano,E.R., and Wilwon,W.B.,*ENDF/B-VI*
5. Ohtsubo,H., Sato,T., and Kobayashi,T., *private communication* (1993)
6. Igashira,M., and Masuda,K., *private communication* (1994)
7. Senoo,K., Nagai,Y., Shima,T. Ohsaki,T., and Igashira,M.,*Nucl, Instr, and Meth.*A**339**,556(1994)

Search for low-energy resonances in the ^{3}He(^{3}He, 2p)^{4}He reaction

P. Descouvemont

*Physique Nucléaire Théorique et Physique Mathématique, CP229
Université Libre de Bruxelles, B1050 Bruxelles - Belgium*

Abstract. The three-cluster Generator Coordinate Method is used to investigate the ^{3}He(^{3}He, 2p)^{4}He and ^{3}H(^{3}H, 2n)^{4}He reactions at low energies. The three-body exit channels are simulated by distortion effects in the wave functions. We test the model through the experimental ^{3}He + ^{3}He elastic cross sections, and discuss the different contributions to the S-factors. We show that, in the ^{3}He(^{3}He, 2p)^{4}He cross section, the (α+p)+p sequential breakup dominates the α+(p+p) process. A systematic search for resonances in ^{6}Be and ^{6}He is carried out, but we do not find any evidence for a new resonance.

INTRODUCTION

The ^{3}He(^{3}He, 2p)^{4}He low-energy reaction is expected to be the dominant source of ^{4}He in low-mass stars (1). The ^{3}He + ^{3}He branch (chain I) represents about 86 % of the *pp* chain and therefore significantly dominates the ^{3}He + ^{4}He branch (chains II and III, $\simeq$ 14 %). In addition, it is well known that the observed Solar-neutrino flux sensitively depends on the ^{3}He(^{3}He, 2p)^{4}He cross section. More than twenty years ago, it has been suggested (2) that a narrow resonance, located close to the ^{3}He + ^{3}He threshold, might be present in the ^{6}Be spectrum. If so, this expected resonance might enhance the ^{3}He(^{3}He, 2p)^{4}He reaction rate, and provide an explanation for the long-standing Solar-neutrino problem.

Several experiments have been devoted to the measurement of the low-energy cross section, and to the search for a narrow resonance near the ^{3}He + ^{3}He threshold (3-5). However, if recent experiments succeeded in measuring the cross section down to stellar energies ($\approx$ 20 keV for the Sun), all attempts to observe a resonance failed up to now. It has been suggested (5) that this resonance might be located below the experimental lower limit (24.5 keV) and, consequently, that its width would be much lower than the energy resolution of current experiments.

In the present work, we use the Generator Coordinate Method (GCM - see Ref. (6) and refs. therein) to investigate the nuclear component of the ^{3}He(^{3}He, 2p)^{4}He and ^{3}H(^{3}H, 2n)^{4}He low-energy S-factors. Screening effects are not taken into account here. A microscopic model, such as the GCM, presents a fairly strong predictive power since, except in the nucleon-nucleon interaction, there is no free parameter. Experimental data are not necessary, but are used as a validity test of the theory. The main characteristics of our work are the following: (i) the 3-body exit channel is simulated by different $\alpha + p + p$ three-cluster configurations. (ii) The hamiltonian of the system reads:

$$H = \sum_{i=1}^{6} T_i + \sum_{i<j=1}^{6} V_{ij} \qquad (1)$$

where T_i is the kinetic energy of nucleon i, and V_{ij} the nucleon-nucleon interaction. This two-body interaction is adjusted on the experimental $\alpha + p$ elastic phase shifts (7) and ^{3}He + ^{3}He elastic cross sections (8) at different angles; this procedure makes the ^{3}He(^{3}He, 2p)^{4}He cross section independent of parameters. Details are given in Ref. (9). (iii) The mirror reaction ^{3}H(^{3}H, 2n)^{4}He is investigated simultaneously, and is used as a consistency test of the theory. Let us stress that the present three-cluster model does not aim at providing the "best" ^{3}He(^{3}He, 2p)^{4}He cross section, but we focus here on the search for possible resonances close to the ^{3}He + ^{3}He threshold in ^{6}Be.

RESULTS

We present in Figure 1 the GCM ^{3}H(^{3}H, 2n)^{4}He S-factor, where partial waves up to $J = 4$ are included. In our approach, the total S-factor results from different two-channel contributions which, together, are expected to simulate three-body effects. In such a model, it is therefore possible to distinguish between a sequential breakup into n+^{5}He followed by the ^{5}He $\rightarrow$ ^{4}He+n decay, and a sequential breakup into (2n)+^{4}He with the subsequent decay (2n) $\rightarrow$ n+n. In Figure 1, we display the contributions of the α +(n+n), $(\alpha+n)^{3/2^-}$ +n and $(\alpha + n)^{1/2^-}$ +n channels. It turns out that all of them have similar energy dependencies. This is not surprising since the energy dependence is mainly given by the ^{3}H + ^{3}H entrance channel. We find that the principal contribution arises from the $(\alpha + n)^{1/2^-}$ +n exit channel; the α +(n+n) channel represents about 1/3 of the total cross section. The $(\alpha + n)^{3/2^-}$ +n contribution is negligible in our model. The GCM total S-factor is fairly consistent with the recent data (10,11), but we slightly overestimate the normalization. The comparison of the theoretical S-factor with older data (12,13) is less good, but these data have been suggested to have large systematic errors (14).

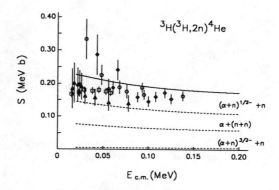

FIGURE 1. ^{3}H(^{3}H, 2n)^{4}He S-factor (full curve), with partial contributions (dashed curves). The data are from Ref.(10) (triangles), Ref.(11) (squares), Ref.(12) (rhombi) and Ref.(13) (circles).

Let us now discuss the ^{3}He(^{3}He, 2p)^{4}He S-factor, presented in Figure 2. Except for the Coulomb force, the ingredients of the model are identical to those of the ^{3}H(^{3}H, 2n)^{4}He investigation. Here also, the main contribution arises from a sequential breakup into $(\alpha + p)^{1/2^-}$ +p, which represents about 3/4 of the total cross section. At astrophysical energies, only the α +(p+p) channel is also significant. The $(\alpha+p)^{3/2^-}$ +p contribution amounts to 10 % beyond 1 MeV only. In the ^{3}He(^{3}He, 2p)^{4}He reaction, the first p+p pseudostate, referred to as (p+p)*, may contribute beyond 0.2 MeV. This contribution, as well as the $(\alpha+p)^{3/2^-}$ +p contribution yields a change in the energy dependence of the total S-factor; from about 0.6 MeV, the slope of the GCM S-factor is reduced, as it is observed in the data of Dwarakanath and Winkler (15).

After these tests of the microscopic model, we have searched for resonances in ^{6}He and ^{6}Be. We use an extension of the microscopic R-matrix (MRM) method, based on an iterative procedure (16). This method has been used in many two-cluster or three-cluster systems, and has been shown to be efficient and accurate. The best accuracy is obtained for narrow resonances, but this procedure can be reliably used for resonances whose width reaches about 1 MeV. Both for ^{6}He and ^{6}Be, we have investigated partial waves from $J = 0$ to $J = 4$ for positive and negative parities. As expected, we found the 0^+ ground state, and the low-energy 2^+ resonance in ^{6}He and ^{6}Be. However, there is no indication for a resonance near the ^{3}H + ^{3}H threshold in ^{6}He or near the ^{3}He + ^{3}He threshold in ^{6}Be. This result has been confirmed by a careful analysis of the GCM phase shifts. They are found quite monotonic, and characteristic from a pure non-resonant process, in good agreement with a recent measurement of the Tübingen group (17).

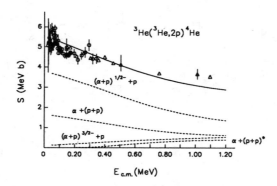

FIGURE 2. ^{3}He(^{3}He, 2p)^{4}He S-factor (full curves), with partial contributions (dashed curves). The data are from Ref.(15) (triangles) and Ref.(5) (circles).

REFERENCES

1. Rolfs, C., and Rodney, W.S., "Cauldrons in the Cosmos", (Chicago, London), 1 (1988).
2. Fetisov, V.N., and Kopysov, Y.S., Phys. Lett. **B40**, 602 (1972).
3. Dwarakanath, M.R., Phys. Rev. **C9**, 805 (1974).
4. McDonald, A.B., et al. Nucl. Phys. **A288**, 529 (1977).
5. Krauss, A., et al. Nucl. Phys. **467**, 273 (1987).
6. Descouvemont, P., J. Phys. **G19**, S141 (1993).
7. Ajzenberg-Selove, F., Nucl. Phys. **A490**, 1 (1988).
8. Tombrello, T.A., and Bacher, A.D., Phys. Rev. **130**, 1108 (1963).
9. Descouvemont, P., to be published.
10. Serov, V.I., et al. Sov. J. At. Energy **42**, 66 (1977).
11. Brown, R.E. and Jarmie, N., Radiation Effects **92**, 45 (1986).
12. Agnew, H.M., et al. Phys. Rev **84**, 862 (1951).
13. Govorov, A.M., et al. Sov. Phys. JEPT **15**, 266 (1962).
14. Jarmie, N., Nucl. Sci. Eng. **78**, 78 (1981).
15. Dwarakanath, M.R., and Winkler, H., Phys. Rev. **C4**, 1532 (1971).
16. Descouvemont, P., and Vincke, M., Phys. Rev. **A42**, 3835 (1990).
17. Staudt, G., private communication.

The PIAFE Project and Its Possible Implications in r-Process Studies

Gabriele Fioni[1] and the PIAFE Collaboration[2]

[1]Institut Max von Laue - Paul Langevin, BP 156, 38042 Grenoble, France
[2]c/o Institut des Sciences Nucléaires, 53 Av. des Martyrs, 38026 Grenoble, France

Abstract. In the framework of the intense worldwide interest in Radioactive Nuclear Beams, the PIAFE project was proposed and this is now in an advanced phase of study. A fission source will be placed near to the core of the ILL high-flux nuclear reactor; the fission fragments will be extracted, ionized, pre-accelerated, mass separated and transferred to the SARA heavy ions accelerator complex of the ISN through a 400 m long beam line. The expected beam intensities for many of the produced neutron rich nuclei are orders of magnitude higher than at any other existing or planned facility. This will enable accurate nuclear spectroscopy experiments to determine the characteristics of unknown n-rich nuclei along the r-process path.

1. INTRODUCTION

Low energy nuclear fission is probably the easiest and the most efficient way to produce n-rich nuclei and several facilities were built around the world over the last 20 years (1, 2, 3). In Grenoble, two major installations are separated by a distance of about 400 m, the ILL High Flux Reactor with a thermal neutron flux of about 10^{15} n/s/cm^2 and the SARA heavy ions accelerators (at energies from 2 to 20 MeV/amu) of the Institut des Sciences Nucléaires (ISN). The PIAFE project (in French "Project d' Ionization et d' Accélération de Faisceaux Exotiques") proposes to couple them in order to get low and medium energy radioactive beams (4, 5). The realization will be in two main stages: the first, called PIAFE-Phase 1, concerns the ILL part of the facility which will provide very intense low energy mass separated radioactive beams, suitable for basic nuclear spectroscopy experiments and possible solid state applications. In a second stage, called PIAFE-Phase 2, the mass separated beam will be transferred to the SARA cyclotron complex through a 400 m long beam line to be accelerated. A wide range of possible applications, from nuclear physics to medical sciences, are foreseen (6). The interest of this project in nuclear astrophysics is mainly in relation with the nuclear spectroscopy experiments connected to the Phase 1, and it will be discussed in the following.

2. GENERAL LAYOUT OF THE FACILITY

A source of the same type as the one used at the OSIRIS facility (7) will be placed near to the core of the ILL reactor in a thermal neutron flux of $5 \cdot 10^{14}$ n/s/cm^2 using the beam tube of the Lohengrin mass separator. It consists of a porous graphite matrix of cylindrical shape, impregnated with a solution of neutral uranium nitrate containing about 1 g of ^{235}U. The estimated yield of $2.2 \cdot 10^{14}$ fissions/s will keep the temperature of the source at about 2200 °C, allowing fast diffusion times for most of the fission fragments. A Rhenium plate placed near to the extraction hole of the source will induce surface ionization to 1^+ for alkali metals with an efficiency of about 100%, while for alkali earths and lanthanides efficiencies of few percent are expected. Other elements can be extracted from the source by laser or by electron-impact ionization schemes.

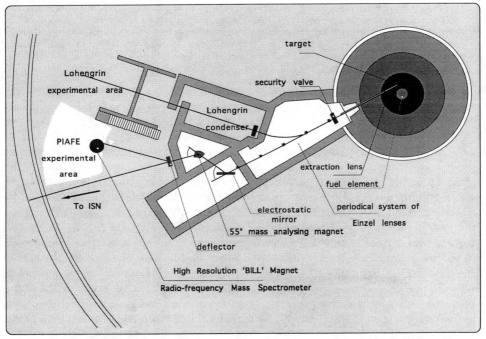

FIGURE 1. Schematic view of the PIAFE-phase 1 project in the ILL reactor experimental hall.

The ions extracted from the source will be accelerated by an electric potential of 30 kV and will be transported over 15 m by a system of 10 electrostatic Einzel lenses (8) up to an electrostatic mirror which will deflect the fission fragments out of the neutron beam. After some beam matching optical elements, the beam will enter in a 55 degrees double focusing dipole magnet like the one used in Mainz (3 and 9) where it will be mass separated with a resolution of $A/\Delta A \approx 600$. The beam can then be either switched to an experimental area in the ILL reactor hall, or sent to the ISN where it will be further accelerated after having changed its charge state by using an ECR source. A schematic view of the ILL part of the project is shown in Figure 1. In the experimental area, additional mass separators will be used. The double focussing "BILL" magnet (10) with a mass resolution up to $A/\Delta A \approx 7000$ will be installed to provide a high purity beam for investigation of rare n-rich nuclei by Q_β, half-life and ß-delayed neutron emission measurements, and by nuclear and laser spectroscopy experiments. High accuracy direct mass determinations will be possible using a Smith type Radio-Frequency Mass Spectrometer (11), with a mass resolution of $A/\Delta A > 50000$, where measurements at very low count-rates (< 1 cps) are feasible.

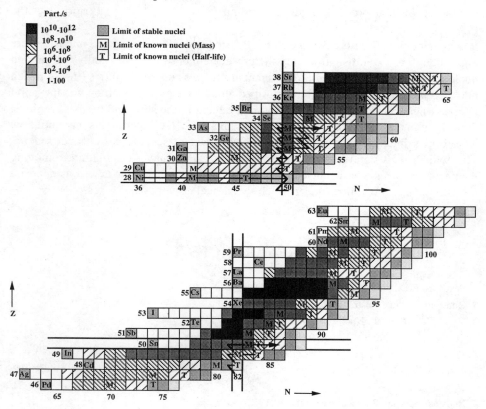

FIGURE 2. Estimated intensities at the PIAFE mass separator for the light and heavy fission fragments. A possible r-process path (15) is shown.

3. INTENSITIES ALONG THE r-PROCESS PATH

A conservative estimate of the intensities expected after the mass separator was made by A. Pinston (12) mainly on the basis of a description of H. Faust (13) for far-asymmetric light fission products and of A. Wahl (14) for other mass regions. Diffusion rates and ionization probabilities were deduced from the available experimental data from the OSTIS mass spectrometer (7).

The results for the light and heavy fission fragments are shown in Figure 2, where a possible r-process path (15) is shown. Refractory elements, from Zr to Rh are not included, as they can not diffuse out of the source. For very light n-rich elements from He to Si which are produced at low yields in ternary nuclear fission, no systematic evaluation of their production yield is yet made as it is not of first priority for the PIAFE project.

It is evident from Figure 2 that for all nuclides concerned, the expected intensities are high enough to extend significantly the present limits of experimentally determined nuclear masses, half-lifes and ß-delayed neutron emission probabilities (P_n) of n-rich nuclei. Systematic studies can be performed along the r-process path on waiting-point nuclei at neutron magic numbers N=50 and 82 and in the surrounding mass regions. Especially in the region of the double-magic nucleus ^{78}Ni, experiments that until now were impossible, can be performed with count-rates of several particles per second. This will strongly contribute to enhance r-process abundancies calculations removing the uncertainty in the basic nuclear data and giving at the same time new constraints on the astrophysical conditions under which stellar nucleosynthesis operates.

REFERENCES

1. Wuensch K.D., Nucl. Instr. Meth., 186(1981)89.
2. Rudstam G., Nucl. Instr. Meth., 139(1976)239.
3. Von Reisky L. et al., Nucl. Instr. Meth., 172(1980)423.
4. Status report of the PIAFE project, ISN - Piafe Collaboration Report, (1993).
5. Faust H. et al., ILL report 93FA9T (1993).
6. The PIAFE Collaboration, PIAFE project: Physics Case, Nifernecker H. editor, sub. to Annales de Physique.
7. Rudstam G. et al., Radiochimica Acta, 49(1990)155.
8. Fioni G. et al., ISN - Piafe Collaboration Report, in press.
9. Von Reisky L., Thesis, Univesität Mainz, (1978).
10. Mampe W. et al., Nucl. Instr. Meth., 154(1978)127.
11. De Saint Simon et al., Nucl. Instr. Meth., B70(1992)459.
12. Fogelberg B. et al., PIAFE project: Physics Case, Nifernecker H. editor, sub. to Annales de Physique.
13. Faust H., ILL report 94FA1T (1993).
14. Wahl A.C., Atomic and Nuclear Data Tables, 39(1988)1.
15. Kratz K.-L., Inst. Phys. Conf. Ser. No. 88/J. Phys. G., 14(1988)S331.

Measurements of (n,γ) Cross Sections with Small Samples

Paul E. Koehler[a] and Franz Käppeler[b]

[a]*Los Alamos National Laboratory, Los Alamos, New Mexico 87545;*
Present address: Oak Ridge National Laboratory, Oak Ridge, Tennessee 37830
Managed by Martin Marietta Energy Systems, Inc. under contract
DE-AC05-84OR21400 with the U.S. Department of Energy
[b]*KfK Karlsruhe, Institut für Kernphysik, D-76021 Karlsruhe, Germany*

Abstract. Neutron capture cross section data for certain isotopes of very small natural abundance are crucial for a better understanding of the s- and p-processes of nucleosynthesis. Also, recent work has shown that many previous (n,γ) measurements need to be extended to lower neutron energies and that the accuracy of some previous data need to be improved. At Los Alamos we have developed a system for measuring (n,γ) cross sections on samples as small as 1 mg. We give examples of measurements made with this apparatus and discuss the nuclear astrophysics motivation for these and future measurements.

INTRODUCTION

Recent developments in stellar models and improvements in abundance determinations have demonstrated that further improvement in our understanding of the s-process is dependent on new neutron capture cross section measurements. Recently, for example, for the first time good agreement was obtained between the data and a dynamical s-process calculation based on a stellar model (1). However, firm conclusions concerning the validity of the stellar model were obscured by uncertainties in the input neutron capture cross sections. Furthermore, these results and others indicate that the predominant neutron exposure during the main component of the s-process occurs at a temperature, $kT \approx 12$ keV, which is significantly lower than the "canonical" s-process temperature of $kT=30$ keV. Hence, many of the older (n,γ) measurements which had a lower energy cutoff of 3.5 keV need to be extended to lower energy in order to determine accurately the Maxwellian-averaged cross section at 12 keV. Recent measurements for the barium isotopes (2) have dramatically illustrated the importance of extending the older measurements to lower energies.

In another example of the need for better (n,γ) data, recent abundance determinations obtained from meteorites (3-5) have apparently detected isotopic signatures from the s-process. Again, however, the authors concluded that a good understanding of the meteoritic data is hampered by the need for better neutron capture data and that some of the currently accepted cross sections are probably incorrect.

Many of the proton rich, intermediate and heavy elements are bypassed by the s-process and instead are thought to originate in the p-process (6). In the p-process these nuclei are thought to be produced mainly through photodisintegration of heavier nuclei. For some nuclei however, the (n,γ) reaction is calculated to compete with and, in some cases, dominate photo-induced reactions, so accurate (n,γ) cross sections are important in determining the final calculated abundances. Most of the relevant cross sections have not been measured because the involved isotopes have very small natural abundances so that the necessary separated samples are prohibitively expensive. Currently p-process calculations have to rely on statistical model estimates for many of the cross sections. As a result, the parameters of the stellar site for the p-process are not very well determined.

In this paper we give examples of some recent measurements which demonstrate the ability to obtain many of the needed data. We conclude with a brief discussion of possible future measurements.

MEASUREMENT TECHNIQUE

Typically, (n,γ) measurements have required samples in the few grams to tens of grams range. Because the neutron flux at LANSCE is as much as several orders of magnitude larger than at other neutron sources, the required sample size can be reduced to the mg range and the cost of the separated isotopic samples is no longer prohibitive. The (n,γ) apparatus we have developed at LANSCE was deployed on the same flight path we have used in the recent past for several (n,p) and (n,α) measurements (7). Major changes made included installing improved collimation and shielding which greatly reduced the background in the γ-ray detectors from the "halo" of the neutron beam and from unshielded neutrons from other nearby LANSCE flight paths.

As in many previous measurements at other facilities, a pair of deuterated benzene scintillators were used to detect the neutron capture γ-rays. In contrast to previous measurements, no pulse-height weighting was used in the conversion of the measured yields to cross sections. Instead, a procedure very much like the one we have used in previous (n,p) and (n,α) measurements was employed. The γ-ray yield as a function of energy was measured down to thermal energy and the measured thermal cross sections were used to normalize the yields to absolute cross sections.

As a test of this procedure a 2.4 mg sample of gold wire was used to measure the ^{197}Au(n,γ)^{198}Au cross section between thermal energy and approximately 50 keV. The total beam time for the experiment was about 1 day.

Good agreement between the LANSCE data (8) and the latest ENDF evaluation was obtained across the entire energy range after the ENDF cross sections had been averaged over the LANSCE resolution.

Next, the ^{191}Ir(n,γ)^{192}Ir and ^{193}Ir(n,γ)^{194}Ir cross sections were measured using 1 mg and 2 mg samples, respectively. The samples consisted of metallic powder which had been "super-glued" to 0.8 μm-thick aluminum backing foils. Again, the total beam time was about 1 day for each sample. The motivation for the measurements was that previous results differed by about 40%.

The resulting ^{191,193}Ir(n,γ)^{192,194}Ir cross sections above 1 keV are compared to previous data in Fig. 1. The data below 1 keV reveal many resonances (8) but are not shown here due to space limitations. The measured resonance energies are in good agreement with those previously reported. The previous data of Macklin et al. (9) are shown as "steps" representing the average cross section across the range of energies reported. For ^{191}Ir(n,γ)^{192}Ir there have been no other data reported. For ^{193}Ir(n,γ)^{194}Ir, both Chaubey and Sehgal (10) and Dovbenko et al. (11) have reported activation measurements. The data of Dovbenko et al. are approximately 40% larger than Macklin et al. whereas Chaubey and Sehgal are roughly 40% smaller than Macklin et al. at 25 keV. Our data for both isotopes are in good agreement with those of Macklin et al. at the lower end of their energy range, but above approximately 6 keV our data are significantly larger. On the other hand, our data for ^{193}Ir(n,γ)^{194}Ir appear to be in fair agreement with those of Dovbenko et al. Because our data for both isotopes show the same general difference with the data of Macklin et al. at higher energies, we are exploring possible reasons for such a systematic deviation and thus consider these results preliminary. One argument against such systematic problems in our data is the fact that the ^{197}Au(n,γ)^{198}Au data (8), which were taken with the same system, do not show a similar deviation from the ENDF evaluation. We are currently analyzing data taken with mg-sized ^{144}Nd

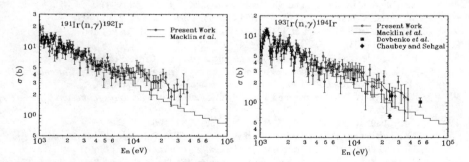

FIGURE 1. The ^{191,193}Ir(n,γ)^{192,194}Ir cross sections from 1 to 100 keV. Our new LANSCE data are shown as circles with a smooth curve to guide the eye. The data of Macklin et al. are shown as steps representing their energy-averaged data whereas the data of Chaubey and Sehgal and Dovbenko et al. are shown as a diamond and squares, respectively.

and ^{113,115}In samples during the same LANSCE run cycle.

FUTURE POSSIBILITIES

For measurements on stable samples of interest to the s- and p- processes several complementary facilities exist. First, very precise measurements can be made using the Karlsruhe barium fluoride detector (12). Because this facility is typically limited to neutron energies greater than approximately 5 keV, additional measurements at other facilities may be required to accurately determine the astrophysically relevant average cross section at kT=12 keV. For example, Voss *et al.* (12) have recently estimated that for ^{135}Ba and ^{137}Ba, the data below 5 keV are responsible for almost 40% of the Maxwellian-averaged (n,γ) cross section at 12 keV. When sufficiently large samples are available complementary measurements can be made at facilities such as GELINA or ORELA. For samples that would be prohibitively expensive to obtain in the relatively large sizes required at these facilities, our data demonstrate that measurements can be made on mg-sized samples at LANSCE. Also, recent work (8) has shown that (n,γ) measurements can be made for some important radioisotopes by using very small radioactive samples. The measurement of all the important isotopes of interest will require the continued availability of all of these facilities.

REFERENCES

1. F. Käppeler, R. Gallino, M. Busso, G. Picchio, and C.M. Raiteri, *Astrophys. J.* **354**, 630 (1990).
2. H. Beer, F. Corvi, and K Athanassopulos, "The Stellar Neutron Capture Rate of the s-Process Monitor ^{138}Ba", *Proc. Symposium on Capture Gamma-Ray Spectroscopy and Related Topics*, edited by J. Kern (World Scientific, Singapore, 1994), p. 698.
3. E. Zinner, S. Amari, and R.S. Lewis, *Astrophys. J.* **382**, L47 (1991).
4. C.A. Prombo, F.A. Podosek, S. Amari, and R.S. Lewis, *Astrophys. J.* **410**, 393 (1993).
5. R. Gallino, C.M. Raiteri, and M. Busso, *Astrophys. J.* **410**, 400 (1993).
6. M. Rayet, N. Prantzos, and M. Arnould, *Astron. Astrophys.* **227**, 271 (1990).
7. P.E. Koehler, S.M. Graff, H.A. O'Brien, Yu.M. Gledenov and Yu. P. Popov, *Phys. Rev. C* **47**, 2107 (1993).
8. P.E. Koehler and F. Käppeler, "Measurements of (n,γ) Cross Sections for Very Small Stable and Radioactive Samples of Interest to the s- and p-process", *Proc. International Conference on Nucl. Data for Science and Technology, 1994*, in press.
9. R.L. Macklin, D.M. Drake, and J.J. Malanify, *Los Alamos Scientific Laboratory Report LA-7479-MS*, (1978).
10. A.K. Chaubey and M.L. Sehgal, *Phys. Rev.* **152**, 1055 (1966).
11. A.G. Dovbenko, V.E. Kolesov, V.P. Koroleva, and V.A. Tolstikov, *Soviet Atomic Energy* **27**, 1185 (1969) (English translation of *Atomnaja Energija* **27**, 406 (1969)).
12. F. Voss, K. Wisshak, P. Guber, and F. Käppeler, "Stellar (n,γ) Cross Sections of ^{134}Ba, ^{135}Ba, ^{136}Ba and ^{137}Ba - Measurements and Astrophysical Results", *KfK Karlsruhe, Institut für Kernphysik Report 5253* (1994).

Neutron Capture in Barium and Gadolinium Isotopes: Implications for the s-Process.

K. Wisshak, F. Voss, F. Käppeler

Kernforschungszentrum Karlsruhe, IK-III, Postfach 3640, D-76021 Karlsruhe, Germany

Abstract.
Neutron capture cross sections of barium and gadolinium isotopes have been measured using the Karlsruhe 4π Barium Fluoride Detector. Severe discrepancies were found compared to previous experiments. The results allow to improve the analysis of branchigs in the s-process path. The barium data are also used for the interpretation of isotopic annomalies and to separate the r- and s- contributions of the element barium. There are indications that the solar barium abundance is larger than given in the literature.

Key words: nucleosynthesis, s-process, neutron capture, Ba and Gd isotopes

1. INTRODUCTION

Since the last conference Nuclei in the Cosmos'92 in Karlsruhe the experimental program with the Karlsruhe 4π Barium Fluoride Detector has been continued. Its main aim is the accurate determination of neutron capture cross sections of s-only isotopes. This is an essential prerequisite in order to derive information on the physical conditions in Red Giant stars from the analysis of branchings in the s-process path. First results on tellurium and samarium isotopes were discussed in the proceedings of the previous conference (1) and in more detail in Refs.(2) and (3). Here, we present our results on barium (4) and preliminary data for the gadolinium isotopes.

2. EXPERIMENT

The neutron capture cross sections of ^{134}Ba, ^{135}Ba, ^{136}Ba, ^{137}Ba and ^{152}Gd, ^{154}Gd, ^{155}Gd, ^{156}Gd, ^{157}Gd, ^{158}Gd have been measured in the neutron energy range from 5 to 225 keV. Capture events were registered with the Karlsruhe 4π BaF$_2$ Detector (5). The experimental method has been described in detail in Ref.(6). The measurements on barium were complicated by the combined effect of the comparably small cross sections and high scattering yields of the Ba$_2$CO$_3$ compound that had to be used as sample material. Therefore, only an uncertainty of $\sim$3% was reached which, however, represents still an improvement by factors of five to eight compared to previous experiments. In case of the gadolinium isotopes a further complication had to be faced for the first time. The natural abundances of the s-only isotopes ^{152}Gd and ^{154}Gd are only 0.2 and 2.2%, respectively. Therefore, highly enriched samples were not

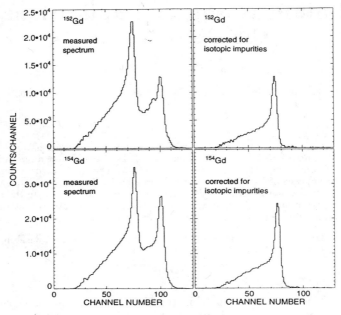

Fig. 1.— Sum energy spectra of capture events in the Karlsruhe 4π BaF$_2$ Detector demonstrating the reliable correction of isotopic impurities in the ^{152}Gd and ^{154}Gd samples

available. The rather low enrichments of 32.5% and 57% required significant corrections for isotopic impurities. This is demonstrated in Fig.1, which shows the gamma-ray spectra for capture events before and after this correction. Note that the neutron binding energies of the even Gd isotopes are between 6 and 6.4 MeV, but 8 to 8.5 MeV for the odd isotopes. Thus, capture in the odd isotopes is only partly disturbing the measured effect since most of these events are located at high energies (above channel 80) in a region that is not used for data evaluation of the even isotopes. This represents an important improvement compared to conventional techniques, where no information on the sum energy of the capture events is obtained. In such measurements up to 75% of the observed effect had to be subtracted to correct for isotopic impurities.

3. RESULTS FOR THE MAXWELLIAN–AVERAGED CROSS SECTIONS

From the experimental results stellar cross sections were derived for temperatures from kT=10 to 100 keV. The 30 keV values for the s–only isotopes are compared in Table 1 with recent evaluations (7) and (8). Severe discrepancies were found for ^{134}Ba and ^{154}Gd.

Table 1: Stellar neutron capture cross sections of the s–only barium and gadolinium isotopes at kT=30keV

Isotope	$<\sigma>$[mb]		
	(this work)	(Bao and Käppeler)	(Beer, Voss and Winters)
^{134}Ba	176.3±5.6 [1]	222±35	221±32
^{136}Ba	62.0±2.1 [1]	69±10	60±10
^{152}Gd	1017 [2]	985±61	1045±65
^{154}Gd	1022 [2]	1278±102	878±27

[1] The 1.5% uncertainty of the standard is not considered since it cancels out in most applications of relevance for nuclear astrophysics.
[2] Preliminary results with an uncertainty of about 2%.

4. IMPLICATIONS FOR s–PROCESS MODELS

From the barium experiment the following conclusions were drawn as described in Ref.(4):

- Together with the data on tellurium and samarium an improved description of the $N_s <\sigma>$ systematics was derived. The calculation shown in Fig. 2 is normalized at the s–only isotopes ^{124}Te and ^{152}Sm. The Ba isotopes are significantly overproduced compared to the empirical values if the solar abundances from Ref.(9) are used (open circles in Fig.2). This discrepancy is removed if the solar barium abundance is normalized by a factor of 1.22 (filled circles in Fig.2).

- For the first time the s–process branching at ^{134}Cs could be defined quantitatively. An analysis with the classical approach indicates a lower mean temperature around 17 keV than was estimated from other branchings. In order to confirm these findings, an improved value of the critical log ft value for the β–decay of ^{134}Cs is needed.

- The new data allow for an improved decomposition of the isotopic and elemental barium abundances into the respective s– and r–process contributions. The r–process contribution to elemental barium is estimated to 15±6%.

- The new cross sections confirm the renormalization factors of Gallino et al. (10) which were suggested to explain the isotopic anomalies in silicon carbide grains.

The gadolinium data allowed a first analysis of the branchings at ^{151}Sm and ^{154}Eu, which yielded the following indications: A consistent description of both branchings seems possible but the temperature required is significantly

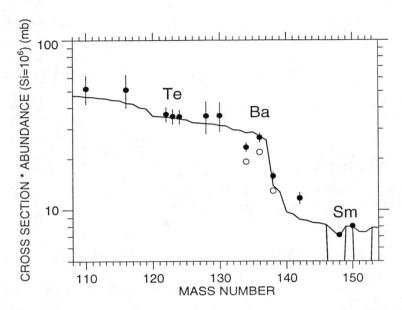

Fig. 2.— The $N_s < \sigma >$-curve calculated in the mass region $120<A<155$ compared to the empirical $N_s < \sigma >$-values

larger than in the barium case. The p– process contributions, esspecially to ^{152}Gd, are smaller than assumed previously (11). This holds also for the capture cross section of the radioactive branch point nucleus ^{154}Eu, where improved calculations are urgently required.

REFERENCES

(1) K. Wisshak, F. Voss, K. Guber, and F. Käppeler, in Nuclei in the Cosmos, Eds. F. Käppeler and K. Wisshak, Institute of Physics Publishing, Bristol and Philadelphia 1993 p.203.

(2) K. Wisshak, F. Voss, F. Käppeler, and G. Reffo, Phys. Rev. C. **45** (1992) 2470.

(3) K. Wisshak, K. Guber, F. Voss, F. Käppeler, and G. Reffo, Phys. Rev. C. **48** (1993) 1401.

(4) F. Voss, K. Wisshak, K. Guber, F. Käppeler, and G. Reffo, report KfK–5253, Kernforschungszentrum Karlsruhe, and Phys. Rev. C. (submitted).

(5) K. Wisshak, K. Guber, F. Käppeler, J. Krisch, H. Müller, G. Rupp, and F. Voss, Nucl. Instr. Meth. A. **292** (1990) 595.

(6) K. Wisshak, F. Voss, F. Käppeler, and G. Reffo, Phys. Rev. C. **42** (1990) 1731.

(7) Z.Y. Bao and F. Käppeler, Atomic Data Nucl. Data Tables **36** (1987) 411.

(8) H. Beer, F. Voss, and R.R. Winters Astrophys. J. Suppl. **80** (1992) 403.

(9) E. Anders and N. Grevesse, Geochim. Cosmochim. Acta **53** (1989) 197.

(10) R. Gallino, C.M. Raiteri, and M. Busso, Astrophys. J. **410** (1993) 400.

(11) H. Beer and R.L. Macklin Astrophys. J. **331** (1988) 1047.

The Stellar Capture Rate of ^{208}Pb

F. Corvi, P. Mutti, K. Athanassopulos

CEC, JRC, Institute for Reference Materials and Measurements, Retieseweg, B-2440 Geel, Belgium

H. Beer

Kernforschungszentrum IK III, P.O. Box 3640, D-76021 Karlsruhe, Germany

Abstract. High resolution neutron capture measurements of a highly enriched ^{208}Pb sample were carried out at the GELINA pulsed neutron facility and the capture areas of twelve resonances in the range 1-400 keV were determined. From these values the Maxwellian-averaged cross section $<\sigma>$ was calculated vs kT from 5 to 100 keV. After adding the estimated direct capture contribution, our results agree with those of previous activation measurements but are a factor of two lower than the Oak Ridge neutron capture data. The reason for such a discrepancy was singled out.

INTRODUCTION

It is well known that the s-process of stellar nucleosynthesis terminates at the isotopes of lead and bismuth since all further neutron captures lead to α-unstable nuclei which are then cycled back to the main lead isotopes. In order to explain the abundances of these isotopes, a so-called strong s-process component was introduced. To study its characteristics, the capture cross sections of all concerned isotopes, and in particular of ^{208}Pb is needed. Being double magic, this nucleus has the smallest Maxwellian-averaged capture (MAC) cross section $<\sigma>$ of all the heavier elements and as such it introduces the most stringent limitations for the s-process flow at the termination of the synthesis. Unfortunately a great uncertainty exists about the value of $<\sigma>$ at 30 keV since the one derived at ORELA from neutron capture in resolved resonances (1) is a factor of two larger than the results of activation measurements performed at Oak Ridge (2) and more recently at Karlsruhe (3). In an effort to solve this discrepancy, the cross section of ^{208}Pb (n,γ) was remeasured with high resolution at the Geel linac in the energy range 1-900 keV

EXPERIMENT

The measurements were performed at the GELINA pulsed neutron source facility using the time-of-flight (TOF) technique. The Geel linac and associated compressing magnet were operated to provide electron bursts of 100 MeV average energy and 1 ns width at a repetition frequency of 800 Hz and an average beam current of 50 µA. Neutrons produced by bremsstrahlung gamma-rays inside the rotating U target were subsequently moderated in two 4 cm thick water slabs canned in beryllium. The sample, consisting of an 8 cm diameter lead disc enriched to 99.75% ^{208}Pb, of thickness 0.00591 at/b., was placed inside an evacuated thin carbon fibre pipe at a flight distance of 59.42 m. It was viewed by four cylindrical C_6D_6-based liquid scintillators placed perpendicularly to the neutron direction. The pulse-height weighting function, applied to the data in order to make the detector efficiency proportional only to the total γ-energy emitted, was that obtained by us in an earlier experiment with similar geometrical conditions (4). The relative neutron flux was measured at the same time as capture by a gas-flow ionization chamber placed upstream in the beam and containing three back-to-back ^{10}B deposits of 9 cm diameter and 40 µg/cm^2 thickness. The effective measurement time was about 530 hours.

The data were independently normalized to gold capture at 4.9 eV and silver capture at 5.2 eV using the saturated resonance method: the results of the two normalizations differed by 5% and their average was adopted. Additionally, the γ-ray energy loss in the lead sample was measured by comparing capture at 5.2 eV in a thin Ag foil to that in a Pb-Ag alloy containing an equivalent quantity of silver and the same quantity of lead as the ^{208}Pb sample. It was found that the reduction in the "weighted" counting rate of the thicker sample was 13%. Assuming that the energy loss for ^{208}Pb(n,γ) is the same as for ^{109}Ag(n,γ), we introduced this correction in the normalization procedure.

RESULTS

After subtraction of a smooth background function, correction for the neutron flux shape and normalization, the amplitude weighted counting rate was reduced to a neutron capture yield which was subsequently analysed with the R-matrix shape fitting code FANAC (5). In the region below 400 keV we observed twelve resonances whose energies are listed in the first column of Table 1: they were already known from previous works except the one at 350.43 keV. On the other hand, we were unable to see some of the resonances visible in transmission measurements (1,6). In fact, detection of weak capture levels is hindered by the presence, in the yield spectrum, of an underlying structure due to prompt capture of scattered neutrons mainly in the aluminium canning of the detectors or in other surrounding material.

To check whether all observed resonances belong to ^{208}Pb(n,γ), we derived also a TOF spectrum of those events delivering more than 4.6

MeV energy in the scintillators: since the neutron separation energy of ^{208}Pb is only 3.936 MeV, all ^{208}Pb levels should completely disappear in such a spectrum. This was verified within the available statistics for all observed resonances except for the broad one at 77.85 keV where a peak, shifted towards larger TOF values, was seen. We attributed this remaining structure to prompt capture in the surrounding materials of resonance scattered neutrons: such a background is visible in this case because of the large Γ_n value of the 77.85 keV resonance. We succeeded in calculating a correction for this effect by normalizing this remaining area to the corresponding yield in the prominent 5.9 keV aluminium peak which dominates the low-energy capture. In this way it was found that about one third of the area of the 77.85 keV resonance is due to this type of background.

The values of the neutron widths Γ_n as well as of l and J given in Table 1 are the results of an R-matrix fit of total and differential elastic scattering data of Horen et al. (6). Keeping these values fix in our analysis, we could derive the capture width Γ_γ and the area $g\Gamma_n\Gamma_\gamma/\Gamma$. In case of resonances for which these transmission data were not available, only the capture area was given. The errors quoted for the capture widths and areas are only the statistical ones except for $E_0 = 77.85$ keV where an uncertainty equal to 50% of the correction for resonance scattering was also considered.

In the last column of Table 1 are reported the corresponding areas from Ref. 1. One may notice that these last values, while being in reasonable agreement with ours for resonances with relatively small Γ_n, are definitely much larger for those levels with large Γ_n/Γ_γ ratio. We ascribe this discrepancy to an insufficient, if any, correction for the prompt background from resonance neutron scattering.

Table 1: Resonance parameters and capture areas of ^{208}Pb + n in the range 1-400 keV

E_0 (keV)	l	J	Γ_n (eV)	Present Work		Macklin et al. (1)
				Γ_γ (meV)	$g\Gamma_n\Gamma_\gamma/\Gamma$ (meV)	$g\Gamma_n\Gamma_\gamma/\Gamma$ (meV)
43.34	--	--	--	--	26.5 ± 0.8	30.5 ± 3.2
47.33	--	--	--	--	38.5 ± 1.2	38.0 ± 4.3
71.21	1	3/2	101 ± 5	12.4 ± 2.0	24.8 ± 4.0	70 ± 30
77.85	1	3/2	958 ± 10	125 ± 30	250 ± 60	1076 ± 41
86.58	1	1/2	75.4 ± 3.0	15.2 ± 6.0	15.2 ± 6.0	<25
116.78	1	3/2	317 ± 6	27 ± 10	55.0 ± 20	276 ± 30
130.25	2	5/2	9.7 ± 0.9	101.7 ± 4.0	302 ± 12	374 ± 22
153.31	1	3/2	10.5 ± 1.0	26.7 ± 4.5	53.2 ± 9.0	--
169.48	1	3/2	21.9 ± 1.6	73.9 ± 3.0	147 ± 6	170 ± 26
193.69	--	--	--	--	277 ± 10	414 ± 29
350.43	--	--	--	--	319 ± 34	--
359.14	--	--	--	--	253 ± 44	276 ± 140

THE MAC CROSS SECTION $<\sigma>$

The MAC cross section $<\sigma>$ calculated from the data of Table 1 is listed in the second column of Table 2 for kT values in the range 5-100 keV. The quoted errors stem from the quadratic composition of the error on $<\sigma>$ due to the statistical uncertainties of the capture areas and a 6% estimated relative error in the normalization procedure. In order to get the total $<\sigma>$ one should add to the resonance part the contribution from direct capture. This part, mainly due to p-wave capture, was recently calculated by H. Oberhummer (7) with an estimated relative uncertainty of 20%, and is given in column 3. The values of total $<\sigma>$ so derived (column 4) are about a factor of two lower than the ORELA data (1) but are in excellent agreement with the 30 keV activation result of ref. (3) : $<\sigma>$ = 0.36 ± 0.03 mb.

Table 2 : The resonance and direct capture component of the MAC cross section $<\sigma>$ as a function of stellar temperature.

kT (keV)	$<\sigma>$ (mb)			kT (keV)	$<\sigma>$ (mb)		
	Present Work	Direct a) Capture	Total		Present Work	Direct a) Capture	Total
5	0.0015±.0001	0.056±.011	0.058±.011	35	0.235±.029	0.145±.029	0.380±.041
8	0.018±.001	0.070±.014	0.088±.014	40	0.241±.029	0.153±.031	0.394±.042
10	0.039±.003	0.079±.016	0.118±.016	45	0.243±.025	0.162±.032	0.405±.041
12	0.063±.005	0.086±.017	0.149±.018	50	0.241±.028	0.169±.034	0.410±.044
15	0.102±.009	0.096±.019	0.198±.021	60	0.231±.026	0.180±.036	0.411±.044
17	0.126±.012	0.102±.020	0.228±.023	70	0.216±.023	0.189±.038	0.405±.044
20	0.157±.017	0.111±.022	0.268±.028	80	0.201±.021	0.194±.039	0.395±.044
25	0.196±.023	0.124±.025	0.320±.034	90	0.186±.015	0.195±.039	0.381±.042
30	0.221±.027	0.135±.027	0.356±.038	100	0.171±.017	0.195±.039	0.366±.043

a) data from ref. (7)

REFERENCES

1. Macklin, R.L., Halperin, J., and Winters R.R., Ap. J. **217**, 222-226 (1977).
2. Macklin, R.L., and Gibbons, J.H., Phys. Rev. **181**, 1639-1642 (1969).
3. Ratzel, U., KfK Karlsruhe, Private Communication (1988).
4. Corvi, F., Fioni, G., Gasperini, F., and Smith, P.B., Nucl. Sci. Eng. **107**, 272-283 (1991)
5. Fröhner, F.H., Report KFK-2145 (1976).
6. Horen, D.J., et al., Phys. Rev. C **34**, 429-442 (1986).
7. Oberhummer, H., Private communication (1994).

Recent (n,p)- and (n,alpha)-Measurements of Relevance to Astrophysics

Cyriel Wagemans[1], Steven Druyts[1,2] and Robert Barthélémy[2]

[1] Dept. of Subatomic and Radiation Physics, RUG,
Proeftuinstraat 86, B-9000 Gent, Belgium

[2] CEC, JRC, Institute for Reference Materials and Measurements,
Retieseweg, B-2440 Geel, Belgium

Abstract. (n,p)- and (n,α)- reactions on ^{35}Cl, ^{36}Cl and ^{41}Ca have been investigated with thermal and resonance neutrons. From these data, stellar reaction rates relevant to the s-process nucleosynthesis are calculated.

INTRODUCTION

The discrepancy between calculated and observed abundances of rare isotopes such as ^{36}S and ^{46}Ca is a longstanding problem (1). One of the reasons for this discrepancy was the lack or even the complete absence of experimental data on (n,p)- and (n,α)- reactions occurring during the nucleosynthesis process. These reactions can however strongly modify the nucleosynthesis path, especially in cases where $\sigma(n,p)$ or $\sigma(n,\alpha)$ exceeds the corresponding $\sigma(n,\gamma)$.

So an experimental programme was set up to investigate (n,p)- and (n,α)- reactions and to determine their cross-sections for neutron energies from thermal up to about 500 keV. The measurements with thermal neutrons were performed at the high flux reactor of the ILL (Grenoble, France); those with resonance neutrons are being done at the GELINA facility of the IRMM (Geel, Belgium).

EXPERIMENTAL METHOD AND RESULTS

Measurements with Thermal Neutrons

The measurements with thermal neutrons had to be temporarily interrupted due to the renovation of the high flux reactor, which is expected to be completed this summer.

© 1995 American Institute of Physics

Measurements with Resonance Neutrons

The measurements with resonance neutrons were performed at the pulsed linear electron accelerator GELINA (IRMM, Geel, Belgium), which produces neutrons by bombarding a massive uranium target with 100 MeV electrons. After moderation of the neutrons in water, a broad energy spectrum (from thermal up to a few MeV) can be obtained.

The present experiments were performed at flight-path lengths of 9m and 30m, using a gridded ionization chamber coupled to a multiparameter data acquisition system. The neutron flux was determined via the $^{10}B(n,\alpha)$- and $^{235}U(n,f)$-reactions. Results have been obtained for $^{35}Cl(n,p)$, $^{36}Cl(n,p)$, $^{36}Cl(n,\alpha)$, $^{41}Ca(n,\alpha)$ and $^{41}Ca(n,p)$.

$$^{35}Cl(n,p)^{35}S$$

A detailed report was published last month (2).

$$^{36}Cl(n,p)^{36}S \text{ and } ^{36}Cl(n,\alpha)^{33}P$$

Two samples with an enrichment in ^{36}Cl of 41.65% were prepared by evaporating AgCl on a 20 µm thick aluminium backing foil coated with 400 µg/cm² of platinum. In this way, two homogeneous and stable layers were obtained with thicknesses of 68 and 174 µg AgCl/cm².

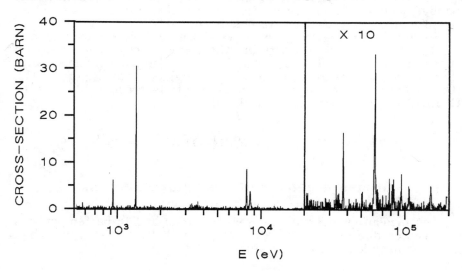

FIGURE 1. Sum of the $^{36}Cl(n,p)$ + $^{36}Cl(n,\alpha)$ cross-sections versus neutron energy.

Fig. 1 shows the results of a high resolution measurement with the thickest sample on a 30 m flight-path. Due to the energy loss in the sample, the 2.2 MeV $^{36}Cl(n,\alpha)$-particles were not perfectly separated from the 1.9 MeV $^{36}Cl(n,p)$-particles. The data shown in fig. 1 correspond to the sum of both reaction cross-sections. Except for the resonance at 930 eV, the $^{36}Cl(n,p)$-reaction is by far dominating. Additional measurements will be performed with the thinner sample in order to achieve an unambiguous separation of both processes.

$^{41}Ca(n,p)^{41}K$ and $^{41}Ca(n,\alpha)^{38}Ar$

In addition to their astrophysical importance, neutron induced reactions on ^{41}Ca are very interesting from a spectroscopic point of view. Indeed, besides the $^{41}Ca(n,p)$-transition, also the $^{41}Ca(n,\alpha_0)$-transition to the ground-state of ^{38}Ar, the $^{41}Ca(n,\alpha_1)$-transition to the first excited level and $^{41}Ca(n,\gamma\alpha)$-transitions occur. A detailed investigation of all these reactions is being performed up to 500 keV neutron energy.

During a first series of measurements, a $CaCO_3$ sample with a thickness of $\approx$ 27 µg/cm² and an enrichment in ^{41}Ca of 63.38% was viewed in an almost 2π-geometry by a 2000 mm² surface barrier detector. In a second series of experiments, the same sample was mounted in a gridded ionization chamber. Fig. 2 shows the $^{41}Ca(n,\alpha_0)$-cross-section determined at a 30m flight-path. The uncertainty on these data is mainly due to a 15% uncertainty on the number of ^{41}Ca-atoms.

At present, measurements are being performed with a CaF_2 sample with a thickness of about 2 µg/cm² and an enrichment in ^{41}Ca of 81.69%, mounted in a similar ionization chamber.

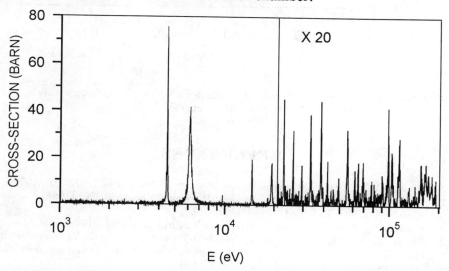

FIGURE 2. $^{41}Ca(n,\alpha_0)$ cross-section obtained at 30 m with an ionization chamber.

DISCUSSION

Koehler et al. recently reported ^{35}Cl(n,p)-, ^{36}Cl(n,p)- and ^{36}Cl(n,α)- measurements (3,4) which were performed, however, with a much poorer neutron energy resolution. For ^{41}Ca, no other data are available so far. Based on these experimental data, Maxwellian averaged cross-sections are calculated as a function of the stellar temperature. The results for ^{36}Cl(n,p)+(n,α) are shown in fig.3, the (n,α) contribution being about 8%. Also the ^{41}Ca(n,α$_0$) data are displayed in the same figure. The ^{41}Ca(n,α$_1$) + (n,γα) as well as the ^{41}Ca(n,p) cross-sections roughly amount to 5-10% of the corresponding ^{41}Ca(n,α$_0$)-values. The astrophysical relevance of these cross-sections is discussed in refs. 2-5.

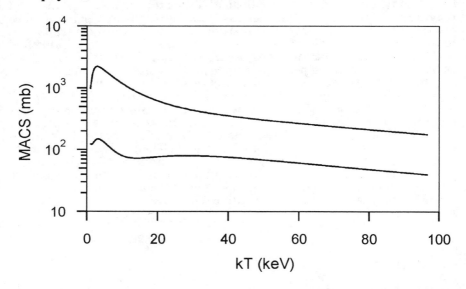

FIGURE 3. Maxwellian averaged cross-sections for ^{41}Ca(n,α$_0$) (upper curve) and ^{36}Cl(n,p) + (n,α) (lower curve).

REFERENCES

1. Howard,W., Arnett,W., Clayton,D., and Woosley,S., Astrophys.J. **175**, 201 (1972).
2. Druyts,S., Wagemans,C., and Geltenport,P., Nucl. Phys. **A573**, 291 (1994).
3. Koehler,P., Phys. Rev. **C44**, 1675 (1991).
4. Koehler,P., Graff,S., O'Brien,H. and Popov,Y., Phys. Rev. **C47**, 2107 (1993).
5. Takahashi,K., and Hillebrandt,W., "On the nβ-process and the Ca-Ti anomalies in the Allende meteorite", in *Proceedings of the 7th Workshop on Nuclear Astrophysics, Ringberg Castle*, 1993, pp. 99-106.

Recent Results of Measurements of the $^{14}N(n,p)^{14}C$, $^{35}Cl(n,p)^{35}S$, $^{36}Cl(n,p)^{36}S$ and $^{36}Cl(n,\alpha)^{33}P$ Reaction Cross Sections

Yu.M.Gledenov*, V.I.Salatski*, P.V.Sedyshev*,
M.V.Sedysheva*, P.E.Koehler♣, V.A.Vesna♦, I.S.Okunev♦.

*Frank Laboratory of Neutron Physics, Joint Institute for Nuclear Research,
141980 Dubna, Russia,
♣ Los Alamos National Laboratory, Los Alamos, New Mexico 87545, USA,
♦ Petersburg Nuclear Physics Institute, 188350 Gatchina, Russia.

Abstract. Experiments are reported for measuring the cross section of the $^{14}N(n,p)^{14}C$ reaction over the neutron energy range from thermal energy to 150 keV at the IBR-30 pulsed booster at JINR, Dubna and the WWR-M reactor at INR, Kiev. The reaction cross section values were found for the thermal energy and for the neutron energies of 24 keV, 54 keV, 144 keV. The $^{36}Cl(n,p)^{36}S$ cross section was measured for the neutron energies from thermal energy to approximately 800 keV at the neutron source of LANSCE, Los Alamos. The contributions of the $^{36}Cl(n,p)^{36}S$ and $^{36}Cl(n,\alpha)^{33}P$ reactions to resonances at 0.9 keV and 1.3 keV were identified. Also, at the WWR-M reactor of PINR, Gatchina, preliminary measurements of the $^{36}Cl(n,p)^{36}S$ cross section at the thermal neutron energy were conducted. The $^{35}Cl(n,p)^{35}S$ reaction cross section was measured at the IBR-30 pulsed booster.

Data obtained as a result of investigations of the (n,p) and (n,α) reactions may be very important for a better understanding of many scenarios of nucleosynthesis. The recent experimental results, presented below, apply to the processes of explosive nucleosynthesis and the s-process. So, depending on the $^{14}N(n,p)^{14}C$ reaction cross section, ^{14}N may act as a strong neutron "poison" during the chain of reactions involving the $^{13}C(\alpha,n)O^{16}$ neutron source in the s-process [1]. Also, the ^{14}N isotope may play a critical role in the nucleosynthesis of ^{19}F. The ^{35}Cl and ^{36}Cl isotopes are associated with production of the rare ^{36}S isotope in explosive carbon burning [2] or in the s-process [3]. Up to now, various astrophysical calculations appear to overproduce ^{36}S. Recently, however, some data have been obtained on the $^{36}Cl(n,p)^{36}S$ cross section [4-7], and the theoretical estimations were used in the calculations.

Detector. Most of our measurements were performed using a double grid ionization chamber (DGIC). Our detector represents a double section plate ionization chamber, which is actually two DGIC with a common cathode (see

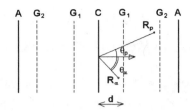

FIGURE 1. Scheme of the double grid ionization chamber.

Fig.1). The investigated and standard targets are placed in the center of the cathode in the "back-to-back" geometry. As a rule these targets are the same size. The cathode and anode signals are registered in the coincidence mode. The pulse-height values from charged particles are described by the well known expressions:

$$V_a = E, \quad V_c = \Delta E\left(1 - \frac{X}{d}\cos\theta\right) \quad (1)$$

where V_a and V_c are the signals from the anode and cathode respectively, E is the energy of the particle, ΔE is the part of energy which the particle loses into the C-G1 volume, X is the center of gravity of the particle track limited by this volume, θ is the angle between the normal of the cathode and the particle trace. After digital conversion the signals are recorded by the measurement system as an integral spectra and as a file of consecutive codes (V_a, V_c) (or (V_a, V_c, t) in the case of measurements by the time-of-flight method) for "off-line" data proceeding. Our method allows identification and selection of different kinds of similar energy particles using the cathode amplitude distribution. This is important for experiments with slow neutrons, where the cross sections of the ^{6}Li and ^{10}B microadmixtures are very large. Also, it allows reduction of the background from the neutron beam and of γ-quanta from the neutron source and the radioactive sample, because the additional G1 grid reduces the sensitive volume in which the background emission pulses could be registered. These advantages of the DGIC can be shown in the example of measurements with the ^{36}Cl isotope.

$^{36}Cl(n,p)^{36}S$ and $^{36}Cl(n,\alpha)^{33}P$ reactions. The preliminary measurements with the ^{36}Cl isotope were performed at the high intensity thermal neutron channel of the WWR-M reactor of PINR, Gatchina. The vertical neutron channel provides thermal neutrons with a flux of $2 \cdot 10^{10} s^{-1}$ (~$5 \cdot 10^8$ 1/cm^{2}s) [8] with a negligible amount of γ-raises and fast and epithermal neutrons. The 25 mm diameter ^{36}Cl target was made by vacuum evaporation onto an aluminium foil 8.5 μm thick. The chlorine was enriched to about 60% with ^{36}Cl and the sample activity was 5.31 μCi. The neutron beam was limited to 30 mm in diameter by the ^{6}LiF-collimator. The neutron flux was determined by gold sample activation.

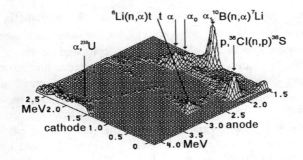

FIGURE 2. An anode-cathode matrix of the ^{36}Cl(n,p)^{36}S measurement for thermal neutrons.

Fig.2 represents a matrix of anode-cathode events from these measurements. Here one can clearly see that the proton distribution and α-particles from the reactions on ^{6}Li and ^{10}B are well separated. The contribution of products from the reaction on ^{36}Ar was determined during measurements with another section of the chamber without a sample and the backing was empty. We obtained the following result for the ^{36}Cl(n$_{th}$,p)^{36}S reaction cross section: σ_{th}=23.7±8.7 mb. Our value is by a factor of about 2 less than the result given by Wagemans et al. [9]. However, it should be noticed that our result is a preliminary one, because we were not successful in eliminating all the system errors in the experiment.

The measurements of the ^{36}Cl(n,p)^{36}S cross section in the neutron energy range from thermal energies to 800 keV were carried out on the neutron source of LANSCE at Los Alamos using the semiconductor detector. Parameters of 8 resonances were determined [10]. Using the DGIC we succeeded in finding the α-channel of decay for the resonances of 902.6 eV and 1297.3 eV. A fragment of the time-of-flight spectrum read out during a run of these measurements is shown in Fig.3. The upper spectrum is from amplitude windows corresponding to protons (Ep=1.87 MeV).

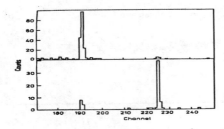

FIGURE 3. Time-of-flight spectra of the ^{36}Cl measurement for resonance neutrons.

The lower one is for α-particles from the $^{36}Cl(n,\alpha)^{33}P$ reaction (E_α=2.10 MeV). At the resonance of 1297.3 eV only protons, for the most part, were registered, but at the resonance of 902.6 eV, α-particles consisted of about 75% of the registrations. We are presently planning to do precision measurements of the $^{36}Cl(n,p)^{36}S$ and $^{36}Cl(n,\alpha)^{33}P$ reaction cross sections for thermal and resonance neutrons. The separation of the α and p-channels may reduce the $^{36}Cl(n,p)^{36}S$ reaction rate [10].

The $^{14}N(n,p)^{14}C$ measurements. These experiments were carried out on a neutron beam of the IBR-30 pulsed booster in Dubna [11] and on the filtered neutrons from the WWR-M reactor of INR in Kiev [12,13]. We measured the $^{14}N(n,p)^{14}C$ reaction cross section at thermal neutrons and neutron energies of 24.5 keV, 53.5 keV and 144 keV to be 1.83±0.07 b, 2.04±0.16 mb, 2.08±0.23 mb and 2.07±0.27 mb, respectively. In Fig.4 our values are compared with data taken from [14-20]. It can be seen that our results agree with the cross section values reported in ref.[9,14-16,18,19]. Our measurements support the reaction rate calculated by Bahcall and Fowler [20] over the threefold reduction in this rate recommended by Brehm et al. [17]. Therefore, our experiment confirms the conclusion of ref.[18,19], that ^{14}N is potentially a much stronger neutron poison during the possible operation of a $^{13}C(n,\alpha)^{16}O$ s-process neutron source, than the data of Brehm et al. imply. Also, it seems to rule out the s-process as a source of significant amounts of ^{15}N.

The $^{35}Cl(n,p)^{35}S$ measurements. Currently in Dubna we are carrying out the measurement of this reaction having the aim to connect the thermal cross section with the lower resonances of ^{35}Cl. We have a preliminary result for the $^{35}Cl(n,p)^{35}S$ cross section for thermal neutrons: σ_{th}=540±40 mb. In the future we hope to reduce the uncertainty of this value.

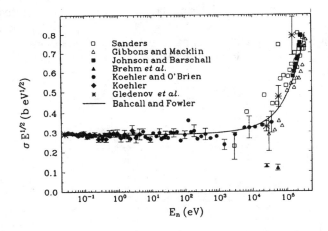

FIGURE 4. The $^{14}N(n,p)^{14}C$ reduced cross section vs. laboratory neutron energy.

We will soon finish the modernization of our measurement apparatus. We have plans to continue measuring reactions of importance to astrophysics. The next step in our research will be measuring the angular distribution of the reaction products and the experiments with gas targets. The start of operation of the new neutron source (IREN) in Dubna will be in the near future. IREN will have a neutron energy range an order of magnitude greater than IBR-30 has now. Then it will be possible to realise a wide research program for the study of reactions of importance to nuclear astrophysics.

REFERENCES

1. Käppeler, F., et al., Rep.Prog.Phys. **52**, 945 (1988).
2. Howard, W., Astrophys.Journ. **175**, 201 (1972).
3. Beer, H., and Penzhorn, R.-D., Astron.Astrophys. **174**, 323 (1987).
4. Andrzejewski, J., et. al., Report of JINR P3-87-319, Dubna, (1987).
5. Gledenov, Y.M., et al., Z.Phys.A **322**, 685 (1985).
6. Koehler, P.E., and O'Brien, H.A., NIM B **40/41**, 494 (1989).
7. Druyts, S., et al., Ann.Geophys. **9**(suppl.), C336 (1991).
8. Altarev, I.S., et al., Pizma v JETF, **44**, 269 (1986).
9. Wagemans, C., et al., in *Proc. Intern.Conf. on Nucl. Data for Sci. and Tech.*, 1992, p.638.
10. Koehler, P.E., et al., Phys.Rev.C **47**, 2107-2112 (1993).
11. Gledenov, Y.M., et al., Z.Phys.A **364**, 307-308 (1993).
12. Gledenov, Y.M., et al., Z.Phys.A (submitted).
13. Gledenov, Y.M., et al., in *Proc. of the 8 Intern.Symp. on Capture γ-ray Spec.*, 1993, p.584.
14. Sanders, M., Phys.Rev. **104**, 1434 (1956).
15. Gibbons, J.H., and Macklin, R.L., Phys.Rev. **114**, 571 (1959).
16. Jonson, C.H., and Barschall, H.H., Phys.Rev. **80**, 818 (1950).
17. Brehm, K., et al., Z.Phys.A **330**, 167 (1988).
18. Koehler, P.E., and O'Brien, H.A., Phys.Rev.C **39**, 1655 (1989).
19. Koehler, P.E., Phys.Rev.C **48**, 439-440 (1993).
20. Bahcall, N.A., and Fowler, W.A., Astrophys.J. **157**, 659 (1969).

Far from stability: evaluation of properties of nuclear excited states

A. Mengoni[1], G. Maino[1,2] and Y. Nakajima[3]

[1] *ENEA, Settore Fisica Applicata, Bologna, Italy*
[2] *INFN, Sezione di Firenze, Firenze, Italy*
[3] *JAERI, Nuclear Data Center, Tokai-mura, Japan*

Abstract. We propose an integrated technique for evaluation of nuclear excitation properties using a model (based on the Fermi Gas with pairing and shell correction) for the level density in the continuum and the Interacting Boson Model for the calculation of the discrete spectra of nuclei far from the stability line. An example is shown relevant to the calculation of the excitation energies of the Cd isotopes in the N = 50-82 shell.

Key words: Nuclear level density, Fermi Gas Model, Interacting Boson Model, neutron-rich nuclei

1. Introduction

The calculation of nuclear bulk properties and ground state energies of nuclei far from stability has a history as old as the von Weiszäcker mass formula. In the recent years, in addition to full microscopic calculations (rarely applied on wide mass regions), alternative approaches based on the combination of macroscopic-phenomenological models, integrated with microscopic calculations for basic nuclear structure effects have been proposed. A typical example of this integration is given by the so-called macroscopic-microscopic model (latest version in Möller *et al.* 1994) for the evaluation of nuclear masses and other properties of nuclear ground states. The good degree of predictability for nuclear ground-state properties achieved using such methods gave us the idea of a similar integration for the evaluation of nuclear excited states properties.

For applications in nuclear astrophysics and in particular for the calculation of nuclear reaction rates it is indispensable to provide the nuclear level density in the full energy range, from the ground state up to excitation energies of several MeV. We have been recently involved in the construction of a new parameter systematics for the calculation of nuclear level density based on the Fermi Gas (Bethe formalization) which takes into account of fundamental nuclear structure effects such as pairing interaction, deformation and shell inhomogeneities (Mengoni and Nakajima 1994). In addition, it has been recently proposed (Mengoni *et al.* 1994, and references therein) to integrate the FGM and use the Interacting Boson Model (IBM) to evaluate the contribution of collective degrees of freedom in nuclear excitation which have shown to play a fundamental role at excitation energies corresponding to the neutron binding energy and higher. Because the IBM provides the low-lying *discrete* energy levels of medium-mass and heavy nuclei with different shapes and ground state structures, it is possible to use the IBM not only for the evaluation of the contribution in the continuum but also for the calculation of the low-lying bands. Following these developments, we propose here to extend the evaluation of nuclear excitations to nuclei far from the stability line.

2. The Fermi Gas Model and its parametrization

A recent parametrization of the standard FGM which takes into account pairing correlations and shell inhomogeneities resulting in a systematics of the FGM parameter a (proportional to the density of single-particle states at the Fermi energy) or of its shell corrected analog $a(*)$, provides

$$a(U) = a(*)[1 + \frac{E_{sh}}{U}(1 - e^{-\gamma U})] \qquad (1)$$

where U is the excitation energy and $\gamma = 0.4 A^{-1/3}$ MeV^{-1} a damping parameter. The shell correction energy, E_{sh} can be either evaluated from the knowledge of the experimental nuclear mass or from microscopic shell model calculation.

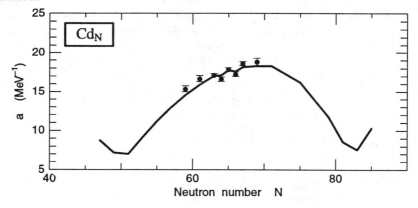

FIGURE 1. Level density parameter a vs the neutron number N for Cd isotopes. Experimental values are derived as described in the text.

Details of the methods used are given in the references (Mengoni and Nakajima 1994, Mengoni and Nakajima 1993). Here, we have calculated the level density parameter a for the Cadmium isotopes in the N = 50-82 shell. Several experimental values of the average spacings of s-waves neutron resonances are available for the Cadmium isotopes, thus allowing for a determination of a for $107 \leq A \leq 117$; they are reported in Figure 1 together with the calculated values for the other isotopes within the shell. Only nuclei with an odd number of neutrons are considered in the extrapolated regions. For the isotopes where the experimental values are available the agreement with the calculation is good allowing for confidence in the extrapolated regions. A comparison of the whole systematics for a vs N confirms the behavior in the extrapolated regions.

3. The Interacting Boson Model and the low-lying spectra

The IBM is a model which provides a description of the low-lying collective spectra of medium-mass and heavy nuclei in terms of excitations represented by scalar $s(l^\pi = 0^+)$ and quadrupole $d(l^\pi = 2^+)$ bosons. Several versions of the model are available for even-even, odd-mass and odd-odd nuclei. Here we only use the model for even-even nuclei, in which neutron and proton degrees of freedom are taken into account (IBM-2). The Hamiltonian can be written as

$$\hat{H} = \hat{h}_0 + \epsilon(\hat{n}_{d_\nu} + \hat{n}_{d_\pi}) + \kappa \hat{Q}^{(2)}_\nu \cdot \hat{Q}^{(2)}_\pi + \hat{V}_{\nu\nu} + \hat{V}_{\pi\pi} + \hat{H}_M. \qquad (2)$$

The definition of all the operators and parameters in Equation (2) can be found in the literature (Iachello and Arima 1987).

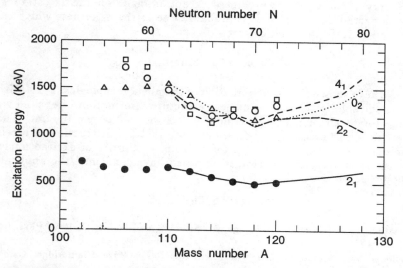

FIGURE 2. Low-lying energy levels of Cd isotopes. The various lines are the results of IBM-2 calculations.

The IBM provides a unified description of low-lying collective spectra for nuclei with different ground-state properties and deformations. In particular, its three dynamical symmetry limits furnish a full description of spherical, axisymmetric and γ-unstable nuclei in a unified and consistent fashion. Moreover, transitional nuclei can as well be treated in a similar simple picture, provided that a suitable choice of the Hamiltonian parameters is adopted. As far as our problem is concerned, we are here interested in the calculation of the low-lying spectra of the even-even Cadmium isotopes whose structure is close to the IBM U(5) dynamical symmetry limit. We can use a parametrization of the Hamiltonian for Cd isotopes recently derived by Giannatiempo et al. 1991 where all the parameters vary smoothly as a function of the boson number N_B. In particular, the only parameter which varies linearly with N_B is χ_ν in the definition of the quadrupole operator

$$\hat{Q}^{(2)}_\nu(\chi_\nu) = [d^\dagger \times \tilde{s} + s^\dagger \times \tilde{d}]^{(2)} + \chi_\nu [d^\dagger \times \tilde{d}]^{(2)}. \qquad (3)$$

It is therefore possible to extrapolate the calculation to mass regions where experimental information is not available. In our case, we have calculated the low-lying spectra of the same set of Cd isotopes selected in the previous paragraph for the evaluation of the FGM parameters. The results are shown in Figure 2. There, the energies of the first 2^+ state and of the triplet $0^+{}_2, 2^+{}_2, 4^+{}_1$ are shown in comparison with the available set of experimental data (Kumpulainen et al. 1992). It should be stressed here that we are showing only these few levels in the figure, but the full collective spectra are provided by the IBM-2 calculations. The agreement is reasonable for experimentally accessible nuclei. For the neutron rich nuclei, close to the shell closure an increase of the energy of the triplet is predicted exception made for the $2^+{}_2$ state which shows a partial decrease. This is probably due to the γ-softness which can be enhanced close to the shell closure.

4. Conclusions

We have shown that a combination of the phenomenological FGM with the IBM is suitable for obtaining nuclear structure information of excited states, for the low excitation energy region as well as in the continuum. Though this work is to be considered as a first step in the construction of a reliable technique and a parametrization for treating properties of excited nuclei far from the stability line, the results obtained have confirmed the applicability of the proposed methodology.

REFERENCES

Giannatiempo, A., Nannini, A., Perego, A., Sona, P., and Maino, G. 1991, Phys. Rev. C, 44, 1508

Iachello, F., and Arima, A. 1987, *The Interacting Boson Model* (Cambridge: Cambridge University Press)

Kumpulainen, J., Julin, R., Passoja, A., Trzaska, W. H., Verho, E., and Väärämäki, J. 1992, Phys. Rev. C, 45, 640

Mengoni, A., Maino, G., Ventura, A., and Nakajima, Y. 1994, *Proc. Internat. Conf. on Perspectives of the Interacting Boson Model, Padova, June 1994*, (Singapore: World Scientific), in press

Mengoni, A., and Nakajima, Y. 1993, JAERI-M report, 93-177

Mengoni, A., and Nakajima, Y. 1994, J. Nucl. Sci. Tech., 31, 151

Möller, P., Nix, J. R., Myers, W. D., and Swiatecki, W. J. 1994, Atomic Data Nucl. Data Tables, in press

Shell Effects in Neutron Capture on Pb

T. Rauscher*, R. Bieber†, S. Lingner†, H. Oberhummer†

*Institut für Kernchemie, Universität Mainz, Germany
†Institut für Kernphysik, Technische Universität, Vienna, Austria

> We calculate direct neutron capture on $^{210-228}$Pb with energy levels and masses taken from two different models. A sharp drop in the cross sections due to the lack of low spin states is seen with RMFT levels but cannot be found with the folded-Yukawa shell model.

I. INTRODUCTION

Over the last years it has been realized that the direct reaction mechanism cannot only dominate the reaction cross sections involving light elements. Also for intermediate and heavy nuclei it can become important near shell closures or for neutron rich isotopes when the level density becomes too low for the compound nucleus mechanism. Recently, we calculated direct capture (DC) cross sections for a number of cases (?,?). In this work we want to focus on (n,γ) reactions on the even-even isotopes $^{210-228}$Pb. They are of importance in the r-process path going to very heavy elements.

II. DIRECT CAPTURE AND FOLDING PROCEDURE

The theoretical cross section σ^{th} is derived from the DC cross section σ^{DC} given by (?)

$$\sigma^{\text{th}} = \sum_i C_i^2 S_i \sigma_i^{\text{DC}} \qquad (1)$$

The sum extends over all possible final states (ground state and excited states) in the residual nucleus. The isospin Clebsch-Gordan coefficients and spectroscopic factors are denoted by C_i and S_i, respectively. The DC cross sections σ_i^{DC} are essentially determined by the overlap of the scattering wave function in the entrance channel, the bound-state wave function in the exit channel, and the multipole transition operator. For the computation of the DC cross section we used the direct capture code TEDCA (?).

For determining the nucleon-nucleus potential the folding procedure was employed, a method already successfully applied in the description of many systems. In this approach the nuclear target density ρ_T (derived from experimental charge distributions or from theory) is folded with an energy and density dependent nucleon-nucleon interaction v_{eff} (?):

$$V(R) = \lambda V_{\rm F}(R) = \lambda \int \rho_{\rm T}(\vec{r}) v_{\rm eff}(E, \rho_T, |\vec{R} - \vec{r}|) d\vec{r} \quad , \tag{2}$$

with $\vec{R}$ being the separation of the centers of mass of the two colliding nuclei. The normalization factor λ accounts for effects of antisymmetrization and is close to unity. The potential obtained in this way ensures the correct behavior of the wave functions in the nuclear exterior. At the low energies considered in astrophysical events the imaginary parts of the optical potentials are small.

III. ENERGY LEVELS AND MASSES

The levels (and masses) needed as input for the DC calculation were taken from two different approaches. The first one was the relativistic mean field theory (RMFT) which has turned out to be a successful tool for the description of many nuclear properties (?). The RMFT describes the nucleus as a system of Dirac nucleons interacting via various meson fields. There are six parameters which are usually obtained by fits to finite nuclear properties. For our calculations we have used the parameter set NLSH (?,?). Also the density distributions needed for the determination of the folding potentials were derived from RMFT wave functions.

The second method is the shell model developed by Möller et al. (?). In this model folded-Yukawa potentials are used in the determination of the single-particle potentials. For pairing the Lipkin-Nogami pairing model (?) is employed. This shell model proved to be very successful in reproducing ground state spins along magic numbers (?) and has been used in QRPA calculations of β-decay half lives (?) and as microscopic part in nuclear mass determinations (?).

To be able to compare the predictions from both models the nuclei were considered to be spherically symmetric.

IV. RESULTS AND DISCUSSION

Having obtained the relevant spins and Q-values (masses) as discussed above we still had to determine the strength parameter λ of the scattering potentials (see Equation (2)). This was done by adjusting λ so that experimental scattering data for ^{208}Pb+n at low energies were reproduced. In the potentials for each of the other isotopes a factor λ was chosen giving the same volume integral as for the fitted ^{208}Pb-potential. For the bound state potentials λ is fixed by the requirement of correct reproduction of the binding energies. The spectroscopic factors were assumed to be unity for all transitions considered.

The results of our calculations are summarized in Fig. 1. The most striking feature is the sudden drop over several orders of magnitude in the cross sections calculated with the RMFT levels. This is due to the lack of low spin levels which are cut off by the decreasing neutron separation energy. For

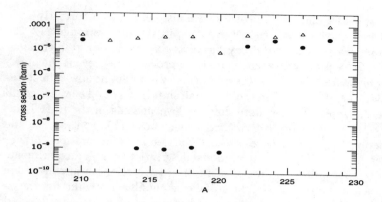

FIG. 1. 30 keV cross sections for different Pb isotopes. Levels and masses are calculated with models by Sharma et al. (?) (dots) and Möller et al. (?) (triangles).

comparison, the levels from both models are shown for ^{211}Pb and ^{215}Pb in Table 1. Only after the $1i_{11/2}$ shell (which forms the state at lowest energy in the RMFT) has been filled completely at ^{222}Pb the cross section is increasing because low spin states become available again.

The values resulting from the Möller model exhibit a smoother behavior. Only a slight dip is visible for ^{220}Pb(n,γ) since the previously accessible $1/2^+$ and $3/2^+$ states have become unbound in ^{221}Pb. Beyond ^{223}Pb the $2g_{9/2}$ shell (which is at lower energy than the $11/2^+$ level in this model) has been filled and at least one of the low spin states can be populated again. The known

TABLE 1. Comparison of theoretical bound energy levels (only the even parity states are shown)

^{211}Pb				^{215}Pb			
RMFT (?)		Möller (?)		RMFT (?)		Möller (?)	
E_x [MeV]	J^π	E_x [MeV]	J^π	E_x [MeV]	J^π	E_x [MeV]	J^π
0.000	11/2+	0.000	9/2+	0.000	11/2+	0.000	9/2+
1.317	9/2+	0.520	11/2+	1.479	9/2+	0.290	11/2+
3.139	5/2+	2.260	5/2+			1.224	5/2+
3.277	7/2+	3.218	1/2+			2.885	1/2+
3.327	1/2+	3.431	7/2+			3.029	7/2+

ground state spins for the lighter isotopes are also reproduced correctly.

If the r-process path crosses the Pb-isotopes in the masses range 212–220 due to the comparatively long β-decay half lives the sudden drop in the neutron capture cross section would terminate the r-process and prevent the build-up of heavier elements. Since in some cases there are unbound low spin states ($J^\pi = 1/2^+, 3/2^+$) close to the threshold a small shift in the level energies could already close the gap. However, note that the level spacing in the RMFT has the tendency to increase towards neutron rich nuclei (?), contrary to the Möller prediction (and contrary to a recent Hartree-Fock calculation (?)).

Nevertheless, even if improved shell model descriptions and/or experiments have to clarify this point, the mechanism of cross section suppression due to the lack of low spin states could still be a possible restraint for the r-process path towards even more neutron rich isotopes and for nuclei beyond Pb. Because of the low level density the compound nucleus model would not be applicable in those cases.

V. ACKNOWLEDGEMENTS

We want to thank P. Möller and M.M. Sharma for making their computer codes available to us. This work was supported by the Austrian Science Foundation (project P8806–PHY). TR is an Alexander von Humboldt fellow.

REFERENCES

1. Balogh, W., Bieber, R., Oberhummer, H., Rauscher, T., Kratz, K.-L., Mohr, P., Staudt, G., and Sharma, M.M., in *Proc. Europ. Workshop on Heavy Element Nucleosynthesis*, eds. E. Somorjai and Z. Fülöp, Debrecen: Inst. Nucl. Res., 1994, p. 67.
2. Grün, K., Pichler, R., and Oberhummer, H., in *Proc. 4th Int. Conf. Appl. Nucl. Techn.*, in print.
3. Kim, K.H., Park, M.H., and Kim, B.T., *Phys. Rev. C* **23**, 363 (1987).
4. Krauss, H., computer code TEDCA, TU Wien, 1992, unpublished.
5. Kobos, A.M., Brown, B.A., Lindsay, R., and Satchler, G.R., *Nucl. Phys.* **A425**, 205 (1984).
6. Gambhir, Y.K., Ring, P., and Thimet, A., *Ann. Phys. (NY)* **198**, 132 (1990).
7. Sharma, M.M., Nagarajan, M.A., and Ring, P., *Phys. Lett. B* **312**, 377 (1993).
8. Sharma, M.M., Lalazissis, G.A., and Ring, P., *Phys. Lett. B* **317**, 9 (1993).
9. Möller, P. and Nix, J.R., *Nucl. Phys.* **A361**, 117 (1981).
10. Pradhan, H.C., Nogami, Y., and Law, J., *Nucl. Phys.* **A201**, 357 (1973).
11. Möller, P. and Randrup, J., *Nucl. Phys.* **A514**, 1 (1990).
12. Möller, P., Nix, J.R., Myers, W.D., and Swiatecky, W.J., *At. Data Nucl. Data Tables*, in print.
13. Sharma, M.M., Lalazissis, G.A., Hillebrandt, W., and Ring, P., *Phys. Rev. Lett.* **72**, 1431 (1994).
14. Dobaczewski, J., et al., *Phys. Rev. Lett.* **72**, 981 (1994).

G-T strength distribution in f-p shell nuclei and ν_e from stellar collapse

F. K. Sutaria and A. Ray

Tata Institute of Fundamental Research, Bombay-400 005, INDIA

Abstract. The Gamow-Teller strength distribution in f-p shell nuclei is important in determining e^--capture and β-decay rates on neutron rich nuclei in presupernova and collapsing cores of massive stars. We use the recently available data for p-n and n-p reactions on some of these nuclei to generate a generalised relation for the centroid (in energy) of the G-T distribution in terms of nuclear mass number and neutron excess. We then calculate the energy spectrum of the electron-capture neutrinos which stream freely out of the collapsing star *without further interactions* (i.e. the pre-trapping neutrinos). The number and spectra of such neutrinos detectable through the charged current reaction on Deuterium in the Sudbury Neutrino Observatory has been computed to be $\simeq 50$ for a supernova explosion at 500 pc.

1. Introduction

The G-T strength in nuclei is distributed over a broad region of excitation energies known as the Gamow-Teller Giant Resonance (GTGR) and contains a large fraction of the total weak-interaction strength. At temperatures and densities ($T \simeq 1 MeV$ and $\rho \simeq 10^{10} g/cm^3$) characteristic of the presupernova core, the main contribution to the rates will come from the GTGR region. Using the experimentally obtained G-T strength distribution with respect to the daughter excitation energy (which is available only for a few nuclei of interest) it is possible to map out the strength disribution in energy, in terms of a few shape parameters. These parameters can be related to nuclear properties like the nuclear isospin and the mass number and a corresponding approximate distribution can be constructed for any nucleus of astrophysical interest in that shell. We consider here nuclei in the f-p shell since in the post-Si-burning phase the core of a massive star has a substantial abundance of neutron rich nuclei $A \geq 60$. Further, during core collapse upto densities of 10^{11} g/cm^3, the neutrinos emitted due to the e^--captures on free protons and nuclei stream freely out of the star without scattering. The spectrum of these neutrinos will be characteristic of the nuclear species which emitted them and thus of the physical conditions in the core. Here we compute the energy spectrum of the unscattered pre-bounce neutrinos and their expected number of detections from a close by Supernova explosion.

2. The Gamow-Teller Strength Distribution

The charge-exchange (p,n) and (n,p) reactions are iso-spin and spin-dependent over a wide range of projectile energies. The weak interaction strength distribution (B_{GT}^+ or B_{GT}^-) is propotional to the $\Delta L = 0$, forward scattering cross-section in the high projectile energy, low momentum transfer limit. This can be seen by comparing the corresponding operators[1]. Several attempts to calculate the G-T distribution from direct shell model calculations using a truncated shell model basis space have been made, (e.g. Aufderheide et al)[2]. Since the number of basis space states involved for mid fp-shell nuclei is very large, the direct shell model calculations, despite substantial computational efforts, have remained somewhat approximate. Here, following an earlier work[3] we have used a statistical approach instead, and applied the Spectral Distribution Theory. According to this theory, for a many particle space of large dimensionality, the smoothed-out eigenvalue distribution of a (2+1) body Hamiltonian is approximately Gaussian. A skewed Gaussian of the form:

$$B_{GT} = A_0[1 + \gamma_1(x^3 - 3x)/6 + \gamma_1^2(x^6 - 15x^4 + 45x^2 - 15)/72]\exp(-x^2/2)$$

$$x = (E_{ex} - E_{GT})/\sigma$$

is used here. The parameters γ_1 (the skewness factor), E_{GT} (the 'energy centroid'), σ (the effective half-width) and A_0 (normalisation factor) are obtained for each nucleus by fitting the above formula to the experimentally obtained strength. These parameters will be reported elsewhere. Here we report the relation developed for the energy centroid E_{GT}. In making these fits, greater weightage has been given to the main peak of the G-T distribution, which generally lies below 10 MeV, because we are primarily interested in the presupernova conditions, where the typical e^- chemical potentials are $\simeq$ few MeV. Comparing the change in the energy of a single particle level under the action of either G-T or Fermi operators it is found that the difference $E_{GT^-} - E_F$ is expected to depend on the spin-orbit splitting term[4] ($\sim A^{-1/3}$), and on the isospin dependant Lane potential term[5] ($\sim (N-Z)/A$). The Isobaric Analog State E_{IAS} (where the Fermi centroid is located) is obtained either experimentally, or from a theoretical relation[5], which is observed to agree very well with the experimental data used in this work. This gives

$$E_{GT^-} - E_{IAS} = 44.16 A^{-1/3} - 76.1(N-Z)/A$$

The correlation between the theoretical relation developed here and the experimental values is shown in Fig. 1. A similar relation was derived earlier[4] using data for nuclei beyond f-p shell and the corresponding coefficients differ widely from those obtained for the f-p shell nuclei. The data used in the present work came from refs. 10 ($^{54,56}Fe$, $^{58,60}Ni$), 11 (^{81}Kr), 12 (^{51}V) and 13 (^{71}Ga).

Since there is no IAS state in the GT^+ direction (for the ground state to excited state transition), E_{GT^+} is expected to depend on the spin-orbit splitting term, the Lane potential term and, for odd-A nuclei on the pairing energy term as well. The correlation between the fitted value

$$E_{GT^+} = 14.15 A^{-1/3} - 3.3(N-Z)/A + 12 A^{-1/2}\delta_{A_{odd}}$$

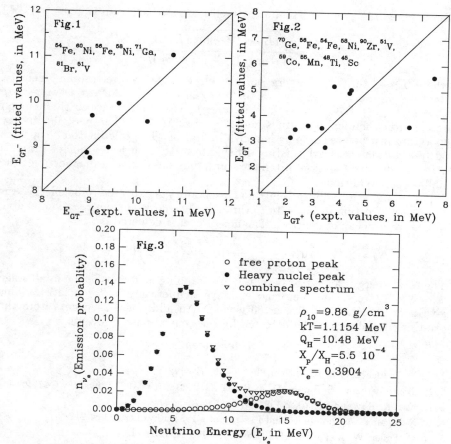

and the experimental value is shown in Fig. 2. The nuclei used in both figures are stated in the order of increasing E_{GT}(experimental). For the latter fit, the data was obtained from refs. 14 (^{45}Sc), 15 (^{56}Fe, ^{58}Ni, ^{55}Mn), 16 (^{51}V, ^{59}Co), 17 (^{70}Ge), 18 (^{90}Zr), 19 (^{48}Ti).

3. Neutrino detection and spectroscopy in SNO

The dominant nuclear species at each instant of collapse is obtained from a one-zone collapse calculation[9] with Saha equilibrum conditions. The neutrino trapping criterion was defined as the point at which the mean-free-path[6] equalled the radius of the homologously collapsing core. This gives trapping density $\rho = 10^{11} g/cm^3$. The unscattered neutrino spectrum is given by:

$$n_{\nu_e} dE = \frac{\sqrt{(E+Q)^2 - m_e^2} (E+Q) E^2 dE}{(1 + \exp((E+Q) - E_F)/kT) . ft}$$

where E_F is the Fermi energy of the electrons at a pre-trapping density. Because the nuclei are in a 'sea' of free neutrons, the nuclear Q-values are $Q = \mu_n - \mu_p + 1.297 + E_{GT-}$ where $\mu_n - \mu_p$ is the difference in the neutron and proton chemical potentials and $E_{GT-} \simeq 3 MeV$ [7]. The neutrino energy spectrum just before trapping is shown in Fig. 3. The peak due to e^--captures on free protons is clearly distinguishable from that on heavy nuclei and this being a sensitive function of temperature, may be used as a diagnostic of the stellar core conditions.

The planned Sudbury Neutrino Observatory is expected to be energy sensitive and direction sensitive to ν_e through the charge-current reaction leading to inverse β-decay of deuteron. The no. of events detected is[8]:

$$N_d = 2.49 \times 10^3 \left(\frac{M}{Kt}\right) \left(\frac{D}{Kpc}\right)^{-2} \left[0.7 \left(\frac{E_{\nu_e}}{10^{52}}\right) \left(\frac{T_{\nu_e}}{4 MeV}\right)^{1.49}\right]$$

The total energy carried away by the pre-trapping neutrinos in the infall stage is $\sim 3 \times 10^{50}$ ergs and neutrino effective temperature is $T_{\nu_e} \simeq 1.24$. This gives, for a SN at 500 pcs, $\simeq 50$ events. An additional 25 events from the pre-trapping phase will be detected by the neutral current events.

References

1. T. N. Taddeucci et al 1987, Nucl.Phys.A**469**, 125-172
2. M. B. Aufderheide, S. Bloom, D. A. Resler, G. J. Mathews 1993, Phys.Rev.C**47**, 2961
3. K. Kar, A. Ray and S. Sarkar 1994, ApJ, Oct. 20
4. K. Nakayama, A. Pio Galleõ and F. Krmpotić 1982, Phys.Lett.B**114**,217
5. G. M. Fuller, W. A. Fowler, M. J. Newman 1982, ApJ**252**,715
6. G. E. Brown, H. A. Bethe and G. Baym, 1982, Nucl.Phys.A**375**, 481
7. H. A. Bethe, G. E. Brown, J. Applegate, J. M. Lattimer 1979, Nucl.Phys.A**324**, 487
8. A. S. Burrows 1990, Supernova, Ed. A. G. Petschek, Springer-Verlag
9. A. Ray, S. M. Chitre and K. Kar 1984, ApJ**285**, 766
10. J. Rapaport et al 1983, Nucl.Phys.A**410**, 371
11. D. Krofcheck et al 1987, Phys.Lett.B**189**, 299
12. J. Rapaport et al 1984, Nucl.Phys.A**427**, 332
13. D. Krofcheck et al 1985, Phys.Rev.Lett**55**, 1051
14. W. P. Alford et al 1991, Nucl.Phys.A**531**, 97
15. S. El-Kateb et al 1994, Phys.Rev.C**49**, 3128
16. W. P. Alford et al 1993, Phys.Rev.C**48**, 2818
17. M. C. Vetterli et al 1992, Phys.Rev.C**45**, 997
18. K. J. Raywood et al 1990, Phys.Rev.C**41**, 2836
19. W. P. Alford et al 1990, Nucl.Phys.A**514**, 49

Beta-Decay Studies of Nuclei Far From Stability near N = 28

O. Sorlin [1], R. Anne [2], L. Axelsson [1], D. Bazin [2], W. Böhmer [3], V. Borrel [1], D. Guillemaud-Mueller [1], Y. Jading [3], H. Keller [1], K-L. Kratz [3], M. Lewitowicz [2], S.M. Lukyanov [4], T. Mehren [3], A.C. Mueller [1], Yu. E. Penionzhkevich [4], B. Pfeiffer [3], F. Pougheon [1], M.G. Saint-Laurent [2], V.S. Salamatin [4], S. Schoedder [3], A. Wöhr [3].

[1] *Institut de Physique Nucléaire, F-91406 ORSAY, France*
[2] *Grand Accélérateur National d'Ions Lourds (GANIL), BP-5027, F-14021 CAEN, France*
[3] *Institut für Kernchemie, Universität Mainz, D-6500 MAINZ, Germany*
[4] *Laboratory of Nuclear Reactions, JINR, POB 79, DUBNA, Russia*

Abstract. Beta-decay half-lives and β-delayed neutron-emission probabilities of the very neutron-rich nuclei ^{43}P, ^{42,44,45}S and $^{44-46}$Cl, ^{47}Ar have been recently measured. These isotopes, which lie at or close to the N=28 magic shell were produced at GANIL in interactions of a 60 MeV/u ^{48}Ca beam with a ^{64}Ni target, and were separated by the doubly achromatic spectrometer LISE3. Their decay was studied by a β or β-n time correlation measurement. The results are compared to recent model predictions and indicate a rapid weakening of the N=28 shell below $^{48}_{20}$Ca$_{28}$. The nuclear structure effects reflected in the decay properties of the exotic S and Cl isotopes may be the clue for the astrophysical understanding of the unusual ^{48}Ca/^{46}Ca abundance ratio measured in the solar system, or observed in EK-inclusions of the Allende meteorite.

1. Introduction

The influence of shell effects is one of the most fascinating challenges of studying nuclei far from stability. For the neutron-magic number N=20, a new region of deformation was found in the very neutron-rich Na and Mg nuclei. The study of these isotopes is still actively persued, both experimentally and theoretically. Concerning the N=28 neutron-magic number, only very little experimental information was available "south" of ^{48}Ca. The present contribution reports on the first study of P to Ar isotopes in the vicinity of N=28.

2. Experimental method of production and selection of nuclei.

Unstable nuclei of interest in the vicinity of N = 28 were produced by fragmentation of a 60 MeV/u ^{48}Ca beam delivered by the GANIL accelerator impinging at onto a ^{64}Ni target (116 mg/cm^2). Both, beam and target are neutron rich and were chosen in order to maximize the production of neutron-rich isotopes. The selection of the desired nuclei among all fragments was achieved by the LISE3 spectrometer with a magnetic analysis and an energy-loss selection with a wedge-shaped degrader (1). Different settings of the magnetic rigidity (Bρ) were carefully chosen in order to optimize the production rates of the isotopes under study. The selected nuclei were detected in a telescope composed of three silicon detectors (300μm, 300μm, and 5000μm). The third detector was a Si(Li) diode which served for the implantation of the nuclei and for the detection of their correlated β decay. The identification of the nuclei transmitted has been achieved by combined informations of TOF, energy-losses in the two first silicon detectors, and the remaining energy in the implantation one. Depending on the settings, a rate of 3-50 nuclei per minute was obtained.

3 Determination of decay properties.

Each time a nucleus hit the first detector, the primary beam was switched off. The cut-off duration lasted about five times the expected β half-life of the implanted nucleus. During this time, a second data acquisition recorded coincidences between a β particle and a neutron at low background. These neutrons were detected in a nearly 4π neutron counter surrounding the telescope. This device was composed of three concentric rings consisting of 60 ^{3}He-proportional counters embedded in a polyethylene moderator matrix. The detector has a nearly energy independent efficiency of ε_n = 31±2% up to $E_n \approx$ 2 MeV and has no low-energy cut-off, as do scintillators. With this set-up, almost all β-delayed neutron coincidences could be directly correlated to implanted precursor nuclei. The triple coincidence (nucleus-β-neutron) led to a very low background rate. Half-lives were deduced from constructing a time histogram of the β-n coincidences detected after the identification of the corresponding nucleus. For nuclei with very low β-delayed neutron emission probabilities, the β-events correlated in time with the implantation of a nucleus were taken to determine the half-lives.

TABLE 1 : Determination of $T_{1/2}$ and P_n for some nuclei. Nuclei quoted with a star in subscript have been already published (2) in 1992, whereas the others are new. The number of β-neutron coincidences or of betas correlated with the implanted nucleus is indicated in the second column.

Nucleus analyzed in 1992* or 1994	Total number of $N_{\beta c}$ a) or $N_{\beta n}$ b) detected	$T_{1/2}$ (ms)	P_n (%)
^{43}P	151 b)	33 ± 3	100
^{42}S	356 a)	560 ± 60	< 4
^{44}S*	575 b)	123 ± 10	18 ± 3
^{45}S	111 b)	82 ± 13	54
^{44}Cl	465 a)	434 ± 60	< 8
^{45}Cl*	880 b)	400 ± 43	24 ± 4
^{46}Cl*	285 b)	223 ± 37	60 ± 9
^{47}Ar	85 a)	≈ 700	< 1

The detection system was calibrated with the isotope ^{15}B which has precisely determined values of $T_{1/2}$ and P_n. This led to a β-efficiency of ε_β = 65 ± 5 %. Table 1 combines the results of two experiments made in 1992 (2) and in 1994. The previous measurements have been confirmed in the 1994 data. The results obtained for ^{43}P, ^{42,45}S, ^{44}Cl, and ^{47}Ar in the 1994 run are preliminary, error bars are still tentative in some cases.

4. Comparison with theoretical calculations.

A Comparison of the experimental results with the 1986 TDA predictions of (3) is given in Figure. 1, summarizing the situations in 1986 (upper part) and 1994 (lower part). We find that the $T_{1/2}$ of ^{44}S and ^{45}Cl with neutron number N = 28 are shorter by factors of 3 and 4 than the predictions. Apart from the "key role" of these nuclei in the nucleosynthesis of ^{46}Ca (see §5), nuclear studies around this neutron closed shell offer a good test of the models and may bring a wealth information on a new zone of deformed nuclei below ^{48}Ca. Because of the predicted "soft" potential-energy surfaces in this mass region (5) in the vicinity of the N = 28 shell, none of the predicted g.s shapes can be excluded a priori. As an example, we used the recent QRPA model of Möller and Randrup (4) with a simple residual GT interaction for the study of the decay of ^{45}S. The Nilsson model is applied for determining the wave functions of the initial and final nucleus. The Lipkin-Nogami approximation (6) has been applied to calculate pairing correlations. Theoretical GT strength functions have been determined for three "typical" shapes of ^{45}S, prolate with $\varepsilon_2 = 0.158$, spherical and oblate with $\varepsilon_2 = -0.158$. As the influence of Q_β is of minor importance in this case, the differences in $T_{1/2}$ and P_n must be due to the different nuclear structures reflected in the S_{GT} pattern. From this calculation it is evident, that the experimental value can only be reproduced when assuming that the g.s of the ^{45}S is not spherical but presumably <u>oblate</u> deformed. Oblate g.s deformation has also been deduced in the case of ^{44}S, and is furthermore required to fit the decay properties of ^{43}P. So far, no detailed spectroscopic information on N ≈ 28 $_{15}$P to $_{17}$Cl isotopes is available, to unambiguously prove nuclear deformation in this mass region. But the integral quantities $T_{1/2}$ and P_n clearly reflect a rapid quenching of the N = 28 shell below ^{48}Ca. This new finding is of considerable interest for our understanding of nuclear structure very far from β-stability in the A ≈ 45 region.

5. Astrophysical interest.

The rapid neutron capture process (r-process) occuring in the high entropy bubble (7) of type II supernovae (SN) produces most of the neutron-rich <u>stable</u> isotopes observed in the solar system. In this scenario, neutron densities ranging from 10^{19} to 10^{24} n.cm^{-3} are obtained for several seconds of the SN explosion. The nucleosynthesis "path" is then driven very far from stability where short-lived species are encountered. If, in a non-equilibrium r-component, the neutron capture time ($T_{1/2}$ (n)) for a certain isotope is longer than its beta-decay time ($T_{1/2}$ (β)), it acts as an r-process turning points. During the freeze-out phase, when the available neutrons have been consumed, beta-decays back to stability will modulate the initial abundances in the r-process path by delayed neutron emission to result in the final abundances of the stable isotopes. A typical example for the importance of such turning points and their β-decay properties is revealed in the isotopic ratio of ^{48}Ca/^{46}Ca in the solar system (53) and in the EK-inclusions (250). The underabundance of ^{46}Ca in the Cosmos is one of the longstanding mysteries that now seems to have been solved with the combined informations on nuclear physics (improved neutron-capture cross section calculation (8), and experimental β decays) and of astrophysics (dynamic of stellar explosions). Compared with earlier theoretical predictions of β decay properties, the most striking difference arises from the shorter half-life of ^{45}Cl and the higher P_n value of ^{46}Cl, which together explain the low production of ^{46}Ca.

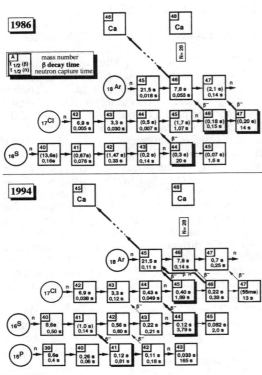

FIGURE 1. Neutron capture path in the $_{16}$S to $_{18}$Ar chains for a stellar temperature of $8 \cdot 10^8$ K and a neutron density of $5 \cdot 10^{-5}$ mol cm^{-3}; upper part: status in 1986, lower part: present status. With the new experimental data obtained at LISE3/GANIL facility, at both N=28 "turning point" isotopes, ^{44}S and ^{45}Cl, β-decay starts to dominate over further neutron capture. Hence, the possible A=46, 47 progenitors of ^{46}Ca will be produced in small amounts only. Even if the neutron capture path reaches ^{46}Cl, its high P$_n$-values (experimental : 60%) largely contributes to the reduction of the ^{46}Ca production. With this, large ^{48}Ca/ ^{46}Ca ratios can be obtained, as required to explain the observed abundances. The half-lives in parenthesis are taken from TDA (1986) or QRPA (1994) calculations, neutron-capture cross sections from (8,9).

This work has also provided constraints on neutron fluxes that are necessary to produce A = 46 and 48 progenitors, which - after β-decay - produce the observed ^{48}Ca/^{46}Ca abundance ratio. Furthermore, the wide study of nuclei in this region will bring substantial contraints on nuclear physics models (see §4), necessary for future, more reliable, predictions. It is important to note that the observed overabundances of ^{48}Ca, ^{50}Ti and ^{54}Cr seem to be correlated with those of ^{58}Fe, ^{64}Ni, and ^{66}Zn. Hence, astrophysical models have to show how they can produce all correlated abundances in a realistic and selfconsistent way. This emphasizes the growing complementarity between nuclear physics and astrophysics and strengthens the necessity of studying the decay properties of the expected progenitor isotopes of the above neutron-rich stable nuclei. This is foreseen with the next experiment in this context at GANIL/LISE3.

References

1. Mueller A. C. and Anne R. , *Nucl. Instr. Meth.* **B56/57**, 559 (1991).
2. Sorlin O. et al., *Phys. Rev.* **C 47**, 2941 (1993).
3. Klapdor H. V. et al., *At. Data Nucl. Data Tables* **31**, 81 (1984).
4. Möller P. and Randrup, *Nucl. Phys.* **A541**, 1 (1990).
5. Möller P. et al., *At. Data Nucl. Data Tables* (1994), in print.
6. Möller P. , Nix, J. R. *Nucl. Phys.* **A536**, 20 (1992).
7. Meyer B. S. et al., *Ap. J.* **399**, 656 (1992).
8. Wöhr A. et al., *Proc. of the 8th Int. Symp. on Capture Gamma-Ray Spectroscopy*, Fribourg, Switzerland, Sept. 20-24, 1993, Ed. J. Kern, World Scientific, p 762.
9. Thielemann F-K, PhD Thesis, Technische Hochschule Darmstadt 1980.

Measurement of the $\Gamma_\gamma/\Gamma_{tot}$ of the $E_x = 2.646$ MeV state in ^{20}Na

M.A. Hofstee, J.C. Blackmon, A.E. Champagne[1], N.P.T. Bateman,
Y. Butt, P.D. Parker, S. Utku, K. Yildiz[2], M.S. Smith[3], R.B. Vogelaar[4],
A.J. Howard[5].

[1] *Department of Physics and Astronomy, University of North Carolina at Chapel Hill, Chapel Hill, NC 27599 and Triangle Universities Nuclear Laboratory, Duke University, Durham, NC 27706,* [2] *Department of Physics, Yale University, New Haven, CT 06511,* [3] *Physics Division, Oak Ridge National Laboratory, Oak Ridge, TN 37831,* [4] *Department of Physics, Princeton University, Princeton, NJ 08544,* [5] *Department of Physics and Astronomy, Trinity College, Hartford, CT 06106*

Abstract. In energetic nova events, breakout of the hot CNO cycle may proceed via the reaction chain ^{15}O$(\alpha,\gamma)^{19}$Ne$(p,\gamma)^{20}$Na. This triggers a sequence of rapid proton captures and beta decays which can process CNO nuclei to heavier elements (rp-process). The rate of the ^{19}Ne$(p,\gamma)^{20}$Na reaction is sensitive to resonances in the ^{19}Ne + p system. The first excited state above the proton threshold of 2.199 MeV in ^{20}Na (at $E_x = 2.646$ MeV) is expected to be the most important in determining the reaction rate. The gamma width (Γ_γ) of this state is still unknown and the J^π assignment of this state is currently the subject of debate (1). We have performed ^{20}Ne(^{3}He,tγ) experiments to determine the branching ration Γ_γ/Γ of the $E_x = 2.646$ MeV excited state in ^{20}Na. Some preliminary results of these experiments will be presented.

1. INTRODUCTION

Over the past decade systematic spectroscopic observations have revealed a large class of novae which exhibit overabundance of Neon and Magnesium in their ejecta(2). Although the source of these overabundances has not been determined, one possible scenario involves breakout from the Hot CNO cycle on the surface of a CO white dwarf(3). To determine the feasibility of this process one needs to know the rates of the reactions which form the breakout path from the Hot-CNO cycle to the rp-process. One of these steps, the ^{19}Ne$(p,\gamma)^{20}$Na reaction, is the subject of the current investigation.

1.1. The $E_x = 2.646$ MeV state in ^{20}Na

Resonances in the ^{19}Ne + p system above the proton emission threshold (at $E_x = 2.199$ MeV) are expected to dominate the reaction rate. Investigations of the level scheme of ^{20}Na (4–7) have revealed a number of states in this region. For $E_p \leq 1$ MeV the excitation energies have been accurately measured, but their J^π assignments are still subject of debate. In this work, the excitation energies of M.S. Smith *et al.* (6) have been used. Large uncertainties exist in estimates of resonance strength ($\omega\gamma$), caused both by uncertain J^π assignments and by unknown partial widths.

The first excited state above the proton emission threshold is found at $E_x = 2.646$ MeV. In earlier publications a 1^+ assignment for this state was inferred from the angular distribution in charge exchange reactions (7), making it the analog of the 3.173 MeV 1^+ state in ^{20}F. However, comparison of ^{20}Ne(^{3}He,t) and (t,^{3}He) measurements (5) have undermined this assignment. Searches for G-T strength (8) have not found any evidence for a 1^+ state at this excitation energy. These and other experimental results have led to a proposed 3^+ assignment for this state (1), making it the analog of the 3^+ ($E_x = 2.966$ MeV) in ^{20}F. This implies a Γ_γ of about 0.12 eV and thereby makes a measurement of $\Gamma_\gamma/\Gamma_{tot}(= 0.09)$ feasible.

2. THE EXPERIMENT

We produced the 2.646 MeV excited state in ^{20}Na via the ^{20}Ne(^{3}He,t) reaction. For this experiment a $E_{^3\text{He}} = 27$ MeV beam from the Yale ESTU Tandem Accelerator was directed onto a Neon-gas cell, designed for this experiment. The charged particle reaction products were measured with two 1-cm^2 position-sensitive Si detectors. The Si-detectors consisted of a 100 μm thick position sensitive ΔE-detector (Left and Right readout), a 1000 μm thick E detector, followed by a 1000 μm E-detector used as a veto. The detectors were mounted, at 5 cm distance from the reaction center at an angle of 35°, on a base with the gas target. The horizontal opening angle of 3° was defined by Tu-alloy wedges, also functioning as shielding.

Gammas were measured in coincidence with the charged particles by a large (5" × 5") NaI detector. Tu-alloy shielding was put directly on top of the gas cell to provide adequate shielding from background gammas (mostly 511 keV) coming from the thin (0.3 mil) Kapton entrance- and exit- windows. Gas pressure was kept at about 1/4 atm for most of the experiment.

The detector geometry and shielding reduced spurious coincidences as well as real coincidences from the windows and other background sources. On average, a beam intensity of 10-15 pnA at 1/4 atm pressure resulted in a countrate of about 80 kHz in the NaI detector.

In the replay the triton energy was corrected (with a linear correction) for kinematics, optimized for $E_x = 3$ MeV. The overall energy resolution was about 200 keV, mainly as a result of the large horizontal opening angle.

In a particle identification (PID) spectrum, a gate was set on the tritons, with adjacent gates to determine the shape of the background. The position and width of the gate was chosen such that as little background was included without excluding any tritons. Both the ^{3}He and deuteron contamination (except for the elastically scattered ^{3}He) could be estimated from the inclusion of the $E_\gamma = 1.6$ MeV line in the NaI spectrum. This line is produced by the deexcitation of the lowest excited state in ^{20}Ne, and is populated at these deuteron energies by the reaction ^{20}Ne*(^{3}He,d)^{21}Na*(,p)^{20}Ne*(γ).

The energy scale of the NaI spectrum has been calibrated both by using lines from sources with beam off (countrate about 20 kHz), and by using known lines from ^{20}Ne* coincident with inelastic scattered ^{3}He. Both calibrations agree with one another, indicating that the energy calibration is independent of the countrate in the detector. The lines from calibrated gamma-sources and other lines of which the (relative) intensity are known can also be used for the (energy dependent) efficiency calibration of the NaI detector.

3. RESULTS

From the triton spectrum of real (i.e. prompt - random) triton-γ coincidences in figiure 1 it can be seen that the coincidence yield for states above the proton emission threshold is low, while the background is considerable. This makes the extraction of $(\Gamma_\gamma/\Gamma_{tot})$ rather difficult. Several methods have been employed to estimate $(\Gamma_\gamma/\Gamma_{tot})$ by comparison of the relative cross section with the states around $E_x = 2$ MeV (below the proton emission threshold).

- Fitting the peaks in the singles spectrum and the analysing the coincidence spectrum (lower statistics), keeping widths and position of the states fixed. The background was either drawn by hand, or the average of the gated spectra with gates on the PID below (light) and above (dark) the triton band were used (subtracted from the spectrum).
- The sum of the total number of counts in the region of the peaks was taken with a straight-line background contribution subtracted.

A very preliminary analysis seems to indicate that $\Gamma_\gamma/\Gamma_{tot}$ is of the order of 10%, in agreement with the predictions of Brown et al. (1). Because of the large background contribution, the error bars on the extracted $\Gamma_\gamma/\Gamma_{tot}$ are rather large. A re-analysis is in progress, aimed at reducing the background and optimizing the energy resolution of the triton spectra. In a final analysis the extracted branching ratio will also have to be corrected for the gamma energy dependence of the NaI detector efficiency.

REFERENCES

[1.] Brown, B.A., Champagne, A.E., Fortune, H.T., and Sherr, R., *Phys. Rev. C* **48**, 1456(1993).

[2.] Williams, R.E., Phillips, M.M. and Hamuy, H., *Ap.J.Suppl.* **90**, 297–316 (1994).

[3.] Champagne, A.E. and Wiescher, M., *Annu. Rev. Nucl. Part. Sci.* **42**, 39-76 (1992).

[4.] Lamm, L.O., Browne, C.P., Görres, J., Graff, S.M., Wiescher, M., and Rollefson, A.A., *Nucl. Phys.* **A510**, 503 (1990).

[5.] Clarke, N.M., Hayes, P.R., Becha, M.B., Pinder, C.N., and Roman, S., *J. Phys. G* **16**, 1547 (1990).

[6.] Smith, M.S., Magnus, P.V., Hahn, K.I., Howard, A.J., Parker, P.D., Champagne, A.E., and Mao, Z.Q., *Nucl. Phys.* **A 536**, 333 (1992).

[7.] Kubono, S., et al., *Z. Phys. A* **331**, 359 (1989); *Ap. J.* **344**, 460 (1989).

[8.] Anderson, B.D., et al., *Phys. Rev. C* **43**, 50-58 (1991).

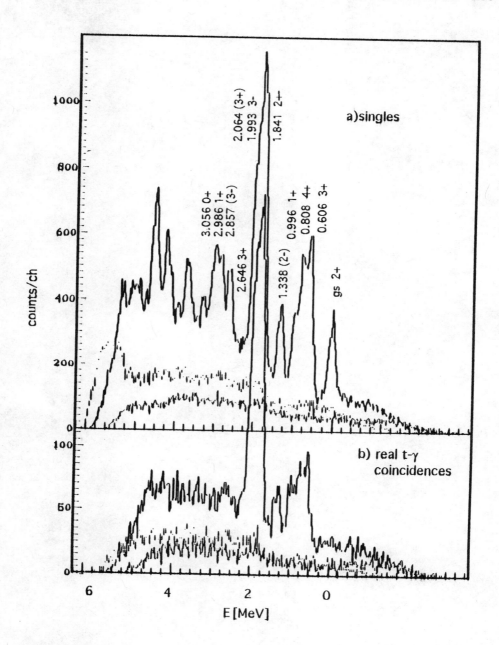

Fig. 1.— Triton spectra for singles (a) and real t-γ coincidences (b).

Is ^{16}O a Strong Neutron Poison?

Y.Nagai*, M.Igashira+, T.Shima*, K.Masuda+, T.Ohsaki+ and H.Kitazawa++

Department of Applied Physics, + Research Laboratory for Nuclear Reactors and ++ Department of Electrical and Electronical Engineering, Tokyo Institute of Technology, O-okayama, Meguro, Tokyo 152, Japan

Abstract. The cross section of a ^{16}O(n,γ)^{17}O reaction was measured in the neutron energies of between 10 and 80 keV. The Maxwellian-averaged cross section at kT=30 keV is 38±4µb; 200 times larger than the value measured previously. The results may have significant effects for the nucleosynthesis theories of s- and p-processes in metal-deficient stars due to a strong neutron poison of ^{16}O. The large cross section is attributed to a non-resonant p-wave neutron capture.

In order to construct models of the chemical evolution of the Galaxy[1], observational efforts of the s-process isotopes for a wide range of metallicities and theoretical efforts to calculate yields of these isotopes as functions of metallicity and stellar mass[2] are being paid. So far, in the models of the chemical evolution of the s-isotopes the relationship of the yields of these isotopes vs the abundance of either ^{56}Fe(seed) nuclei or ^{16}O(source) nuclei was suggested to be linear[3]. However, it was shown to be non-linear for low-metallicity massive stars[2]. The non-linearity is considered to be due to the role played by light nuclei as neutron poisons. If the light nuclei absorb neutrons, the yields of p-nuclei also decrease, since the s-process nuclei are considered to be the immediate predessors of p-process nuclei[4].

Therefore, in order to construct models to predict the s-and p-isotope productions as a function of metallicity it is very important to know the neutron capture cross sections of light nuclei at stellar neutron energy, since these cross sections are poorly known.

So far, the cross section of a ^{16}O(n,γ)^{17}O reaction was measured at stellar neutron energy by Allen and Macklin[5]. They obtained the small Maxwellian-averaged capture cross section of 0.2(1) µbarn at kT=30 keV, and therefore they concluded that ^{16}O plays a minor role as a neutron poison.

In this study the cross section of the ^{16}O(n,γ)^{17}O reaction was measured at astrophysically important neutron energies between 10 and 80 keV.

© 1995 American Institute of Physics

cross sections. Assuming no-resonance in the neutron energies of between 80 and 280 keV, the partial capture cross sections were obtained as $\sigma_0(E)=0.97E^{\frac{1}{2}}$ and $\sigma_1(E)=3.51E^{\frac{1}{2}}$, respectively, where E is in the keV unit. A total capture cross section at the neutron energy E was obtained by adding these partial cross sections to the estimated value obtained from the s-wave thermal neutron capture cross section by assuming a $1/v$ law as

$$\sigma(E) = 1.0E^{-\frac{1}{2}} + 4.61E^{\frac{1}{2}} \quad (2)$$

Therefore, the Maxwellian-averaged capture cross section at the temperature T, defined as $<\sigma>=<\sigma v>/v_T = 2/\sqrt{\pi}$, is given as

$$<\sigma> = 1.0(kT)^{-\frac{1}{2}} + 6.9(kT)^{\frac{1}{2}} \quad (3)$$

Here v_T is given as $v_T = \sqrt{2kT/m}$, where m is the reduced mass of a neutron and a nucleus. The Maxwellian-averaged capture cross section at 30 keV is 38(4) μb; 200 times larger than the previously measured value of 0.2(1) μbarn. The astrophysical S-factor, defined as $S = <\sigma> v_T$, is obtained as

$$S = (4.5 + 2700T_9) \times 10^{-23} \quad (4)$$

Therefore, the reaction rate, defined as $N_A S$, is calculated as

$$N_A S = 27 + 16200T_9 \; cm^3 s^{-1} mole^{-1} \quad (5)$$

Here N_A is the Avogadro constant and T_9 is the temperature in unit of 10^9 K. The present result of the large reaction rate has important implications for nucleosynthesis theories of s- and p-processes, especially in metal deficient massive stars as a very important neutron poison. Therefore, it would cause the large non-linearity of the relationship of the s-isotope yields vs the abundance of either ^{16}O(source) nuclei or ^{56}Fe(seed) nuclei. The large capture cross section obtained here can be explained by a non-resonant p-wave capture. The interpretation is based on the facts that the observed γ-ray branching ratio from a captured state to low-lying states is quite different from the value obtained by using thermal neutrons and that each partial cross section is proportional to $E^{\frac{1}{2}}$, not $E^{-\frac{1}{2}}$. The non-resonant p-wave capture was found also in the measurement of the ^{12}C(n,γ)^{13}C reaction cross section[9]. In these experiments, a prompt γ-ray detection method combined with a good energy resolution NaI(Tl) spectrometer enabled us to observe a discrete γ-ray from a captured state to a low-lying state and thus to assign the p-wave neutron capture assuming an E1 character for the capture γ-transition.

The measurement was done by using pulsed neutrons and detecting prompt γ-rays from a captured state. A bunched proton beam, provided from the 3MV Pelletron Accelerator of the Research Laboratory for Nuclear Reactors at the Tokyo Institute of Technology, was used to produce the neutrons by a ^{7}Li(p,n)^{7}Be reaction. The samples of an enriched ^{7}Li$_2$O and a gold (Au) were used. The absolute capture cross section was obtained by normalizing the γ-ray yields of the O sample with those of Au, since the cross section of Au has been well known. The prompt γ-rays were detected by an anti-Compton NaI(Tl) spectrometer. A background subtracted γ-ray pulse height spectrum is shown in fig.1, where the γ-rays from the captured state to $\frac{5}{2}^+$ (ground) and $\frac{1}{2}^+$ (first) states in ^{17}O were clearly seen. However, cascade γ-rays from the captured state to the $\frac{1}{2}^+$ state via a $\frac{1}{2}^-$ (second) state were not observed clearly, although the γ-ray intensities were reported to be strongest in the ^{16}O(n,γ)^{17}O reaction measured by thermal neutrons[6]. The intensities of discrete γ-rays were analyzed by a stripping method, using the response function of the γ-ray spectrometer. The partial capture cross section, σ_i, corresponding to the γ-transition from a captured state to a low-lying state, i, is given by using the γ-ray intensity as

$$\sigma_i = \frac{C_{Au}}{C_O} \frac{n_{Au}}{n_O} \frac{\phi_{Au}}{\phi_O} \frac{Y_O}{Y_{Au}} \frac{\Omega_{Au}}{\Omega_O} \sigma_{Au} \qquad (1)$$

In equation (1) Ω is the solid angle of the sample to the neutron source and n is the thickness of the sample. ϕ and Y_O (Y_{Au}) are the neutron yield, measured by a ^{6}Li-glass scintillation detector, and the γ-ray yield of the O (Au) sample, respectively. The capture cross section of Au is indicated by σ_{Au}. C is the correction factor for the multiple scattering effect and the shielding of incident neutrons in a sample, respectively. They were obtained by using the Monte Carlo code TIME-MULTI[7]. Finally, the partial capture cross sections of $\sigma_0(E)$ and $\sigma_1(E)$, corresponding to the γ-transitions from a captured state to the $\frac{5}{2}^+$(ground) and $\frac{1}{2}^+$(first) states, at a neutron energy E were obtained, respectively. Here it should be mentioned that a $p_{\frac{3}{2}}$- wave neutron resonance was observed at 4552 keV in ^{17}O, 408.4 keV above the neutron threshold, with a total radiative width of 3.65(49) eV[8]. It is therefore expected that the resonance contributes to the cross sections measured for the neutron energy between 10 and 80 keV. The contribution was calculated by using the Breit-Wigner one level formula as 0.8, 1.3 and 1.8 μb for the average neutron energies of 20, 40 and 60 keV, respectively. Here experimental values of 3.65 eV and 45 keV for γ-ray and neutron widths were used. Consequently, the non-resonance cross sections were obtained by subtracting the resonance effects from the measured cross sections. The results were combined with the measured non-resonance cross sections at E=280 keV, 30(30) μb and 55(30) μb for the σ_0 and σ_1, respectively[8], to obtain the energy dependence of the

ACKNOWLEDGEMENTS

Finally, we thank Dr M.Hashimoto for useful discussions. The present work was supported partly by the Toray Science Foundation, and by a Grant-in-Aid of the Japan Ministry of Education, Science and Culture.

REFERENCES

1. Wheeler,J.C., Sneden,C., and Truran,J.W.,*Annu. Rev. Astron. Astrophys.***27**,279 (1989).
2. Prantzos,N., Hashimoto,M., and Nomoto,K., *Astron.and Astrophys.***234**,211(1990). Baraffe,I., El Eid,M.F., and Prantzos,N., *Ap.J.***258**,357 (1992).
3. Tinsley,B.,*Fund.Cosm.Phys.***5**,287 (1980).
4. Lambert,D.,*Astron. and Astrophys. Rev.***3**,201 (1992)
5. Allen,B.J., and Macklin,R.L.,*Phys. Rev.***C3**,1737 (1971)
6. Mcdonald,A.B., Earle,E.D., Lone,M.A., Khanna,F.C., and Lee,H.C.,*Nucl. Phys.***A281**,325(1977)
7. Senoo,K., Nagai,Y., Shima,T., Ohsaki,T., and Igashira,M.,*Nucl. Instr. and Meth.***A339**,556(1994)
8. Igashira,M., Kitazawa,H., and Takaura,K.,*Nucl. Phys.***A536**,285(1992)
9. Ohsaki,T., Nagai,Y., Igashira,M., Shima,T., Takeda, K., Seino,S., and Irie,T.,*Ap.J.***422**,912(1994)

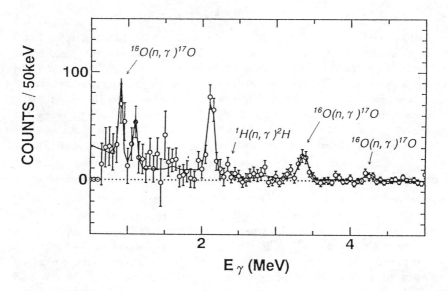

FIGURE 1. Background subtracted γ-ray spectrum at the average neutron energy of 40 keV.

New Measurement of the ^{14}N(n,p)^{14}C Cross Section at Stellar Energy with a Drift Chamber

T. Shima, K. Watanabe[1], H. Sato, T. Kii, Y. Nagai
and M. Igashira*

Department of Applied Physics, Tokyo Institute of Technology,
O-okayama, Meguro-ku, Tokyo 152, Japan
*Research Laboratory for Nuclear Reactors, Tokyo Institute of Technology,
O-okayama, Meguro-ku, Tokyo 152, Japan

Abstract. The ^{14}N(n,p)^{14}C reaction is crucial for nucleosynthesis in stars, since it may act as a neutron poison for stellar s-process. We made a new measurement of its cross section by using a newly developed gas scintillation drift chamber, which contained N$_2$ gas as an active target in addition to the chamber gas. A preliminary value of 1.07±0.31(stat.)±0.19(syst.) mb was obtained at $\bar{E}_n = 49.7$ keV.

INTRODUCTION

A slow-neutron capture process (s-process) occuring at helium shells in stars is considered to be one of main processes of the nucleosynthesis beyond iron [1,2]. As for the neutron souce, two reactions of the ^{13}C(α,n)^{16}O in low-mass stars and the ^{22}Ne(α,n)^{25}Mg in intermediate-mass stars have been suggested [1,2]. However, since the helium shell contains a large amount of ^{14}N produced by the CNO hydrogen burning, ^{14}N may impede s-process by consuming neutrons (neutron poison) [3], if the cross section of the ^{14}N(n,p)^{14}C reaction is large at the stellar temperature. Until now, the cross section at the keV region has been measured by two groups. Brehm et al. obtained 0.81±0.05 mb and 0.52±0.06 mb at $kT = 25.0$ keV and 52.4 keV, respectively [4]. These values are factors of two or three smaller than the values estimated by using the data of reverse reactions [5], indicating ^{14}N is not a poison for s-process. On the other hand, the cross sections measured by Koehler and O'Brien are consistent with the estimated values in the energy range from 61 meV to 34.6 keV [6], and they claimed ^{14}N to be a poison for s-process. Since the difference

---
[1]Present address : Nippon Oil Company, Minato-ku, Tokyo

between the above two results is quite serious, it is neccessary to measure the cross section again with an independent experimental method.

EXPERIMENTAL METHOD AND RESULT

The measurement was carried out at the Research Laboratory for Nuclear Reactors (RLNR) at the Tokyo Institute of Technology. A pulsed neutron beam, provided from the 3.2 MV Pelletron Accelerator of RLNR, was introduced into the gas scintillation drift chamber (GSDC) [7]. Schematic drawings of the GSDC are shown in fig. 1. A vertical drift chamber is contained in a gastight box filled with the mixed gas of He and N_2. When a $^{14}N(n,p)^{14}C$ reaction event occurs, the produced proton moves both exciting and ionizing the gas atoms. Four photomultiplier (PM) tubes (Hamamatsu H1584, $\phi 5$") are used to detect the primary scintillation photons emitted from the excited gases, in order to determine the (n,p) reaction time. The ionized electrons are drifted via the uniform electric field in the drift region, and are detected by the multi-wire proportional counter (MWPC) placed behind the drift region. Energy losses and the trajectories of the charged particles emitted in the (n,p) reactions are observed by the MWPC. The GSDC has the following merits.
(i) N_2 works as an active target, leading to a very large detection efficiency.
(ii) Energies of the neutrons can be determined by the time-of-flight method, since a (n,p) reaction time can be known by detecting primary photons.
(iii) True (n,p) events can be separated from background ones by observing tracks of the emitted protons.

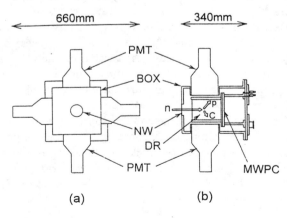

FIGURE 1. Schematic drawings of (a): a front view and (b): a cross sectional side view. n: neutron, p: proton, C: ^{14}C nucleus produced by a $^{14}N(n,p)^{14}C$ reaction, PMT: PM tubes, BOX: gastight box, NW: neutron window, DR: drift region, MWPC: multi-wire proportional counter.

The absolute value of the cross section of the ^{14}N(n,p)^{14}C reaction was determined relative to that of the ^{3}He(n,p)^{3}H reaction, which has been well known at the keV region [8]. In order to compare these two cross sections, two types run A and run B of measurements were carried out. Run A was made by using the sample gas of N_2 120 Torr + natural He 480 Torr, and run B was by using the sample gas of N_2 120 Torr + 0.2% enriched ^{3}He 480 Torr. Fig. 2 shows a typical pulse height spectrum of the charged particles produced in the ^{14}N(n,p)^{14}C reaction. True (n,p) events were separated by the proton energies from the background ones.

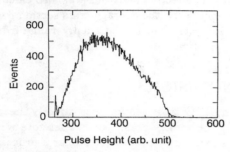

FIGURE 2. Pulse height spectrum of the charged partcles emitted in the (n,p) reactions.

Fig. 3 shows the time-of-flight spectra of neutrons from run A and run B. The events due to the (n,p) reactions with keV neutron can be found around the region from 60 nsec to 250 nsec. Subtracting the contributions of the thermal neutron, the net counting rates R_A and R_B of the (n,p) reaction events with keV neutron were obtained as 9.6×10^{-3} cps and 0.21 cps, respectively. Then the counting rates R_N and R_{He} of the ^{14}N(n,p)^{14}C reaction and the ^{3}He(n,p)^{3}H reaction were obtained as $R_N = R_A = 9.6 \times 10^{-3}$ cps and $R_{He} = R_B - R_A \cdot (\phi_B/\phi_A) = 0.20$ cps, respectively, where the ratio of neutron intensities ϕ_B/ϕ_A was measured by a Li-glass scintillation counter.

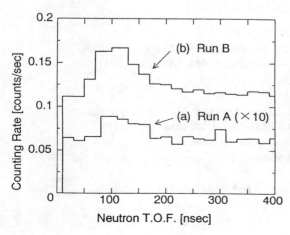

FIGURE 3. Time-of-flight spectra; (a): for run A and (b): for run B.
In the spectrum (a), values ten times as much as the observed counting rates are shown.

Then σ_N is given as

$$\sigma_N = \frac{R_N}{\varepsilon_N \cdot n_N \cdot \phi} = \frac{\varepsilon_{He}}{\varepsilon_N} \cdot \frac{n_{He}}{n_N} \cdot \frac{\phi_A}{\phi_B} \cdot \frac{R_N}{R_{He}} \cdot \sigma_{He} \ . \tag{1}$$

Here σ, ε and n stand for the cross sections, the detector efficiencies and the number densities of the effective nuclei, respectively. N and He denote the variables associated with the $^{14}N(n,p)^{14}C$ events and the $^3He(n,p)^3H$ events, respectively. ε_N and ε_{He} are determined as 0.64 and 0.85, by simulating both events of neutron scattering and (n,p) reactions with the Monte Carlo cord TIME-MULTI [9]. Mean energy of the effective neutrons was also calculated by the simulation as 49.7 keV. Consequently, the average cross section $\bar{\sigma}_N$ was obtained as 1.07±0.31(stat.) ±0.19(syst.) mb in the energy range from 10 keV to 100 keV ($\bar{E}_n = 49.7$ keV).

CONCLUSION

In this work, the average $^{14}N(n,p)^{14}C$ cross section was directly measured to be 1.07±0.31(stat.)±0.19(syst.) mb at $\bar{E}_n = 49.7$ keV. Therefore we may conclude that ^{14}N is important as a neutron poison for stellar s-process. In addition, the production of ^{15}N by the $^{14}N(\alpha,\gamma)^{18}F(n,\alpha)^{15}N$ reaction chain may be hindered by the $^{14}N(n,p)^{14}C$ reaction. It must be noted that the $^{14}N(n,p)^{14}C$ cross section is also important to study origins of ^{14}C both in stars and on the Earth.

ACKNOWLEDGMENTS

We wish to thank Prof. T. Yamazaki and Prof. H. Ejiri for kind encouragements. This work was supported by the Yamada Science Foundation.

REFERENCES

1. Cameron, A.G.W., *Phys. Rev.* **93**, 932 (1954)
2. Iben, I.Jr., *Astrophys. J.* **196**, 525–547 and 549–558 (1975)
3. Iben, I.Jr., and Renzini, A., *Astrophys. J. Lett.* **263**, L23–L27 (1982)
4. Brehm, K., et al., *Z. Phys.* **A330**, 167–172 (1988)
5. Sanders, M., *Phys. Rev.* **104**, 1434–1440 (1956),
   Gibbons, J.H., and Macklin, R.L., *Phys. Rev.* **114**, 571–580 (1959)
6. Koehler, P.E., and O'Brien, H.A., *Phys. Rev.* **C39**, 1655–1657 (1989)
7. Shima, T., et al., *submitted to Nucl. Instr. and Meth. A.*
8. Smith, M.S., Kawano, L.H., and Malaney, R.A., *Astrophys. J. (Suppl.)* **85**, 219–247 (1993)
9. Senoo, K., et al., *Nucl. Instr. Meth.* **A339**, 556–563 (1994)

Charge-Changing Cross Sections near the Δ-Isobar Resonance

P. Doll and H.J. Crawford [+]

*Kernforschungszentrum Karlsruhe, Institut für Kernphysik 1,
P.O. Box 3640, D - 76021 Karlsruhe*
[+] *Space Science Laboratory, University of California, Berkeley, California 94720*

Abstract. Investigations of the charge-changing fragmentation cross sections for nuclei ranging from ^{12}C to ^{58}Ni and energy / nucleon from 300 to 2100 MeV on hydrogen emphasize the influence of nucleon isobar excitations.

INTRODUCTION

Secondary-to-primary ratios studied in cosmic rays (1) at an energy of around 1 GeV/nucleon provide a sensitive test of cosmic ray propagation calculations. Such calculations take into account the energy dependence of the path-length-distribution (PLD) and solar modulation. From the analysis of the Boron/Carbon (B/C) ratio e.g., it is concluded (1) that the mean-free-path in Galactic propagation must be fully energy dependent. For the understanding of this energy dependence the interaction cross sections of nuclei with the hydrogen in the interstellar medium must be studied. Therefore, the 'transport collaboration' (2, 6) made it his business to measure and investigate charge-changing (CC) cross sections for nuclei from ^{12}C to ^{58}Ni interacting with a hydrogen target in the energy range from 390 to 910 MeV/nucleon.

CALCULATION

The present model (3) calculates in the framework of a multiple scattering mechanism the total CC cross sections for protons on the nuclei mentioned above. The intranuclear nucleon-nucleon (NN) cascade is dominated by the total NN cross sections σ_{pp} and σ_{pn}. The NN cross sections determine the mean-free-path inside the nucleus and in its outer hemisphere, where most of the fragmentation takes place. Already the total CC cross sections exhibit a rise near the bombarding energy (600 - 900 MeV/nucleon) where a Δ-resonance can be excited in the individual NN collision. This finding is very similar to the excitation of the Δresonances in charge exchange experiments (4) where the structure of these resonances are found to decay not only into pion and nucleon but also into two nucleons. The latter decay mode appears only in nuclei. We therefore modify the mean-free-path in nuclei by adding the decay of the Δ-hole state into 2 particle - 2 hole states. Such states may correspond to highly excited Gamow-Teller (GT) components.

This additional Δ-GT coupling is parametrized in a Breit-Wigner form, in which the positions and widths of the GT- and Δ-resonances are taken as free parameters in the fits to the partial CC data.

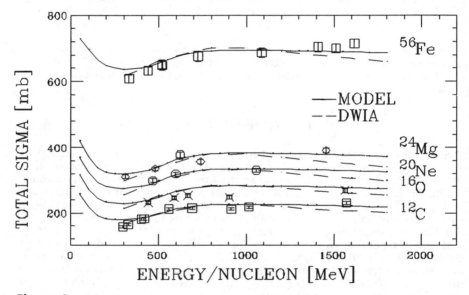

Figure 1. Total charge-changing cross sections for various nuclei and calculations in the model described in the text (3).

Calculating the partial cross sections, we assume that the energy of a Δ-resonance is distributed among the excitation energy of various fragments and their energy of relative motion. In the framework of the multiple scattering mechanism, the excitation energy spectrum for a specific fragment determines the weight of the corresponding partial cross section (see e.g. ref. 8).This excitation spectrum is assumed to depend exponentially on the binding energy difference between the target and the fragment. This energy difference exhibits a characteristic dependence on the charge Z and neutron-proton asymmetries N-Z of the target and fragment, respectively. We correct the binding energies for the energies of the corresponding giant GT-resonances.

The figures show the measured and calculated spallation cross sections. Figure 1 shows the total CC cross sections for various nuclei (^{12}C, ^{16}O, ^{20}Ne, ^{24}Mg and ^{56}Fe) described by the present model (solid curves) and distorted-wave-impulse-approximation (DWIA) calculations (3) (dashed curves). The data for ^{56}Fe were taken from refs. (5) and (7) and work quoted therein. The data for the other nuclei were mainly taken from the compilation by P. Ferrando et al. (5).

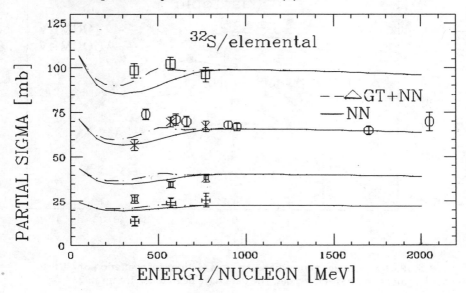

Figure 2. Partial elemental charge-changing cross sections for spallation of ^{32}S into various fragment charges (for details see text).

Figure 2 shows the results for ^{32}S. The partial elemental cross sections for spallation of ^{32}S on hydrogen are shown for fragment charges of 14, 12, 10 and 8 from top to bottom. Most of the excitation functions are only measured at three energies (2, 6) emphasizing the need for more experiments and a good description (9) of such data. The staggering of the cross sections with fragment charge emphasizes an $\exp(\alpha \cdot \Delta Z)$ dependence (5,9). The single and double charge transfer exhibit pronounced Δ-GT mixing.

Figure 3 shows the pronounced dependence of the partial isotopic yields on the neutron-proton asymmetry N - Z which is contained in the asymmetry energy of the liquid-drop model. Data (2, 6) and calculations were taken at energies (400 and 600 MeV) where Δ-GT mixing is most pronounced. The figure demonstrates the predictive power of the present calculations even for isotopic yields far away from neutron-proton symmetry in the fragment nuclei.

CONCLUSIONS

It is the purpose of the presented calculations to provide a physics based model which can predict spallation cross sections for many nuclei and energies around the Δ-resonances. With respect to the semi-empiri-

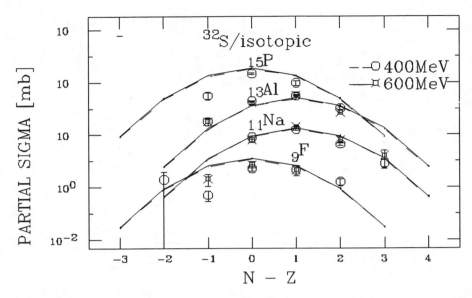

Figure 3. Partial isotopic charge-changing cross sections for spallation of ^{32}S into various fragment charges and masses. The data have been taken from ref. 2.

cal calculations by Silberberg and Tsao (9) which cover a wider charge and mass transfer and energy range, the present calculations concentrate on small charge and mass transfer. We hope to improve the knowledge of cross sections and of the PLD needed around the maximum of the cosmic ray B/C ratio. Thereafter, more detailed calculations on the influence of solar modulation on, e.g. the B/C ratio can be carried out.

REFERENCES

1. M. Garcia-Munoz, J.A. Simpson, T.G. Guzik, J.P. Wefel and S.H. Margolis, *Astrophys. Jour. Suppl. Series* **64, 269** (1987)
2. C.E. Tull et al., *Contr.23.Int.*, Cosmic Ray Conf., Calgary, 1993, **2**, 163
3. P. Doll and H.J. Crawford, *Contr.Int.Symp.*, Nuclei in the Cosmos, Karlsruhe, 1992, IOP, Publishing Ltd.1993, 417 and to be published
4. T. Hennino et al., *Phys.Lett.* **283B, 42** (1992)
5. P. Ferrando, W.R. Webber, P. Goret, J.C. Kish, D.A. Schrier, A. Soutoul and O. Testard, *Phys. Rev.* **37C, 1490** (1988)
6. C.N. Knott et al., *Contr.23.Int.*, Cosmic Ray Conf., Calgary, 1993, **2, 187**, and priv. communication
7. W.R. Webber and D.A. Brautigam, *Astrophys. Jour.* **260, 894** (1982)
8. J. Hüfner, U. Schäfer and B. Schürmann, *Phys.Rev.* **12C, 1888** (1975)
9. R. Silberberg and C.H. Tsao, *Phys.Rep.* **191, 351** (1990)

S-Factor in the Framework of the "Shadow" Model

A.Scalia

Istituto Nazionale di Fisica Nucleare, Corso Italia 57, 95129 Catania, Italy

Dipartimento di Fisica dell' Università, Corso Italia 57, 95129 Catania, Italy

Abstract. *The S(E) factor is obtained in the framework of the "shadow" model. The analytical expression of the S(E) function is not compatible with a S-factor which is a slow varying function of the energy.*

The fusion process between charged particles was investigated by using the "shadow" model [1].In this model the fusion process is considered as the shadow of the Rutherford scattering so that the particles which fuse are those that in Rutherford scattering are detected in the "shadow" region.By using this point of view a regularity was found by analysing fusion data relative to systems with $24 \leq (A_1 + A_2) \leq 194$ and a new expression for the fusion cross-section was suggested [1].The "shadow" model is based on this observed regularity and it is able to reproduce the experimental values of fusion cross-section for about 100 reaction at energies below the Coulomb barrier.For light systems $3 \leq (A_1 + A_2) \leq 8$ and at energies far below the Coulomb barrier observed regularity disappears and the "shadow" point of view is not able to describe the fusion process.However it is possible to generalize the "shadow" model so that it is able to reproduce the experimental values of fusion cross-section for light systems at very low energy [2].To judge the success of the "shadow" model in reproducing the experimental data we used a logarithmic scale to plot the rapidly varying cross-section [1],[2] but it is more easily to judge the success of the model if we transform the fusion cross-section as calculated from the "shadow" model equations into the astrophysical S(E) factor [3]

$$S(E) = E\sigma_f(E)\exp(2\pi\eta) =$$

$$\frac{\pi(Z_1 Z_2 e^2)^2}{E} G(y)(1+G(y))(1-g_1(y))(1-g_2(y))(1-g_3(y))......(1-g_j(y))\exp(2\pi\eta)$$

where E is the centre of mass energy, η is the Coulomb parameter $\sigma_f(E)$ is the fusion cross-section obtained in the framework of the "shadow" model,

$$y = \frac{E_B - E}{E_S},$$

$$G(y) = \exp(-\exp(\exp(y)))$$

$$g_i(y) = \exp(-2.789(\frac{d-y}{d-y_i})^{\gamma_i}) = \exp(-2.789(\frac{E}{E_i})^{\gamma_i}), i = 1,2,3,....j,...$$

$$d = \frac{E_B}{E_S}$$

E_B is the energy of the Coulomb barrier, E_S is a scale parameter expressed in MeV, $E_1, E_2, E_3, \ E_j,$ are the "door" energies, $\gamma_1, \gamma_2, \gamma_3, \ \gamma_j,$ are determined by fitting experimental data [1],[2]. In figures a comparison among the values of S(E) obtained by using the experimental values of fusion cross-section, those obtained by using the "shadow" model and the values of S(E) obtained by using a polinomial fit is shown for different reactions. The values of parameters are reported in table 1. By inspection of figures we can assert that at very low energy the astrophysical factor S(E) obtained in the framework of the "shadow" model shows an oscillating behavior for $E_n \leq E \leq E_B$, E_n is smaller value of "door" energies, for energies smaller than E_n it shows an exponential increase.

References

[1] A.Scalia,Nuovo Cimento A104 (1991) 1467;A105 (1992) 233.

[2] A.Scalia,P.Figuera,Phys.Rev. C46 (1992) 2610;A.Scalia, Nuovo Cimento A105 (1993) 855.

[3] A.Scalia,Phys.Rev.C49 (1994) 2847.

[4] A.Krauss,H.W.Becker,H.P.Trautvetter,C.Rolf,Nucl.Phys. A467 (1987) 273.

[5] B.W.Filippone,A.J.Elwyn,C.N.Davids,D.D.Koetke,Phys.Rev. C28 (1983)2222.

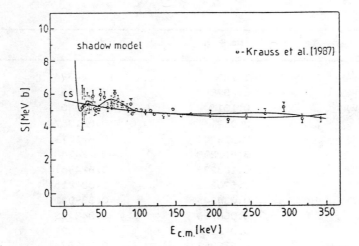

FIG. 1. $S(E)$ function for the reaction $^3\text{He}(^3\text{He},2p)^4\text{He}$. Experimental data are from Ref. [4]. The shadow model line is the energy dependence predicted by the "shadow" model by using the experimental data of Ref. [4]. The cs line is the energy dependence predicted by the classical $S(E)$ function obtained by using a polynomial fit to the data

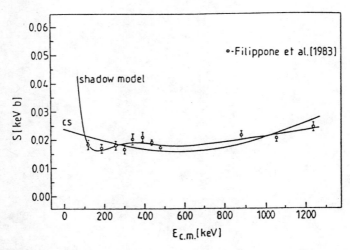

FIG. 2 Same as Fig. 1 for the reaction $^7\text{Be}(p,\gamma)^8\text{B}$

TABLE 1

SYSTEM		$^3\text{He} + {}^3\text{He}$	$p + {}^7\text{Be}$
E_B	(MeV)	1.285352	11.134794
E_S	(MeV)	1.297857	10.348685
E_1	(MeV)	0.1513	0.279802
γ_1		3.0191	0.581357
E_2	(MeV)	0.319229	0.871671
γ_2		3.660290	−2.124748
E_3	(MeV)	0.136238	0.924961
γ_3		1.670239	1.827215
E_4	(MeV)	0.03434	0.389921
γ_4		5.068369	3.048433
E_5	(MeV)	0.071170	0.256891
γ_5		4.212952	0.287673

An improved experimental Approach to the $^{12}C(\alpha,\gamma)^{16}O$ Problem *

G. Roters, C. Rolfs, and H.P. Trautvetter

Institut für Physik mit Ionenstrahlen, Ruhr-Universität Bochum
44780 Bochum, Germany

Abstract. The capture reaction $^4He(^{12}C,\gamma)^{16}O$ as well as the elastic scattering process $^4He(^{12}C,^{12}C)^4He$ have been investigated in the energy range of E_{cm} = 0.9 - 3.4 MeV. An extended, windowless and recirculating ^4He - gas target system with large apertures was bombarded by an intense ^{12}C - beam. Angle integrated gamma ray yield (E1 + E2 component) was observed by using a 4 in. ∅ × 4 in. BGO detector in close geometry. The E1 component was observed by a 2 in. ∅ × 2 in. BGO detector in far geometry, where the E2 contribution was significantly reduced. Five large area plastic detectors in anticoincidence to the BGO detectors provided an active shield against γ-ray background induced by cosmic rays.

INTRODUCTION

The predicted abundance distribution of elements (between carbon and iron) as well as the evolution and the final evolutionary state of massive stars (neutron stars and black holes) depend critically upon the rate of the radiative capture reaction $^{12}C(\alpha,\gamma)^{16}O$ [1,2]. Although this reaction has been the subject of intense experimental and theoretical efforts over the past twenty years, there is still considerable uncertainty in the value of the astrophysical S(E) - factor at stellar helium burning energies (E_{cm} = 0.3 MeV). S(E) is defined as S(E) = σ(E) E exp(2πη), where η is the Sommerfeldparameter. The cross section σ(E) at E_{cm} = 0.3 MeV is expected to be dominated by the p-wave (E1) and d-wave (E2) capture to the (J^π = 0⁺) ^{16}O ground state. Two bound states, at 6.92 MeV (J^π = 2⁺) and 7.12 MeV (J^π = 1⁻), which correspond to subthreshold resonances at E_R = - 245 and - 45 keV, appear to provide the bulk of capture strength through their finite widths that extend into the continuum [3]. The determination of the S(E) - factor at stellar energies requires a considerable extrapolation from the experimentally accessible energies of $E_{cm} \cong$ 1 MeV, where the cross section is dominated by a broad E1 resonance (J^π = 1⁻) at E_R = 2.39 MeV. The cross section σ(E) of this E1 resonance varies from about 40

---

* Supported in part by the Deutsche Forschungsgemeinschaft (Ro 429/21-3)

nb on top of the resonance to a few pb off resonant ($E_{cm} \cong 1$ MeV). The major problem in these measurements arises from the combination of a low γ-ray capture yield and a high γ-ray background, induced by the sensitivity of the γ-ray detectors to neutrons. In order to avoid neutron-induced γ-ray background, mainly from the $^{13}C(\alpha,n)^{16}O$ reaction due to ^{13}C contaminants in the target and on the surface of beam-defining collimators, the role of projectiles and target has been interchanged. A windowless and recirculating ^{4}He gas target system with large apertures has been constructed and was bombarded by an intense ^{12}C beam. Large apertures are of crucial importance in order to avoid neutron-induced γ-ray background arising mainly from the $^{12}C(^{12}C,n)^{23}Mg$ reaction by hitting ^{12}C contaminants on the surface of the apertures. The measurements were carried out over a wide energy range varying from E_{cm} = 0.9 to 3.4 MeV. To obtain information on the E1- as well as the E2 -component of the γ-radiation a special detector set-up of BGO detectors was used. The cosmic-ray induced background was suppressed with the use of an anti-coincidence shielding consisting of five NE 102A plastic scintillators.

EXPERIMENTAL EQUIPMENT AND SET-UP

The Accelerator

The 4 MV Dynamitron tandem accelerator at Bochum [4] provided a ^{12}C beam of up to 20 µA particle current at the target with energies of E_{lab} = 3.7 - 13.6 MeV. Measurements of the energy calibration of the analysing magnet as well as of the beam energy resolution for heavy ions have been carried out previously [5].

The Gas Target System

A windowless and recirculating gas target system with large apertures has been constructed [6]. Schematic diagrams of the relevant parts of this system and of the target chamber are shown in Fig. 1. The gas target system consists of 5 pumping stages, where roots blowers were used in the first and second stages and turbomolecular pumps in the third, fourth, and fifth stages. The beam enters the disc shaped target chamber through the apertures F to A and is stopped in a shielded beam dump (Faraday cup) at a distance of about 7 m from the target chamber. The beam dump is surrounded by paraffin and is positioned in a completely separated hall in order to reduce neutron-induced γ-ray background from the beam stop. The diameter of the apertures, which are long pipes with high gas flow impedances, vary from 8 mm (A) to 40 mm (F). The 2 mm aperture indicated in Fig.1 is a removable, water-cooled aperture which is of crucial importance for good focusing. It is installed in front of aperture A and used to guide the beam into the target chamber without hitting any apertures. The 2 mm aperture is removed after minimising the beam current on it. This procedure was repeated every 2 - 3 hours to check the beam focusing. The gas pressure in the target chamber was measured with a baratron capacitance manometer to an accuracy of ± 4 %. The ^{4}He target gas pressure was chosen to be 9 Torr (chemical purity 99.9999%) and the corresponding pressure reduction at the fifth pumping stage was about 10^{-6} Torr. The purity of the gas as well as the number of projectiles can be monitored via elastic scattering yields from the target nuclids, using well collimated surface barrier detectors placed at θ_{lab} = 45° and 75° to the beam axis.

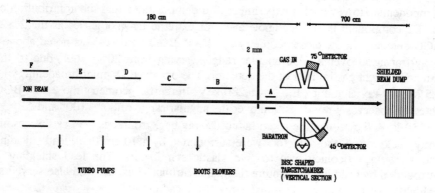

FIGURE 1. Schematic diagram of the relevant parts of the windowless gas target system. The beam enters the disc shaped target chamber through the apertures F to A and is stopped in a shielded beam dump (Faraday cup) at a distance of about 7 m from the target chamber.

Gamma-ray Detection

The γ-ray transitions were observed by two BGO detectors ($Bi_4Ge_3O_{12}$). BGO was used instead of NaI(Tl) because of the higher γ-ray efficiency (a factor of ≅ 2.5). This expectation was verified by spectra obtained on top of the broad J^π = 1$^-$ resonance at E_{cm} = 2.39 MeV: a 4 in. ⌀ × 4 in. BGO detector (in the following 4"× 4" BGO) and a 4"×4" NaI(Tl) detector concurrently detected the 9.55 MeV γ-ray transition to the ground state. Another advantage of BGO compared to NaI(Tl) is the lower sensitivity to neutrons [7]. A specific arrangement of the two detectors was chosen in order to separate the two radiation patterns E1 and E2. The 4"×4" BGO was placed in close geometry at a distance of 2 cm from the beam axis, where the observation γ-ray angle varied from 20° to 160° relative to the beam axis. In such geometry it was shown [8] that nearly all angle integrated information is obtained for both E1 and E2 radiation patterns. The 2"×2" BGO was placed in far geometry at a distance of 24.5 cm from the beam axis. Such a geometry should be mainly sensitive to the E1 component of the radiation since the E2 radiation pattern exhibits zero yield at 90° whereas the E1 component has its maximum at this angle. Only the finite solid angle of the 2"×2" BGO gives rise to a small E2 contribution to the measured yield. This expectation was tested at the $J^\pi = 2^+$, $E_{cm} = 2.68$ MeV resonance exhibiting pure E2 radiation. A reduction in the yield of about a factor 6 was observed in the 2"×2" BGO compared to the 4"×4" BGO after proper normalisation due to different distances and efficiencies for the two detectors. With such a detector set-up it is possible to distinguish between the two radiation components. However, this experiment also demonstrates that the contribution of the E2 component is not negligible even for extreme far geometries as the 2"×2" BGO detector. In the close geometry the 4"×4" BGO observes - as noted above - γ-ray transition over a wide angular range, varying from 20° - 160°, due to the extended target gas in front of the detector. The observed Doppler broadening of a 9.5 MeV γ-ray is about $\Delta E_{Dopp} \cong \pm 280$ keV. In the far geometry the 2"×2" BGO detector observes γ-ray transition over the angular range of 80° - 100° and $\Delta E_{Dopp} \cong \pm 50$ keV. To reduce the room induced γ-ray background as well as the cosmic γ-ray background, the detectors were surrounded by a 10 cm thick lead shielding and were in anticoincidence to five plastic scintillators. The lead shielding is surrounded by cadmium to capture thermal neutrons. Thus, the lead also served as a shield against the neutron capture-γ-rays from cadmium.

GAMMA-RAY SPECTRA

In Fig.4 gamma-ray spectra observed by the two BGO detectors on top of the broad $J^\pi = 1^-$ resonance are shown. The spectra are dominated by a 9.5 MeV γ-ray peak coming from the transition to the ground state. On the low energy part of the spectrum some background peaks are visible and on the high energy part not completely suppressed cosmic background is seen (suppression factor of about 10 for the 4"×4" BGO and about 25 for the 2"×2" BGO for the γ-energy region of 8 to 11 MeV). Due to the better energy resolution and the far geometry of the 2"×2" BGO, the full energy-, the single escape- and the double escape peak of the 9.5 MeV γ-ray transition could be resolved. Another detector set-up was used were the 2"×2" BGO was replaced by a high purity Ge detector in order to get some information about target impurities not coming from the target gas but from hitting any apertures, beam pipes or the target chamber. The Ge and BGO detectors were placed in close geometry face to face. Any γ-ray background not coming from target gas impurities (no Doppler shift) should be seen as a sharp peak in the Ge detector spectrum but no sharp peaks above $E_\gamma = 5$ MeV were visible. Further background studies are in progress.

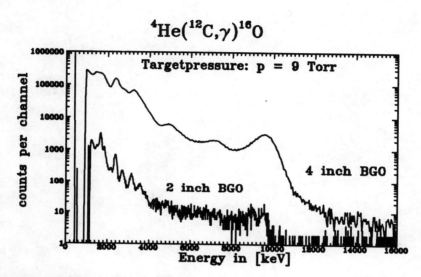

FIGURE 2. Gamma-ray spectra observed on top of the broad $J^\pi = 1^-$, $E_{cm} = 2.39$ MeV resonance for the 4"×4" BGO and for the 2"×2" BGO. The spectra are dominated by a 9.5 MeV γ-ray peak coming from the ground state transition.

REFERENCES

1. W. D. Arnett and F.-K. Thielemann, *Astrophys. J* **295** (1985) 589, 604.
2. S. E. Woosley and T. A. Weaver, *Nucleosynthesis and its Implication for Nuclear and Particle Physics*, edited by J. Audouze and N. Mathieu (Reidel, Dordrecht, 1986) 145.
3. W. A. Fowler, *Rev. Mod. Phys.* **56** (1984) 149.
4. K. Brand, *Nucl. Instrum. Methods* **141** (1977) 519.
5. M. Bahr, L. Borucki, H. W. Becker, M. Berheide, M. Buschmann, C. Rolfs, G. Roters, S. Schmidt, W. H. Schulte, G. E. Mitchell, J. S. Schweitzer *Zeitschrift für Physik,* (1994) in progress.
6. G. Roters, *Diplomarbeit, Universität Münster* (1990).
7. O. Häusser, M. A. Lone, T. K. Alexander, S. A. Kushneriuk and J. Gascon, *Nucl. Instr. and Meth.* **213** (1982) 301.
8. K. U. Kettner, *Thesis, Universität Münster* (1982).

Approximate Penetration Factor for a Diffuse Edge Well with Coulomb Barrier. Astrophysical Applications.

Cornelia Grama, N.Grama, I.Zamfirescu

Institute of Atomic Physics, Bucharest, MG-6, Romania

The cross sections of the charged particle induced reactions at low energies, in particular those of astrophysical interest, are determined by the penetration through the Coulomb and centrifugal barriers. The astrophysical reaction rates at a certain temperature are expressed by the strengths of the resonances in the Gamow window. There are cases when the reaction strengths cannot be experimentally measured and consequently have to be theoretically evaluated. This imposes the calculation of the partial widths for the formation and decay of the given resonant state. In the frame of the R matrix theory a partial width for a given channel can be separated into two factors : a penetration factor that depends on the external barrier and on the energy and a reduced width which is a function of the nuclear configuration describing the state. The penetration factor for the diffuse edge potential could be obtained by the numerical integration of the Schrödinger equation. However in those cases where a large amount of calculation is necessary an approximation could become very useful.

In the frame of the equivalent square well (ESW) model [2], [3], [4] the diffuse edge potential is replaced by an equivalent square well potential that has the same resonance properties. The parameters of the ESW are chosen in such a way that both the diffuse edge and ESW have resonances at the same energies and the same reduced widths at the corresponding matching radii. The difference in the reflection properties is taken into account by a reflection factor f that multiplies the penetration factor of the ESW so that $P_D = fP_{ESW}$.

However both the numerical calculation and the equivalent square well model have the disadvantage that they do not stress the dependence on the parameters of the diffuse edge potential, on the incident energy and on the reduced mass of the particles in a given channel. In practice as e.g. in the evaluation of the thermonuclear reaction rates in the frame of the statistical theory of nuclear reactions [1] the ESW model is used taking ΔR and f constant with respect to the incident energy and target mass.

However, as it was shown in [4] there is a moderate energy dependence of the reflection factor and a rather strong dependence of ΔR and f on the mass of the target for the light-medium target nuclei.

In the following an alternative treatment of the penetration factor of the Woods-Saxon well plus Coulomb barrier is presented. It has the advantage that it stresses the dependence on the parameters. As the Woods-Saxon potential $U(r) = -U_0\{1 + exp((r-R)/a)\}^{-1}$ has a small layer at r=R in which the potential falls off rapidly varying the semiclassical approximation, valid if only $|\rho = [kU(r)]^{-1}dU/dr| \ll 1$, cannot be used. The anticlassical approximation of the wave function and of the penetration factor for the Woods-Saxon well with Coulomb and centrifugal barrier are obtained by taking diffuseness a as a small parameter and using the method of matched asymptotic expansions.

We look for an asymptotic expansion for $\epsilon = a/R \to 0$ of the solution of the Schrödinger equation that behaves asymptotically as an outgoing wave when $r \to \infty$. For $\epsilon \to 0$ the Woods-Saxon potential converges nonuniformly at r=R due to the exponential term. The Schrödinger equation is expressed in terms of a new variable $y = \{1 + exp[(r-1)/\epsilon]\}^{-1}$. It results a singular perturbation problem and no asymptotic expansion is uniformly valid on the whole range of the variable $y \in (0,1]$. There are two layers at $y \approx 0$ and $y \approx 1$ where the solution has not the correct asymptotic behaviour. In order to obtain a uniform asymptotic expansion on the whole interval the matching of the asymptotic expansions is used [6],[7]. The approximate solution is not a single expansion in terms of a single scale. An expansion valid on $y \in (0,1]$ except $y \approx 0$ and $y \approx 1$ represents the "outer" solution. On the boundary layers $y \approx 0$ and $y \approx 1$ the "inner" solutions are constructed by changing the scale, i.e. by introducing inner variables so that the solution has the right asymptotic behaviour. The expansions that represent asymptotically the solutions in different regions are then matched according to the asymptotical matching principle. The inner expansions include exponentially small terms that are essential in the matching. In other words a high precision asymptotics of an unusual kind, which includes exponentially small terms is necessary in order to obtain the matching of the inner and outer solutions.

The approximate solutions are then used in order to obtain the penetration factor. According to the definition the penetration factor is the imaginary part of the logarithmic derivative of the outgoing wave function at the channel radius. One obtains

$$P_D \approx \frac{k}{H_l^+(k)H_l^-(k)\left\{1 - \frac{\epsilon^2 U_0 \pi^2}{12}\right\}^2} \qquad (1)$$

Table 1: Comparison of the exact and approximate penetration factors and of the exact and approximate reflection factors respectively

m	V_0R^2	cR	ϵ	$\epsilon^2 U_0$	l	kR	E/E_B	P_{ex}	P_{app}	f_{ex}	f_{app}
1.	500.	0.	0.05	0.06	0	0.2		0.22E0	0.22E0	1.11	1.10
						1.0		0.11E1	0.11E1	1.10	
			0.1	0.24	0	0.2		0.31E0	0.28E0	1.54	1.40
						1.0		0.15E1	0.14E0	1.49	
		1.0	0.05	0.06	0	0.4	0.11	0.47E-4	0.47E-4	1.09	1.10
						1.0	0.68	0.26E0	0.26E0	1.10	
			0.1	0.24	0	0.4	0.13	0.60E-4	0.60E-4	1.34	1.34
						1.0	0.84	0.34E0	0.33E0	1.45	
3.	1000.	10.	0.05	0.36	0	2.0	0.27	0.42E-5	0.42E-5	1.57	1.59
						3.0	0.60	0.52E-1	0.51E-1	1.64	
			0.08	0.92	0	2.0	0.30	0.76E-5	0.67E-5	2.87	2.52
						3.0	0.67	0.11E0	0.81E-1	3.41	
6.	1500.	40.	0.05	1.08	0	4.5	0.33	0.89E-8	0.86E-8	2.90	2.78
						6.0	0.59	0.26E-2	0.21E-2	3.46	
					10	7.0	0.36	0.96E-6	0.53E-6	2.41	2.78
						9.0	0.60	0.36E-2	0.37E-2	2.70	

where $H_l^{\pm}(k) = G_l(k) \pm i F_l(k)$. Because $P_S = \dfrac{k}{H_l^+(k) H_l^-(k)}$ is the penetration factor for the square well potential it results

$$P_D \approx P_S(1 + \frac{1}{6}\epsilon^2\pi^2 U_0) \qquad (2)$$

provided that $|\epsilon^2 U_0| \ll 1$. In the above expression P_S and P_D are the penetration factors of the diffuse edge and square well potentials respectively, calculated at the same radius.

The quality of the approximation and its range of validity have been checked by comparison of the approximate penetration factor P_{app} given by (2) to the "exact" numerically calculated penetration factor P_{ex} for a large range of values of k and of the parameters $m, U_0 = 2mV_0R^2/\hbar^2, c = Z_1 Z_2 e^2 MR/\hbar^2, \epsilon = a/R, l$. The diffuse edge well penetration factor P_{ex} was calculated by the numerical integration of the Schrödinger equation. The square well penetration factor P_S was obtained by using the RCWFN subroutine that calculates the regular and irregular Coulomb wave functions F_l and G_l [8]. In table 1 the comparison is done for values of the parameters c, m, V_0 and R that could correspond to physical systems. For example $m \approx 1$ represent the proton ($cR \approx 1$) and neutron ($cR = 0$) channels. For the nuclear system $^{12}C + ^{12}C$ the reduced mass is $m = 6$ and $cR \approx 40$. The intermediate value $m = 3$, $cR = 10$ would illustrate some α +nucleus systems (e.g. for α +^{16}O, cR=8.44 and $m = 3$ amu for a radius calculated according to the formula $R = 1.25A^{1/3} + 1.6\ fm$. The chosen values of the parametr $V_0 R^2$ are obtained for $V_0 \approx 50 MeV$ and resonable values of R. As long as the condition of validity of the approximation $|\epsilon^2 U_0 \ll 1|$ is fulfilled there is a good agreement of P_{ex} and P_{app}.

One can see from Equation (2) that the effect of the diffuseness is to increase the penetration factor. The factor that multiplies the penetration factor for the square well plays the role of the reflection factor in the ESW method. However in the ESW the square well penetration factor is calculated at a matching radius that differs from the radius of the Woods-Saxon potential by ΔR, where ΔR can represent tens of percents of the radius itself. Consequently a numerical comparison cannot be done directly. The reflection factor introduced in the present work contains the explicit dependence on the parameters of the Woods-Saxon well and shows the quadratic dependence on the diffuseness, as it was claimed in the ESW method for neutron channels [2]. From Equation (2) one can see that this quadratic dependence on the diffuseness a is valid for the charged particle channels too. Moreover the approximate reflection factor contains the dependence on the nuclear system through the reduced mass m and radius R, but it does not depend on the energy, similar to the reflection factor defined in the ESW method.

References

[1] Woosley S.E., and Fowler W.A., *At. Data Nucl. Data Tables* **22** 371 (1978).

[2] Vogt E.,*Rev. Mod. Phys.* **34**, 723 (1962).

[3] Vogt E., Michaud G. and Reaves H., *Phys. Lett.* **19**, 570 (1965).

[4] Michaud G., Scherk L. and Vogt E., *Phys. Rev.* **C1**, 864 (1970).

[5] Kolkunov,V. A., *Teor. Math. Fiz.* **3** 72 (1970).

[6] Van Dyke M.,*Perturbation methods in fluid mechanics, Annotated edition* Stanford:Parabolic Press, 1975, p. 29.

[7] Eckhaus W.,*Asymptotic Analysis of Singular Perturbations*, Amsterdam: North Holland Publ. Comp., 1979.

[8] Barnet A.R, Feng D.H., Steed J.W., Goldfarb L.J.B., *Comp. Phys. Comm.* **8**, 377 (1974).

The s–Process Cross Sections of the Tin Isotopes

Ch. Theis(1), K. Wisshak(1), K. Guber(1), F. Voß(1), F. Käppeler(1),
L. Kazakov(2), and N. Kornilov(2)

*(1) Kernforschungszentrum Karlsruhe, IK-III, Postfach 3640,
D-76021 Karlsruhe, Germany
(2) Institute of Physics and Power Engineering, Obninsk, Russia*

Abstract. A recent investigation of the abundance pattern in the Cd–In–Sn region has shown that the s– and r–process contributions to the rare isotopes ^{113}In, ^{114}Sn, and ^{115}Sn are not sufficient to account for the observed stellar abundances. Furthermore, the s–process contributions may also depend on temperature due to branchings at low lying isomers. For the reliable determination of the s–process abundances, the stellar (n,γ) cross sections have been measured for $^{114-118}$Sn and ^{120}Sn using the Karlsruhe 4π BaF$_2$ detector.

Key words: nucleosynthesis — neutron capture — s–process

1. INTRODUCTION

Of the three rare tin isotopes ^{112}Sn, ^{114}Sn, and ^{115}Sn only the even isotopes are significantly produced by the p–process (1, 2), while ^{115}Sn is assumed to originate mainly from the s– and r–processes. A recent investigation (3) of the s– and r–process contributions to these isotopes showed that the r–process can contribute only about 60 % to the solar ^{115}Sn abundance, in accordance with the results of Beer *et al.* (4). Since the main s–process path proceeds via ^{114}Cd – ^{115}Cd – ^{115}In – ^{116}In – ^{116}Sn, the s–process yield of ^{115}Sn is restricted to a small branching due to the decay of the 335 keV isomer in ^{115}In. At low s–process temperatures, this isomer is sufficiently populated to yield a significant ^{115}Sn abundance. At higher temperatures, however, isomer and ground state in ^{115}In are thermally equilibrated, resulting in a strong depopulation of the isomer. Accordingly, the classical approach ($T_s \sim 3 \times 10^8$ K) yields less than 1 % s–process contribution to ^{115}Sn. The alternative scenario, helium shell burning in low mass stars, starts with a low s–process temperature and, hence, a more efficient branching to ^{115}Sn. However, the temperature increase at the end of the helium shell burning episodes leads to a rearrangement of the abundance pattern, thus limiting the net production of ^{115}Sn to about 8 %.

The quantitative discussion of this problem is hampered by the lack of accurate cross sections. In the neutron energy range above 20 keV, Timokhov *et al.* (5) reported (n,γ) cross sections for the tin isotopes with uncertainties of ∼5 % for ^{115}Sn and ∼7 % for the s–only isotope ^{116}Sn. At lower energies, which are relevant in stellar models, only theoretical estimates exist for ^{115}Sn. In view of this situation, new measurements were carried out over a larger energy range.

© 1995 American Institute of Physics

2. MEASUREMENT

The (n,γ) cross sections for ^{114}Sn, ^{115}Sn, ^{116}Sn, ^{117}Sn, ^{118}Sn, and ^{120}Sn have been measured between 5 and 220 keV using the Karlsruhe 4π BaF$_2$ detector. For a description of the experimental setup see the contribution of Wisshak et al. to this conference (6) and the references given therein.

Isotopically enriched metallic tin samples were provided by IPPE, Obninsk. The characteristics summarized in Table 1 indicate that the isotopic enrichment of the ^{114}Sn and ^{115}Sn samples was relatively low. For the correction of the respective isotopic impurities it was important that the entire sequence of all tin isotopes was measured in the same experiment. The impurities due to the missing isotope ^{119}Sn were always less than 2 %. The corresponding corrections could be approximated by the properly normalized spectra of ^{117}Sn, resulting in uncertainties of ≤ 0.3 % in all cases.

The measurements were carried out relative to gold as a standard, and a graphite sample was used to determine the background due to neutrons scattered from the samples into the detector.

Table 1: Sample Characteristics

Sample	Mass (g)	Enrichment (%)	Binding energy (MeV)
^{114}Sn	2.5884	70.2	7.546
^{115}Sn	0.7544	45.3	9.562
^{116}Sn	7.5284	97.6	6.944
^{116}Sn	2.4462	97.6	6.944
^{117}Sn	0.9177	92.1	9.326
^{118}Sn	4.7372	98.5	6.484
^{120}Sn	10.2791	99.6	6.171

Two runs with maximal neutron energies of 100 keV and 200 keV were performed with the 4π BaF$_2$ detector operated as a γ-ray calorimeter by recording the sum energies of the capture γ-ray cascades. In a third run, the individual γ-rays of each cascade were measured using an ADC system coupled to the 40 modules of the 4π detector. In this way, the fraction of unobserved capture events could be determined on the basis of experimental information.

3. DATA ANALYSIS

The detected capture events were sorted in two-dimensional spectra containing the γ-ray sum energy versus time of flight (TOF) with the cascade multiplicity as an additional parameter. The data analysis was performed using the full set of recorded events as well as a reduced set, which was obtained by rejecting those events where only neighboring detector modules contributed. These events were considered as being due to the background from sample–scattered neutrons (see below). The sample–independent background was determined from the spectrum measured with an empty position. A small time–independent background was removed by means of events observed at long flight times. Then, the spectra were corrected for isotopic impurities by subtracting the correspondingly normalized spectra of the impurity isotopes.

Scattered neutrons from the samples were partly captured by the barium of the scintillator and caused a significant background. The binding energies of the even tin isotopes and of ^{197}Au are well below those of the odd barium isotopes ^{135,137}Ba. Therefore, events observed in the spectra of the even isotopes at sum energies of ~ 9 MeV are only due to scattered neutrons, allowing an easy determination of the scattering correction. The binding energies of the odd tin isotopes are slightly higher than of the odd barium isotopes. In these cases, the scattering corrections were determined via captures of scattered neutrons in the even barium isotopes ^{134,136}Ba. The uncertainties of this important correction are rather small compared to other detection techniques, since the related information can be derived directly from the experimental spectra as a function of TOF.

The background corrected TOF spectra are eventually used to calculate the capture cross sections. A detailed description of this procedure is given by Wisshak et al. (7), including the corrections for unobserved events, and for multiple scattering and self–absorption effects in the sample.

Preliminary cross sections for the even isotopes are compared in Figure 1 with the results of Timokhov et al. (5) and of Macklin et al. (8). Around 30 keV, the uncertainties of the present data are typically 1 %. In general, good agreement was obtained with the previous measurements. Comparison of the Maxwellian averaged cross sections for kT = 30 keV in Table 2 shows, however, that the uncertainties of the stellar cross sections could be reduced by a factor of 3.

Table 2: Preliminary Maxwellian averaged cross sections for kT = 30 keV compared with the results of Timokhov et al. [5]

Isotope	This work	Ref. [5]
^{114}Sn	137 ± 2.4	149 ± 8
^{116}Sn	96 ± 1.1	104 ± 4
^{118}Sn	66.5 ± 1.2	76 ± 4
^{120}Sn	41.7 ± 0.9	41.0 ± 3.3

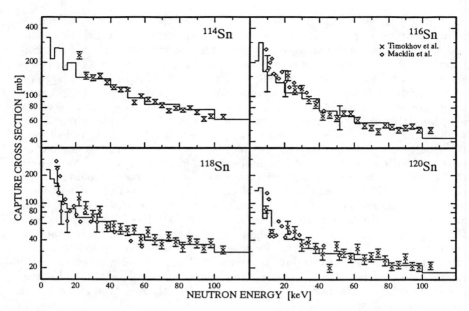

Fig. 1.— Comparison of the present capture cross sections (histograms) with previous data

4. CONCLUSION

New and more accurate cross sections data are now available for a series of stable tin isotopes. These data will allow a more precise investigation of the s-process flow in the Cd-In-Sn region, in particular of the possible s-process branching at ^{115}Cd.

REFERENCES

(1) M. Rayet, N. Prantzos, and M. Arnould 1990, *Astron. Astrophys.* **227**, 271

(2) W. Howard, B. Meyer, and S. Woosley 1991, *Ap. J.* **373**, L5

(3) Zs. Németh, F. Käppeler, Ch. Theis, T. Belgya and S. Yates 1994, *Ap. J.* **426**, 357

(4) H. Beer, G. Walter, and F. Käppeler 1989, *Astron. Astrophys.* **211**, 245

(5) V. Timokhov, M. Bokhovko, A. Isakov, L. Kazakov, V. Kononov, G. Mantorov, E. Poletaev and V. Pronyaev 1989, *Sov. J. Nucl. Phys.* **50**, 375

(6) K. Wisshak, F. Voß, and F. Käppeler 1994, *contribution to this conference*

(7) K. Wisshak, K. Guber, F. Voß, F. Käppeler and G. Reffo 1993, *Phys. Rev. C* **48**, 1401

(8) R. Macklin, T. Inada and J. Gibbons 1962, *AEC, Washington Reports to the NCSAG* **1041**, 30 (Data from EXFOR 11981)

The Stellar (n,γ) Cross Sections of Unstable s–Process Nuclei

Stefan Jaag and F. Käppeler

Kernforschungszentrum Karlsruhe, IK-III, Postfach 3640, 76021 Karlsruhe, Germany

Abstract. When the s–process nucleosynthesis path encounters an unstable isotope with a half–life comparable to the neutron capture time, the resulting branching of the reaction flow leads to an abundance pattern that carries information on the physical parameters of the stellar plasma. The (n,γ) cross section of these isotopes are important clues for the interpretation of these abundance patterns. While the neutron capture cross sections of the stable isotopes are known with good accuracy, there is a complete lack of experimental data for the radioactive branch point isotopes. This contribution describes a method for measurements on unstable targets, and presents the results on ^{155}Eu and ^{163}Ho.

Key words: nuclear astrophysics, neutron capture, s–process

1. INTRODUCTION

There are two different methods to measure stellar (n,γ) cross sections in the keV–region: detection of the prompt capture gamma–rays in combination with the time–of–flight (TOF) method as well as the activation technique, which is based on counting of the induced activity.

If the relevant unstable branch point nucleus is followed by a stable isotope, one is forced to use the TOF–method. Since this method requires sample masses of a few mg the γ–ray background related to the sample activity presents a serious problem. In those cases, where this background can not be sufficiently suppressed, it is practically impossible to identify the true capture events.

The better sensitiyity of the activation technique offers the advantage that very small quantities of sample material are sufficient. However, the number of astrophysically interesting isotopes is limited due to the requirement that also the 'produced' nuclei must be unstable (1).

This technique has been first applied to the determination of the stellar cross sections of the unstable isotopes ^{155}Eu and ^{163}Ho.

2. THE STELLAR (n,γ) CROSS SECTION OF ^{155}Eu

The branchings at ^{154}Eu ($t_{1/2} = 8.8\,yr$) and ^{155}Eu ($t_{1/2} = 4.68\,yr$) can both be considered as s–prosess thermometers. While the ^{154}Eu branching is characterized by the s–only isotope ^{154}Gd, the ^{155}Eu branching can not be defined by an s–only nucleus since the ^{155}Gd abundance received an r–process contribution as well. For experimental reasons, there is almost no chance to measure the ^{154}Eu cross section directly. Therefore, the motivation of the present measurement was to use the experimental value for ^{155}Eu for improving the theoretical calculation of the more important ^{154}Eu cross section. On the other hand, it will also be possible to analyze the ^{155}Eu–branching directly, as soon as reliable s–abundance data for the Gd–isotopes become available, e.g. via the discovery of the corresponding isotopic anomalies.

2.1. Sample preparation

In order to achieve the required isotopic purity, the ^{155}Eu–sample was breeded in a reactor from an enriched batch of $1.55\,g$ ^{154}Sm. Before, this batch had to be purified chemically in order to reduce small but disturbing impurities from other rare earth elements to an acceptable level. In this respect, an 0.013 % Eu contamination was the most critical point, since this gives rise to an interfering background from ^{152}Eu and ^{154}Eu activities in the later activation. According to the estimated effect of this background, more than 99 % of the Eu impurity had to be removed before the actual cross section measurement.

The purification was carried out via the ion exchange technique, using the strongly basic anion exchange resin AMBERLITE IRA 400 CG (200–400 mesh, nitrate form). The eluant was an aqueous solution of 80 % methanol, 1.5 N ammonium nitrate and 0.01 N nitric acid. The column had a volume of $\approx 280\,cm^3$, the flow rate of the eluant was $9\,ml/min$ and the temperature $50°C$. After the separation, which took about 8 hours, the purified Sm was washed out and precipitated with oxalic acid. In this way, the Eu impurity could be reduced by a factor 400.

The purified batch of ^{154}Sm was exposed to a thermal neutron flux of $10^{13}\,sec^{-1}\,cm^{-2}$ in the reactor of the DKFZ Heidelberg for $90\,min$. The so produced ^{155}Eu was then separated in a second step, using the same ion exchange technique. The final sample of $(3.34 \pm 0.13) \cdot 10^{14}$ atoms of ^{155}Eu was characterized by its activity. This ^{155}Eu was imbedded in a remaining macroscopic quantity of Sm_2O_3, which was pressed to a pellet and canned in a thin walled graphite box. Before the final activation, a waiting time of 220 days was observed to allow all of the coproduced ^{156}Eu to decay ($t_{1/2} = 15.6\,d$).

2.2. Activation, Counting and Results

The activation was performed at the Karlsruhe Van de Graaff accelerator using a quasi–stellar neutron spectrum of the ^{7}Li(p,n)^{7}Be reaction (2). The ^{155}Eu–sample was sandwiched between two gold foils for flux normalization and mounted directly on the Li–target. The irradiation lasted for 17 days at an average neutron flux of $3 \cdot 10^9\,sec^{-1}\,cm^{-2}$, and the induced ^{156}Eu activity was observed for 29 days using a shielded $175\,cm^3$ HPGe detector.

The resulting γ–ray spectrum contained more than 200 lines, mainly from the remaining ^{152}Eu, ^{154}Eu activities and from natural background. From the observed ^{156}Eu transitions, 11 lines could be used for determining the induced activity. Because of the close counting geometry, the high sample activity, and the finite sample thickness, corrections had to be applied for cascade summing, pile–up, dead time, self shielding and for the spatial distribution of the activity within the sample.

The resulting relative cross section for the quasi–stellar spectrum is

$$\frac{<\sigma(^{155}\text{Eu})>}{<\sigma(^{197}\text{Au})>} = 2.28 \pm 0.13,$$

corresponding to a Maxwellian averaged neutron capture cross section of the ^{155}Eu ground state

$$<\sigma(^{155}\text{Eu})>_{30\,keV} = 1320 \pm 84\,mb.$$

Comparison with theoretical calculations of Harris (3) and of Holmes et al (4) (2088 mb and 1730 mb, respectively) shows that the experimental result is significantly smaller. The effect of excited states in the stellar plasma had also been considered in the above calculations yielding corrections of only a few %, but of opposite sign. Therefore, no corrections were made for this effect.

3. THE STELLAR (n,γ) CROSS SECTION OF ^{163}Ho

Under s–process conditions the terrestrially stable ^{163}Dy becomes unstable (*bound state decay*). Due to the resulting branching of the synthesis path, ^{164}Er is mainly of s–process origin. The analysis of this branching allows to discuss the electron density in the stellar interior, since the half–lives of the involved isotopes depend strongly on this parameter.

3.1. Sample preparation

The ^{163}Ho–sample ($t_{1/2} = 4570\,yr$) was a fraction of a sample breeded in Los Alamos from enriched ^{152}Er, which had been chemically concentrated using the ion exchange technique by the ISOLDE group (α–HIB as an eluant and the cation exchange resin Aminex A 5, (5)).

This sample, which still contained some Er and Dy, was adsorbed at 5 mg graphite and then pressed to a pellet. The elemental composition was determined quantitatively using the X–ray flourescence technique. The isotopic composition was measured by mass spectroscopy, yielding a total of $N(^{163}\text{Ho}) = (6.62 \pm 0.1) \cdot 10^{17}$ atoms.

3.2. Activation, Counting and Results

Six activations of 1^h duration were performed at the Van de Graaff accelerator (same method as in the ^{155}Eu–measurement). The induced ^{164}Ho activity was counted with a 40 cm^3 HPGe–detector. The deconvolution of the mixed X–ray multiplets and a measurement of the β–spectrum allowed also to reevaluate the gamma–ray line intensities and the half–life of ^{164}Ho. As a by–product, the stellar ^{162}Eu cross section could be determined as well:

$$< \sigma(^{163}\text{Ho}) >_{30 keV} = 2125 \pm 95 \, mb$$

$$< \sigma(^{162}\text{Er}) >_{30 keV} = 1624 \pm 124 \, mb$$

Both results are in excellent agreemet with the theoretical predictions of Harris (3) and Holmes et al (4), who give values of $2264\,mb$ and $2880\,mb$ for ^{163}Ho and $1670\,mb$ and $1534\,mb$ for ^{162}Er, respectively.

The analysis using the classical s–process approach and adopting the latest results on neutron density (6) and temperature (7) yields the following estimate for the electron density during the s–process: $10 \cdot 10^{26} < n_e < 30 \cdot 10^{26} \, cm^{-3}$. Compared to the previous result of (8) (which was obtained with different neutron density and temperature parameters), the electron density range could be reduced by ~60%.

Apart from the ^{164}Er production, the new capture cross section provides a reliable ^{163}Ho abundance. This is important since it decays after the neutron irradiation and contributes significantly to the s–abundance of ^{163}Dy. Due to the difference in the stellar cross sections of ^{163}Ho and ^{163}Dy, the final s–abundance of ^{163}Dy depends on the branching factor, and could, therefore, also be used for deriving the electron density. In fact, recent studies of isotopic anomalies in meteoritical SiC–grains (9) seem to indicate an undisturbed s–process pattern for the dysprosium isotopes. If these first results can be confirmed, a corresponding independent determination of the electron density would be possible.

REFERENCES

1. Käppeler F., *Radioactive Nuclear Beams (Louvain–la–Neuve)*, Bristol: Adam Hilger, 1991, 305.
2. Ratynski W. and Käppeler F., *Phys.Rev.C*, **37**, 595 (1988).
3. Harris M.J., *Ap&SS*, **77**, 357 (1981).
4. Holmes J.A. et al., *Atomic Data and Nuclear Data Tables*, **18**, 305 (1976).
5. Beyer G.J. and Ravn H.L., private communications.
6. Wisshak K. et al., *Report KfK 5067*, (1992).
7. Kppeler F. et al., *ApJ*, **410**, 370 (1993).
8. Beer H. et al., *Capture Gamma–Ray Spectroscopy and Related Topics (Knoxville)*, New York: AIP, 1984, 778.
9. Richter S. and Ott U., private communications.

β-Delayed Neutron Decay in ^{17}B and ^{19}C

G. Raimann(1), A. Ozawa(2), R.N. Boyd(1,3), F.R. Chloupek(1),
M. Fujimaki(2), K. Kimura(4), T. Kobayashi(2), J.J. Kolata(5),
S. Kubono(6), I. Tanihata(2), Y. Watanabe(2), K. Yoshida(2)

(1) Department of Physics, The Ohio State University, Columbus, OH 43210, USA

(2) The Institute of Physical and Chemical Research (RIKEN), Wako-shi, Saitama 351-01, Japan

(3) Department of Astronomy, The Ohio State University, Columbus, OH 43210, USA

(4) Nagasaki Institute of Applied Science, Nagasaki, Nagasaki 851-01, Japan

(5) Department of Physics, University of Notre Dame, Notre Dame, IN 46556, USA

(6) Institute of Nuclear Study, University of Tokyo, Tanashi, Tokyo 188, Japan

> The β-delayed neutron decays of ^{17}B and ^{19}C were studied using radioactive ion beams. The neutron energies, measured via time-of-flight, give information on states above the neutron decay threshold in ^{17}C and ^{19}N, respectively. These low lying states are of possible interest for Big Bang nucleosynthesis.

ASTROPHYSICAL MOTIVATION

The abundances calculated within the framework of the Inhomogeneous Models (IM) of primordial nucleosynthesis differ from those obtained within the Standard Model, especially for the elements ^{7}Li, ^{9}Be and ^{11}B (1,2). Comparison of the predicted primordial abundances with the observed ones should clarify the level to which inhomogeneities existed in the early universe. The abundance of ^{7}Li, however, may not be a good measure (3). The primordial abundances of ^{9}Be (4) and ^{11}B (5) are somewhat less sensitive to unknown factors, but neither is easy to observe. The IMs also predict (6) a considerably higher abundance of elements with mass 12 and greater than does the Standard Model. Thus it would be desirable to have an independent way of testing the predictions of the IMs by looking at slightly heavier nuclides. In order to establish accurate predictions of elemental abundances based on the IMs, accurate knowledge of nuclear reaction rates between the intervening nuclides is essential.

Most of the nucleosynthesis in the IMs, however, occurs just beyond the neutron rich side of stability, and thus involves short-lived nuclides. Recent studies (7,8) on the pathway of nucleosynthesis anticipated in IMs have identified some of the key branch points in that flow.

In this work we investigated levels above the neutron threshold in ^{17}C and ^{19}N, which play a crucial role for the ^{16}C$(n,\gamma)^{17}$C and ^{18}N$(n,\gamma)^{19}$N reactions that would lead on to yet higher masses.

EXPERIMENTAL SETUP AND RESULTS

The experiment was performed at the recoil ion facility (RIPS) of the Institute of Physical and Chemical Research (RIKEN). A pulsed primary beam of ^{22}Ne with an energy of about 112 MeV/nucleon hit a ^{12}C production target. The recoil ions were analysed by two dipole magnets. The beam finally passed through a thin mylar window and through a rotatable Al-absorber which downgraded the energy of the ions sufficiently for them to be stopped in a sandwich of plastic scintillators that detected the β-decay of ^{17}B ($T_{1/2} = 5$ ms) or ^{19}C ($T_{1/2} = 50$ ms). With a probability of $p_{1n} = 63$ % (9) for ^{17}B and 47 % (9) for ^{19}C the β-decay was followed by a one-neutron decay. The energy of the emitted neutron could be determined by time-of-flight (TOF), with the β-decay providing the start signal, and large neutron walls yielding the stop signal. In addition, NaI counters detected γ-events from decays of excited states of the residual nucleus in case the neutron decay was not to the ground state.

Detection of the neutrons was achieved by three large neutron walls, each having an active surface area of about 110 × 100 cm, that were positioned at a distance of 125 cm from the sandwich of β-detectors. The energy resolution of the neutron walls was about 13% (for 1 MeV neutrons), and the time resolution below 1 ns.

The signals from the neutron walls were corrected for electronic effects and for different path lengths of the neutrons. An energy calibration was performed using the known energies of β-delayed neutrons from ^{17}N. Since the $E_n = 0.38$ MeV neutron peak was quite strong, this indicates that the neutron detection threshold was well below that energy. The efficiency of the neutron walls was determined via a Monte Carlo simulation.

The β-delayed ^{17}B neutron spectrum (not shown) exhibits 3 peaks, one of which has a broad shoulder. The peaks were fitted simultaneously with 4 Gaussians, and a broad Gaussian was used for the background. An investigation of possible daughter decays reveals that the lowest-energy peak ($E_n = 0.82$ MeV) corresponds to the β-delayed neutron breakup of ^{16}C into ^{15}N + n. This identification is strongly supported by the analysis of the β energy spectrum. In the absence of prominent γ-peaks the data are consistent with neutron breakups of ^{17}C that leave the residual ^{16}C nucleus in its ground state. Table 1 summarizes the results.

TABLE 1. Preliminary energy assignments, branching ratios (normalized to the known probability for 1n-breakup) and log ft-values for observed β-delayed neutrons from ^{17}B and ^{19}C. The peaks are listed as a function of decreasing energy; see Fig. 1.

beam	E_n (MeV)	E_x (MeV)	in	decay to	branching (%)	log ft
^{17}B	2.95(10)	3.86(10)	^{17}C	g.s.	26(2)[a]	4.7(2)
^{17}B	1.81(5)	2.65(5)	^{17}C	g.s.	37(3)[a]	4.7(1)
^{17}B	0.82(2)	3.37(2)	^{16}N[b]	g.s.		
^{19}C	^{17}B peak					
^{19}C	2.08(10)	6.15(10)	^{19}O[c]	g.s.		
^{19}C	^{17}B peak					
^{19}C	1.51(7)	7.03(7)	^{19}N	1st. exc.	14(1)[d]	4.9(1)
^{19}C	1.02(4)	6.51(4)	^{19}N	1st. exc.	21(2)[d]	4.8(1)
^{19}C	^{17}B peak					
^{19}C	0.62(3)	4.61(3)	^{19}O[e]	g.s.		
^{19}C	0.45(2)	6.38(2)	^{19}N	2nd. exc.	12(2)[d]	5.1(1)

[a] Normalized to 63(1) % (9)
[b] $E_x = 3.36$ (10)
[c] $E_x = 6.06$ from ^{18}O(n,n)^{18}O (11)
[d] Normalized to 47(3) % (9)
[e] $E_x = 4.59$ from ^{18}O(n,n)^{18}O (11)

Fig. 1 shows the β-delayed ^{19}C neutron spectrum. The peaks were fitted simultaneously with 8 Gaussians, and again a broad Gaussian was used for the background.

Due to a ^{17}B beam contamination in the ^{19}C beam, the previously identified boron peaks were also found in the ^{19}C spectrum. Two weak peaks were assigned to β-delayed neutron decay of ^{19}N into ^{18}O + n. No literature data were found about branching ratios of ^{19}N β-decay, but the observed energies closely match those from ^{19}O levels seen in ^{18}O(n,n)^{18}O elastic neutron scattering (11). The γ-spectrum shows several peaks. All three neutron groups are strongly coincident with 115 keV γ-rays from the first excited state (12) in ^{18}O, and the 0.45 MeV peak is coincident also with 460 keV γ's, indicating a neutron decay to the second excited state in ^{18}O (587 keV (12)). Table 1 summarizes the results.

In the present study, several new levels in ^{17}C and ^{19}N were observed, but no evidence was found for levels close to the neutron threshold.

ACKNOWLEDGMENTS

We wish to thank the staff of the RIKEN cyclotron for the smooth and reliable operation of all equipment. GR acknowledges support through an

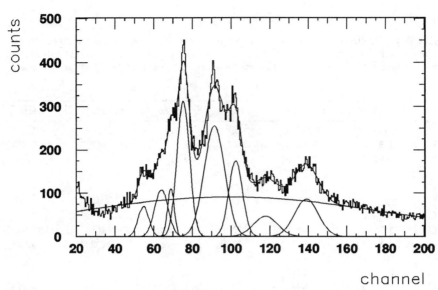

FIG. 1. Neutron time-of-flight spectrum for β-delayed neutron decay of ^{19}C.

Ohio State University postdoctoral fellowship. This work was supported in part by NSF grant PHY-9221669 and INT-9218241.

REFERENCES

1. Kajino, T., and Boyd, R.N., *Ap. J.* **359**, 267 (1990).
2. Thomas, D., Schramm, D.N., Olive, K.A., and Fields, B.D., *Ap. J.* **406**, 569 (1993).
3. Jedamzik, K., and Fuller, G.M., *submitted to Ap. J.* (1994).
4. Gilmore, G., Edvardsson, B., and Nissen, P.E., *Ap. J.* **378**, 17 (1991).
5. Duncan, D., Lambert, D.L., and Lemke, M., *Ap. J.* **401**, 584 (1992).
6. Jedamzik, K., Fuller, G.M., Mathews, G.J., and Kajino, T., *Ap. J.* **422**, 423 (1994).
7. Rauscher, T., Applegate, J.H., Cowan, J.J., Thielemann, F.K., and Wiescher, M., *Ap. J.* in press (1994).
8. Kajino, T., Mathews, G.J., and Fuller, G.M., *Ap. J.* **364**, 7 (1990).
9. Dufour, J.-P., Del Moral, R., Hubert, F., Jean, D., Pravikoff, M.S., Fleury, A., Mueller, A.C., Schmidt, K.-H., Sümmerer, K., Hanelt, E., Fréhaut, J., Beau, M., and Giraudet, G., *Phys. Lett. B* **206**, 195 (1988).
10. Ajzenberg-Selove, F., *Nucl. Phys.* **A 460**, 1 (1986).
11. Ajzenberg-Selove, F., *Nucl. Phys.* **A 475**, 1 (1987).
12. Pravikoff, M.S., Hubert, F., Del Moral, R., Dufour, J.-P., Fleury, A., Jean, D., Mueller, A.C., Schmidt, K.-H., Sümmerer, K., Hanelt, E., Fréhaut, J., Beau, M., Giraudet, G., *Nucl. Phys.* **A528**, 225 (1991).

Shapes of nuclei in the inner crust of a neutron star

Kazuhiro Oyamatsu* and Masami Yamada†

*Department of Energy Engineering and Science, Nagoya University, Furo-cho, Chikusa-ku, Nagoya 464-01 Japan
†Advanced Research Center for Science and Engineering, Waseda University, Okubo, Shinjuku-ku, Tokyo 169 Japan

Abstract. Nuclear shapes in the inner crust of a neutron star is discussed in the compressible liquid-drop model and the Thomas-Fermi model. We also calculate the shell energies of non-spherical nuclei. We conclude that non-spherical nuclei exist in the density region $(1.0-1.5) \times 10^{14}$ g/cm^3.

1. INTRODUCTION

About ten years ago two groups [1,2] pointed out, with the compressible liquid-drop model, that the stable nuclear shape may no longer be spherical at subnuclear densities just below the normal nuclear density. These liquid-drop studies suggested that, with increase of the matter density, the shape of the stable nucleus may change successively from sphere to cylinder, slab, cylindrical hole and spherical hole before going into uniform matter. Our group has been studying these non-spherical nuclei in the neutron-star matter, while the result of our early liquid-drop studies [2,3] is general enough to be also applied to the supernova matter.

We consider β-stable and charge-neutral matter composed of protons, neutrons and electrons at 0 K. In the neutron-star crust, nuclei form lattice to reduce the long-range Coulomb energy. Since the β-stability is achieved by making the matter extremely neutron rich, some neutrons drip out of nuclei. In this paper, we define "nucleus" as the domain occupied by protons. The nuclear shapes considered in this paper are sphere, cylinder, slab, cylindrical hole, and spherical hole. These shapes can be considered as the one-, two- or three-dimensional spheres which minimize the surface energies.

In the compressible liquid-drop model, a sharp nuclear surface is assumed, and each of the local proton and neutron number densities inside the nucleus and the neutron number density outside the nucleus is assumed to be constant. The nucleon distributions can be characterized by the following five parameters:
n^{in} : total nucleon number density inside the nucleus,

x : proton number fraction inside the nucleus,
n^{out} : nucleon (neutron) number density outside the nucleus,
u : volume fraction of the nucleus in a cell,
a : lattice constant.
We write the average energy density as

$$\frac{W}{a^3} = w_b(u, n^{in}, x, n^{out}) + \sigma(n^{in}, x, n^{out}) g(u, shape) a^{-1} + (e\, n^{in}\, x)^2 w_{Cl}(u, shape, lattice) a^2. \quad (1)$$

Here, W is the energy of a cell, w_b is the average bulk energy density in a cell. and σ is the surface tension. The last two terms in Eq. (1) are the surface and the Coulomb energies and depend on the nuclear shape. The values of the parameters in the nucleon distributions are chosen so as to minimize this energy density for each nuclear shape and lattice type. The stable nuclear shape is the one that gives the lowest energy among all nuclear shapes. By minimizing the average energy density with respect to the lattice constant a and then eliminating a, Eq. (1) can be written as

$$\frac{W}{a^3} = w_b(u, n^{in}, x, n^{out}) + \frac{3}{2}\sqrt[3]{2}\left[en^{in}x\sigma(n^{in}, x, n^{out})\right]^{2/3} G(u, shape, lattice), \quad (2)$$

with

$$G(u, shape, lattice) = g(u, shape)^{2/3} w_{Cl}(u, shape, lattice)^{1/3}. \quad (3)$$

Therefore the stable nuclear shape and lattice type are those that minimize the geometric factor G. The term "geometric factor" was used by Cooperstein and Baron [4] although the definition (3) is slightly different from and more general than theirs. By choosing the lattice type which minimizes the Coulomb energy for each nuclear shape, the stable nuclear shape is determined as a function of the volume fraction u as shown in FIGURE 1.

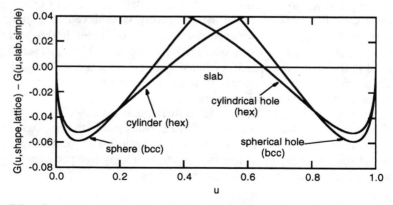

FIGURE 1. Geometric factor relative to that of slab. The stable nuclear shape is the one that gives the lowest G value.

2. THOMAS-FERMI CALCULATION

In order to go beyond the liquid-drop model, we performed a realistic Thomas-Fermi calculation for the neutron-star matter [5]. In this calculation, we paid special attention to the surface energy; calculations were performed with four different energy-density functions, and the neutron and proton distributions were parametrized independently of each other. In all the four models, we found that the stable nuclear shape changes successively as in the liquid-drop model in the density region $(1.0-1.5) \times 10^{14}$ g/cm^3 (0.06-0.09 fm^{-3}) although the nuclear surfaces are appreciably diffused as shown in FIGURE 2.

3. SHELL EFFECTS

It is known that shell effects are important in normal nuclei. Masses of normal nuclei are well represented by the sum of semiclassical energies and shell energies; we take the same strategy for the non-spherical nuclei and calculate shell energies of the non-spherical nuclei [6]. The non-spherical single-particle potential is constructed by making full use of the results of the Thomas-Fermi calculation. The spin-orbit splittings in the calculated single-particle energies are found to be much smaller than those of normal nuclei because the spin-orbit

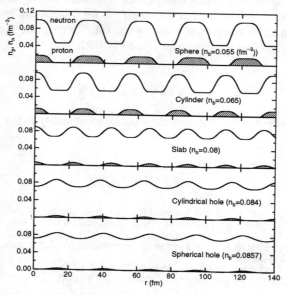

FIGURE 2. The neutron (upper) and proton (lower, shaded) distributions along the straight lines joining the centers of the nearest nuclei or hole. Calculations are made with model I, but the other models give similar results.

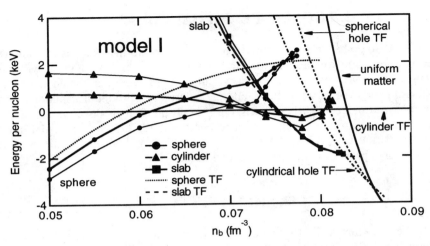

FIGURE 3. Energy per nucleon relative to the Thomas-Fermi value of the cylindrical nucleus in model I of Refs. 5 and 6. The thick lines correspond to the standard case while the thin lines correspond to the case with extremely large shell energies. TF stands for the Thomas-Fermi value.

potential, which is propotional to the gradient of nucleon distributions, is small. The shell energies are extracted from the single-particle energies in the single-particle potential. Neutron shell energies are one-order of magnitude smaller than proton shell energies. The band structure effects for protons are found to be small, and the proton shell energies are somewhat smaller than the energy difference between the successive nuclear shapes (see FIGURE 3).

4. CONCLUSION

We conclude that the non-spherical nuclei exist stably in the inner crust of a neutron star.

REFERENCES

1. Ravenhall, D.G., Pethick, C.J. and Wilson, J.R., Phys. Rev. Lett. **50**, 2066–2069 (1983).
2. Hashimoto, M., Seki, H. and Yamada, M., Prog. Theor. Phys. **71**, 320–326 (1984).
3. Oyamatsu, K., Hashimoto, M., and Yamada, M.,, Prog. Theor. Phys. **72** 373–375 (1984).
4. Cooperstein, J and Baron, E.A., in Supernovae, ed. Petschek, A. G. (Springer-Verlag, New York, 1990) pp. 213–266.
5. Oyamatsu, K., Nucl. Phys. **A561**, 431–452 (1993).
6. Oyamatsu, K. and Yamada, M., to be published in Nucl. Phys. **A**.

Capture Reactions at Astrophysically Relevant Energies

P. Mohr[†], V. Kölle[†], S. Wilmes[†], U. Atzrott[†], F. Hoyler[†],
C. Engelmann[†], H. Grollmuss[†], G. Staudt[†], H. Oberhummer[*]

[†] *Universität Tübingen, Physikalisches Institut, Tübingen, Germany*
[*] *Technische Universität, Institut für Kernphysik, Vienna, Austria*

> The Direct Capture model using double–folded α–nucleus potentials is applied to the study of several (α,γ) capture reactions.

THE POTENTIAL MODEL: DOUBLE–FOLDED α POTENTIALS

The large uncertainty in available cross section data of the $^{12}C(\alpha,\gamma)^{16}O$ reaction is one of the central problems in Nuclear Astrophysics. We present calculations for α capture cross sections for some light nuclei which have been done in the framework of a simple potential model. In these calculations double–folded α–nucleus potentials (1,2) are used being extracted from the study of quasi-molecular α-cluster states in the nuclei $^{16,17,18}O$, ^{19}F, and ^{20}Ne or from a fit to elastic phase shift data. In the excitation functions for elastic α scattering the quasi–bound α–cluster states are observed as potential resonances. Therefore, our calculations consistently describe the resonant and non-resonant behaviour of the (α,γ) cross sections. For that reason some certainty can be expected in the $^{12}C(\alpha,\gamma)^{16}O$ results.

ELASTIC SCATTERING DATA AND B(E$\mathcal{L}$)–VALUES

The α–nucleus potential is given by the Coulomb and the folding potential: $U(r) = V_C(r) + \lambda V_F(r)$. The fit parameter λ is used to adjust the strength of the potential to the energies of quasi-molecular α–cluster states or to experimental phase shift data. Wave functions for these α-cluster states are determined by the number of nodes N and the angular momentum number L which are correlated to the oscillator quanta Q: $Q = 2N + L = \sum_{i=1}^{4}(2n_i + l_i) = \sum_{i=1}^{4} q_i$. In $^{16,17,18}O$, ^{19}F, and ^{20}Ne two quasi-molecular rotational bands with $Q = 8$ and $Q = 9$ can be observed. In the case of $^{17}O = {}^{13}C \otimes \alpha$ and $^{19}F = {}^{15}N \otimes \alpha$ a small spin-orbit potential in the usual Thomas form ($\sim 1/r \cdot dV_F/dr$) has been added whose strength has been adjusted to reproduce the spin–orbit splitting. A weak parity dependence is observed for the Q=8 and Q=9 quasi-molecular bands in the nuclei $^{16,17,18}O$, ^{19}F, and ^{20}Ne. The folding potentials $\lambda V_F(r)$ for $^{16}O = {}^{12}C \otimes \alpha$

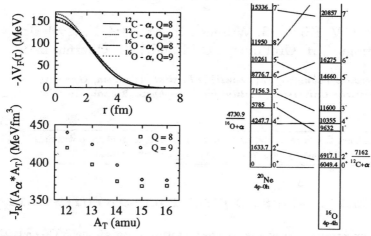

FIG. 1. Potentials and volume integrals of double–folded potentials for the nuclei ^{16,17,18}O, ^{19}F, and ^{20}Ne (left side); partial level scheme of ^{16}O and ^{20}Ne (right side)

and ^{20}Ne=^{16}O⊗ α and the systematic behaviour of the volume integrals per interacting nucleon pair $J_R/(A_\alpha \cdot A_T) = \lambda \int V_F(r) d\vec{r}/(A_\alpha \cdot A_T)$ are shown in Fig. 1. The level scheme for ^{16}O and ^{20}Ne is given in Fig. 1.

Using these potentials the following properties can be calculated:
1) Elastic scattering cross sections and phase shifts (s. Fig. 2).
2) reduced transition strengths $B(E\mathcal{L}) \sim |<u_{N_i L_i}(r)|r^\mathcal{L}|u_{N_f L_f}(r)>|^2$.
The experimental (3) and calculated $B(E\mathcal{L})$–values are given in Table 1.

TABLE 1. Experimental (3) and calculated $B(E\mathcal{L})$-values

	E_i (keV)	E_f (keV)	L_i	L_f	$\mathcal{L}$	$B_{exp.}(E\mathcal{L})$ (W.u.)	$B_{th.}(E\mathcal{L})$ (W.u.)
^{20}Ne	5785	0	1^-	0^+	1	(8.0±3.0) -6	3.93 -6
^{20}Ne	1634	0	2^+	0^+	2	(1.8±0.2) 1	1.47 1
^{20}Ne	4248	1634	4^+	2^+	2	(2.1±0.2) 1	2.10 1
^{20}Ne	8777	4248	6^+	4^+	2	(2.0±0.2) 1	1.97 1
^{20}Ne	11950	8777	8^+	6^+	2	(9.0±1.0) 0	9.50 0
^{20}Ne	7156	5785	3^-	1^-	2	(4.7±0.8) 1	4.17 1
^{20}Ne	4248	0	4^+	0^+	4	(2.3±0.5) 1	3.18 1
^{16}O	7117	0	1^-	0^+	1	(3.4±0.3) -4	3.35 -7
^{16}O	9632	0	1^-	0^+	1	(6.0±0.8) -5	4.13 -8
^{16}O	6917	0	2^+	0^+	2	(3.5±0.4) 0	3.29 0
^{16}O	6917	6049	2^+	0^+	2	(3.1±0.4) 1	2.75 1
^{16}O	10355	6917	4^+	2^+	2	(6.2±0.8) 1	4.03 1
^{16}O	10355	0	4^+	0^+	4	(6.0±2.0) 0	9.89 0

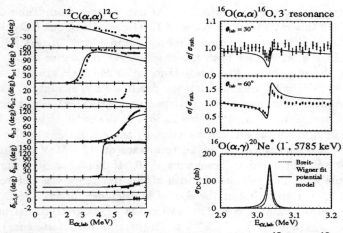

FIG. 2. Experimental and theoretical phase shifts of ^{12}C$(\alpha,\alpha)^{12}$C (4) (left side) and cross sections in the 3^- resonance of ^{16}O$(\alpha,\alpha)^{16}$O (5) (right side)

DIRECT CAPTURE (DC) CROSS SECTIONS

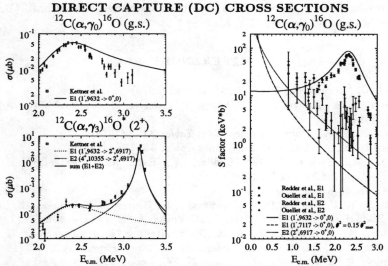

FIG. 3. Calculated DC cross section for ^{12}C$(\alpha,\gamma)^{16}$O compared to experimental data from refs. (6–8)

Furthermore, if the potentials are known, DC cross sections $\sigma^{DC} \sim C^2 S \cdot |<u_{bound}(r)|T_{E\mathcal{L}/M\mathcal{L}}|\chi_{scatt.}(r)>|^2$ can be calculated (9,10). But there are two problems: 1) E1 transitions are isospin-suppressed in N=Z nuclei. The transition strengths depend strongly on small T=1 admixtures. 2) Only few spectroscopic factors are available between the $1p$ and $2s1d$ shell.

Therefore, we have normalized the calculated capture cross sections with the ratio of the experimental to the calculated $B(E\mathcal{L})$ values which are often available from electron scattering experiments. As can be seen from Table 1, for all E2 transitions this ratio is in the order of 1. The results of our DC calculations, normalized in this way, are shown in Figs. 2 and 3 for ^{16}O$(\alpha,\gamma)^{20}$Ne and ^{12}C$(\alpha,\gamma)^{16}$O. In the latter case both the observed 4$^+$ resonance at 3.1 MeV and the tail of the sub–threshold 2$^+$ resonance at -0.245 MeV are well reproduced. $S_{E2}(300\ \mathrm{keV}) = 82.6\pm10$ keVb is the result of this calculation. For E1 transitions the ratio mentioned above is a measure for the T=1 admixtures in the wave functions. For ^{12}C$(\alpha,\gamma)^{16}$O this ratio is in the order of 10^3 (s. Table 1). The normalization of the DC cross section by the aid of the B(E1)–values gives good results with respect to the 1$^-$ resonance at $\sim$2.4 MeV (s. Fig. 3). The tail of the sub–threshold 1$^-$ resonance at -0.045 MeV has been normalized in the same way. The calculated width of this resonance was reduced by $\theta_\alpha^2 = 0.15\pm0.1$ (7). Depending on the sign of interference for the two 1$^-$ resonances one obtains a value of $S_{E1}(300\ \mathrm{keV}) = 128.4\pm45$ keVb (constructive interference) or $S_{E1}(300\ \mathrm{keV}) = 18.6^{+45}_{-18}$ keVb (destructive interference).

ACKNOWLEDGMENTS

We would like to thank the Deutsche Forschungsgemeinschaft (DFG–projects Sta290/2-2, Sta290/3-1, Mu705/3-1).

REFERENCES

1. Abele, H., Staudt, G., *Phys. Rev.* **C47**, 742 (1993)
2. Kobos, A. M., Brown, B. A., Lindsay, R., Satchler, G. R., *Nucl. Phys.* **A425**, 205 (1984)
3. Endt, P. M., *At. Data and Nucl. Data Tables* **23**, 3 (1979)
   Endt, P. M., *At. Data and Nucl. Data Tables* **55**, 171 (1993)
4. Plaga, R., Becker, H. W., Redder, A., Rolfs, C., Trautvetter, H. P., *Nucl. Phys.* **A465**, 291 (1987)
5. Knee, H., to be published
6. Kettner, K. U., Becker, H. W., Buchmann, L., Görres, J., Kräwinkel, H., Rolfs, C., Schmalbrock, P., Trautvetter, H. P., Vlieks, A., *Z. Phys.* **A308**, 73 (1982)
7. Redder, A., Becker, H. W., Rolfs, C., Trautvetter, H. P., Donoghue, T. R., Rinckel, T. C., Hammer, J. W., Langanke, K., *Nucl. Phys.* **A462**, 385 (1987)
8. Ouellet, J. M. L., Evans, H. C., Lee, H. W., Leslie, J. R., MacArthur, J. D., Latchie, W., Mak, H.-B., Skensved, P., Whitton, J. L., Zhao, X., Alexander, T. K., *Phys. Rev. Lett.* **69**, 1896 (1992)
9. Mohr, P., Abele, H., Zwiebel, R., Staudt, G., Krauss, H., Oberhummer, H., Denker, A., Hammer, J. W., Wolf, G., *Phys. Rev.* **C48**, 1420 (1993)
10. Kim, K. H., Park, M. H., Kim, B. T., *Phys. Rev.* **C23**, 363 (1987)

Solar Abundances and s-Process by Combined Burning of Two Neutron Sources

Hermann Beer*, B. Spettel+, H. Palme++

* Kernforschungszentrum Karlsruhe, IK III, P.O. Box 3640, 76021 Karlsruhe
+ Max-Planck-Institut für Chemie, Saarstraße 23, 55128 Mainz
++ Universität Köln, Mineralogisch-Petrol. Inst., Zülpicherstr. 49b, 50674 Köln

Abstract. A new set of solar abundances is presented. With these input data refined s-process model calculations were carried out based on the combined burning of two neutron sources. The s-, r-, and p-process abundance distributions are improved. The s-Ba detected in SiC grains of meteorites was found to be similar to solar system s-material.

INTRODUCTION

The solar abundances of the chemical elements are indispensable as carriers of signatures of nucleosynthetic processes in stellar environments. This is especially true for the elements heavier than iron. Therefore there is a special interest in the decomposition of heavy solar abundances into s-, r-, and p-process components. Improvements are achieved by a refined understanding of s-process synthesis.

SOLAR ABUNDANCES

For the new compilation of solar abunances (1) the most important source of information was found from new analyses of the CI carbonaceous chondrites Orgueil, Ivuna, Alais (Table 1). Results from other sources are given in italics. Compared with the previous compilation of Anders and Grevesse (2) the largest changes concern S(-19%), Hg(17%) Se(12%), P(-11%), Au(8%), and Sr(-7.8%). The changes in Se and Sr lead to more stringent constraints in the weak s-process analysis, the change in S makes s-process synthesis of ^{36}S more difficult, and in Hg allows for realistic estimates of the p-abundances of ^{196,198}Hg.

The heavy elements (A>56) were decomposed into s-process and residual r-process abundances. The plots of the s-process σN curve and the r- process residuals showed deficiencies and irregularities at six different mass regions (1). The deficiencies arise from the uncertainties in subtracting a major s-process component from the solar abundance at the mass numbers 63-71, 86-94, and 137-138 and require a refinement of the weak and main s-process models with improved cross sections (3), especially an improved ^{138}Ba value (4), and solar abundances, especially improved Se, and Sr values. The irregularities are a manifestation of a significant p-process component.

Table 1. Elemental abundances relative to $N(Si)=10^6$, mainly from CI meteorites

Element		N	Accuracy (%)	Element		N	Accuracy (%)
1	H	2.79E+10		44	Ru	1.86	10
2	He	2.72E+09		45	Rh	0.342	20
3	Li	56.4	10	46	Pd	1.37	10
4	Be	0.73	10	47	Ag	0.480	10
5	B	21.2	10	48	Cd	1.59	10
6	C	1.01E+07		49	In	0.178	10
7	N	3.13E+06		50	Sn	3.72	10
8	O	2.38E+07		51	Sb	0.287	10
9	F	805	15	52	Te	4.68	10
10	Ne	3.44E+06		53	I	0.90	20
11	Na	5.70E+04	5	54	Xe	4.1	
12	Mg	1.04E+06	3	55	Cs	0.372	5
13	Al	8.43E+04	3	56	Ba	4.61	5
14	Si	1.00E+06	3	57	La	0.464	5
15	P	9.38E+03	10	58	Ce	1.20	5
16	S	4.31E+05	10	59	Pr	0.180	10
17	Cl	5176	15	60	Nd	0.864	5
18	Ar	1.01E+05		62	Sm	0.269	5
19	K	3658	5	63	Eu	0.100	5
20	Ca	6.24E+04	3	64	Gd	0.341	5
21	Sc	34.5	3	65	Tb	0.0620	10
22	Ti	2421	5	66	Dy	0.411	5
23	V	280	5	67	Ho	0.0904	10
24	Cr	1.34E+04	3	68	Er	0.261	5
25	Mn	9251	3	69	Tm	0.0398	10
26	Fe	8.58E+05	3	70	Yb	0.251	5
27	Co	2257	3	71	Lu	0.0382	5
28	Ni	4.82E+04	3	72	Hf	0.158	5
29	Cu	542	10	73	Ta	0.020	10
30	Zn	1299	10	74	W	0.136	7
31	Ga	36.6	5	75	Re	0.0541	7
32	Ge	118	10	76	Os	0.672	5
33	As	6.35	5	77	Ir	0.628	3
34	Se	70.8	5	78	Pt	1.34	10
35	Br	11.5	10	79	Au	0.203	5
36	Kr	56		80	Hg	0.41	20
37	Rb	7.14	5	81	Tl	0.184	10
38	Sr	21.8	5	82	Pb	3.21	10
39	Y	4.64	5	83	Bi	0.140	15
40	Zr	11.2	5	90	Th	0.0338	5
41	Nb	0.696	5	92	U	0.0086	10
42	Mo	2.54	5				

Table 2. Weak and main s-process parameters. Besides fraction of iron seed and exposure, temperature kT, neutron density n_n, exposure per pulse $\Delta\tau$, and overlap factor $r=\exp(-\Delta\tau/\tau_0)$ are given.

Weak component	Fraction of iron seed	Exposure τ (mbarn^{-1})	kT (keV)	n_n (cm^{-3})		
	0.00313	0.110	25	10^6		
	0.00151	0.131	25	10^6		
		0.135	86	$2\ 10^{10}$		

Main Component	Fraction of iron seed	Av. exposure τ_0 (mbarn^{-1})	kT (keV)	n_n (cm^{-3})	$\Delta\tau$ (mbarn^{-1})	r
CALC.1	0.001816	0.131	12	$7.5\ 10^8$	0.143	0.33
		$4.15\ 10^{-4}$	26	$5\ 10^8$	$4.54\ 10^{-4}$	
CALC.2	0.001384	0.131	12	$1.5\ 10^8$	0.0287	0.80
		$4.15\ 10^{-4}$	26	$2\ 10^8$	$0.909\ 10^{-4}$	

WEAK AND MAIN s-PROCESS

The weak s-process was studied as a superposition of two synthesis processes: combined exposures at low temperature, kT, and neutron density, n_n, and at high kT and n_n plus an exposure at low kT and n_n alone (Table 2). The astrophysical parameters were derived from the s-only nuclei but also from the abundance residuals of non s-only isotopes. The final results are shown in Fig. 1. and given in Table 2. The oversubtractions at A=63-71 are removed. It was found that ^{70}Ge must have a p-process contribution of $\approx 15\%$.

The main s-process was formulated as the combined pulsed burning of two neutron sources. The model is a special case of a more general double pulse model (5). In Fig. 2 two calculations in comparison with the empirical data are shown. The calculations are parameter fits (Table 2) to the s-only isotopes on the unique synthesis path and to s-only isotopes on the branched path which are sensitive to duration and strength of the neutron pulse and additionally often to temperature as well. Especially sensitive is the synthesis of ^{152}Gd, e.g. in CALC.1 after irradiation 1 and 2(unpulsed) only 1.8%, then 10% of the solar ^{152}Gd is formed but taking into account the pulse conditions we arrive at 50%. The improvement of the new calculations compared with old calculations is obvious (Fig. 2). The s-only ^{136}Ba isotope is now correctly reproduced and for isotopes ^{137,138}Ba the s-process overproduction is avoided. The s-process material detected in SiC grains of meteorites is found to be similar to solar s-process. The evidence is provided by s-Ba, especially by the ratio ^{138}Ba/^{136}Ba which is sensitive to the

Table 3. Ba isotope ratios and s-Ba from SiC grains

	CALC. 1	CALC. 2	SiC grains Ref.(6)	SiC grains Ref.(7)
134/136	0.17	0.30	0.33	0.34
135/136	0.20	0.17	0.16	0.13
137/136	0.79	0.78	0.72	0.68
138/136	6.69	6.68	6.48	6.22

average exposure (Table 3). The ^{136}Xe/^{130}Xe ratio bears similar evidence: $3.1\ 10^{-3}$ and $2.3\ 10^{-3}$ for CALC.1 and 2, respectively, compared with $\approx 7.1\ 10^{-3}$ from SiC grains.

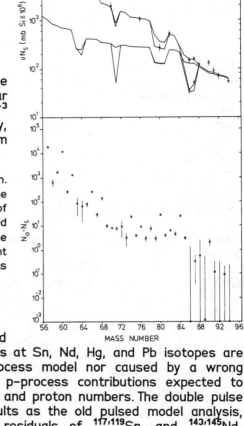

Fig. 1. Weak s-process calculation. The σN curve is shown above. The higher curve is the composite of weak and main s-process adjusted to the indicated empirical data. The lower curve is the main component only. Below are shown the residuals after s-process subtraction.

p-PROCESS DISTRIBUTION

The irregularities in the s- and r-process abundance distributions at Sn, Nd, Hg, and Pb isotopes are neither dependent on the s-process model nor caused by a wrong solar abundance but significant p-process contributions expected to be pronounced at magic neutron and proton numbers. The double pulse model analysis gives similar results as the old pulsed model analysis, and the s-process subtracted residuals of ^{117,119}Sn, and ^{143,145}Nd, where a negligible p-contribution is expected (< ^{113}In, ^{115}Sn abundance), show no inconsistencies. The p-process distribution found is shown in Fig. 3. At Pb the assumption of a significant p-contribution in ^{204}Pb is as important as the double pulse model for the analysis of the s-process termination and cosmochronology with radiogenic ^{235,238}U in Pb (8).

CONCLUSIONS

The present average neutron exposure of the main component $\tau_0 = 0.131$ mbarn^{-1} is significantly smaller than the corresponding value

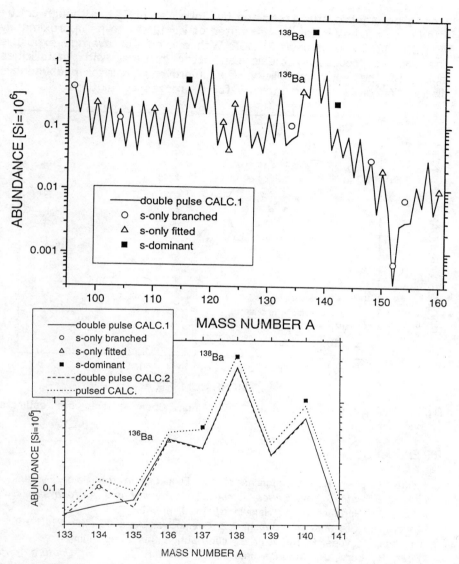

Fig. 2 Main s-process calculation by a double pulse model. The adjustment is to the s-only isotopes. It is remarkable how well s-only ^{136}Ba is reproduced. It is a consequence of the new double pulse model, burning at kT=12 keV and improved cross sections (3,4). On an extended scale (below) the peak at Ba is shown in comparison with the previous old pulsed model in which ^{136}Ba as well as ^{137,138}Ba of mixed s- and r-process origin are overproduced. ^{134}Ba on the branched synthesis path requires neutron densities as used in CALC.2.

from a parametric study reported by Gallino et al. (9). Although calculated at kT=30 keV it can be estimated at kT=12 keV to be approximately $\tau_0 \approx 0.28 \sqrt{12/30} = 0.18$ mbarn^{-1} (10). With our smaller average exposure the solar s-process synthesis can occur in stars with metallicities closer to the solar metallicity. Our improved empirical p-abundance distribution gives new constraints for p-process calculations.

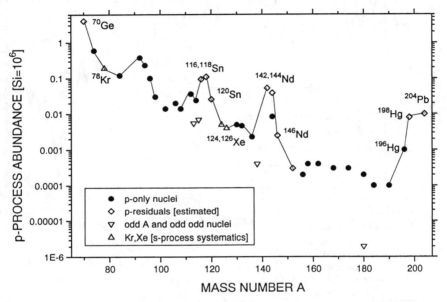

Fig. 3 The p-process abundance distribution. Open symbols are estimated abundances.

REFERENCES

1. Palme, H., Beer, H., "Abundances of the Elements in the Solar System" in *Landolt Börnstein, New Series, Group VI, Astronomy and Astrophysics*, Subvolume 3a, Springer Verlag Heidelberg, 1993, pp. 196-221
2. Anders, E., Grevesse, N., Geochim. & Cosmochim. Acta **53**, 197-214(1989)
3. Wisshak, K., Voss, F., Käppeler, F., contribution this symposium
4. Beer, H., Corvi, F., Athanassopulos, K., *8th Int. Symp. on Capture γ-Ray Spectr. and Related Topics*, Fribourg, 20-24 Sept. 1993, ed. J. Kern, World Sientific, Singapore 1994, pp. 698-705
5. Beer, H., *Ap. J.* **379**, 409-419(1991)
6. Ott, U., Begemann, F., *Ap. J.* **353**, L57-L60(1990)
7. Prombo, C. A., et al., *Ap. J.* **410**, 393-399(1993)
8. Corvi, F., Mutti, P., Athanassopulos, K., Beer, H., contribution this symposium
9. Gallino, R., Raiteri, C. M., Busso, M., *Ap. J.* **410**, 400-411(1993)
10. Gallino, R., private communication

A Direct Measurement of the ^{19}Ne(p,γ)^{20}Na Reaction

R.D. Page[1], G. Vancraeynest[2], A.C. Shotter[1], M. Huyse[2],
C.R. Bain[1], F. Binon[3], R. Coszach[4], T. Davinson[1], P. Decrock[2],
Th. Delbar[4], P. Duhamel[3], M. Gaelens[2], W. Galster[4], P. Leleux[4],
I. Licot[4], E. Liénard[4], P. Lipnik[4], C. Michotte[4], A. Ninane[4],
P.J. Sellin[1], Cs. Sükösd[5], P. Van Duppen[2], J. Vanhorenbeeck[3],
J. Vervier[4], M. Wiescher[6] and P.J. Woods[1]

[1]Department of Physics and Astronomy, University of Edinburgh, The King's Buildings, Edinburgh EH9 3JZ, United Kingdom
[2]Instituut voor Kern- en Stralingsfysika, Katholieke Universiteit Leuven, Leuven, B-3001 Belgium
[3]Institut d'Astronomie, d'Astrophysique et de Géophysique, Université Libre de Bruxelles, B-1050 Bruxelles, Belgium
[4]Institut de Physique Nucléaire, Université Catholique de Louvain, Louvain-la-Neuve, B-1348 Belgium
[5]Institute of Nuclear Techniques, Technical University of Budapest, H-1521 Budapest, Hungary
[6]Department of Physics, University of Notre Dame, Notre Dame, Indiana 46556, USA

Abstract. The ^{19}Ne(p,γ)^{20}Na reaction may be of considerable influence for the reaction flow between the CNO and the NeNa mass region in high temperature hydrogen burning conditions. The proposed low energy resonance at 0.447 MeV has been measured in inverse kinematics with the Louvain-la-Neuve ^{19}Ne beam, using novel activation techniques. An upper limit to the resonance strength has been determined and the implications for the spin assignment of the 2.646 MeV state in ^{20}Na as well as for the stellar reaction rate are discussed.

In a previous contribution to a conference (1), the motivation of this work (2) and the experimental techniques have been described at length. Only the latter will be briefly recalled in this contribution. Intense ^{19}Ne beams of 11.4 MeV energy, up to 150 ppA, were used to bombard polyethylene targets, about 150 μg/cm^2 thick. Two techniques were used sequentially to detect α-particles from ^{20}Ne levels fed by a 20 % strength β-decay branch of ^{20}Na.

In the first method, ^{20}Na nuclei recoiling from the target are implanted into an aluminium stopping surface mounted on a perspex disc which is rotated by 180° every 870 ms, bringing then the irradiated surface in front of two double-sided silicon strip detectors (3). The strip detectors comprise 48 300 μm x 16 mm strips on each face, their thickness being 34 or 67 μm. Each pixel has thus a very small

small efficiency for the β-particles (up to 10^9 ^{19}Ne nuclei per s are also implanted with the ^{20}Na nuclei). Energy spectra are reconstructed for the different pixels and then summed. The global efficiency of this method, including branching factor, disc rotation time, solid angle of the strip detectors, is about 0.6 %.

In the second method, recoiling ^{20}Na nuclei are slowed down in an aluminium degrader and implanted in a thin aluminium stopper mounted on one arm of a windmill structure which is rotated by 90° every 1.03 s, presenting successively the implanted zone in front of pairs of solid state nuclear track detectors (one on each side). The three pairs of track detectors provide a lifetime signature. The efficiency of the track detectors for α-detection was determined versus energy and angle of incidence using beams of Rutherford backscattered α-particles. Tracks induced by α-particles are revealed when the detectors are subsequently etched (4). Irradiations with sources have demonstrated the insensitivity of these detectors to β-particles. The global efficiency of this method is about 1.6 %.

Both methods were first tested by measuring the ^{19}Ne(d,n)^{20}Na reaction at 11.4 MeV; both agreed on a cross section of 1.4 ± 0.2 mb. Moreover this measurement was needed to subtract the deuterium contribution from the CH$_2$ target. For the ^{19}Ne(p,γ) measurement, each method was able to give a 90 % confidence level upper limit for the 447 keV resonance strength, i.e. 20 meV for the strip detectors and 26 meV for the track detectors. Combining both results gives a 90 % UL of 18 meV, which is inconsistent with a recent calculation (80 meV) based on a 3$^+$ assignment for the resonant level in ^{20}Na (5); the lifetime of the proposed analog state in ^{20}F, at 2.996 MeV, has been recently measured (6), an upper limit of 15 fs (90 % CL) having been determined. This sets a lower limit of 20 meV to the resonance strength of the 3$^+$ state in ^{20}Na if one accepts the proton widths calculated in (5). A 3$^+$ state is thus excluded with more than 90 % confidence by the present measurement. On the other hand, our 18 meV upper limit is consistent wit the 6 meV strength calculated by Lamm et al (7) on the basis of a 1$^+$ spin and an intruder character for the state.

Before a significant stellar reaction rate can be calculated for the ^{19}Ne(p,γ)^{20}Na reaction, a lower limit for the present state as well as the γ-widths of the higher lying levels should be determined. The latter item is presently studied at the Louvain-la-Neuve facility.

REFERENCES

1. Page R.D. et al., "Measurement of the ^{19}Ne(p,γ) reaction rate using radioactive beams", in *Proceedings of the Third Int. Conf. on Radioactive Nuclear Beams, Lansing, Michigan, USA,* 1983, D.J. Morrissey (Ed.), Editions Frontières, Gif-sur-Yvette, France, p. 489.
2. Wallace R.K. and Woosley S.E., *Astrophys. J.* **184**, 493 (1973).
3. Sellin P.J. et al., *Nucl. Instr. and Meth. in Phys. Res.* **A311**, 217 (1992).
4. Vanmarcke H. et al., *Nucl. Tracks* **12**, 689 (1986).
5. Brown B.A. et al., *Phys. Rev.* **C48**, 1456 (1993).
6. Görres J. et al., *Bull. Am. Phys. Soc.* **38**, 1844 (1993).
7. Lamm L.O. et al., *Nucl. Phys.* **A510**, 503 (1990).

Neutron Producing Reactions in Stars*

A. Denker, H.W. Drotleff†, M. Große, H. Knee, R. Kunz, A. Mayer,
R. Seidel, M. Soiné, A. Wöhr‡, G. Wolf and J.W. Hammer

Institut für Strahlenphysik der Universität Stuttgart
† *TÜV Hannover–Sachsen-Anhalt*
‡ *Institut für Kernchemie, Mainz*

Abstract: The nucleosynthesis of elements heavier than iron occurs mainly by neutron capture in the s– and r– process. The possible neutron sources have been investigated within present limits of sensitivity. Cross section, resonance parameters, S–factors and reaction rates have been obtained.

SURVEY

Using methods already described [1, 2, 3, 4, 5], state of the art measurements for $^9\text{Be}(\alpha,\text{n})^{12}\text{C}$, $^{13}\text{C}(\alpha,\text{n})^{16}\text{O}$, $^{17}\text{O}(\alpha,\text{n})^{20}\text{Ne}$, $^{18}\text{O}(\alpha,\text{n})^{21}\text{Ne}$, $^{21}\text{Ne}(\alpha,\text{n})^{24}\text{Mg}$ and $^{22}\text{Ne}(\alpha,\text{n})^{25}\text{Mg}$ have been performed. For the reactions $^{25}\text{Mg}(\alpha,\text{n})^{28}\text{Si}$ and $^{26}\text{Mg}(\alpha,\text{n})^{29}\text{Si}$ further experiments are in preparation. A sensitivity limit at the cross section of 10^{-11} b could be achieved.

The He$^+$ beam of at least 100 μA on the target was provided by the 4 MeV DYNAMITRON of the Institut für Strahlenphysik der Universität Stuttgart [6]. The beam energy was determined before each experiment using known narrow resonances. Solid state targets designed for high beam power were used for ^9Be, ^{13}C, ^{25}Mg and ^{26}Mg. For the isotopes ^{17}O, ^{18}O, ^{21}Ne and ^{22}Ne the Stuttgart gas target facility RHINOCEROS was employed. The background was determined not only by gas in–gas out measurements, but also using the ^{20}Ne isotope as target, which shows realistic background, having no open (α,n) channel at low energies. The detector was designed especially for this experiments with an *absolute* efficiency of up to 38%. The efficiency was determined with a calibrated neutron source and by Monte Carlo calculations.

Resonance parameters have been determined as far as possible [3, 7, 8].

Reaction	Excitation function S–factor	lowest σ	Resonances obs. (new)	$\omega\gamma$
$^9\text{Be}(\alpha,\text{n})^{12}\text{C}$	0.36 MeV $\leq \text{E}_\alpha \leq$ 3.6 MeV	$5\cdot 10^{-7}$ b	8 (3)	6
$^{13}\text{C}(\alpha,\text{n})^{16}\text{O}$	0.35 MeV $\leq \text{E}_\alpha \leq$ 1.4 MeV	$6\cdot 10^{-11}$ b	3 (-)	-
$^{17}\text{O}(\alpha,\text{n})^{20}\text{Ne}$	0.7 MeV $\leq \text{E}_\alpha \leq$ 2.8 MeV	$7\cdot 10^{-10}$ b	25 (18)	17
$^{18}\text{O}(\alpha,\text{n})^{21}\text{Ne}$	0.85 MeV $\leq \text{E}_\alpha \leq$ 2.7 MeV	$2\cdot 10^{-9}$ b	15 (9)	12
$^{21}\text{Ne}(\alpha,\text{n})^{24}\text{Mg}$	0.55 MeV $\leq \text{E}_\alpha \leq$ 2.7 MeV	$2\cdot 10^{-10}$ b	37 (28)	32
$^{22}\text{Ne}(\alpha,\text{n})^{25}\text{Mg}$	0.55 MeV $\leq \text{E}_\alpha \leq$ 2.35 MeV	$4\cdot 10^{-10}$ b	20 (1)	20
$^{25}\text{Mg}(\alpha,\text{n})^{28}\text{Si}$	1.9 MeV $\leq \text{E}_\alpha \leq$ 3.3 MeV	–	- (-)	-
$^{26}\text{Mg}(\alpha,\text{n})^{29}\text{Si}$	1.0 MeV $\leq \text{E}_\alpha \leq$ 3.6 MeV	–	- (-)	-

*supported by the Deutsche Forschungs–Gesellschaft under contract No. Ha 962/12

For brevity only the reaction rates as final results can be given in this paper. For details of the measurements we refer to [1, 2, 3, 4, 5, 7, 8].

REACTION RATES

The astrophysical reaction rates have been calculated by numerical integration using the code CALCRATE [3]. Analytical expressions have been fitted to the data.

$$^9\text{Be}(\alpha,\text{n})^{12}\text{C}$$

Our measurements determine the reaction rate for $0.2 \leq T_9 \leq 4$ [5, 8]. For higher temperatures, cross sections of Schmidt et al. [9] and v.d. Zwan & Geiger [10] were used. The fit is valid for $0.2 \leq T_9 \leq 10$.

$$N_A \langle \sigma v \rangle = \sum_{i=1}^{6} \frac{a_i}{T_9^{3/2}} \exp\left(\frac{-b_i}{T_9}\right)$$

The coefficients are:

i	1	2	3	4	5	6
a_i	70.01	$8.853 \cdot 10^4$	$7.706 \cdot 10^5$	$5.517 \cdot 10^8$	$4.412 \cdot 10^9$	$5.188 \cdot 10^{10}$
b_i	2.447	4.138	6.230	13.18	18.64	37.37

$$^{13}\text{C}(\alpha,\text{n})^{16}\text{O}$$

For $T_9 \leq 0.4$, the reaction rate had to be extrapolated [5, 7], the following fit gives the most probable value. The fit is valid for $0.01 \leq T_9 \leq 1$.

$$N_A \langle \sigma v \rangle = \frac{1016.988}{T_9^{3/2}} \cdot \exp\left(-\frac{6.259}{T_9}\right) + \frac{3.474 \cdot 10^5}{T_9^{3/2}} \cdot \exp\left(-\frac{8.430}{T_9}\right)$$
$$+ \frac{6.788 \cdot 10^{15}}{T_9^2} \cdot \exp\left(-\frac{33.093}{T_9^{1/3}}\right) \left(1 + 0.485\, T_9^{1/3} - 7.948\, T_9^{2/3} + 10.725\, T_9\right)$$

$$^{17}\text{O}(\alpha,\text{n})^{20}\text{Ne}$$

Our measurements determine the reaction rate for $0.35 \leq T_9 \leq 2.5$ [3, 4]. For higher temperatures the reaction rates of CF88 [11] were used. Below $T_9 \leq 0.35$ extrapolations were necessary, the fit gives the most probable value. The fit is valid for $0.01 \leq T_9 \leq 10$.

$$N_A \langle \sigma v \rangle = \sum_{i=1}^{9} \frac{a_i}{T_9^{3/2}} \exp\left(\frac{-b_i}{T_9}\right)$$

with the coefficients:

i	1	2	3	4	5
a_i	$3.5693 \cdot 10^{-21}$	$1 \cdot 10^{-13}$	$3.325 \cdot 10^{-8}$	$1.3 \cdot 10^{-3}$	15.6
b_i	1.276	2.03	2.975	4.521	6.91
i	6	7	8	9	
a_i	3893	$2.274 \cdot 10^6$	$2.06 \cdot 10^8$	$1.9 \cdot 10^9$	
b_i	8.88	14.51	21.4	29	

$^{18}O(\alpha,n)^{21}Ne$

Our measurements determine the reaction rate from threshold up to $T_9 = 2.5$ [3, 4]. To calculate reaction rates for higher temperatures, the cross section was extrapolated using data of Bair and Willard [12] and Hansen et al. [13]. The fit is valid for $0.025 \leq T_9 \leq 10$. A new ratio of (α,γ) to (α,n) has also been calculated.

$$N_A \langle \sigma v \rangle = \sum_{i=1}^{7} \frac{a_i}{T_9^{3/2}} \exp\left(\frac{-b_i}{T_9}\right)$$

The coefficients are:

i	1	2	3	4	5	6	7
a_i	1.814	104.6	$3.225 \cdot 10^4$	$2.158 \cdot 10^6$	$7.8 \cdot 10^8$	$1.1 \cdot 10^{10}$	$5 \cdot 10^{10}$
b_i	8.2083	9.017	11.15	14.73	23.61	36.1	48.5

$^{21}Ne(\alpha,n)^{24}Mg$

Our experiments determine the reaction rate for $0.6 \leq T_9 \leq 2$ [3, 4]. To higher temperatures the reaction rates of CF88 [11] were used. To lower temperatures extrapolations were made, the fit gives the most probable value and is valid for $0.025 \leq T_9 \leq 10$.

$$N_A \langle \sigma v \rangle = \sum_{i=1}^{9} \frac{a_i}{T_9^{3/2}} \exp\left(\frac{-b_i}{T_9}\right)$$

with the coefficients:

i	1	2	3	4	5
a_i	$2.85 \cdot 10^{-24}$	$2 \cdot 10^{-15}$	$3 \cdot 10^{-8}$	0.4	3170
b_i	1.7	2.705	4.4	7.98	12.423
i	6	7	8	9	
a_i	$1.852 \cdot 10^6$	$3.162 \cdot 10^8$	$3.49 \cdot 10^{10}$	$6.126 \cdot 10^{11}$	
b_i	17.62	24.749	41.406	67.69	

^{22}Ne$(\alpha,n)^{25}$Mg

For temperatures below $T_9 = 0.5$ extrapolations were necessary [1, 7]. The fit gives the most probable value and is valid for $0.15 \leq T_9 \leq 1.6$. There was no evidence for a resonance at 630 keV within the limits of our data.

$$N_A\langle\sigma v\rangle = 1.266 \cdot 10^{21} \cdot \frac{t^{5/6}}{g \cdot T_9^{3/2}} \cdot \exp\left(-\left(\frac{0.203}{t}\right)^{17.384}\right) \cdot \exp\left(-\frac{50.899}{t^{1/3}}\right)$$
$$+ \frac{7.425 \cdot 10^{-3}}{g} \cdot \exp\left(-\frac{3.989}{T_9^{4/3}}\right) - \frac{1.969 \cdot 10^{-14}}{g} \cdot \exp\left(-\frac{7.902}{T_9^{3/4}}\right)$$
$$- \frac{1.094 \cdot 10^4}{g} \cdot \exp\left(-\frac{35.934}{T_9^{3/2}}\right) + \frac{2.624}{g} \cdot \exp\left(-\frac{7.654}{T_9}\right)$$

with

$$g = 1 + 129.22 \cdot \exp\left(-\frac{5.056}{T_9}\right) \;;\; t = \frac{T_9}{1 - 0.027 \cdot T_9}$$

REFERENCES

[1] H.W. Drotleff, PhD thesis, Universität Bochum und Stuttgart, 1992
[2] A. Denker, H.W. Drotleff, M. Große, J.W. Hammer, H. Knee, R. Seidel, G. Wolf, *Reaction Rates of* $^{21}Ne(\alpha,n)^{24}Mg$, $^{17}O(\alpha,n)^{20}Ne$, $^{18}O(\alpha,n)^{21}Ne$ in Proceedings of the 7th workshop on Nuclear Astrophysics, pp123-131, Eds. W. Hillebrand & E. Müller, MPA/P7, München, 1993
[3] A. Denker, PhD thesis, Universität Stuttgart, 1994
[4] A. Denker, H.W. Drotleff, M. Große, J.W. Hammer, H. Knee, R. Kunz, G. Wolf, *Reaction Rates of* $^{21}Ne(\alpha,n)^{24}Mg$, $^{17}O(\alpha,n)^{20}Ne$ and $^{18}O(\alpha,n)^{21}Ne$ in Proceedings of the European Workshop on Heavy Element Nucleosynthesis, pp 145-150, Eds. Zs. Samorjai & Zs. Fülöp, Budapest, 1994
[5] R. Kunz, A. Denker, H.W. Drotleff, M. Große, H. Knee, S. Küchler, R. Seidel, M. Soiné, J.W. Hammer, *Neutron Sources in Nuclear Astrophysics* in Proceedings of the 4th International Conference on Applications of Nuclear Techniques: Neutrons and their Applications, in press
[6] J.W. Hammer et al., Nucl. Inst. Meth. **161**, 189 (1979)
[7] H.W. Drotleff, A. Denker, H. Knee, M. Soiné, G. Wolf, J.W. Hammer, U. Greife, C.Rolfs, H.P. Trautvetter, Astrophys. J. **414**, 735-739 (1993)
[8] R. Kunz, S. Barth, A. Denker, H.W. Drotleff, J.W. Hammer, H. Knee, A. Mayer, *Reaction Rates of* $^9Be(\alpha,n)^{12}C$ in Proceedings of the European Workshop on Heavy Element Nucleosynthesis, pp 134-139, Eds. Zs. Samorjai & Zs. Fülöp, Budapest, 1994
[9] D. Schmidt, R. Böttger, H. Klein, R. Nolte, Report PTB-N-7 and Report PTB-N-8, Physikalisch - Technische Bundesanstalt, 1992
[10] L.v.d. Zwan, K.W.Geiger, Nucl. Phys. **A 152**, 481 (1970)
[11] G.R. Caughlan, W.A. Fowler, At. Data Nucl. Data Tables **40** No.2, 284-334 (1988)
[12] J.K. Bair, H.B. Willard, Phys. Rev. **128**, 299-304 (1962)
[13] L.F. Hansen, J.D. Anderson, J.W. McClure, B.A. Pohl, J. Nucl. Phys., **98**, 25 (1967)

The Reaction Rate for ^{31}S(p,γ)^{32}Cl and its Influence on the SiP-Cycle in Hot Stellar Hydrogen Burning

S. Vouzoukas, C.P. Browne, J. Görres, H. Herndl[*], C. Iliadis[†], J. Meißner, J.G. Ross[‡], L. van Wormer[§], M. Wiescher

University of Notre Dame, Department of Physics, Notre Dame, IN 46556, USA

A. Lefébvre, P. Aguer, A. Coc, J.P. Thibaud

Centre de Spectrométrie Nucléaire et de Sprectrométrie de Masse, Orsay 91405, France

I. INTRODUCTION

The rp-process has been identified as the dominant nucleosynthesis process in explosive hydrogen burning [1] expected for example in novae and x-ray bursts. An impedance, however, is caused by cyclic reaction sequences [2]. In these cases the cycle may be closed by a (p,α) reaction (see NeNa-, SiP-, SCl-cycle etc.) and nuclear material is stored in the cycle for a certain period of time τ_{cycl}. For higher temperature scenarios a break-out may also occur by a proton capture on a long-lived isotope in the cycle, e.g. ^{31}S(p,γ).

In the present work we attempted to determine the excitation energies of the proton unbound states in ^{32}Cl with high accuracy. These states may contribute as resonances to the ^{31}S(p,γ) break-out reaction channel and will determine the reaction rate.

II. EXPERIMENTAL PROCEDURES AND RESULTS

The experiment was performed with the FN-Tandem accelerator and the 100 cm broad-range magnetic spectograph at the Nuclear Structure Laboratory of the University of Notre Dame. The ^{32}S(^{3}He,t)^{32}Cl reaction (Q=-12.706 MeV) was investigated at various ^{3}He beam energies between 22.5 and 25 MeV with beam intensities ranging between 0.1 and 0.5 $p\mu$A on target. Data were taken only at forward angles, 7.5° $\leq \Theta_{lab} \leq$ 20°. The Notre Dame

---

[*]permanent address, T U Wien, Institut für Kernphysik, Wien, Austria

[†]present address, TRIUMF, Vancouver, Canada

[‡]present address, University of Central Arkansas, Conway, AR

[§]present address, Hiram College, Hiram, OH

SNICS ion source was used to implant 70 keV ^{32}S$^-$ ions into 40 μg/cm^2 carbon foils. A typical triton spectrum obtained using these targets is shown in figure 1. In the kinematic analysis of the observed levels, excitation energies were measured relative to that of the well known state at 1.168 MeV [6]. The weighted averages for the excitation energies of those levels observed with the implanted target are listed in Table I. The results of previous work are also listed, as are the compilation values [7] and the results from a test run that we did in Orsay.

The resonance strength, $\omega\gamma$, of the observed ^{32}Cl unbound states in the ^{31}S(p,γ)^{32}Cl reaction channel can be calculated directly from the proton- and γ-partial widths, Γ_p and Γ_γ, respectively. The proton partial width can be calculated as a function of the penetrability through the Coulomb barrier $P_\ell(E_p)$ and the single particle spectroscopic factor C^2S of the resonance level and the γ-partial width can be calculated in terms of the reduced transition probability B(ΠL) for the γ-decay with multipolarity L. These parametes were determined from shell model calculations (for details see reference [9]). Table I also lists the resonance strengths from the experimental input parameters. The error of $\omega\gamma_{exp}$ reflects only the uncertainty of the resonance energy which influences the penetrability.

The stellar reaction rate of ^{31}S(p,γ)^{32}Cl is determined mainly by the resonant reaction contributions. The weak 3$^+$ resonance at 162 keV contributes only at low temperatures, $T_9 \leq 0.25$, while for higher temperatures the reaction rate is dominated by the 3$^+$ resonance at 556 keV. At very high temperatures, $T_9 \geq 0.7$, the 1$^+$ resonance at 640 keV determines the total rate.

Figure 2 shows the reaction flow in the Si to Ar mass range calculated for four different temperature and density conditions, to demonstrate the formation and the leakage of a SiP-cycle. Part a) of the figure indicates the flux for a fairly low temperature and density scenario typical for novae, at these conditions the SiP-cycle operates, there is no leakage via the ^{31}S(p,γ) reaction but some leakage occurs via the ^{31}P(p,γ) reaction [5,6]. Part b) shows the flux at temperature and density conditions, $T = 4 \cdot 10^8$K, $\rho = 10^4$g/cm^2. The flow pattern is characterized by a continuous rp-process flux because a considerable leak from the SiP-cycle occurs via ^{31}S(p,γ). For high temperature and low density scenarios as expected in Thorne Zytkow objects [8], however, the cyclic flow pattern reemerges (part c) because the large ^{31}S(p,γ) rate is considerably reduced by the inverse photodisintegration rate. For higher densities anticipated in x-ray burst events, the flow again bypasses the SiP-cycle (part d).

A collaboration with a team from the Centre de Spectrométrie Nucléaire et de Sprectrométrie de Masse and the Institute de Physique Nucléaire is underway to measure the branching ratios and γ-partial widths of these states via the study of the ^{32}S(^{3}He,tγ)^{32}Cl reaction. A test run of that experiment yielded enough statistics to allow an accurate determination of the energies of some of the peaks. These preliminary results from the test run are listed on Table I and are in excellent agreement with the present work.

Acknowledgment: This work was supported by the U.S. National Science Foundation Grant No. PHY 88-03035 and PHY 91-00708.

[1] W.K. Wallace, S.E. Woosley, Ap.J.Suppl. **45**, 389 (1981)

[2] A.E. Champagne, M. Wiescher, Ann.Rev.Nucl.Part.Sci. **42**, 39 (1992)

[3] C. Iliadis, U. Giesen, J. Görres, S. Graff, M. Wiescher, R.E. Azuma, M. Buckby, J. King, C.A. Barnes, T.R. Wang, Nucl.Phys. **A533**, 153 (1991)

[4] C. Iliadis, J. Görres, J.G. Ross, K.W. Scheller, M. Wiescher, C. Grama, Th. Schange, H.P. Trautvetter, H.C. Ewans, Nucl.Phys. **A559**, 83 (1993)

[5] C. Jeanperrin, L.H. Rosier, B. Ramstein, E.I. Obiajunwa Nucl.Phys. **A503**, 77 (1989)

[6] T. Björnstad, M.J.G. Borge, P. Dessagne, R.-D. von Dincklage, G.T. Ewan, P.G. Hansen, A. Huck, B. Jonson, G. Klotz, A. Knipper, P.O. Larson, G. Nyman, H.L. Ravn, C. Richard-Serre, K. Riisager, D. Schardt, G. Walter, Nucl.Phys. **A443**, 283 (1985)

[7] P.M. Endt, Nucl.Phys. **A521**, 1 (1990)

[8] G.T. Biehle, Ap.J. **380**, 167 (1991)

[9] S. Vouzoukas, C.P. Browne, U. Giesen, J.Görres, S.M. Graff, C. Iliadis, L.O. Lamm, J. Meißner, J.G. Ross, K. Scheller, L. van Wormer, M. Wiescher, A.A Rollefson, H. Herndl , Phys. Rev. C in print (1994)

TABLE I. Excitation energies and spin and parities of the observed levels in ^{32}Cl and ^{32}P

^{32}S(^{3}He,t)^{32}Cl		^{32}S(^{3}He,tγ)^{32}Cl	^{32}S(^{3}He,t)^{32}Cl		^{32}P	
E_x [keV]	$\omega\gamma_{exp}$ [eV]	E_x [keV]	E_x [keV]	J^π	E_x [keV]	J^π
present		Orsay test	Jeanperrin [5]		Endt [7]	
1168.5^a		1167±2	1157±5	1$^+$	1150±3	1$^+$
1329±3		1328±2	1326±5	2$^+$	1323±2	2$^+$
1735±3	$2.3^{+5.8}_{-1.7}\cdot 10^{-10}$	1735±1	1719±4	3$^+$	1755±10	3$^+$
2129±3	$5.5\pm1.2\cdot 10^{-3}$	2127±1	2122±7	(3-5)$^+$	2177±2	3$^+$
2213±3	$1.3\pm0.8\cdot 10^{-2}$		2193±7	(1,2)$^+$	2230±6	1$^+$
2281±3	$3.5\pm0.6\cdot 10^{-3}$		2270±5	(1,2)$^+$	2218±11	2$^+$

^{a}reference value (Björnstad [6])

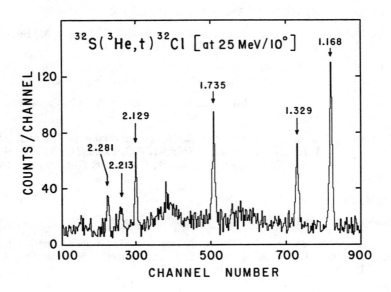

FIG. 1. A ^{32}S(^{3}He,t)^{32}Cl triton spectrum measured at $\Theta = 10°$.

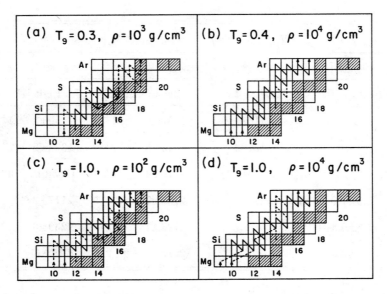

FIG. 2. Flow pattern of nuclear reactions in the mass range A=22-36 for different temperature and densities, calculated for a time period of t = 10^3 s.

Sequential proton capture reactions in the rp-process

L. Van Wormer

Hiram College, Dept. of Physics, Hiram, Ohio, USA

J. Görres, H. Herndl, M. Wiescher

University of Notre Dame, Dept. of Physics, Notre Dame, Indiana, USA

B. A. Brown

Michigan State University, Dept. of Physics, East Lansing, Michigan, USA

F. K. Thielemann

Universität Basel, Institut für Physik, Basel, Switzerland

Detailed shell shell model calculations have been performed to study the structure of unbound excited states in the neutron deficient T=3/2 and T=2 nuclei in the mass range A=20-30. Excitation energies, single particle structure as well as γ transition strengths were computed. On the basis of these results new reaction rates were calculated for the two proton capture sequences, ^{22}Mg(p,γ)^{23}Al(p,γ)^{24}Si and ^{26}Si(p,γ)^{27}P(p,γ)^{28}S. The consequences of these new rates for the reaction flow and the rp-process nucleosynthesis at different stellar temperature and density conditions will be discussed.

I. INTRODUCTION

Explosive hydrogen burning in novae is considered a possible source for the production of the long-lived γ-emitters ^{22}Na and ^{26}Al. At temperature conditions T$\geq$0.15 these isotopes are mainly produced by the β-decay of the progenitors ^{22}Mg and ^{26}Si (1). A detailed study of the depleting proton capture rates of these isotopes is therefore necessary.

Proton capture on even-even T=1 nuclei like ^{22}Mg and ^{26}Si have small Q-values and are strongly hindered at high temperature conditions by the inverse photodisintegration of the T=3/2 compound nuclei ^{23}Al and ^{27}P. This will cause an enrichment of ^{22}Mg and ^{26}Si before β-decay relative to the abundances of ^{23}Al and ^{27}P. A fast destruction of these T=3/2 nuclei, however, either via (p,γ) or β-decay will also result in a rapid depletion of the T=1 waiting point nuclei via sequential two-proton capture or proton capture followed by β-decay (1,2).

II. REACTION RATES

Previously the proton capture rates for the reaction sequences ^{22}Mg(p,γ)^{23}Al(p,γ)^{24}Si and ^{26}Si(p,γ)^{27}P(p,γ)^{28}S have been estimated from the level structure of the compound nuclei and the respective mirror nuclei (1,3,4). The main uncertainties for the proton capture on the T=1 and T=3/2 nuclei are the excitation energies and the γ- and proton partial-widths of the resonance levels. To improve the previous estimates extended shell model calculations have been performed with the OXBASH code (5) using the Wildenthal-interaction for pure sd-shell configurations (6). Coulomb- and Thomas-Ehrman shifts were taken into account for the excitation energies. The γ-widths were derived from the electromagnetic transition rates, the proton-widths were calculated from the single particle spectroscopic factors of the unbound states. Details of the calculations will be discussed elsewhere (7). These results have been used to calculate the resonant and nonresonant reaction rates of ^{23}Al(p,γ)^{24}Si, ^{26}Si(p,γ)^{27}P and ^{27}P(p,γ)^{28}S as described for the case of ^{22}Mg(p,γ)^{23}Al (4). Figure 1 shows the present rates based on the shell model calculations. While the rates of ^{23}Al(p,γ)^{24}Si and ^{26}Si(p,γ)^{27}P

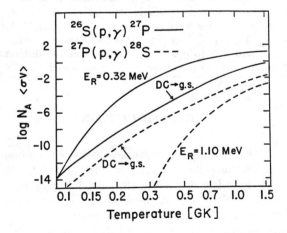

FIG. 1. Reaction rate contributions of ^{26}Si(p,γ)^{27}P (solid line) and ^{27}P(p,γ)^{28}Si (dashed line).

are dominated by a low energy resonance, the rate of ^{27}P(p,γ)^{28}S is mainly determined by the direct capture to the groundstate. Comparison with the previously estimated rates show significant differences for ^{26}Si(p,γ)^{27}P due to an additional resonance, and for ^{27}P(p,γ)^{28}S owing to much stronger direct capture contributions as previously anticipated.

III. REACTION NETWORK CALCULATIONS

Network calculations have been performed for the temperatures T=0.15, 0.2, 0.25 and 0.3 GK at a density of 10^4 g/cm^3 using an initial solar abundance distribution to investigate the influence of the new rates on the reaction flow and the abundance of the involved isotopes.

At T=0.15 and 0.2 GK, no changes in the reaction path or in the flux strengths along the path were discernable. Correspondingly, the abundance versus time plots showed no differences with the new shell model rates.

The modifications in the rate for ^{23}Al(p,γ)^{24}Si at low temperatures do not show any change for the reaction flux and abundances in that mass region. The modified rates of ^{26}Si(p,γ)^{27}P and ^{27}P(p,γ)^{28}S however cause a significant change as temperature increases. At T=0.25 and 0.3 GK, ^{26}Si will undergo a proton capture instead a β-decay when the shell model based rates are applied. Figure 2 shows the corresponding reaction path for a temperature of T=0.3 GK. The solid lines indicate the main reaction flow integrated over 1000 s, the dashed lines represent a 1 to 10 % and the dot-dashed lines a 0.1 to 1 % flow of the maximum flow, respectively.

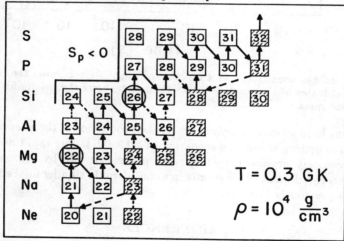

FIG. 2. Time integrated reaction flux in the mass A=20 to 40 region, for details see text.

This indicates that the flux bypasses the relatively long β-decay of ^{26}Si via ^{26}Si(p,γ)^{27}P($\beta^+\nu$)^{27}Si which will influence the abundance distribution in this mass range. Figures 3 shows abundance versus time plots for selected waiting point isotopes at T=0.3 GK using the previous rates (3,1) and the present shell model based rates, respectively.

Figure 3 indicates that after 0.1 s the initial solar abundances of stable isotopes (A$\geq$20) are converted and stored in the waiting point isotopes ^{22}Mg, ^{26}Si and ^{30}S for a time period of $\approx$20 s. Applying the shell model rates causes a significant change for the waiting point isotope ^{26}Si and consequently its daughter nucleus ^{26}Al, produced by β-decay. The new, largely enhanced ^{26}Si(p,γ) rate causes a rapid processing of ^{26}Si towards ^{27}P and further to ^{30}S thus bypassing ^{26}Al. This is reflected in the rapid decrease of the ^{26}Si and ^{26}Al abundance after 0.1 s. Not affected are the abundances of the two other waiting points ^{22}Mg and ^{30}S which remain significantly enriched for $\approx$20 s until further processing shifts the abundance distribution towards higher masses (1).

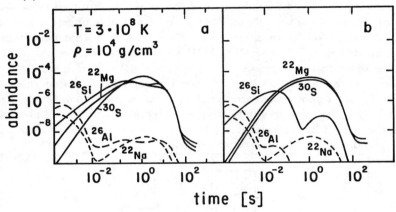

FIG. 3. Isotopic abundances as a function of time. Part a shows the results for previous estimates of the reaction rates (1), part b shows the results for the shell model based rates.

This may have serious consequences for the production of the γ- emitter ^{26}Al in explosive hydrogen burning scenarios (T$\geq$0.2 GK). The rapid destruction of the previously proposed waiting point progenitor ^{26}Si by proton capture prevents its enrichment. This results in a considerably smaller final abundance of ^{26}Al as previously thought.

REFERENCES

1. L. van Wormer, J. Görres, C. Iliadis, M. Wiescher, F.K. Thielemann, Ap.J. (1994) in press
2. A. Champagne, M. Wiescher, Annu.Rev.Nucl.Part.Sci.**42**, 39 (1992)
3. M. Wiescher, J. Görres, F.K. Thielemann, H. Ritter, Astr.Ap. **160**, 56 (1986)
4. M. Wiescher, J. Görres, B. Sherrill, M. Mohar, J. Winfield, B.A. Brown, Nucl.Phys. A **484**, 90 (1988)
5. B.A. Brown, A. Etchegoyen, W.D.M. Rae, N.S. Godwin, OXBASH (1984) unpublished
6. B.H. Wildenthal, Prog.Part.Nucl.Phys. **11**, 5 (1984)
7. H. Herndl, J. Görres, M. Wiescher, B.A. Brown, L. van Wormer, (1994) to be published

Nuclear recoil from atomic electrons and resulting effects on low-energy fusion cross sections

Timothy D. Shoppa

Kellogg Radiation Lab, California Institute of Technology, Pasadena CA 91125

Abstract. Recent measurements of low energy fusion cross sections in the lab using atomic and molecular targets have shown apparent atomic screening effects in excess of what can be explained using the adiabatic limit. Here the effect of the nuclear momentum distribution caused by the momentum-space wavefunction of a hydrogenic atomic target is folded with a "bare" nuclear cross-section. Because of the very long tail of the atomic momentum distribution, laboratory fusion cross sections below a few keV are strongly enhanced beyond what is predicted by the adiabatic limit. Several limitations of the simple folding procedure used here are discussed.

1. INTRODUCTION

The nucleus of an atom acquires a momentum distribution as a result of recoil from the momentum distribution of bound atomic electrons. In particular, for the case of a one-electron atom in the hydrogenic 1s state, the momentum space wave function is given by

$$\phi(\mathbf{p}) = (8Z^5\pi^{-2})^{1/2}(Z^2 + p^2)^{-2} \qquad (1)$$

where $\mathbf{p}$, the relative momentum between the nucleus and the electron, is measured in units of $m_e \alpha c$. This wave function ignores relativistic effects and assumes a point-like nuclear charge distribution. If one assumes that the center of mass of the atom is at rest in the lab, then the normalized probability distribution for finding the nucleus with a momentum $\mathbf{p}$ in the lab is $g(\mathbf{p}) = 8|\psi(2\mathbf{p})|^2$.

The effect of this nuclear momentum distribution in atomic targets has been discussed before in the context of narrow resonances (1). Here I consider the effect in the case of nuclear fusion. Because of the Coulomb barrier, for nuclear center-of-mass energies $E_{\rm cm}$ less than the height of the coulomb barrier, there is an exponential rise in cross section as the relative energies between the nuclei increases. These fusion cross sections usually are expressed with the fastest-varying parts factored out:

$$\sigma_{\rm bare}(E_{\rm cm}) = \frac{S(E_{\rm cm})}{E_{\rm cm}} \exp(-2\pi\eta(E_{\rm cm})) \qquad (2)$$

where all of the effects of the nuclear physics are embodied in the "S-factor" $S(E_{\rm cm})$, which is (barring any sharp resonances) a much more slowly varying function of energy than $\sigma_{\text{bare}}(E_{\text{cm}})$. Here $\eta(E_{\text{cm}})$ is the Sommerfeld parameter for the initial nuclei: for incident and target nuclei of charge Z_i, Z_t, mass m_i, m_t, and reduced mass μ, we have $2\pi\eta(E_{\text{cm}}) = Z_i Z_t \alpha (2E_{\text{cm}}/\mu c^2)^{\frac{1}{2}}$.

A simple way to take the distribution of the target nucleus momentum into account is to write the cross section for a beam of incident nuclei with laboratory energy $E_{\text{beam}} = (P_{\text{beam}})^2/2m_i$ as

$$\sigma_{\text{lab}}(E_{\text{beam}}) = \int dp_z g(p_z) \sigma_{\text{bare}}((P_{\text{beam}} + p_z)^2/2\mu) \qquad (3)$$

where $g(p_z)$ is the probability of finding the target nucleus with momentum p_z in the direction of the incoming beam. First I give an example of this folding, and then I discuss some limitations of this simple treatment.

2. EXAMPLE

As a specific example, consider the case of the d(^{3}He,p)^{4}He reaction, where (for simplicity) the target ^{3}He nucleus is assumed to have only one electron bound in the ground state. The normalized target nucleus momentum distribution in the direction of the beam is given by

$$g(p_z) = \frac{16}{3\pi} Z_t^5 (Z_t^2 + (2p_z)^2)^{-3} \qquad (4)$$

Using the $n = 4$ parameterization given by Chulick et. al. (2) of the d(^{3}He,p)^{4}He cross section and performing the folding indicated by Eq. [3], one obtains the enhancement factor $f = \sigma_{\text{lab}}/\sigma_{\text{bare}}$ shown in Figure 1 as the dashed line. For comparison, the enhancement factor predicted for electron screening in the adiabatic limit (3) for an atomic ^3He target ($U_e = 119$ eV) is shown as the dot-dashed line. The product of the two enhancements is shown as the solid line.

3. DISCUSSION

The simple folding performed here is far from a full quantum-mechanical treatment of the problem. The results of the folding do not agree with the apparent excess screening seen in experiment (4). In particular, the enhancement predicted using Eq. [3] does not become important until E_{cm} is below 3 keV, while the experiment does not go below 6 keV. Several effects that have been ignored in the simple treatment given above are discussed below.

The most obvious shorcoming of the example above is that it assumes that the momentum distribution for large nuclear separations is the one that should be used in the folding. At the energies at which electron screening in fusion reactions has been observed, the electron wavefunction adapts itself nearly adiabatically to the nuclear charge distribution (5). As a result, the atomic wave function when the nuclei are near each other is probably better

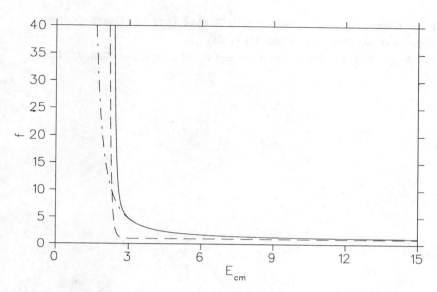

Fig. 1.— Figure 1. Enhancement factors for the adiabatic approximation (dot-dashed line), the momentum distribution folding (dashed line), and their product (solid line).

described in terms of the atomic states of the compound nucleus. Because the compound nucleus will have a higher effective Z, the momentum distribution will be broadened compared to what was assumed above, and this broader distribution will result in a greater enhancement of fusion cross sections in this simple model.

On the other hand, as the two nuclei near each other, they will share the recoil momentum. If the two nuclei have nearly identical charge to mass ratios, their recoil velocities when near each other will be nearly identical, and there will be very little enhancement due to the recoil effect. So in this limit, one would expect to see recoil effect enhancements in reactions like $d(^3He,^4He)p$ and $^6Li(p,\alpha)^3He$ but not in reactions such as $d(d,p)t$, $^3He(^3He,2p)^4He$, or $^6Li(d,\alpha)^4He$

The example given above only considers the case of one electron. Most laboratory targets have more than one bound electron; if this is the case, the recoil momentum may be absorbed by other electrons instead of the nucleus. A better treatment of this aspect of the problem requires an atomic wave function that accounts for the detailed correlations of the electrons out to very high momenta.

REFERENCES

1. J.S. Briggs and A. M. Lane, Phys. Lett. **B106**, 436 (1981).
2. G.S. Chulick, Y.E. Kim, R.A. Rice, and M. Rabinowitz, Nucl. Phys. **A551**, 255 (1993).

3. H.J. Assenbaum, K. Langanke, and C. Rolfs, Z. Phys **A327**, 461 (1987).
4. S. Engstler *et. al.*, Phys. Lett. **B202**, 179 (1988).
5. T. D. Shoppa, S. E. Koonin, K. Langanke, and R. Seki, Phys. Rev. **C48**, 837 (1993).

Subthreshold Resonance Effect in the ^{6}Li(d,α)^{4}He Reaction

H.Bucka, K.Czerski, P.Heide, A.Huke, G.Ruprecht and B.Unrau

Institut für Strahlungs- und Kernphysik, Technische Universität Berlin, D-10623 Berlin

Abstract. The reaction ^{6}Li(d,α)^{4}He has been measured for deuteron energies between 50 and 180 keV. In addition to the direct reaction component, which was calculated in the frame of the DWBA formalism, a significant contribution of a broad 2^+ subthreshold resonance in the compound nucleus ^{8}Be to the reaction cross section was identified. The results of the present analysis are compared with those obtained for the ^{6}Li(d,p)^{7}Li and ^{6}Li(d,n)^{7}Be reactions. Implications on the astrophysical S-factor value at zero deuteron energy and on the electron screening potential are discussed.

1. Introduction

In a recent study of the mirror reactions ^{6}Li(d,p)^{7}Li and ^{6}Li(d,n)^{7}Be it has been shown [1] that the decrease of the (d,n) to (d,p) cross section ratio below 200 keV [2] can be explained by a broad 2^+ subthreshold resonance in the compound nucleus ^{8}Be, this resonance being composed of two isospin mixed states. One of these states with an isospin 0 should have a large α-width and should be observed in the ^{6}Li(d,α)^{4}He reaction. So far the cross section for this reaction has been described either within R-matrix theory due to contributions from distant resonances (energy-independent R-matrix parametrization) [3] or within DWBA theory via a direct reaction mechanism [4].

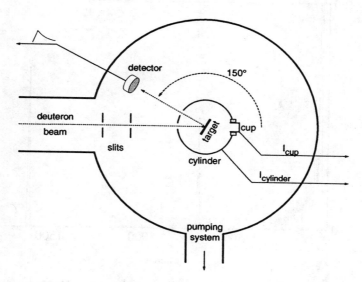

Figure 1. Experimental set-up.

In the present work we have measured the ^{6}Li(d,α)^{4}He reaction at deuteron energies between 50 and 180 keV, showing that the low energy cross section for this reaction is a sum of two components: a direct contribution and a subthreshold resonance contribution. The evaluation of the resonance contribution allows to determine the astrophysical S-factor at zero deuteron energy, also yielding an appropriate value for the electron screening potential.

2. Experiment

The experimental set-up is shown in Fig.1. Deuterons from the ISKP cascade accelerator impinged on a thin ^{6}LiF target (10 μg/cm^2 on a 10 μg/cm^2 carbon-backing, corresponding to 2-3 keV deuteron energy loss). The α-particles were detected by a 100 mm^2 Canberra PIPS Si-detector being placed at an angle of 150° with respect to the beam in 10 cm distance from the target. Since the experiment was designed to also measure the recoil nuclei from the ^{6}Li(d,p)^{7}Li and ^{6}Li(d,n)^{7}Be stripping reactions, there was no protective foil in front of the detector. To avoid pile-ups from elastically scattered deuterons we used for spectroscopic purpose a fast timing amplifier together with a stretcher. A typical spectrum is shown in Fig.2. The details of the experimental set-up will be published elsewhere. For a correct measurement of the beam current it was necessary to take care of deuterons (up to 50 %) that have changed their charge state when traversing the target. Therefore the target was electrically connected to a surrounding cylinder box. The true beam current is then the sum of the Faraday cup plus the cylinder current (see Fig.1).

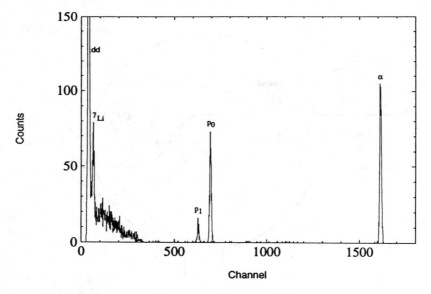

Figure 2. Charged particle spectrum measured at $E_d^{lab} = 160$ keV. The peaks labeled p_0 and p_1 are from the ^{6}Li(d,p)^{7}Li reaction and ^{7}Li stands for the recoil nucleus. The dd-line corresponds to elastic deuteron scattering.

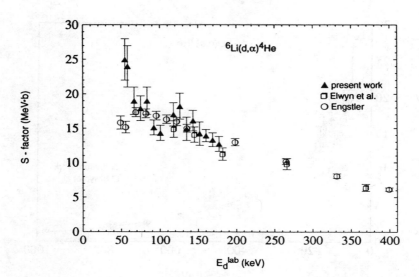

Figure 3. Astrophysical S-factors for the ^{6}Li(d,α)^{4}He reaction from the present work and from previous measurements [5], [6].

3. Results

Astrophysical S-factors have been determined assuming an isotropic angular distribution [5] for the ^{6}Li(d,α)^{4}He reaction at deuteron energies below 200 keV. The results are presented in Fig.3 together with experimental data from other authors [5], [6]. All results are in good agreement and indicate an increase of the S-factors with decreasing deuteron energy. As in the case of the ^{6}Li(d,p)^{7}Li and ^{6}Li(d,n)^{7}Be reactions [1] we expect a significant contribution of a broad 2$^+$ subthreshold resonance in this energy region. This resonance, according to Ajzenberg-Selove [7], has a rather large width of 800 keV and lies 80 keV below the reaction threshold. The direct reaction contribution should dominate at higher deuteron energies. Its magnitude we have calculated in the frame of zero range DWBA with conventional finite range and nonlocality parameters [8]. Two angular momentum transfers were considered ($l = 0,2$) with the corresponding spectroscopic factors $C^2S = 1.032$ and 0.087, respectively [4]. The optical model parameters for the deuteron channel are from [9] and for the α-α channel from [10]. With one normalization factor $D_0^2 = (2.1\pm0.1)10^3$ MeV^{2}fm^3 (so called zero range strength factor) it was possible to describe the experimental angular distributions [5] for six deuteron energies between 370 keV and 975 keV; however, in all cases some isotropic contribution corresponding to the resonance component had to be added. Subtracting the calculated direct reaction contribution from the measured S-factor values one gets the compound resonance component which can be fitted by a Lorentz curve (s-wave resonance). With a fixed value of the resonance width $\Gamma = 800$ keV we obtained as a result of the fitting procedure the resonance energy $E_R = (-340 \pm 190)$ keV. In Fig.4 the DWBA calculation result for the direct component together with the resonance contribution is presented.

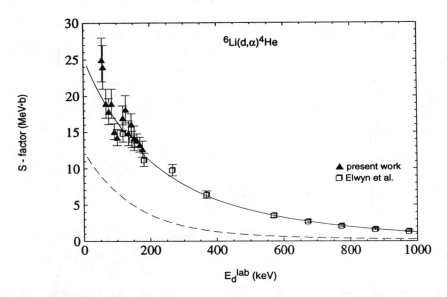

Figure 4. Comparison of the experimental (from the present work and from [5]) and theoretical S-factors for the ^{6}Li(d,α)^{4}He reaction. The dashed line represents the DWBA calculation of the direct reaction component. The solid line is the S-factor predicted taking an additional resonance contribution into account.

4. Discussion

It has been shown that the subthreshold resonance plays an important role for understanding the excitation function of the ^{6}Li(d,α)^{4}He reaction at low deuteron energies. In Fig.5 the cross sections for this reaction are shown. A comparison with the S-factor curves (Fig.4) allows us to conclude that a prominent structure in the cross section at deuteron energies around 600 keV is a result of the subthreshold resonance, the strength of which is shifted to higher deuteron energies due to the Coulomb penetration effect. The derived resonance energy $E_R = (-340 \pm 190)$ keV is significantly smaller than the value $E_R = (-100 \pm 80)$ keV which we got from the ^{6}Li(d,p)^{7}Li and ^{6}Li(d,n)^{7}Be reactions [1]. This supports the hypothesis that the resonance state in ^{8}Be at an excitation energy of 22.2 MeV does in fact consist of two close lying isospin mixed 2^+ states where the lower one has a large α-width. For astrophysical applications it is important to extrapolate the measured S-factor values to zero deuteron energy, $S(0)$. However, the direct experimental determination of this value can be very difficult because of the growing importance of the electron screening effect at deuteron energies below 50 keV. Taking the resonance contribution into account, we get an S-factor value $S(0) = 25$ MeV·b compared to $S(0) = 19.5$ MeV·b from [11]. This difference has of course consequences for the evaluation of the electron screening potential U_e. In the work [11] the authors obtained a value of $U_e \approx 350$ eV in comparison with the theoretical value of about 200 eV. The recognition of the subthreshold resonance contribution

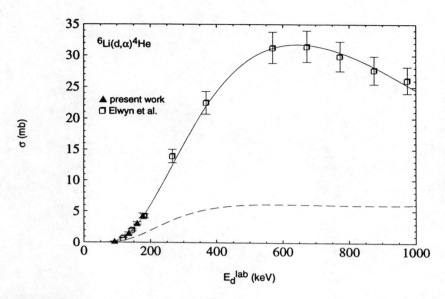

Figure 5. Total cross sections for the ^{6}Li(d,α)^{4}He reaction. The description corresponds to that given for Fig.4.

reduces the experimental value for U_e leading to a better agreement with theory.

REFERENCES

[1] Czerski, K., Bucka, H., Heide, P. and Makubire, T., *Phys.Lett.* **B307**, 20 (1993)
[2] Cecil, F.E., Peterson, R.J. and Kunz, P.D., *Nucl.Phys.* **A441**, 477 (1985)
[3] Elwyn, A.J. and Monahan, J.E., *Phys.Rev.* **C19**, 2114 (1979)
[4] Raimann, G., PhD thesis, Universität Münster 1991
[5] Elwyn, A.J., Holland, R.E., Davids, C.N., Meyer-Schutzmeister, L., Monahan, J.E., Mooring, F.P. and Ray, W. Jun., *Phys.Rev.* **C16**, 1744 (1977)
[6] Engstler, S., PhD thesis, Universität Münster 1991
[7] Ajzenberg-Selove, F., *Nucl.Phys.* **A490**, 1 (1988)
[8] Kunz, P.D., code DWUCK4, Colorado University, unpublished
[9] Satchler, G.R., *Nucl.Phys.* **85**, 183 (1966)
[10] Marquet, L., *Phys.Rev.* **C28**, 2525 (1983)
[11] Engstler, S., Raimann, G., Angulo, C., Greife, U., Rolfs, C., Schröder, U., Somorjai, E., Kirch, B. and Langanke, H., *Phys.Lett.* **B279**, 20 (1992)

The Reaction ^{70}Ge$(\alpha,\gamma)^{74}$Se (p–process)

Zs. Fülöp[†], Á.Z. Kiss[†], E. Somorjai[†], C.E. Rolfs[*], H.P. Trautvetter[*], T. Rauscher[**] and H. Oberhummer[***]

[†] *Institute of Nuclear Research, Debrecen, Hungary*
[*] *Institut für Experimentalphysik III, Ruhr–Universität, Bochum, Germany*
[**] *Institut für Kernchemie, Universität Mainz, Germany*
[***] *Institut für Kernphysik, Technische Universität, Vienna, Austria*

The cross section of the reaction ^{70}Ge$(\alpha,\gamma)^{74}$Se is measured in the bombarding energy range of $4.90 < E_\alpha < 7.53$ MeV. The experimental S–factor values are in good agreement with theoretical calculations. Reaction rates for the inverse (γ,α) reaction are determined.

INTRODUCTION

The stable neutron–deficient isotopes of the elements with charge number $Z \geq 34$ are classically referred to as the p–nuclei. They were observed only in the solar system and here they represent 0.1% to 1% of the bulk elemental abundances, made predominantly of the more neutron–rich s– and r–nuclei. The stellar process synthesizing the p–nuclei is called the p–process. Several models have been developed for describing the p–process (1–3). There are differences in the details (astrophysical site, temperature, time scale, reactions involved, etc.), however, the generally accepted main process are subsequent (γ,n) reactions, which start from s– and r–nuclei and drive the nuclei towards the neutron–deficient region. Along this isotopic path, the binding energy of neutrons is gradually larger. As a result, the reaction flow slows down and, at some point, is deflected by (γ,α) or (γ,p) reactions. At these branching points the matter tends to accumulate. The branching points are the p–nuclei themselves (light ones) or their progenitors (heavy ones). In fact, (n,γ) reactions can be also important and compete with photon–induced reactions.

Above Si, theoretical p–process studies (4,5) have to rely mainly on reaction rates predicted in the framework of the statistical model. For charged particle reactions practically no experimental data is available above the Fe region.

Of crucial importance is the investigation of the branchings. The first candidate is ^{74}Se, the lightest p–nucleus (lowest Coulomb barrier) sitting at such a branching point. Here, the ejection of an alpha particle is favoured relative to the protons (3). Up to now, the ^{70}Ge$(\alpha,\gamma)^{74}$Se reaction has been investigated for energies $E_\alpha > 12$ MeV (6), i.e. above the Gamow window, Coulomb barrier and neutron threshold.

© 1995 American Institute of Physics

EXPERIMENTS AND RESULTS

The experiment on the $^{70}\text{Ge}(\alpha,\gamma)^{74}\text{Se}$ reaction has been performed at the cyclotron in Debrecen and the Dynamitron Tandem accelerator in Bochum from E_α=4.9 MeV up to 7.8 MeV (E_{cm}=4.635–7.378 MeV), using a He^{++}- beam. In the first run, with the target on thick backing, only one gamma-transition from ^{74}Se, namely the decay of the first excited state to the ground state (E_γ=635 keV) was seen. However, the yield was low and the background was very high. High beam current, long measuring time and improved detection efficiency are necessary to compensate the small reaction cross section. For reducing the background yield (predominantly caused by high energy gamma-rays and neutrons from the $^{13}\text{C}(\alpha,\text{n})^{16}\text{O}$ reaction) the thick Ta backing —which integrates the background yield— was replaced by a thin (0.1 μm) gold foil. With this it was not possible to use a high beam current. However, the reaction yield to background ratio improved. The resulted loss in yield has been compensated by using a high efficiency Ge detector (116% HPGe) in close geometry and the background has been further lowered by the proper choice of geometry and shielding. An enriched (86.5%) ^{70}Ge target was used with an averaged target current of 500 particle-nA, avoiding any target deterioration. The target stability was monitored by the strong Coulomb excitation peak of the ^{70}Ge target nucleus (1040 keV). Because of the increased yield to background ratio not only the peak at E_γ=635 keV but also others from ^{74}Se have appeared in the spectrum taken with relatively good statistics at E_{cm}=6.514 MeV, namely the E_γ=985 keV and 219 keV ones belonging to a cascade going through the first excited state. In the same spectrum very weak primary transitions to the ground and first excited states have also appeared. The total cross section of the $^{70}\text{Ge}(\alpha,\gamma)^{74}\text{Se}$ reaction was determined using only the intensity of the E_γ=635 keV peak. In principle this is only a lower limit, however, in the spectra all the peaks mentioned above belonging to the decay of ^{74}Se represent transitions decaying directly or after a few steps through the first excited state to the ground state. The only exception is the extremely weak (see later) primary transition to the ground state. It is clearly demonstrated by these experimental facts (see also Ref. (6)) together with a statistical survey of the decay of the known 53 levels in ^{74}Se (7) (which shows that more than 90% of all the decays —as an average— go through the first excited state), that the 635 keV peak represents the total cross section (within the error limit). To convert the peak area to the cross section, the absolute efficiency of the detector was measured, and the number of target atoms/cm^2 has been determined by the RBS technique. The E_γ=635 keV transition is fed by more transitions, therefore no correction for angular distribution is necessary. The errors of the resulted cross sections vary between 10 and 40% (except at the lowest energy).

The measured cross sections are in excellent agreement with the statistical model calculation performed with the code SMOKER (5). We used the most recent mass table (8) giving a Q-value of 4.077 MeV. The measured and cal-

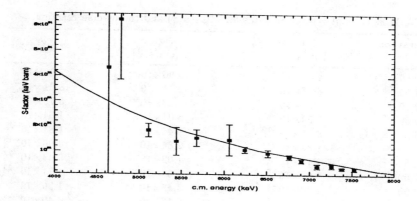

FIG. 1. The S-factor of the reaction $^{70}\text{Ge}(\alpha,\gamma)^{74}\text{Se}$ from the transition between the first excited state (E_x=6438 keV; J^π=2$^+$) and ground state of the residual nucleus.

culated S-factors are shown in Fig. 1. With the help of additional theoretical values at low energies we calculated reaction rates for the forward (capture) reaction as well as for the inverse (photodisintegration) reaction using detailed balance (9). The rates for thermally excited targets are given in Table 1, as well as the ratio of the rate factors. The experimental data contribute to the rates significantly only for $T_9 \geq 1.0$.

The possible contribution of the direct process has also been calculated and was found being negligible.

For the mentioned primary transitions, the cross sections at E_{cm}=6.514 MeV are: σ=1.0±0.6 μbarn and 2.0±0.7 μbarn to the ground state and first excited states, respectively. The values are close to our detection limits and negligible (here the error of the total cross section itself is 23 μbarn).

The work is in progress, the future tasks are to improve the statistics, to perform coincidence measurements for identifying the peaks connected to the 635 keV transition on a reduced background and to determine the cascade corrections.

ACKNOWLEDGEMENTS

We want to thank F.-K. Thielemann for discussions and for the permission to use his code SMOKER. This work was supported by the Hungarian

TABLE 1. Stellar reaction rates for $^{70}\text{Ge}(\alpha,\gamma)$ and the inverse reaction.

Temperature 10^9 K	$N_A <\sigma v>_\alpha$ [cm^{3}s^{-1}mole^{-1}]	$<\sigma v>_\gamma$ [cm^{3}s^{-1}]	$\frac{<\sigma v>_\alpha}{<\sigma v>_\gamma}$
0.50	4.36×10^{-29}	9.06×10^{-60}	7.99×10^{6}
0.60	8.87×10^{-26}	1.71×10^{-49}	8.61×10^{-1}
0.70	3.82×10^{-23}	7.24×10^{-42}	8.76×10^{-6}
0.80	5.59×10^{-21}	6.04×10^{-36}	1.54×10^{-9}
0.90	3.74×10^{-19}	3.44×10^{-31}	1.81×10^{-12}
1.00	1.38×10^{-17}	2.84×10^{-27}	8.06×10^{-15}
1.50	4.17×10^{-12}	1.08×10^{-14}	6.43×10^{-22}
2.00	1.07×10^{-8}	1.04×10^{-7}	1.71×10^{-25}
2.50	2.50×10^{-6}	3.45×10^{-3}	1.20×10^{-27}
3.00	1.36×10^{-4}	5.13×10^{0}	4.40×10^{-29}
3.50	2.83×10^{-3}	1.14×10^{3}	4.12×10^{-30}
4.00	3.05×10^{-2}	7.48×10^{4}	6.78×10^{-31}
4.50	2.04×10^{-1}	2.09×10^{6}	1.62×10^{-31}
5.00	9.48×10^{-1}	3.19×10^{7}	4.94×10^{-32}
6.00	9.15×10^{0}	1.98×10^{9}	7.67×10^{-33}
7.00	4.26×10^{1}	3.74×10^{10}	1.89×10^{-33}
8.00	1.24×10^{2}	3.17×10^{11}	6.49×10^{-34}
9.00	2.65×10^{2}	1.56×10^{12}	2.83×10^{-34}
10.00	4.53×10^{2}	5.13×10^{12}	1.47×10^{-34}

National Science Fund (grant no. 3008) and the Hungarian–German and – Austrian Exchange program (OMFB grant nos. 67 and A.1). TR is an Alexander von Humboldt fellow.

REFERENCES

1. Audouze, J., and Truran, J. *Ap. J.* **202**, 204 (1975).
2. Arnould, M., *Astron. Ap.* **46**, 117 (1976).
3. Woosley, S.E., and Howard, M., *Ap. J. Suppl.* **36**, 285 (1978).
4. Rayet, M., Prantzos, N., and Arnould, M., *Astron. Astrophys.* **227**, 271 (1990); and references therein.
5. Thielemann, F.-K., Arnould, M., and Truran, J.W., in *Advances in Nuclear Astrophysics*, eds. E. Vangioni–Flam et al., Gif-Sur-Yvette: Editions Frontieres, 1986, p. 525.
6. Zipper, W., Seiffert, F., Grawe, H., and von Brentano, P., *Nucl. Phys.* **A551**, 35 (1993).
7. *Table of Isotopes*, eds. C.M. Lederer and V.S. Shirley, John Wiley & Sons Inc., 1978.
8. Audi, G., and Wapstra, A.H., *Nucl. Phys.* **A565**, 1 (1993).
9. Fowler, W.A., Caughlan, G.E., and Zimmerman, B.A., *Ann. Rev. Astron. Astrophys.* **5**, 525 (1967).

β-β Decay with Majorana Neutrino as Possible Reason for the Lack of Solar Neutrinos

V.I. Tretyak, V.V. Kobychev

*Institute for Nuclear Research of the Ukrainian National Academy of Sciences
Prospect Nauki 47, 252028 Kiev, Ukraine*

Abstract. A new mechanism is considered which could contribute to the lack of Solar neutrinos. If neutrino is a Majorana particle ($\tilde{\nu}_e = \nu_e$) and its mass $m_\nu \neq 0$ or/and right-handed admixtures exist in the weak interactions, the β–β decay between the pairs of single β emitters without emission of neutrinos is possible. The estimations of the effective neutrino mass which could twice decrease the emission rate of Solar ^{8}B and ^{7}Be neutrinos give however the value of $\langle m_\nu \rangle$ more than 10^4–10^7 GeV which is in contradiction with experimental limits $\langle m_\nu \rangle \leq 1$–4 eV founded in direct experiments devoted to search for neutrinoless double beta decay of atomic nuclei.

Key words: Solar neutrinos. — Neutrino mass. — β–β decay.

The phehomenon of neutrinoless β–β decay between the pairs of single β emitters was considered for the first time by A.F. Pacheco [1]. If neutrino is a Majorana particle ($\tilde{\nu}_e = \nu_e$) and its mass $m_\nu \neq 0$ and/or right-handed admixtures contribute to the weak interaction, the process is possible when virtual antineutrino emitted in β$^-$ decay of one nucleus may be absorbed as neutrino in another nucleus in neighbourhood provoking also its β$^-$ decay:

$$(A_1, Z_1) \rightarrow (A_1, Z_1+1) + e^- + \tilde{\nu}_e$$

$$\nu_e + (A_2, Z_2) \rightarrow (A_2, Z_2+1) + e^-. \tag{1}$$

The nonzero neutrino mass or right-handed admixtures are required to flip the neutrino helicity and make possible the second reaction in (1). Analogous equations can be written also for two β$^+$ emitters or for one β$^-$ and one β$^+$ nuclei (electron capture instead β$^+$ decay is possible too). As a result of the process (1), only electrons – no neutrinos – will be emitted with energy $E(e_1) + E(e_2) = Q(\beta_1) + Q(\beta_2)$, where $Q(\beta)$ is energy released in β decay.

The neutrinoless β–β decay between the pairs of single β emitters is analogous to the neutrinoless double beta ($\beta\beta$) decay of atomic nuclei (see [2]). In both of them d-quarks are transformed to u-quarks (as regards (1)). However in $\beta\beta$ decay the quarks are located in one or different nucleons of the same nucleus (Δ-isobar or two-nucleon mechanism) whereas in β–β decay – in different nuclei. In fact, the (A_1,Z_1) and (A_2,Z_2) nuclei of (1) can be located not in the same sample but in different objects bringing into being the interaction of a given nucleus with the whole Universe (or, for the case of nonzero m_ν, with its part limited by Yukava radius). Theoretically it could cause the unavoidable electron background in superlow-background experiments devoted to search fo rare and forbidden processes (such as $0\nu\beta\beta$ decay).

As in process (1) only electrons are emitted, it is interesting to consider this phenomenon in connection with the puzzle of the lack of Solar neutrinos which have to be emitted, in particular, in β^+ decay of ^8B:

$$^8B \rightarrow {}^8Be + e^+ + \nu_e. \qquad (2)$$

In accordance with [3], the observed in experiments flux of neutrinos generated in process (2) is at least two times less than its theoretical value.

Probability of neutrinoless β-β decay between two nuclei $\Gamma \sim \langle m_\nu \rangle^2 / D^2$, where $\langle m_\nu \rangle$ is an effective value of neutrino mass, D - distance between the nuclei [1]. In the sample with N_β nuclei the full rate of β-β processes is proportional to the number of pairs $\Lambda \sim N_{pair} \cdot \langle m_\nu \rangle^2 / R^2$, where R – an effective radius of the sample. Since $N_{pair} \sim N_\beta^2 \sim R^6$, so $\Lambda \sim N_\beta^{4/3} \langle m_\nu \rangle^2$ and in a big sample the process of β–β decay could be non-negligible.

To estimate the value of $\langle m_\nu \rangle$ which is needed for decreasing two times the neutrino emission rate, we use the formula (4) in [1] for Γ modified for the case of two different nuclei (A_i, Z_i), (A_j, Z_j) anchored on the distance D_{ij}:

$$\Gamma_{ij} = \frac{\gamma_{ij}}{D_{ij}^2} = \frac{\pi(ln2)^2 \langle m_\nu \rangle^2}{16 m^5} \cdot \frac{H_i H_j f_{ij}}{(ft)_i (ft)_j} \cdot \frac{1}{D_{ij}^2}, \qquad (3)$$

where m is the mass of electron, $H_i = 2\pi\alpha Z_i'/(1-\exp(-2\pi\alpha Z_i'))$ – Coulomb correction factor, $Z_i' = \mp(Z_i \mp 1)$ – atomic number of daughter nucleus created in $\beta^\pm$ decay, $f_{ij} = t_{ij}(t_{ij}^4 + 10 t_{ij}^3 + 40 t_{ij}^2 + 60 t_{ij} + 30)/30$, $t_{ij} = (Q_i+Q_j)/m$, Q_i – maximal energy of neutrino available in decay of (A_i, Z_i), $\alpha = 1/137.036$.

Full rate of β-β decay in homogenous globe with radius R is:

$$\Lambda_{ij} = \frac{9}{4R^2} N_{pair} \gamma_{ij}, \qquad (4)$$

where $N_{pair} = N_i N_j$ for $i \neq j$ and $N_i(N_i-1)/2$ for $i=j$, N_i – the number of nuclei (A_i, Z_i) in the globe. Supposing that the Sun is approximately homogenous in the deepest layers where fusion occurs, we have the formula for the rate of β-β decay within the Sun:

$$\Lambda_{ij} = 3.2 \cdot 10^{-76} \cdot \left(\frac{\langle m_\nu \rangle}{1eV}\right)^2 \left(\frac{R}{R_\odot}\right)^{-2} N_{pair} \chi_i \chi_j f_{ij} \quad (1/sec), \qquad (5)$$

where $\chi_i = H_i/(ft)_i$ depends on properties of the nucleus (A_i, Z_i) only, whereas f_{ij} depends on properties of both the nuclei. The formula (5) allows to calculate the rates of neutrinoless β–β process between the pairs of existing within the Sun active nuclei. This quantities are presented in the Table. The values of Q_i and $(ft)_i$ were taken from [5], $R=0.1\,R_\odot$ was restricted [3], and N_i was calculated using the data [3] about the rates of (A_i, Z_i) decay:

$$N_i = (dN_i/dt)T_{1/2}/\ln 2 = \Omega(\nu_i)T_{1/2}/\ln 2,$$

where $\Omega(\nu_i)$ – theoretical value of flux of i-th neutrinos [3]. The values of $\Lambda_{ij}/\langle m_\nu \rangle^2$ for ^{13}N (Q_i=1190 keV), ^{15}O (1723 and 2754 keV) and ^{17}F (1739 keV) were calculated also and were found to be equal to 10^4–10^5 s$^{-1}\cdot$eV^{-2} for β–β decay with ^{7}Be and less than 1.0 s$^{-1}\cdot$eV^{-2} – with ^{8}B.

We can see from the Table, that the maximal contribution in decreasing the boron neutrinos' flux is determined by β–β decay between ^{8}B and ^{7}Be: $\Lambda = 60 \langle m_\nu(\text{eV})\rangle^2$ 1/sec. For the decreasing two times the theoretical rate of the ^{8}B neutrino emission we need therefore to take the value of $\langle m_\nu \rangle = 1.2 \cdot 10^7$ GeV. For decreasing two times the flux in neutrino monoline of 862 keV from ^{7}Be K-capture (theoretical rate $1.3 \cdot 10^{37}$ 1/sec) we have to take the value of $\langle m_\nu \rangle = 2.6 \cdot 10^4$ GeV.

Required values of neutrino mass are in manifest contradiction with experimental limits on $\langle m_\nu \rangle$ from direct experiments to search for neutrinoless double beta decay of ^{76}Ge [6], ^{116}Cd [7], ^{136}Xe [8], and ^{150}Nd [9]: $\langle m_\nu \rangle < 1...4$ eV.

So, the neutrinoless β–β decay within the Sun could not give a considerable contribution to the lack of Solar neutrinos. Nevertheless it could play more important role in more compact and dense astronomical objects (such as pre-Supernova core) and in stellar nucleosynthesis.

Nucleus	Q_i, keV	lg ft	N_i	^{7}Be 384	^{7}Be 862	^{8}B 331	^{8}B 14018
^{7}Be	384	3.5	$8.8\cdot 10^{43}$	$2.5\cdot 10^8$	$1.5\cdot 10^9$	–	21
^{7}Be	862	3.3	$8.8\cdot 10^{43}$	$1.5\cdot 10^9$	$6.7\cdot 10^9$	1.1	39
^{8}B	331	2.9	$1.8\cdot 10^{34}$	–	1.1	–	–
^{8}B	14018	5.7	$1.8\cdot 10^{34}$	21	39	–	–

Table 1: $\Lambda_{ij}/\langle m_\nu \rangle^2$, s$^{-1}\cdot$eV^{-2}. Values less than 1.0 are not showed.

REFERENCES

1. A.F. Pacheco, Phys.Rev.Lett. 53 (1984) 979
2. T. Tomoda, Rep.Prog.Phys. 54 (1991) 53
3. J.N. Bahcall and R.K. Ulrich, Rev.Mod.Phys. 60 (1988) 297
4. Y. Kohayama, K. Kubodera, and K. Yazaki, Phys.Lett.B 168 (1986) 21
5. Table of Isotopes, eds. C.M. Lederer and V.S. Shirley (7th ed., Wiley, NY, 1978)
6. H.V. Klapdor-Kleingrothaus, Nucl.Phys.B (Proc.Suppl.) 31 (1993) 72
7. F.A. Danevich et al. Sov.J.Nucl.Phys., to be published
8. J.-L. Vuilleumier et al. Nucl.Phys.B (Proc.Suppl.) 31 (1993) 80
9. M.K. Moe, M.A. Nelson, and M.A. Vilent, Prog.Part.Nucl.Phys. 32 (1994) 247

The Search of 2β Decay of ^{116}Cd with ^{116}CdWO$_4$ Crystal Scintillators

F.A.Danevich, A.Sh.Georgadze, V.V.Kobychev, B.N.Kropivyansky,
V.N.Kuts, A.S.Nikolaiko, V.I.Tretyak and Yu.Zdesenko

*Institute for Nuclear Research of the Ukrainian National Academy of Sciences
Prospect Nauki 47, 252028 Kiev, Ukraine*

Abstract. The 2β decay experiment (about 30000 h) with ^{116}CdWO$_4$ scintillators enriched in ^{116}Cd to 83% has been performed in the Solotvina Underground Laboratory. The limits $T_{1/2}^{0\nu} \geq 2.9 \cdot 10^{22}$ y (90% CL) and $T_{1/2}^{2\nu} \geq 1.8 \cdot 10^{19}$ y (99% CL) have been set for $0\nu 2\beta$ and $2\nu 2\beta$ decay of ^{116}Cd. After subsraction of the spectrum of natural CdWO$_4$ scintillator from the spectrum of enriched ^{116}CdWO$_4$ detector the positive effect has been seen which is in accordance with $2\nu 2\beta$ decay of ^{116}Cd with $T_{1/2}^{2\nu} = \{2.7^{+0.5}_{-0.4}(stat.)^{+0.9}_{-0.6}(syst.)\} \cdot 10^{19}$ y. However the imitation of the effect by the impurities of ^{90}Sr can not be excluded.

Key words: 2β decay.— Neutrino mass.— Scintillators.

1. Introduction

The remarkable results have been obtained in double beta decay investigations due to sharp intensification of research during last decade: the half-life limit more than 10^{24} y for $0\nu 2\beta$ decay of ^{76}Ge [1–3]; the large scale experiment on ^{76}Ge with enriched HPGe detectors [3]; the level of 10^{23} y for lim $T_{1/2}^{0\nu}$ of ^{136}Xe [4] and 10^{22} y for ^{82}Se [5], ^{100}Mo [6], ^{116}Cd [7] and ^{130}Te [8]; low temperature high energy resolution bolometer with TeO$_2$ crystal created for study of 2β decay of ^{130}Te [8]; the observation of $2\nu 2\beta$ decay of ^{76}Ge [3, 9], ^{82}Se [5], ^{100}Mo [10–12], ^{150}Nd [11, 13] in direct counting experiments; progress in theoretical interpretation of the phenomenon [14, 15].

These achievements are related with particular place of 2β investigations among many approaches to study the neutrino properties. Now they represent the only way to measure an effective Majorana mass of the electron neutrino and yield the most stringent limits on this mass as well as on lepton charge nonconservation, right-handed admixtures in the weak interaction, the neutrino-Majoron coupling constant and other parameters of theories beyond the Standard Model. On the other hand such progress would be impossible without advances in experimental technique.

The results of 2β decay investigations of ^{116}Cd since 1987 are summarized in presented paper. The problems of detector's making and high-sensitive measuring technique development were solved during these investigations.

2. Detectors, installation and background measurements

The energy released in transition ^{116}Cd $\rightarrow$ ^{116}Sn is equal to 2802 keV, ^{116}Cd abundance is 7.49(9)% [16]. Theoretical estimations [17] of $T_{1/2}(^{116}$Cd) with respect to $2\nu 2\beta$ decay are in the range of 10^{18}–10^{20} y whereas the calculated value of $T_{1/2}^{0\nu} \cdot \langle m_\nu \rangle^2 = 4.87 \cdot 10^{23}$ y·eV^2 was found to be much more stable. This value of product is one of the lowest among the values for all $2\beta^-$ active nuclides. It is near 4 times lower then values for ^{76}Ge and ^{136}Xe.

In order to increase the sensitivity of research, the ^{116}CdWO$_4$ crystals enriched in ^{116}Cd to 83% were grown [18, 19]. Three of samples of crystals (19.0, 14.0 and 12.5 cm^3)

were used in experiment. The number of ^{116}Cd nuclei in these samples is $2.09 \cdot 10^{23}$, $1.54 \cdot 10^{23}$ and $1.37 \cdot 10^{23}$, respectively. The ^{nat}CdWO$_4$ scintillators of volume 8.0, 9.1 and 56.8 cm^3 were used in the experiments also.

All measurements were carried out in the Solotvina Underground Laboratory of INR built in the salt mine of the depth more than 1000 mwe where cosmic muon flux is suppressed by a factor of 10^4, and natural γ background is near 30–70 times lower than in chambers built of common materials [20]. The data acquisition and stabilization system described in details elsewhere [21].

The detectors' background in the energy interval 2.7–2.9 MeV was reduced successively by more than two orders of magnitude [7, 18, 19, 22] with different installations. These measurements let to clear up the origins of different components of the detectors' background and to construct the installation with the best characteristics. The active and passive shieldings are applied. The passive shielding of OFHC copper (5 cm) and lead (25 cm) surrounds the plastic scintillator which was used as active shielding. The cadmium tungstate crystal is viewed py PMT (FEU-110) through a light-guide 51 cm long. The energy resolution of the detectors is equal to 14.1, 8.2 and 7.1% at energy 662, 1770 and 2615 keV respectively for ^{116}CdWO$_4$ and 12.2, 7.9 and 6.7% – for ^{nat}CdWO$_4$. The energy calibration was carried out with ^{207}Bi weekly and with ^{232}Th once in three weeks.

The background of ^{116}CdWO$_4$ detector in the region of ^{116}Cd $0\nu2\beta$ decay (2.7 - 2.9 MeV) in this installation was reduced down to 0.53 counts/y·kg·keV. This value is similar to the characteristics of the best superlow-background HPGe detectors employed in ^{76}Ge 2β decay investigations [1, 3, 9].

3. Two-neutrino 2β decay of ^{116}Cd

An examination of $2\nu2\beta$ decay is embarrassed by reason that the effect to be found has continuous distribution like a background in contrast to clearly expressed peak in $0\nu2\beta$ decay. An effect-to-background ratio ranged within 1/10–1/7 in the majority of previous experiments where $2\nu2\beta$ decay was revealed [5, 10], and only in three last works it was succeeded to increase this value to 1.3:1 for ^{76}Ge [3], (2–3):1 for ^{100}Mo [11, 12] and (3–4):1 for ^{150}Nd [11, 13].

To estimate the probability of ^{116}Cd $2\nu2\beta$ decay, the results of the last measurements with natural and enriched detectors were used: ^{116}CdWO$_4$ (15.2 cm^3) – 2982 h, ^{nat}CdWO$_4$ (9.1 cm^3) – 1596 h (Fig. 2). The simplest (and very conservative) method to evaluate the limit on $2\nu2\beta$ effect in enriched crystal is to set the model $2\nu2\beta$ distribution equal to the experimental spectrum in some energy region and to take the full model's area as an effect limit. To calculate the efficiency of the detector its responce function for events being sought was simulated by Monte Carlo code [23]. The energy and angular distributions of the electrons in various mechanisms of ^{116}Cd 2β decay, the processes of interaction of electrons with the crystal substance as well as the detector's resolution were taken into account. The calculated in this way responce function to $2\nu2\beta$ events is shown in Fig. 1. It corresponds to the limit $T^{2\nu}_{1/2} = 1.8 \cdot 10^{19}$ y (99% CL).

The measurements with ^{nat}CdWO$_4$ crystal (9.1 cm^3) and experimental data about internal radioactive impurities of the crystals were used then. In this order the model distributions were subtracted from measured spectra. These models correspond to background induced by impurities of ^{40}K ($2.3 \cdot 10^{-3}$ Bk/kg), ^{137}Cs ($1.5 \cdot 10^{-3}$ Bk/kg), ^{226}Ra ($1.5 \cdot 10^{-5}$ Bk/kg), ^{232}Th ($1.7 \cdot 10^{-5}$ Bk/kg), ^{238}U ($1.8 \cdot 10^{-5}$ Bk/kg) in enriched crystal and ^{40}K ($3.6 \cdot 10^{-3}$ Bk/kg), ^{232}Th ($1.2 \cdot 10^{-5}$ Bk/kg) in natural one. It should be noted thad substracted fraction was equal to 13% of original experimental spectrum in the energy region of 1.2–2.4 MeV and 7.9% – in region of 1.4–2.4 MeV for enriched detector and 24% and 14% respectively – for the natural crystal.

The maximal possible contribution of thermal neutrons capture by ^{113}Cd nuclei to the spectrum in region of ^{116}Cd $2\nu2\beta$ decay was estimated also. According to calculations this contribution is negligible (<1%).

The spectrum of ^{nat}CdWO$_4$ crystal (normalysed to the volume and time of measurements) was subtracted from the spectrum of ^{116}CdWO$_4$. The difference is shown in Fig. 2

together with theoretical $2\nu2\beta$ distribution for ^{116}Cd with $T^{2\nu}_{1/2} = 2.7 \cdot 10^{19}$ y. The effect-to-background ratio is equal to 1.1:1 in energy interval 1.4–2.4 Mev. Taking in account the possible errors in amounts of radioactive impurities and some uncertainity of the procedure of normalization to the volume, we derive following value of half-life of ^{116}Cd with respect to $2\nu2\beta$ decay:

$$T^{2\nu}_{1/2} = \{2.7^{+0.5}_{-0.4}(\text{stat.})^{+0.9}_{-0.6}(\text{syst.})\} \cdot 10^{19} \text{ y}.$$

However this effect could be explained also by the impurity of ^{90}Sr/^{90}Y in enriched crystal with activity $2 \cdot 10^{-3}$ Bk/kg (or by the difference for this value in enriched and natural crystals). Our experimental restrictions on ^{90}Sr impurities do not exclude this possibility: ^{116}CdWO$_4$ – <$2.8 \cdot 10^{-3}$ Bk/kg; ^{nat}CdWO$_4$ – <$1.4 \cdot 10^{-3}$ Bk/kg. Therefore the firm conclusion that the observed effect is really $2\nu2\beta$ decay of ^{116}Cd can be made only after more precise measurement of ^{90}Sr impurities in the crystals.

4. Neutrinoless 2β decay of ^{116}Cd

For evaluations of $0\nu2\beta$ results the total spectrum was used merging the measurements with ^{116}CdWO$_4$ crystals in different installations in which the background rate in interval 2.7–2.9 MeV was below 1.0 counts/y·kg·keV. Full time of measurements is equal to 5822 h, the product of the number of ^{116}Cd nuclei by time – $1.08 \cdot 10^{23}$ nuclei·y, mean background rate – 0.55 counts/y·kg·keV (2.7–2.9 MeV). A part of the total spectrum in the energy range 2.2–4.0 MeV is shown in Fig. 3. Since the peak of $0\nu2\beta$ decay is evidently absent, the data were used to estimate the limiting probability of this process. The $T_{1/2}$ limit was calculated by known formula:

$$\lim T_{1/2} = \ln 2 \cdot \varepsilon \cdot t \cdot N_n / \lim S_e,$$

where N_n is the number of nuclei of isotope under study, ε – efficiency of registration, t – time of measurements, $\lim S_e$ – limiting number of effect's events which can be excluded with given confidence level. The responce function of ^{116}CdWO$_4$ detectors for potential events of ^{116}Cd $0\nu2\beta$ decay is a gaussian with center at energy 2802 keV and FWHM = 189 keV (such a distribution is shown in Fig. 3 with $T_{1/2} = 1 \cdot 10^{22}$ y). It was found by simulation [23], that for total spectrum $\varepsilon = 83.5\%$. The value of $\lim S_e$ was evaluated by maximum likelihood method and by standard least-squares fit [24]. It was assumed that the experimental spectrum can be described in the region of 2420–3420 keV by a sum of three functions, one of which corresponds to the effect, while two others (a first-degree polynomial and a gaussian with a center at 2614.5 keV and FWHM = 183 keV – γ line of ^{208}Tl) correspond to background. A maximization of the likelihood function in the region 2420–3420 keV gives for the limiting area of $0\nu2\beta$ peak the values ranged from 0.7 to 2.0 counts with CL = 90%. It corresponds to the limit $T^{0\nu}_{1/2} \geq 3.1 \cdot 10^{22}$ y. Least-square fit by above mentioned functions gave for peak area the value less than 2.2 counts (90% CL) which corresponds to the limit $T^{0\nu}_{1/2} \geq 2.9 \cdot 10^{22}$ y. Thus different methods of evaluation give the close values that confirm the reliability of the results and allow to set the following limit for ^{116}Cd $0\nu2\beta$ decay:

$$T^{0\nu}_{1/2} \geq 2.9 \cdot 10^{22} \text{ y, } 90\% \text{ CL } (5.4 \cdot 10^{22} \text{ y, } 68\% \text{ CL}).$$

Comparing this limit with theoretical calculations [17], we have computed the restrictions on the neutrino mass and right-handed admixtures in the weak interaction: $\langle m_\nu \rangle \leq 4.6$ eV, $\langle \eta \rangle \leq 5.9 \cdot 10^{-8}$, $\langle \lambda \rangle \leq 5.3 \cdot 10^{-6}$. Assuming $\langle \eta \rangle = 0$, $\langle \lambda \rangle = 0$ and using the value [17] $T^{0\nu}_{1/2} \cdot \langle m_\nu \rangle^2 = 4.87 \cdot 10^{23}$ y·eV^2, the restriction $\langle m_\nu \rangle \leq 4.1$ eV (90% CL) can be obtained.

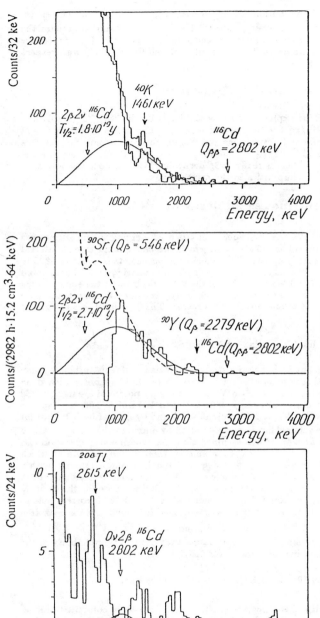

Figure 1. The background of enriched $^{116}CdWO_4$ (15.2 cm^3, 2982 h - single line) and natural $CdWO_4$ (9.1 cm^3, 1596 h - bold line; normalized to the time and volume of $^{116}CdWO_4$) crystals. The smooth curve - theoretical distribution for $2\nu 2\beta$ decay of ^{116}Cd with $T_{1/2}^{2\nu}=1.8\cdot 10^{18}$ y (99% CL limit).

Figure 2. The difference of backgrounds of $^{116}CdWO_4$ and $CdWO_4$ crystals. The spectra are normalized to the volume and time, and the models of radioactive impurities are subtracted. Smooth line corresponds to $2\nu 2\beta$ decay of ^{116}Cd with $T_{1/2}^{2\nu}=2.7\cdot 10^{18}$ y. Dashed line - β spectrum of ^{90}Y with activity 2 mBk/kg.

Figure 3. The total background spectrum of $^{116}CdWO_4$ detectors for 5822 h in the region of ^{116}Cd $0\nu 2\beta$ decay. The theoretical distribution (smooth line) corresponds to $T_{1/2}^{0\nu}=1.0\cdot 10^{22}$ y.

ACKNOWLEDGEMENTS

The research described in this paper was supported in part by Grant No U54000 from the International Scientific Foundation. Authors would like to thank O.A.Ponkratenko for more accurate definition of the influence of bremsstrahlung on the responce function of $CdWO_4$ detectors, and P.P.Berczik for help in preparation of the text.

REFERENCES

1. D.O.Caldwell et al., Nucl. Phys. B (Proc. Suppl.) 13 (1990) 547.
2. I.V.Kirpichnikov, Nucl. Phys. B (Proc. Suppl.) 28A (1992) 210.
3. H.V. Klapdor-Kleingrothhaus (for the Heidelberg-Moscow Collaboration), Nucl. Phys. B (Proc. Suppl.) 31 (1993) 72.
4. J.-C. Vuilleumier et al., Phys. Rev. D 48 (1993) 1009.
5. S.R. Elliot et al., Phys. Rev. C 46 (1992) 1535.
6. M. Alston-Garnjost et al., Phys. Rev. Lett. 71 (!993) 831.
7. F.A. Dahevich et al., Proc. 3rd Int. Symp. on Weak and Electromagn. Interactions in Nuclei WEIN-92 (Dubna, June 1992), ed. Ts.D. Vylov (World Sci. Publ. Co., 1993) p.575.
8. A. Alessandrello et al., Phys. Lett. B 285 (1992) 176;

   A. Alessandrello et al., Proc. 3rd Int. Workshop on Theor. and Phenom. in Astroparticle and Underground Phys. TAUP-93 (Assergi, Sept. 1993), to be published.
9. R.L. Brodzinski et al., Nucl. Phys. B (Proc. Suppl.) 31 (1993) 76.
10. H. Ejiri et al., Nucl. Phys. B (Proc. Suppl.) 28A (1992) 219.
11. S.R. Elliot et al., Nucl. Phys. B (Proc. Suppl.) 31 (1993) 68.
12. NEMO Collaboration, Proc. 3rd Int. Workshop on Theor. and Phenom. in Astroparticle and Underground Phys. TAUP-93 (Assergi, Sept. 1993), to be published.
13. V.A. Artem'ev et al., JETP Lett. 58 (1991) 53.
14. T. Tomoda, Rep. Prog. Phys. 54 (1991) 53.
15. H.V. Klapdor-Kleingrothhaus, Proc. 3rd Int. Symp. on Weak and Electromagn. Interactions in Nuclei WEIN-92 (Dubna, June 1992), ed. Ts.D. Vylov (World Sci. Publ. Co., 1993) p.201.
16. C.M. Lederer and V.S. Shirley, eds., Table of isotopes, 7th ed. (Wiley, New York, 1978).
17. A. Staudt, K. Muto, and H.V. Klapdor-Kleingrothhaus, Europhys. Lett. 13 (1990) 31.
18. F.A. Danevich et al., preprint KINR-88-11 (1988).
19. F.A. Danevich et al., JETP Lett. 49 (1989) 476.
20. Yu.G. Zdesenko et al., Proc. 2nd Int. Symp. on Underground Phys. (Baksan Valley, August 1987), ed. G.V. Domogatsky (moscow, Nauka, 1988) p.291.
21. G.N. Garkusha et al., preprint KINR-86-4 (1986).
22. Yu. Zdesenko, J. Phys. G: Nucl. Part. Phys. 17 (1991) 243.
23. Yu.G. Zdesenko et al., preprints KINR-86-43 (1986), KINR-89-7 (1989), KINR-92-8 (1992).
24. Particle Data Group, Phys. Rev. D 45 (1992), part II.

The keV Neutron Capture Cross Sections of ^{146}Nd, ^{148}Nd, and ^{150}Nd

K.A. Toukan(1), K. Debus(2), and F. Käppeler(2)

(1) College of Engineering and Technology, The University of Jordan, Amman, Jordan
(2) Kernforschungszentrum Karlsruhe, IK-III, Postfach 3640, D-76021 Karlsruhe, Germany

Abstract.

The neutron capture cross sections of ^{146,148,150}Nd have been determined relative to that of gold by means of the activation method. The samples were irradiated in a quasi-stellar neutron spectrum for $kT = 25$ keV using the ^{7}Li(p,n)^{7}Be reaction near threshold. Variation of the experimental conditions in different activations and the use of different samples allowed for the reliable determination of corrections and the evaluation of systematic uncertainties. The resulting stellar cross sections are given with uncertainties around 6%, which corresponds to an improvement by at least a factor of two compared to previous data.

Key words: nuclear astrophysics — neutron capture — s-process

1. INTRODUCTION

Recently, the s-process branchings at ^{147}Nd, ^{147}Pm, ^{148}Pm and ^{149}Pm sketched in Figure 1 have been studied in the light of significantly improved (n,γ) cross sections for the involved s-only isotopes ^{148}Sm and ^{150}Sm (1).

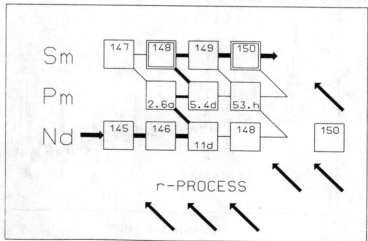

Fig. 1.— The s-process flow through the branchings at A=147, 148, and 149.

In order to optimize the set of neutron capture cross sections required for this analysis, the present work aimed at an improved determination of the relevant stellar cross sections of ^{146}Nd and ^{148}Nd. This study included also ^{150}Nd in order to establish the cross sections for a sequence of isotopes, which is important for comparison with statistical model calculations. Since the previously existing experimental data showed discrepancies up to a factor of 3, the cross section trend with neutron number was completely obscured. This trend represents a sensitive test for the calculated Nd, Pm, and Sm cross sections as the neutron number is increased starting from the magic configuration at N=82, and should allow to reduce the problems for the unstable branch point nuclei in such calculations.

2. EXPERIMENT

For the determination of the stellar neutron capture cross sections of ^{146}Nd, ^{148}Nd, and ^{150}Nd, samples of natural elemental neodymium were irradiated in a quasi-stellar neutron spectrum, which yields the proper cross section average $<\sigma v>/v_T$ for a thermal energy of $kT = 25\pm0.5$ keV (2, 3). The neodymium samples were sandwiched between thin gold foils for normalization to the well known gold cross section of 648 ± 10 mb at $kT = 25$ keV (3).

After the irradiations, the induced activities were determined using a well calibrated HP–Ge detector. Figure 2 shows the gamma-ray spectrum obtained from a 0.1 mm thick Nd sample after an irradiation of 3.5 h.

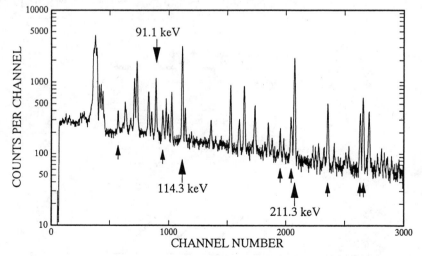

Fig. 2.— The gamma–ray spectrum measured after irradiating a 0.1 mm thick Nd–sample for 3.5 h (0.102 keV per channel). The lines from the decay of ^{147}Nd and ^{149}Nd are indicated by arrows (downward: ^{147}Nd, upward: ^{149}Nd).

Table 1: The 30 keV capture cross sections of ^{146}Nd, ^{148}Nd, and ^{150}Nd (in mbarns) compared with previous data

	^{146}Nd	^{148}Nd	^{150}Nd	
Previous Data				Reference
	150	210±80	240±150	(4)
	105±16	221±40	...	(5)
	113±25	123±20	...	(6)
	152±33	208±26	...	(7)
	205±21	186±19	187±19	(8)
	75±7	99.4±14	...	(9)
	118±14	130±15	153±18	(10)
	104	216	262	(11)
	102	82	147	(12)
This Work				
	87±4	152±9	159±10	

In addition to the numerous lines from the decay of ^{149}Nd ($t_{1/2}$=1.72 h), the spectrum includes also the major line from the decay of the longer-lived ^{147}Nd ($t_{1/2}$=10.98 d, E_γ=91.1 keV) with sufficient statistical significance. This means that both cross sections could be evaluated from the same spectrum. The other lines are partly from the decay of the Pm daughters, partly from the decay of ^{151}Nd. In total, seven activations have been performed.

By systematic variation of the experimental conditions with respect to sample thickness, irradiation time, and neutron flux, the related experimental uncertainties could be studied and the corresponding corrections be determined reliably.

3. RESULTS

The present results for $kT = 30$ keV are compared with the previous data in Table 1. The 6% uncertainty obtained in this work represent an improvement by at least a factor of 2. Consequently, the cross sections of the heavier Nd isotopes are sufficiently well described for the first time.

The data sets with the largest uncertainties (4, 5, 6, and 7) exhibit not only the largest discrepancies but often also very different cross section ratios. The only data set that is in reasonable agreement with the present work is that of Iijima et al. (10). However, even there the cross section trend with neutron number appears too flat. Note that the cross section of ^{146}Nd was significantly overestimated in all previous measurements except from Bradley et al. (9). On the other hand, Bradley et al. (9) find a very low cross section for ^{148}Nd. Since these data were obtained with a filtered reactor beam of 24±2 keV, they may have been affected by the resonance structure in these cross sections close to N=82.

The differences between the present data and the statistical model calculations of Harris (11) and of Holmes *et al.* (12) are compatible within the 50% uncertainty typical of these approaches with *global* parameter sets. However, only the calculation of Harris (11) shows reasonable agreement with respect to the cross section trend versus neutron number. New calculations using an improved parameter set based on a consistent *local* systematics are under way (13).

REFERENCES

(1) Wisshak, K., Guber, K., Voss, F., Käppeler, F., Reffo, G. 1993, *Phys. Rev.*, **C48**, 1401.

(2) Beer, H. and Käppeler, F. 1980, *Phys. Rev.*, **C21**, 534.

(3) Ratynski, W. and Käppeler, F. 1988, *Phys. Rev.*, **C37**, 595.

(4) Allen, B.J., Gibbons, J.H., and Macklin, R.L. 1971, *Adv. Nucl. Phys.*, **4**, 205.

(5) Siddappa, K., Sriramachandra Murty, M., and Rama Rao, J. 1973, *Nuovo Cim.*, **18A**, 48.

(6) Musgrove, A.R. de L., Allen, B.J., Boldeman, J. W., and Macklin, R.L. 1978, *Neutron Physics and Nuclear Data for Reactors and Other Applied Purposes*, (Paris: OECD) p.449.

(7) Nakajima, Y., Asami, A., Kawarasaki, Y., Furuta, Y., Yamamoto, T., and Kandu, Y. 1978, *Neutron Physics and Nuclear Data for Reactors and Other Applied Purposes*, (Paris: OECD) p.438.

(8) Kononov, V.N., Yurlov, B.D., Poletaev, E.D., and Timokhov, V.M. 1978, *Yad. Fiz.*, **27**, 10.

(9) Bradley, T., Parsa, Z., Stelts, M.L., and Chrien, R.E. 1979, *Nuclear Cross Sections for Technology*, eds. J.L. Fowler, C.H. Johnson, and C.D. Bowman (Washington D.C.: NBS), NBS Spec. Publ. 594, p.344.

(10) Iijima, S., Watanabe, T., Yoshida, T., Kikuchi, Y., and Nishimura, H. 1979, *Neutron Cross Sections of Fission Product Nuclei*, eds. C. Coceva and G.C. Panini, Report RIT/FISLDN(80) 1, NEANDC(E) 209"L", C.N.E.N., Bologna, p. 317.

(11) Harris, M.J. 1981, *Ap. Space Sci.*, **77**, 357.

(12) Holmes, J.A., Woosley, S.E., Fowler, W.A., and Zimmerman, B.A. 1976, *Atomic Data Nucl. Data Tables*, **18**, 305.

(13) Reffo, G. 1994, private communication.

Reaction Rate for ^{6}Li(p,α)^{3}He and ^{6}Li(d,α)^{4}He Reactions

J. Szabó

*Institute of Experimental Physics, Kossuth University,
H-4001 Debrecen, P.O.B. 105, Hungary*

Abstract. Thermonuclear reaction rate for ^{6}Li(p,α)^{3}He and ^{6}Li(d,α)^{4}He reactions is given in the range of $T_9 \sim 0.01 - 2.5$ K deduced from direct measured cross sections.

Introduction

The investigation of the ^{6}Li(p,α)^{3}He and the ^{6}Li(d,α)^{4}He reactions in the thermonuclear energy region is important in two aspects.

a.) These reactions are dominant destructive processes for ^{6}Li and ^{7}Li and could be incorporated in explaining the extreme low lithium abundances including the present theories of spallative (GCR) and big-bang nucleosynthesis (BBN) generation of L-elements (1).

b.) The importance of these reactions in the CTR program has been discussed by Feldbacher et al. (2). Among the exotic fuel cycles the ^{6}Li(p,α)^{3}He and ^{6}Li(d,α)^{4}He reactions are relatively clean energy sources as well as high efficiency for energy generation among the exotic fuel elements (Li, Be, B). This can be shown by a reaction-rate ratio for the actual reaction to that of T(d,n)^{4}He. These relative yields at $E_{ion} \sim 25$ keV are appr. 10^{-4} for ^{6}Li + p and 10^{-3} for ^{6}Li + d reactions, respectively, relatively large compared to other exotic fuels.

Both reactions have been investigated in the energy range $E_p < 200$ keV by many authors. As regards ^{6}Li(p,α)^{3}He reaction, Gemeinhardt et al. (3) have measured cross sections in the range $E_p = 50$-190 keV and Fiedler et al. (4) in the range $E_p = 23$-50 keV. Shinozuka et al. (5) and Elwyn et al. (6) determined cross sections at $E_p = 100$-200 keV, while Szabo et al. (7) in the range $E_p = 100$-180 keV. Lately, S. Engstler (8) has made an excellent investigation to obtain cross sections in the range $E_p = 10$-200 keV giving first experimentally evidence of the electron screening effect, too.

For the ^{6}Li(d,α)^{4}He reaction, Sawyer et al. (9) and Lee (10) have measured cross sections at $E_d = 60$-240 and $E_d = 40$-130 keV, respectively. Hirst et al. (11) have determined data in the range $E_d = 60$-200 keV. Szabo et al. (7) have measured cross sections at $E_d = 100$-180 keV, while Elwyn et al. (12) at $E_d = 100$-1000 keV. Golovkov et al. (13) have determined cross section data in the range $E_d = 35$-110 keV and Engstler (8) $E_d = 15$-200 keV.

This work was supported in part by the Hungarian Research Foundation (Contract no: 1734/91).

Experiment

The experimental set-up had been published in detail in our earlier paper (14). The energy calibration of the 200 keV accelerator had been performed by the $^{11}B(p,\gamma)^{12}C$ resonance at E_p = 163.1 keV and non-resonant $D(p,\gamma)^3He$ reaction, in absolute energy scale, using Ge(Li) detector. The energy uncertainty was about 2 keV. For the targets enriched ^6LiF(87.9 %) had been used. For the detection of ^3He and ^4He particles solid-state nuclear track detectors (SSNTD) were used with monitoring Si-detector.

Results

Angular distributions were measured for both reactions at the energies, E_{bomb} = 120 and 180 keV and these were found to be isotropic.

The total cross sections obtained from our experiment are shown in Figs.1, 2 together with other data in the energy range E_{bomb} < 200 keV. We have measured cross sections by direct α - detection using SSNTD of T-Cellit for (p,α) and CA 80-15 for (d,α) reactions, respectively. To distinguish and identify the ^3He and ^4He particles, an alternate etching method was developed. Due to the steep rise with energy no wonder that there is about a factor of 2 (or more) among the cross sections at the same energy measured by different authors.

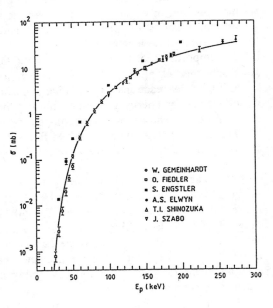

Fig.1. Experimental cross section for the $^6Li(p,\alpha)^3He$ reaction

From the measured cross section data astrophysical S-factors were calculated from the formula (15)

$$S(E) = E \cdot \sigma(E) \cdot \exp(2\pi\eta) \tag{1}$$

with the Sommerfeld-parameters $2\pi\eta = 89.9 \cdot E^{-1/2}$ (E is c·m in keV) for (p,α) and $2\pi\eta = 109.6 \cdot E^{-1/2}$ for (d,α) reactions, respectively. (The calculated numerical data are: 87.23 and 115.32.) The astrophysical S-factors are for (p,α) reaction S = (2.86 ± 0.26)MeV·b and for (d,α) reaction, S = (11.84 ± 0.54)MeV·b.

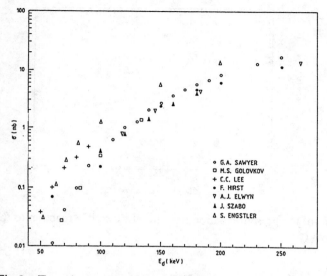

Fig.2. Experimental cross section for the ^{6}Li(d,α)^{4}He reaction

The thermonuclear reaction rate for the non-resonant reaction is given by (15)

$$N_A <\sigma \cdot v> = \frac{N_A}{\mu} \frac{7.2 \cdot 10^{-19}}{Z_1 Z_2} \tau^2 \exp(-\tau) S_0(E_0) \cdot (1+\frac{5}{12\tau}) \qquad (2)$$

where μ is the reduced masses, N_A is the Avogadro's number, $\tau = 3E_0/kT$.

The calculated thermonuclear reaction rates in the temperature range of $T_9 = 0.01 - 2.5$ are shown in Figs. 3 and 4 comparison to the values of Caughlan (15) and Elwyn (6).

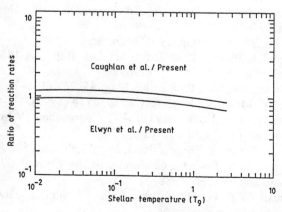

Fig.3. Ratio of the present reaction rates for ^{6}Li(p,α)^{3}He reaction to those of Caughlan (15) and Elwyn (6).

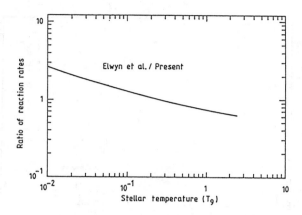

Fig.4. Ratio of the present reaction rates for ^{6}Li(d,α)^{4}He reaction to that of Elwyn (6).

References

1. H. Reeves, Rev. Mod. Phys. 66, 193 (1994).
2. R. Feldbacher, M. Heindler, G. Miley, IAEA-TECDOC-457, Vienna, 1986.
3. W. Gemeinhardt, D. Kamke, C. Rhöneck, Z. Phys. 197, 58 (1966).
4. O. Fiedler, P. Kunze, Nucl. Phys. A 96, 513 (1967).
5. T. Shinozuka, Y. Tanaka, K. Sugiyana, Nucl. Phys. A 326, 47 (1979).
6. A.J. Elwyn, R.E. Holland, C.N. Davids, L. Meyer-Schützmeister, F.P. Mooring, W. Ray, Phys. Rev. C 20, 1984 (1979).
7. J. Szabó, M. Várnagy, Z.T. Bödy, J. Csikai, *Int. Conference Nuclear Data for Science and Technology*, Antwerp, Ed. Böckhoff, K.H., Dordecht 1983, p. 956; unpublished (1990).
8. S. Engstler, Ph.D. Dissertation, 1991, Münster.
9. G.A. Sawyer, J.A. Phillips, Los Alamos Sci. Lab. Rep., LA-1587, 1953.
10. C.C. Lee, J. Korean Phys. Soc. 2,1 (1969).
11. F. Hirst, I. Johnstone, M.J. Poole, Phil. Mag. 45, 762 (1954).
12. A.J. Elwyn, J.E. Monahan, Phys. Rev. C 19, 2114 (1979).
13. M.S. Golovkov, V.S. Kulikanszkasz, V.T. Vorenchev, V.M. Krasnopolsky, V.I. Kukulin, Jad. Fiz. 34, 861 (1981).
14. M. Várnagy, J. Szabó, S. Szegedi, Nucl. Inst. Meth. 154, 557 (1978).
15. C.E. Rolfs and W.S. Rodney, *Cauldrons in the Cosmos*, Univ. of Chicago Press, Chicago and London, 1988.
16. G.R. Caughlan, W.A. Fowler, Atomic Data and Nuclear Data Tables 40, 283 (1988).

Analysis of the Triple–Alpha Process in the Potential Model

H. Krauss[†], K. Grün[†], H. Herndl[†], H. Oberhummer[†], P. Mohr[*], H. Abele[*], G. Staudt[*]

[†] Technische Universität, Institut für Kernphysik, Vienna, Austria
[*] Universität Tübingen, Physikalisches Institut, Tübingen, Germany

> The potential model is applied to the study of the triple–alpha process. The two reaction steps were treated consistently using the folding procedure for the optical and bound-state potentials. The low-temperature reaction rates obtained in the potential model are calculated and compared with earlier calculations.

The bulk of the carbon abundance in the universe is a consequence of the triple–alpha process. This reaction is expected to occur in two steps. The first is the formation of the short living nucleus ^{8}Be ($\tau \approx 10^{-16}$ s). The second step is the production of ^{12}C by the capture reaction ^{8}Be$(\alpha,\gamma)^{12}$C, which reaction rate depends sensitively on the properties of the 0_2^+ state in ^{12}C.

At temperatures corresponding to helium burning in red giants ($T \approx 10^8 - 10^9$ K) the ^{8}Be quantity is in thermal equilibrium with the α-particles. The reaction rate for the triple–alpha process can then be expressed by the number density of ^{8}Be nuclei in equilibrium with the ^{4}He bath multiplied with the reaction rate of ^{8}Be$(\alpha,\gamma)^{12}$C (1).

At lower temperatures (a few times 10^7 K) occuring in accreting white dwarfs (2) and neutron stars (3) this equilibrium does not exist anymore (4). In this case the triple–alpha process proceeds via formation of the ^{8}Be ground-state in the lower wing of the $\alpha+\alpha$ resonance. The triple–alpha reaction-rate now becomes dependent on energies at which ^{8}Be is formed and is given by summing over all these energies.

The low-temperature reaction-rates for the triple–alpha process have been calculated using a Breit-Wigner formula with an energy-dependent width (5). In a subsequent paper a direct contribution using a hard core was added (6). A microscopic calculation in the generator-coordinate method for the second step of the triple–alpha process including the low-temperature reaction rates was carried out by Descouvemont and Baye (7).

In this work we assume that the first and second step of the triple–alpha process proceeds as a direct reaction. Even though in nuclear astrophysics the term *direct* has been used synonymously with *non-resonant*, we think that this is not appropriate. Direct processes can be described in potential models, where sometimes also potential resonances can be observed.

We used the double–folding procedure in calculating the α–α potential in order to describe the elastic scattering data up to energies of about 40 MeV as well as the properties of ^{8}Be. The calculations reproduce excellently the energies and α–widths of the ^{8}Be ground–state (8).

With the knowledge of the ^{8}Be density–distribution obtained from the α–α folding potential (8) we are able to calculate the optical potential for the α–^{8}Be system using again the folding procedure. Since no experimental phase shifts for α–^{8}Be are available, we have fitted the normalization factors of the folding potentials to the relevant bound and resonance energies of ^{12}C.

The theoretical cross section σ^{th} is obtained from the DC cross section

$$\sigma^{\text{th}} = \sum_i C_i^2 S_{\alpha,i} \sigma_i^{\text{DC}} \quad . \tag{1}$$

The sum extends over the ground and first excited state in ^{12}C. The spectroscopic factors $S_{\alpha,i}$ have been taken from Kurath (9) and Kwaśniewicz and Jarczyk (10). The DC cross section σ_i^{DC} is essentially determined by the overlap of the scattering wave function in the entrance channel, which were calculated using the α–^{8}Be potential, the bound–state wave function in the exit channel and the multipole transition–operator. For the computation we used the direct–capture code TEDCA (11).

The results for the DC calculation of ^{8}Be$(\alpha,\gamma)^{12}$C gives values for the α–, γ– and total widths of the important 0_2^+–resonance in ^{12}C which are comparable with the experimental data (8).

We now determine the reaction rates using the astrophysical S–factor calculated with the potential model. We want to present low–temperature rates ($T = 1 \cdot 10^7$ K – $1 \cdot 10^8$ K). Since the astrophysical S–factor consists not only of a resonant and non–resonant part but also of non–neglible interference terms, we prefer to give the low-temperature reaction rates in tabular form. The reaction rate per particle pair is calculated by integrating the astrophysical S–factor multiplied with the Maxwell–Boltzmann velocity–distribution (1):

$$\langle \sigma v \rangle = \frac{8}{\pi \mu} \frac{1}{(kT)^{\frac{3}{2}}} \int_0^\infty S(E) \exp\left[-\frac{E}{kT} - \left(\frac{E_G}{E}\right)^{\frac{1}{2}} \right] dE \quad , \tag{2}$$

with the Gamov energy given by

$$E_G = 2\mu \left(\frac{\pi e^2 Z_1 Z_2}{\hbar} \right)^2 \quad . \tag{3}$$

In the above equations T is the temperature, μ is the reduced mass of the system and $Z_1 e$ and $Z_2 e$ are the charges of the colliding nuclei. For ^{8}Be$(\alpha,\gamma)^{12}$C the Gamow energy $E_G = 167$ keV and the effective mean energy $E_0 = 31$–146 keV corresponding to $T = 1 \cdot 10^7$–$1 \cdot 10^8$ K.

The reaction rate per unit volume for the 3α process is then obtained by

$$r = N_\alpha N_{^8\text{Be}} \left\langle \alpha^8\text{Be} \right\rangle \quad , \tag{4}$$

where N_α and $N_{^8Be}$ are the number densities. The quantity $\langle \alpha^8 Be \rangle$ is the reaction rate per particle pair (Eq. 2) for the $^8Be(\alpha,\gamma)^{12}C$ reaction.

Assuming equilibrium between α-particles and ^8Be-nuclei, the number of ^8Be-nuclei is given by the Saha equation (1):

$$N_{^8Be} = \frac{1}{2} N_\alpha^2 \left(\frac{2\pi}{\mu kT} \right)^{\frac{3}{2}} \hbar^3 \exp\left(-\frac{E_R}{kT} \right) \quad , \tag{5}$$

where $E_R = 92.1$ keV is the resonance energy of the ^8Be ground state.

For the low-temperature rates the triple-alpha process can proceed via forming the ^8Be ground state in the lower wing of the $\alpha+\alpha$ resonance (5,6). In this case the reaction rate $\langle \alpha^8 Be \rangle$ as well as the number density $N_{^8Be}$ become dependent on the energy at which ^8Be is formed. However, as in previous work (6,7), we assume for the calculation of the reaction rate $\langle \alpha^8 Be \rangle$ that ^8Be is formed at the resonance energy. The reaction rates for the triple-alpha process can be calculated by the formulae given by (5), but with the improved $\langle \alpha^8 Be \rangle$ reaction rates given in this work.

TABLE 1. Low-temperature reaction rates for $^8Be(\alpha,\gamma)^{12}C$ in cm^3 s^{-1} for the transition to the 0_1^+- and 2_1^+-state. The square brackets denote powers of ten.

T [10^7 K]	0_1^+	2_1^+	Total
1	2.79[-64]	5.94[-63]	6.22[-63]
2	1.15[-54]	2.76[-53]	2.87[-53]
3	5.24[-50]	1.41[-48]	1.47[-48]
4	4.54[-47]	1.37[-45]	1.42[-43]
5	5.57[-45]	1.88[-43]	1.94[-43]
6	2.18[-43]	8.29[-42]	8.51[-42]
7	4.06[-42]	2.02[-40]	2.06[-40]
8	4.55[-41]	1.07[-38]	1.08[-38]
9	3.51[-40]	7.50[-37]	7.50[-37]
10	2.03[-39]	2.55[-35]	2.55[-35]

As can be seen from Table 1, the transition to the ground-state is very small compared to the transition to the first excited state even in the low-temperature range and is neglible for temperatures above $T = 7 \cdot 10^7$ K. In this region the reaction rate is determined only by the resonant part.

Normalized to the low-temperature reaction rates of the present work the reaction rates calculated by (5) are two times smaller, because these authors considered only the resonant contribution. In the calculations of (6) a resonant contribution was combined with a hard-core direct part. In this calculation the reaction rates are overestimated because the spectroscopic factor of the 2_1^+-state was assumed to be the same as for the ground state. However, shell-model calculations have shown that the spectroscopic factor of the 2_1^+-state is 3.5 times lower (10), because this state has a strong

^{8}Be(2^+)⊗α-configuration. Also in the microscopic calculation by (7) the important ^{8}Be(2^+)⊗α-configuration is missing in the 2_1^+-state of ^{12}C (12). This is probably also reflected by the low B(E2)-value for the transition $0_2^+ \to 2_1^+$ resulting from this microscopic calculation. Therefore in this respect the microscopic calculation suffers from the same deficiency as the calculation using a resonant contribution and a hard-core direct part.

The following advantages in applying the potential model together with the folding procedure to the calculation of the reaction rates for the triple-alpha process are evident. The two reaction steps are treated consistently using the same model, only the α-α cross sections and the energies of the bound and resonance states in ^{12}C are adjusted to the experimental data and the resonant and non-resonant parts of the astrophysical S-factor are obtained simultaneously. These features allow us to calculate the extreme sensitivity of this reaction to variations in the underlying nucleon-nucleon interaction. Such variations shift the resonace energy in the ^{8}Be ground-state as well as the important 0_2^+ state in ^{12}C and change the reaction rates considerably. If the nucleon-nucleon force in our universe would be weaker (stronger) only by 0.1%, the reaction rate of the triple-alpha process would be reduced (enhanced by about a factor of 250 (13). Therefore, these calculations allow an assessment of the importance of the triple-alpha process for the anthropic principle in cosmology.

We want to thank the Fonds zur Förderung der wissenschaftlichen Forschung in Österreich (project P8806-PHY), the Österreichische Nationalbank (project 3924) and the Deutsche Forschungsgemeinschaft (DFG-project Sta290/2) for their support.

REFERENCES

1. Rolfs, C., and Rodney, W. S., *Cauldrons in the Cosmos*, (Chicago, University of Chicago Press 1988)
2. Nomoto, K., *ApJ* **253**, 798 (1982); *ApJ* **257**, 780 (1982)
3. Wallace, R. K., Woosley, S. E., Weaver, T. A., *ApJ* **258**, 696 (1982)
4. Cameron, A. G. W., *ApJ* **149**, 239 (1959)
5. Nomoto, K., Thielemann, F.-K., Hiyaji, S., *A&A* **149**, 239 (1985)
6. Langanke, K., Wiescher, M., Thielemann, F.-K., *Z. Phys.* **A324**, 147 (1986)
7. Descouvemont, P., and Baye, D., *Phys. Rev.* **C36**, 54 (1987)
8. Mohr, P., Abele, H., Kölle, V., Staudt, G., Oberhummer, H., Krauss H., *Z. Phys. A*, 1994, in press
9. Kurath, D., *Phys. Rev.* **C7**, 1390 (1973)
10. Kwaśniewicz, E., and Jarczyk, L., *J. Phys.* **G12**, 697 (1986)
11. Krauss, H., code TEDCA, TU Wien, unpublished, 1992
12. Descouvemont, P., private communication, 1993
13. Oberhummer, H., Krauss, R., Grün, H., Rauscher, T., Abele, H., Mohr, P., Staudt, G., *Z. Phys. A*, 1994, in press

Measurement of the Half-Life of ^{44}Ti

J. Meißner, J. Görres, H. Schatz, S. Vouzoukas, M. Wiescher

Department of Physics, University of Notre Dame, Notre Dame, IN 46556

L. Buchmann

TRIUMF, Vancouver, B.C., Canada

D. Bazin, J. A. Brown, M. Hellström, J. H. Kelley, R. A. Kryger, D. J. Morrissey, M. Steiner

National Superconducting Cyclotron Laboratory, Michigan State University, East Lansing, Michigan 48824

K. W. Scheller

Physics Department, University of Evansville, Evansville, IN 47722

R. N. Boyd

Physics Department, Ohio State University Columbius, OH 43210

I. INTRODUCTION

Previous measurements of the ^{44}Ti half-life display a large spread, and the reported half-lives range from 43 to 67 yrs. Since remnants of supernovae, especially SN 1987A, are expected to be partially powered by the decay of ^{44}Ti it is important to know the half-life to higher precision to predict the abundance of ^{44}Ti. A recent COMPTEL observation of ^{44}Ti in CAS A sets upper and lower limits on the initially produced amount of ^{44}Ti (1). In this case, a tighter constraint on the half-life can considerably decrease the uncertainty. An OSSE observation (2) could not confirm the ^{44}Ti γ-ray emmission in CAS A, but agree that a tighter constraint on the half-life will significantly improve Supernova models. In the present study we have employed a new method (3) to produce and implant on-line a well determined number of ^{44}Ti.

II. EXPERIMENTAL METHOD AND PRELIMINARY RESULTS

The production of a suitable ^{44}Ti sample was carried out at the National Superconducting Cyclotron Laboratory at Michigan State University. With a primary ^{58}Ni beam of E/A= 70 MeV/u and an intensity of about 20 enA a fragmented beam was produced on a ^{9}Be target. Using the A1200 projectile fragment separator ^{44}Ti were separated from most other fragments. During 32 hours several samples were prepared by implanting this beam into Al-foils. The total intensity of all fragments was monitored continously with four PIN-diodes which were placed at various distances and angles around the production target. The transmission of the A1200 was checked by inserting a thin scintillator at the position of the sample every 2 hours. It stayed constant within 4% over the whole experiment. A PIN diode was inserted at the position of the implantation foil before and after the implantation into each sample. This allowed a particle identification from energy loss versus time-of-flight histograms. The resulting ratio of ^{44}Ti versus other fragments stayed constant within 3% over the course of the experiment. The number of implanted ^{44}Ti particles could then be determined from the integrated count rate of the beam monitors and this ratio. Among other fragments one of the main beam contaminants was ^{43}Sc. The number of implanted ^{43}Sc was determined from the time dependent count rate of the monitors and the ratio of ^{43}Sc versus all other fragments. The implantation foils were changed every 6 to 9 hours and were checked for the γ-activity following the ^{43}Sc decay. The measured decay curve is in excellent agreement with the well known half-live of ^{43}Sc ($T_{1/2} = 3.891 \pm 0.012$ hr (4)). The initial ^{43}Sc activity was determined by extrapolation of the decay curve to the time when implantation was stopped. This gave an independent method to determine the number of implanted ^{44}Ti particles using the dead time independent ^{43}Sc/^{44}Ti ratio.

A total number of $(59\pm9)\cdot 10^6$ ^{44}Ti particle were implanted into the three samples.

The half-life was determined by measuring the γ-activity of a stack of all implanted ^{44}Ti-samples. The emission of the 1157 keV γ-ray of the ^{44}Ti(EC)^{44}Sc(β^++EC)^{40}Ca decay cascade was measured over a period of 2 months with two 55% germanium γ-ray detectors. Figure 1 shows the γ-ray spectrum taken with the sample. The efficiency at 1157 keV has been determined with calibrated ^{60}Co, ^{22}Na and ^{137}Cs sources. Corrections for coincidence summing in the ^{60}Co($\beta\nu$)^{60}Ni(γ) cascade have been applied according to the method in (5) and with the total photon absorption cross section from (6).

The analysis of the γ-ray spectra from the ^{44}Ti decay yielded an activity A= $(2.25\pm0.2)\cdot 10^{-2}$ s^{-1}. With

$$T_{1/2} = \ln 2 \frac{N_{^{44}Ti}}{A} \quad (1)$$

a preliminary analysis indicates a half-life of $T_{1/2} = 58$ years with an estimated 1σ uncertainty of 10 years.

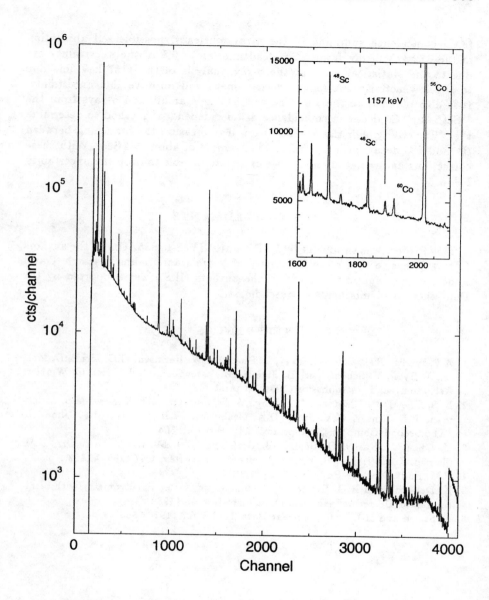

FIG. 1. ^{44}Ti(EC) γ-ray spectrum taken over the course of 2 month

III. OUTLOOK

The large uncertainties in the number of implanted ^{44}Ti particles is mainly due to the the spread of the results for both applied methods. This uncertainty will decrease, as the analysis of the ^{43}Sc decay still lacks final corrections (e.g. coincidence summing in the γ-ray efficiency, possible self absorption of the 372 keV γ-ray from ^{43}Sc, deadtime, etc). 9% of the uncertainty are due to the statistical error in the γ-ray analysis of the 1157 keV line due to the low activity. A longer measurement will improve this uncertainty. It is also planned to measure the two 511 keV anihilation γ-rays from the ^{44}Sc$(\beta^+\nu)^{44}$Ca decay in coincidence as an independent method to determine the ^{44}Ti activity. For this method we will re-measure the branching between β^+ and EC decay with a purified ^{44}Ti sample of about 3 kBeq. With these corrections and remeasurements we expect to be able to drop the uncertainty below 10%.

ACKNOWLEDMENTS

This Project was supportet by NSF grant PHY88-03035. One of the authors (J.M.) wishes to thank the University of Notre Dame Zahm Research Travel Fund for their support and one of the authors (H.S.) was supported by the Deutscher Akademischer Austauschdienst.

REFERENCES

1. A.F. Iyudin, R. Diehl, K. Bennet, H. Bloemen, W. Hermsen, G.G. Lichti, D. Morris, J. Ryan, V. Schönfelder, H. Steinle, M. Varendorff, C. de Vries, C. Winkler, Astronomy and Astrophysics (1994) in print
2. L.-S. The, D.D. Clayton, M. Leising, W.N. Johnson, J.D. Kurfess, M.S. Strickman, R.L. Kinzer, G.V. Jung, D.A. Grabelsky, W.R. Purcell, M.P. Ulmer, in "The Second Compton Symposium" AIP Proceed. 304
3. Y. Chen, E. Kashy, D. Bazin, W. Benenson, D.J. Morrissey, N.A. Orr, B.M. Sherrill, J.A. Winger, B. Young, J.Yurkon, Phys. Rev. C47 (1993) 1462
4. P.M. Endt, Nuclear Physics A521 (1990) 1
5. K. Debertin and R.G. Helmer in "Gamma- and X-Ray Spectrometry with Semiconductor Detectors" North-Holland, Amsterdam (1988)
6. E. Storm and H.I. Israel, Nuclear Data Tables A7 (1970) 565

Measurement of the Neutron Capture Cross Sections of ^{15}N and ^{18}O

J. Meißner, H. Schatz, H. Herndl, M. Wiescher

Department of Physics, University of Notre Dame, Notre Dame, IN 46556

H. Beer, F. Käppeler

Kernforschungszentrum Karlsruhe, Institut für Kernphysik III, P.O. Box 3640, D-76021 Karlsruhe

I. INTRODUCTION

Neutron capture reactions on light nuclei may be of considerable importance for the s-process nucleosynthesis in red giant stars as well as in inhomogeneous big bang scenarios and high entropy supernovae neutrino bubbles. To determine the reaction rates for such different temperature conditions, the cross sections need to be known for a wide energy range.

The reactions ^{15}N(n,γ) and ^{18}O(n,γ) represent two important links in the reaction sequences for the production of heavier isotopes in such scenarios. At high temperature conditions, the cross sections are not only influenced by non resonant s-wave contributions but also by non resonant p-wave contributions, and higher energy resonances. To study these effects the (n,γ) cross sections for both reactions have been measured for different neutron energies using a fast cyclic neutron activation technique. Throughout this work all energies are given in the laboratory system except for resonance energies E_R which are given in the center of mass system.

II. EXPERIMENTAL METHOD

The measurements were performed at the 3.75 MV Van-de-Graaff accelerator at the Kernforschungszentrum Karlsruhe (KfK). The samples consisted of a mixture of ^{nat}Kr and ^{15}N or ^{18}O enriched gases respectivly, confined in stainless steel spheres. Using a fast sample changer (1) the samples were moved continously between irradiation and counting positions. The irradiation time t_b and the counting time t_c were set to 14 s in the case of ^{15}N and 54 s in the case of ^{18}O, representing roughly two half-lifes of the respective sample. Several neutron spectra at laboratory energies of 25, 152, 250 and 371 keV have been been generated by the ^{7}Li(p,n) reaction at proton energies in the range of 1912 keV to 2138 keV. Except for the 25 keV spectrum, which resembles a Maxwell Boltzmann distribution at kT=25 keV (2), the FWHM of the spectra were between 22 keV and 30 keV. Neutron spectra were calculated using the cross section data from (3). At the counting position the decay of the produced isotopes was observed with a well shieldied 100% HPGe γ-ray detector (2 keV resolution).

For the 25 keV runs the ^{18}O(n,γ) and ^{15}N(n,γ) cross sections were measured relative to the ^{86}Kr cross section, which had been determined independently using the same setup and neutron spectrum (4). However, for the higher neutron energies the averaged neutron capture cross section for ^{86}Kr turned out to be too low to use this method. Therefore gold foils were placed into the neutron flux do determine the cross section relative to the well known cross section of the ^{197}Au(n,γ) reaction (5). After each run, the activity of the gold sample was measured using the same detector and counting position as for the gas sample.

The averaged cross sections for the respective neutron spectrum have been determined using the equations derived in (4).

III. PRELIMINARY RESULTS

A. ^{15}N(n,γ)

For comparison with the experimental data we calculated the p-wave direct capture cross section according to (6) and folded the results with the respective neutron spectrum. Spectroscopic factors were determined from shell model calculations for the ground state and the first three excited states. They are C^2S= 0.96, 1.0, 0.98 and 0.98 respectively. The resulting averaged cross section is about 70% higher at 25keV, 46% higher at 150 keV and 26% higher at 370keV. This indicates that the calculated spectroscopic factors might be to big. Using Rauscher's definition for the S-factor (7) we get an averaged $S_n = 30 \pm 7$ μbarn MeV$^{-1/2}$. For the resonance at $E_R = 862$ keV a spectroscopic factor $C^2S = 5.9 \cdot 10^{-3}$ has been determined from an experimental total width $\Gamma = 15$ keV (8) (9) and penetrability calculations. Using the recommended upper limits for the strength of γ-ray transitions (10) a partial width $\Gamma_\gamma = 5.15 \cdot 10^{-2}$ eV was calculated. Using these values we get a resonance strength of $\omega\gamma = 7.7 \cdot 10^{-2}$ eV. The resulting preliminary stellar reaction rate is

$$N_A <\sigma v> = 3.18 + 3.23 \cdot 10^3 T_9 \\ + 1.3 \cdot 10^4 T_9^{-3/2} \exp(-10.0/T_9) [cm^3 s^{-1} mol^{-1}],$$

which is a factor of 1.4 smaller than the theoretical rates previously used by Rauscher et. al. (7) (see Fig 1).

B. $^{18}O(n,\gamma)$

Theoretical p-wave direct capture values for comparison with the experiment were calculated using the same method as for ^{15}N. Spectroscopic factors were taken from (8). Since the data points at 25keV, 129keV and 250keV are in agreement with the calculations it can be concluded that resonant contributions are neglectable at these energies. The resulting averaged S-factor is $S_n = 49 \pm 8$ μbarn MeV$^{-1/2}$. The high experimental averaged cross section for the neutron spectrum around 150 keV shows the contribution of the $J^\pi = 3/2^+$ resonance at $E_R = 152$ keV. Shell model calculations predict a spectroscopic factor of $C^2S = 0.008$ which yields a neutron partial width of $\Gamma_n = 42$ eV. Since this is three orders of magnitudes smaller than the width of the neutron spectrum, we could determine the resonance strength $\omega\gamma$ and Γ_γ directly from the measured averaged cross section. The results are $\omega\gamma = 3.4 \cdot 10^{-2}$ eV and $\Gamma_\gamma = 1.7 \cdot 10^{-2}$ eV. The strength of the resonances at $E_R = 371$ keV, 626 keV and 746 keV has been estimated to be $\omega\gamma = 0.58$ eV, 3.06 eV and 0.72 eV respectivly by using the experimental total width Γ from the literature (8) and typical values for the γ-partial width Γ_γ. For this preliminary analysis the interference term of the p-wave resonance at $E_R = 626$ keV with p-wave direct capture has been neglected. Finally we arrive at a preliminary stellar reaction rate of

$$\begin{aligned}N_A <\sigma v> =\ & 2.12 \cdot 10^1 + 5.273 \cdot 10^3 T_9 \\ & + 5.68 \cdot 10^3 T_9^{-3/2} \exp(-1.764/T_9) \\ & + 9.69 \cdot 10^4 T_9^{-3/2} \exp(-4.305/T_9) \\ & + 5.11 \cdot 10^5 T_9^{-3/2} \exp(-7.265/T_9) \\ & + 1.20 \cdot 10^5 T_9^{-3/2} \exp(-8.657/T_9) [cm^3 s^{-1} mol^{-1}].\end{aligned}$$

Figure 1 shows the stellar reaction rate in comparison with the rates previously used by Rauscher et. al. (7). The difference of up to a factor of 2 between $T_9 = 0.2$ and 0.8 is mainly due to the revised resonance strength of the $E_R = 152$ keV resonance.

IV. ACKNOWLEDGEMENTS

The authors would like to thank W. Rupp and A. Ernst and his crew for the technical support during the experiment. This project was partially funded by NSF grant PHY88-03035. One of the authors (J.M.) was supported by a Grant-in-Aid of Research from the National Academy of Science through Sigma Xi, The Scientific Research Society. On of the authors (H.S.) was supported by the German Academic Exchange Service (DAAD) with a "Doktorandenstipendium aus Mitteln des zweiten Hochschulsonderprogramms".

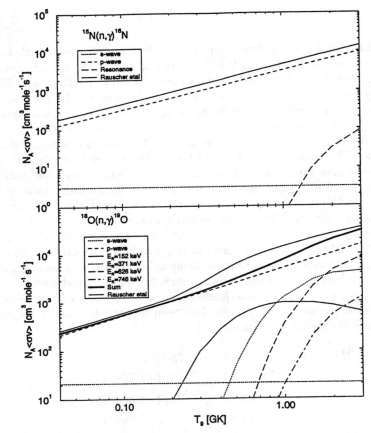

FIG. 1. Stellar reaction rates of ^{15}N(n,γ) and ^{18}O(n,γ) and their contributions. For comparison to the rates from Rauscher *et. al.* see text

REFERENCES

1. H. Beer, G. Rupp, G. Walter, F. Voss, F. Käppeler, NIM A337 (1994) 492
2. W. Ratynsky, F. Käppeler, Phys Rev C 37 (1988) 595
3. H. Liskien, A. Paulsen, At. Data Nucl. Data Tab 15 (1975) 57
4. H. Beer, ApJ 375 (1991) 823
5. R.L. Macklin, private communication (1988)
6. C. Rolfs, Nuclear Physics A217 (1973) 29 565
7. T. Rauscher, J.H. Applegate, J.C. Cowan, F.-K. Thielemann, M. Wiescher, Ap. J. (1993) in print
8. F. Ajzenberg-Selove, Nuc. Phys A 460 (1986) 1
9. S.F. Mughabghab, M. Divadeenam, N.E. Holden, *Neutron Cross Sections, Neutron Resonance Parameters and Thermal Cross Sections* Academic Press (1981)
10. P.M. Endt, Atomic Data and Nuclear Data Tables 55 (1993) 171

Determination of alpha widths in ^{19}F

F. de Oliveira(1), A. Coc(1), P. Aguer(1), C. Angulo(1),
G. Bogaert(1), S. Fortier(2), J. Kiener(1), A. Lefebvre(1),
J.M. Maison(2), L. Rosier(2), G. Rotbard(2), V. Tatischeff(1),
J.P. Thibaud(1) and J. Vernotte(2)

(1) CSNSM, IN2P3-CNRS, 91405 Orsay Campus and
(2) IPN, IN2P3-CNRS, 91406 Orsay Campus

The reaction ^{15}N$(\alpha,\gamma)^{19}$F is of great importance for fluorine nucleosynthesis. At temperatures where ^{19}F is expected to be produced, the narrow resonance associated with the 4.377 MeV level of ^{19}F yields the dominant contribution to cross section. We have extracted its width by using the alpha-transfer reaction ^{15}N$(^7\text{Li},t)^{19}$F and finite range DWBA analysis.

INTRODUCTION

Fluorine nucleosynthesis remains an astrophysical puzzle since no standard stellar evolution model can reproduce the observed abundance (4.05 10^{-7} in mass fraction), in spite of large theoretical and experimental efforts. However, the first reliable measurements of extra-solar fluorine abundances were done on asymptotic giant branch stars (AGB) [1], and it suggests that He-burning shell is the site of fluorine production. Forestini et al. [2] have studied nucleosynthesis in AGB stars and revealed that fluorine can be produced during thermal pulses. In that case (and also for some other scenarios) the reaction ^{15}N$(\alpha,\gamma)^{19}$F is the main contributor to fluorine production. In their latest compilation, Caughlan and Fowler [3] (quoted CF88) used three terms for this rate. The first one is a non resonant term for the lowest temperatures, the second term corresponds to the narrow resonance associated with the 4.377MeV level ($J^\pi = \frac{7^+}{2}$) and the last one is a continuum part for the highest temperatures. Below $T_6 = 500$ (where the helium burning takes place), the rate is dominated by the resonant term. However, this part of the reaction rate is just a crude estimation (0.1 for the reduced width). Since this contribution is very important for fluorine production and could be quite different from the estimation, we measured for the first time the α-width of the 4.377 level. We used the alpha-transfer reaction ^{15}N$(^7\text{Li},t)^{19}$F and a FR-DWBA analysis of the experimental results because the very low cross section hindered a direct measurement.

© 1995 American Institute of Physics

EXPERIMENTAL METHOD

To relieve us from contaminations present in solid targets we have used a ^{15}N confined gas cell (99% enriched) with a pressure close to 100 mbar. The ^{7}Li^{+++} beam was delivered by the I.P.N. Orsay Tandem Accelerator and had an energy of 28 MeV. The intensity of the beam was always higher than 10 nA. The particles were analysed with a magnetic Split Pole spectrograph. An E-ΔE technique gave a clear identification of the outgoing particles in the focal plane of the spectrograph. For normalization purpose, a Si-detector monitored the scattered particles and thus, gave information on the beam intensity and the gas cell pressure. A special off-line treatment was used to correct for the spectrograph aberrations and the variation of energy loss with angle and pressure. Due to the thin entrance window (nickel foil of 0.63μm) and the use of a mylar window (of only 3.5μm thickness) for the outgoing particles, the energy resolution for the reaction products was kept better than σ=50 keV. Angular distributions were obtained from 10 degrees to 60 degrees in the lab system. We analysed the first 16 levels of ^{19}F. The level of interest was well separated from the others as shown in fig. 1.

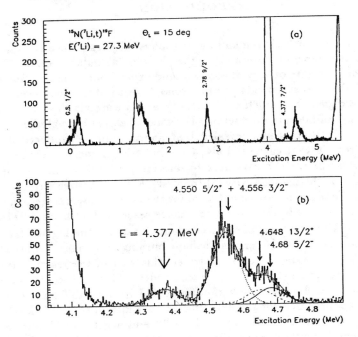

Fig. 1 :(a) The tritium spectrum for an angle of 15° in the lab system. All these peaks could be identified as ^{19}F levels. Only few are labeled. (b) The level of interest for astrophysics (the 4.377 MeV level) is well separated from the other.

FR-DWBA ANALYSES

Cross sections were computed using the DWBA theory and the Ptolemy code (4). The lithium (the fluorine) nucleus is considered as a tritium (nitrogen) core and an alpha cluster with principal number N and orbital angular momentum L. Wave functions were extracted from the nuclear potentials given by Kubo et al. (5) for lithium (N=L=1) and for fluorine (2N+L=7 or 9 for positive parity levels and 2N+L=8 for negative parity levels). These potentials have Woods-Saxon shapes and the well depth is adjusted to reproduce the experimental separation energy of the four nucleons. For unbound levels we have adopted a separation energy of 0.05 MeV. The elastic scattering wave function in the incoming channel $^{15}N(^7Li,^7Li)^{15}N$ was taken from (6) and for the outgoing channel $^{19}F(t,t)^{19}F$ from (7). Assuming the spectroscopic factor of the ^7Li to be unity, the ^{19}F one is extracted from the classical relation : $\left(\frac{d\sigma}{d\Omega}\right)_{exp} = S_{^{19}F} \left(\frac{d\sigma}{d\Omega}\right)_{DWBA} + \left(\frac{d\sigma}{d\Omega}\right)_{HSFB}$ where HSFB refer to Hauser-Feshbach calculation. This part is taken into account for the statistical compound nuclear contribution to the cross section. The comparison of the calculations with the experimental result for the 4.377 MeV level is presented in fig 2.

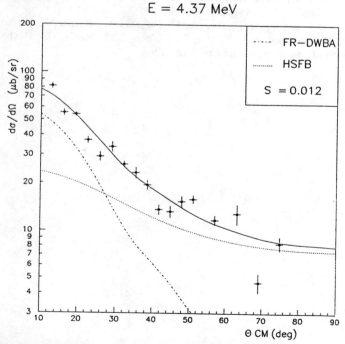

Fig. 2 : Experimental data are compared with a theoretical prediction where the spectroscopic factor C^2S is taken equal to 0.012 for the 4.377 MeV level. Only statistical errors are presented.

RESULTS

From the spectroscopic factor of the first 15 levels in ^{19}F we have deduced the associated reduced α-widths. Where it was possible, the α-width was also calculated for unbound levels. Results were compared with previous studies and we have found a rather good agreement (8). For the level of interest the α-width was found to be $1.5\ 10^{-15}$ MeV, 60 times weaker than the previous value used in the CF88 compilation. That gives for this reaction rate term the new and significantly changed value : $1.6\ 10^{-4}\ T_9^{-\frac{3}{2}}\ exp\ (-\frac{4.232}{T_9})$.

REFERENCES

1. Jorissen, A., Smith, V.V. and Lambert, D.L. ; 1992, A&A 261, 164
2. Forestini, M., Goriely, S., Jorissen, A., and Arnould, M., A&A 261,157
3. Caughlan, G.R. and Fowler, W.A., 1988, ADNDT 40, 283
4. Argonne-Indiana-Stonybrook direct reaction program Ptolemy March 1983 version
5. Kubo, K.I. and Hirata, M., Nucl. Phys. 1972, A187, 186
6. Woods, C.L., Brown, B.A. and Jelley, N.A., 1982, J. Phys. G: Nucl. Phys. 8, 1699
7. Ward, R.P. and Hayes, P.R., 1991, ADNDT 49, 315
8. De Oliveira, F., Coc, A., Aguer, P., Angulo, C., Bogaert, G., Kiener, J., Lefebvre, A., Tatischeff, V., Thibaud, J.P., Fortier, S., Maison, J.M., Rosier, L., Rotbard, G. and Vernotte, J., to be submitted to Nucl. Phys.

Electron Excitations in Nuclear Reactions at Astrophysically Relevant Energies [1]

K. Grün, H. Huber, J. Jank, H. Leeb and H. Oberhummer

Institut für Kernphysik, Technische Universität Wien, A-1040 Vienna, Austria

> A semiclassical method is presented which describes the degree of electron excitations in nucleus–atom collisions at astrophysically relevant energies where neither sudden nor adiabatic conditions are realised. The method is based on the solution of the time dependent Schrödinger equation for the electron wavefunction assuming trajectories for the motion of the nuclei. An application to α–He$^+$ scattering at 92 keV is given, where it provides promising results for the splitting of the astrophysically important resonance in laboratory experiments.

INTRODUCTION

The understanding of stellar evolution and nucleosynthesis processes is closely related to the knowledge of nuclear reaction rates at low energies (1). At these energies direct measurements in laboratory experiments are difficult because there is a strong suppression of charged particle reactions by the Coulomb barrier. Hence experiments are performed at higher energies and the cross sections must be extrapolated to low energies. Unfortunately the laboratory situation is not equivalent to the stellar environment where nuclear reactions take place in a high density electron plasma. In laboratory experiments the collision partners are usually not bare nuclei but contain still part of their electrons in atomic orbits which in turn modify the observed reaction rates at low energies by screening and excitation effects.

The influence of electronic degrees of freedom on the measurements has been observed in several nuclear reactions and there has been great effort to understand the so-called screening effects (2). In this contribution we deal with electron excitations in collision processes at low energies and propose a simple method for their calculation. As a first application of our method we consider the splitting of the ^{8}Be resonance in α–He$^+$ elastic scattering due to electronic excitations. This resonance is of astrophysical importance because it represents the first step of the triple-α process (3) which is the essential link in stellar nucleosynthesis between the light and heavy elements. Already

---

[1] Supported by Austrian *Fonds zur Förderung der wissenschaftlichen Forschung*, Project Number PHY-8807

in 1966 an experimental determination of this s-wave resonance has been reported being the first observation of electronic effects in nuclear reactions (4). At present we have considered only a one-electron system and therefore our example does not correspond to a realistic experimental situation. However, it clearly indicates the procedure how to reach a more quantitative understanding of this resonant process.

THE METHOD

Ion-atom collisions at low energies which are of interest for their astrophysical implications are characterized by the fact that nuclear and electronic degrees of freedom must be considered simultaneously. However, a closer look to the scattering system shows that there are two distinct regions which can be separated in time (5). In the first stage the two nuclei penetrate the electron shells of their collision partners. Because of the Coulomb force generated by the moving nuclei, the electron wave function will be distorted and energy can be exchanged between the electrons and the nuclei (2,5,6). After the penetration of the nuclei through the electron shells we arrive in the second stage of the scattering process. Here the nuclei are strongly decelarated and the nuclear interaction becomes relevant. During this period the displacement on the atomic scale is negligible and therefore the electrons are in stationary states of a Coulomb potential of a point charge corresponding to the sum of charges of the colliding nuclei.

This two-step picture is the basis of our method. In the first step we evaluate the distortion of the electron wave function ψ due to the penetrating nuclei A and B with charges Z_A and Z_B, respectively. We follow the procedure of References (5,6) and solve the time dependent Schrödinger equation assuming the nuclei to move along classical trajectories $\mathbf{r}_A(t)$ and $\mathbf{r}_B(t)$ for central collisions. For simplicity we restrict ourselves to one electron systems. Thus the corresponding Schrödinger equation reduces to

$$i\frac{\partial}{\partial \tau}\psi = -\Big\{\frac{\partial^2}{\partial z^2} + \frac{\partial^2}{\partial \rho^2} + \frac{1}{\rho}\frac{\partial}{\partial \rho} + \frac{Z_A}{\sqrt{(z-z_A(t))^2 + \rho^2}} + \frac{Z_B}{\sqrt{(z-z_B(t))^2 + \rho^2}}\Big\}\psi, \quad (1)$$

where we have used cylindrical coordinates $\mathbf{r} = (\rho, \varphi, z)$ with appropriate scaling. As initial condition we assume the electron to be in the ground state of energy ϵ^A associated with the nucleus A. The Schrödinger equation has to be solved numerically together with the Hamiltonian equations for the positions of the nuclei which are governed by their mutual repulsion and the attraction by the electron.

We enter into the second step of the scattering process when the distance of the nuclei is about 500 fm or below. At such distances the Hamiltonian of

TABLE 1. The occupation probabilities $|a_{n\ell}|^2$ of ^{8}Be one-electron states in α–He$^+$ collisions given in % at $E_{c.m.} = 92$ keV

ϵ_N [eV]	$N = n+\ell+1$	$\ell = 0$	$\ell = 1$	$\ell = 2$	$\ell = 3$	$\sum_\ell |a_{n\ell}|^2$
−217.60	1	36.02	–	–	–	36.02
− 54.40	2	9.78	30.13	–	–	39.90
− 24.18	3	0.72	3.82	2.40	–	6.94
− 13.59	4	0.26	1.20	0.79	0.11	2.35

Equation (1) reduces in good approximation to that of a central Coulomb field of charge $Z_A + Z_B$ and the electrons move in stationary states $\phi_{n\ell}$. Therefore the wave function ψ can be written in terms of the eigenstates $\phi_{n\ell}$

$$\psi = \sum_{n\ell} a_{n\ell}\phi_{n\ell}(\mathbf{r}) \exp(-i\frac{1}{\hbar}\epsilon_N t) \quad , \tag{2}$$

where $|a_{n\ell}|^2$ gives the probability that the electron is in the state $\phi_{n\ell}$ with energy ϵ_N. Because of energy conservation the energy difference of the electron $\epsilon_N - \epsilon^A$ has been transferred to the nuclear motion thus modifying the energy where the nuclear reaction takes place. From this simple picture the measured cross section $\sigma(E)$ is given,

$$\sigma(E) = \sum_{n\ell} |a_{n\ell}|^2 \sigma_{\text{Nucl}}(E + \epsilon^A - \epsilon_N) \quad , \tag{3}$$

where $\sigma_{\text{Nucl}}(E)$ denotes the corresponding cross section of bare nuclei.

After the nuclear reaction takes place there is certainly an outgoing penetration phase where coupling between electronic and nuclear motion can occur. However, this coupling has no effect on the measured cross sections because of the finite resolution of detector systems. Furthermore, the Q-values of nuclear reactions may lead to fast moving ejectiles thus reducing the effect of the coupling of electronic and nuclear motion.

EXAMPLE: THE ^{8}BE RESONANCE

As a first example we consider the splitting of the ^{8}Be resonance in α–α elastic scattering which has been measured in high precision experiments (4,7). Using an implicit algorithm together with a successive overrelaxation method the solution of Equation (1) is straightforward. For the evaluation of the elastic scattering cross section we use the α–α potential of Ref. 8, which leads to a resonance at $E_{\text{res}} = 91.55$ keV. First preliminary results for the electron excitation probability $|a_{n\ell}|^2$ of our calculation neclecting electron screening are given in Table 1. The corresponding differential elastic cross sections at

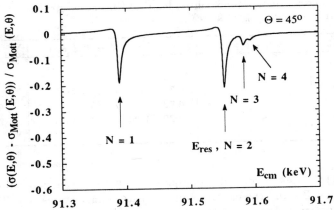

FIG. 1. The splitting of the ^{8}Be resonance due to electron excitations. The occupation probabilities are taken from Tab. 1.

$\Theta_{c.m.} = 45°$ have been evaluated in the vincinity of the ^{8}Be resonance by Equation (3) and are displayed in Figure 1.

Of course this result cannot be compared with existing experimental data, because we have treated only the one electron system (α–He$^+$). However, it clearly indicates the splitting of the resonance due to electron excitations.

SUMMARY

We have proposed a semiclassical method to describe the distortions of the electron wave function and associated excitations during the collision process. The method, although time consuming, is rather simple for one electron systems but becomes tedious for many electrons because of the required projections to determine the excitations. The first preliminary results are promising and further more quantitative work is in progress.

REFERENCES

1. Rolfs, C., and Rodney, W. S., *Cauldrons in the Cosmos*, University of Chicago Press, 1988.
2. Assenbaum, H. J. et al. Z. Phys. **A327**, 461 (1987).
3. Burbridge, E. M. et al. Rev. Mod. Phys. **29**, 547 (1957).
4. Benn, et al. Phys. Lett. **20**, 43 (1966); Nucl. Phys. **A106**, 296 (1968).
5. Bracci, et al. Phys. Lett. **A146**, 128 (1990).
6. Shoppa, T. D. et al. Phys. Rev. **C48**, 837 (1993).
7. Wüstenbecker, S. et al., Z. Phys. **A344**, 205 (1992).
8. Mohr, P., Abele, H., Kölle, V., Staudt, G., Oberhummer, H., and Krauss, H., Z. Phys. A (to be published).

CLASSICAL DISTRIBUTIONS FOR INTERACTING PARTICLES

G.Kaniadakis*, A.Erdas[†], G.Mezzorani[†], P.Quarati*

*Dipartimento di Fisica - Politecnico di Torino, Italy
[†]Dipartimento di Scienze Fisiche - Università di Cagliari, Italy
Istituto Nazionale di Fisica Nucleare - Sezioni di Cagliari e di Torino

Abstract We propose a set of non-Maxwellian statistical distributions, stationary solutions to the Fokker-Planck equation, in which the diffusion and drift coefficients are different from the Brownian case (non interacting particles). These distributions, slightly different from the Maxwellian, describe classical interacting particles and, at equal temperature, have depleted tails compared to the Maxwellian one. This feature could be of great interest in many astrophysical problems such as the solar neutrino production.

Statistical distributions are very important in astrophysics and particularly in the calculation of the astrophysical thermonuclear rates [1]. Usually Maxwellian distribution is assumed in such calculation. We show a particular set of stationary solutions to the Fokker-Planck equation [2], slightly different from the Maxwellian, which could be relevant in this context.

Let us consider the three dimensional Fokker-Planck equation [3]:

$$\frac{\partial n(t,\mathbf{v})}{\partial t} = \nabla\{[\mathbf{J}(t,\mathbf{v}) + \nabla D(t,\mathbf{v})]n(t,\mathbf{v}) + D(t,\mathbf{v})\nabla n(t,\mathbf{v})\} \ . \quad (1)$$

In this equation $n(t,\mathbf{v})$ is the occupational number of the particles in the velocity space, $\mathbf{J}(t,\mathbf{v})$ is the drift coefficient and $D(t,\mathbf{v})$ is the diffusion coefficient. We define also the distribution function $f(t,v)$ by means of

$$f(t,v) = 4\pi v^2 n(t,v) \ , \quad (2)$$

and recall that $f(t,v)dv$ represents the number of particles per unit volume with a speed $v = |\mathbf{v}|$ in the range $[v, v+dv]$. We define also the distribution function in the energy space by means of $p(t,E)dE = f(t,v)dv$; the quantity $p(t,E)dE$ represents the number of particles per unit volume with an energy in the range $[E, E+dE]$. Eq.(1) can be written as a continuity equation in the energy space:

$$\frac{\partial p(t,E)}{\partial t} + \frac{\partial \Phi(t,E)}{\partial E} = 0 \ . \quad (3)$$

The quantity $\Phi(t,E)$ represents the particle current in the energy space and gives the flux of particles outcoming from a sphere of radius $v = \sqrt{2E/m}$ in the velocity space and has the following expression:

$$\Phi(t,E) = -\left(\widehat{J}(t,E) + \frac{\partial \widehat{D}(t,E)}{\partial E}\right)p(t,E) - \widehat{D}(t,E)\frac{\partial p(t,E)}{\partial E} \ , \quad (4)$$

© 1995 American Institute of Physics

$$\widehat{D}(t,E) = 2mED(t,E) \; , \tag{5}$$

$$\widehat{J}(t,E) = 2EJ_s(t,E) - 3mD(t,E) \; . \tag{6}$$

The scalar drift coefficient $J_s(t,E)$ is defined through:

$$\mathbf{J}(t,\mathbf{v}) = \mathbf{v}J_s(t,E) \; . \tag{7}$$

The stationary distributions, solutions of Eq.(3), can be obtained by setting the flux of Eq.(4) equal to zero. We obtain:

$$p(E) = A_1 \frac{\sqrt{E}}{D(E)} \exp\left[-\int \frac{J_s(E)}{mD(E)} dE\right] \; . \tag{8}$$

We have also

$$f(E) = \sqrt{2mE} \, p(E) \tag{9}$$

and taking into account Eq.(8) we find:

$$f(E) = A_2 \frac{E}{D(E)} \exp\left[-\int \frac{J_s(E)}{mD(E)} dE\right] \; . \tag{10}$$

A_1 and A_2 are normalization constants obtained by imposing the condition

$$\int_0^\infty f(t,v) dv = \int_0^\infty p(t,E) dE = \left(\frac{h}{m}\right)^3 \frac{N}{V} \; , \tag{11}$$

where h is Planck's constant and N is the number of particles contained in the volume V. In the following we consider interacting particles. Their interaction can be simulated by means of additional terms in the drift and diffusion coefficients in comparison to the Brownian expressions. We pose

$$J_s(E) = c_0 + c_1 \alpha \epsilon \; , \tag{12}$$

$$D(E) = \frac{c_0}{\beta m} + \frac{c_2}{\beta m} \alpha \epsilon \; , \tag{13}$$

$$\epsilon = \beta E \; , \quad \beta = 1/kT \; . \tag{14}$$

In the case $c_1 = c_2 = 0$, Eqs. (12) and (13) define Brownian motions. We introduce the parameters $\chi_0 = c_0/c_2$ and $\chi_1 = c_1/c_2$, the expression of the distribution $p(E)$ and $f(E)$ becomes

$$p(E) = B_1 \sqrt{\alpha \epsilon} \exp(-\chi_1 \alpha \epsilon) \left(1 + \frac{\alpha \epsilon}{\chi_0}\right)^{\chi_1 \chi_0 - \chi_0 - 1} \; , \tag{15}$$

$$f(E) = B_2 \alpha \epsilon \exp(-\chi_1 \alpha \epsilon) \left(1 + \frac{\alpha \epsilon}{\chi_0}\right)^{\chi_1 \chi_0 - \chi_0 - 1} \; , \tag{16}$$

B_1 and B_2 are normalization constants that can be determined from Eq. (11). Eqs. (15) and (16) represent the main result of our work.

Fig.1 shows the behavior of the function $f(E)$, given by Eq.(16), when the parameters are chosen in such a way to have $\chi_0 = 1, \alpha = 1$, for different values of $\chi_1 (0.5, 1, 3, 10)$. The temperatures of the systems described by the distributions b, c, d, e are less than the temperature of the Maxwellian distribution (curve a). We note a clear depletion in the region of the tails respect to the Maxwellian. As the depletion of the tail increases the temperature of the system decreases.

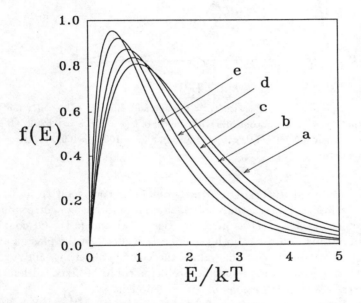

Fig.1 The distribution function $f(E)$ with the parameters $\chi_0 = 1, \alpha = 1$ and different values of χ_1: a (Maxwellian), b ($\chi_1 = 10$), c ($\chi_1 = 3$), d ($\chi_1 = 1$), e ($\chi_1 = 0.5$).

In Fig.2 we report few (b, c, d, e) distributions $f(E)$ together with the Maxwellian (a) having chosen particular values of α (0.620, 0.760, 0.885, 0.995) in order to obtain the same temperature for the different distributions. The different curves correspond to different values of χ. We take $\chi_0 = \chi_1 = \chi = 1.6, 2.5, 5, 100$. We note that the curve $b(\alpha = 0.620, \chi = 1.6)$ is identical to the Maxwellian distribution. We obtain that the same system can be described by distributions corresponding to the same temperature but with different tails.

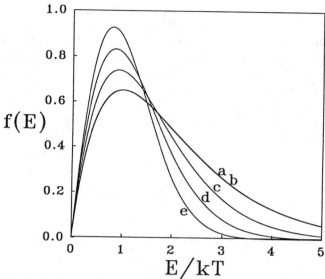

Fig.2 The distribution function $f(E)$ with $\chi_0 = \chi_1 = \chi$ and different values of χ and α. The curve a is the Maxwellian and coincides with the curve b ($\chi = 1.6, \alpha = 0.620$). All the curves a, b, c ($\chi = 2.5, \alpha = 0.760$), d ($\chi = 5, \alpha = 0.885$), e ($\chi = 100, \alpha = 0.995$) correspond to the same temperature.

At present, we are still working on the problem and we want to find the expression of the internal interactions and the external force field corresponding to the coefficients J and D as given by Eqs. (12) and (13). Of course, these interactions must represent both the collective effects of the plasma and the many-particle correlations. The values the coefficients α, χ_0 and χ_1 of Eqs. (15), (16) should assume in specific problems could be fixed ad hoc by inserting the statistical distributions in specific models.

Even with these limitations, we conclude the following. It is well known that the largest contribution to the reaction rates comes from the high-velocity tail of the distribution. The possibility of depleting the tail, without altering the temperature and the pressure could be relevant in the solution of the solar neutrino problem and in other astrophysical processes [4].

References

[1] D. C. Clayton et al, Astroph. J.199 (1975) 399.

[2] G. Kaniadakis and P. Quarati, Physica A 192 (1993) 677.

[3] H. Risken *The Fokker-Planck Equation* (Springer, Heidelberg, 1984).

[4] W. Rodney and C. Rolfs, *Cauldrons in the Cosmos* (Univ. Chicago Press, 1988).

EFFECT OF RADIATIVE RECOMBINATION IN THERMONUCLEAR FUSION RATES

A.Erdas[†], G.Kaniadakis[*], G.Mezzorani[†], P.Quarati[*]
[†]Dipartimento di Scienze Fisiche - Università di Cagliari, Italy
[*]Dipartimento di Fisica - Politecnico di Torino, Italy
Istituto Nazionale di Fisica Nucleare - Sezioni di Cagliari e di Torino

Abstract We show that radiative recombination in dense astrophysical plasma, like the solar interior, can give rise to Coulomb screening effects which affect the thermonuclear reaction rates, and depend on the nuclear electron screening factor, temperature and electron density in the plasma.

Thermonuclear reactions in stars and of course in the sun are affected by electronic screening [1]. This effect, studied also in laboratory nuclear reaction experiments, is described by different authors using various conceptual approaches but a standard treatment, particularly in the case of dense plasma, has not yet been established [2-4]. In the field of electron screening, we want to call the attention on a particular shielding mechanism, which so far has been neglected, which is due to radiative recombination processes [5] in the astrophysical plasma.

We now give an example to explain our previous statement. Let us suppose we have a one-component neutral plasma of electrons and protons in the stellar interior. These protons and electrons recombine radiatively, forming unstable neutral hydrogen atoms of lifetime τ, and this instability is due to the fact that the plasma temperature is higher than the binding energy of such atoms. Before one of these atoms dissociates, a fusion process between the atom and a proton or another such atom can occur, and in this case, the reaction rate is enhanced compared to the case of two bare protons.

We address the question of the importance of this enhancement and the physical conditions that have influence on it. Let us now consider the more general case of a neutral plasma composed by two species of nuclei, a and b, and electrons e^-. We define the fusion rate F_1 of bare nuclei as

$$F_1 = \frac{1}{1+\delta_{a,b}} n_a n_b < \sigma_{ab} v > , \qquad (1)$$

where n_a and n_b are the densities of the nuclei a and b, σ_{ab} is the fusion cross section and v is the relative velocity and $\delta_{a,b}$ is Kronecker's delta. If one takes into consideration the radiative recombination of the nuclei with the electrons in the plasma

$$a + e^- \to (a - e^-)_{atom} + \gamma , \qquad (2)$$

then one obtains a fusion rate F_2 given by

$$F_2 = \frac{1}{1+\delta_{a,b}}\left[(n_a - \epsilon_a)(n_b - \epsilon_b) <\sigma_{ab}v> +\epsilon_a\epsilon_b <\sigma_{(ae^-)(be^-)}v>\right.$$
$$\left.+(n_a-\epsilon_a)\epsilon_b<\sigma_{a(be^-)}v> +\epsilon_a(n_b\epsilon_b)<\sigma_{(ae^-)b}v>\right], \quad (3)$$

where ϵ_a and ϵ_b are the densities of the recombined atomic systems $(a-e^-)$ and $(b-e^-)$ respectively, $\sigma_{a(be^-)}$ is the cross sections for the fusion of the nucleus a with the atomic system $(b-e^-)$, $\sigma_{(ae^-)b}$ is the fusion cross sections for $(a-e^-)$ and b and $\sigma_{(ae^-)(be^-)}$ is the fusion cross sections for $(a-e^-)$ and $(b-e^-)$. Let us suppose that also in a dense plasma the relation between fusion cross section of bare nuclei and fusion cross section of a bare nucleus with a neutral atom is the following:

$$\sigma_{(ae^-)b} = \sigma_{a(be^-)} = f\sigma_{ab}, \quad (4)$$

where f is the enhancement factor due to the Coulomb screening [6]. If we assume that the screening factor f is independent on the relative velocity, we find that the ratio between F_2 and F_1 is

$$\frac{F_2}{F_1} = 1 + \frac{\epsilon_a}{n_a}(f-1) + \frac{\epsilon_b}{n_b}(f-1) + \frac{\epsilon_a}{n_a}\frac{\epsilon_b}{n_b}(f-1)^2. \quad (5)$$

We may write the densities ϵ_a and ϵ_b (for temperatures T much higher than the binding energies) in the form

$$\epsilon_{a,b} = n_{a,b}n_{e^-}^{eff(a,b)}<\sigma_R^{a,b}v>\tau_{(a,b)e^-}, \quad (6)$$

where σ_R^a and σ_R^b are the radiative recombination cross sections for the nuclei a and b, $<\sigma_R^a v>$ and $<\sigma_R^b v>$ are the recombination rates (averaged over a Maxwellian distribution or over stationary non-Maxwellian distributions [7]), τ_{ae^-} and τ_{be^-} are the mean lifetimes of the atomic systems $(a-e^-)$ and $(b-e^-)$ respectively, and $n_{e^-}^{eff(a)}$ and $n_{e^-}^{eff(b)}$ are the "effective" densities of electrons available for the nuclei a and b and are given by $n_{e^-}^{eff(a,b)} = Z_{a,b}n_{a,b}$, where $-Z_a e$ is the charge of the nucleus a and $-Z_b e$ is the charge of the nucleus b. An expression for $\epsilon_{a,b}$ in a neutral plasma that is correct also at low temperatures can be derived considering that $\epsilon_{a,b}$ must satisfy the following differential equation:

$$\frac{d\epsilon_{a,b}}{d\tau} = \frac{\epsilon_{a,b}(\infty) - \epsilon_{a,b}}{\tau_0}, \quad (7)$$

where $\epsilon_{a,b}(\infty) = n_{a,b}$ and τ_0 has the expression

$$\tau_0 = \frac{1}{n_{e^-}^{eff(a,b)}<\sigma_R^{a,b}v>}. \quad (8)$$

Solving Eq.(7), we obtain

$$\epsilon_{a,b} = n_{a,b}\left[1 - \exp\left(-n_{e^-}^{eff(a,b)} <\sigma_R^{a,b}v> \tau\right)\right].\qquad(9)$$

The lifetime τ of an atomic system at temperature T is given by the well-known formula [6]

$$\tau = \left(\frac{2\pi\hbar^2}{\mu kT}\right)^{3/2}\frac{2J_3+1}{(2J_1+1)(2J_2+1)}\frac{\exp(-Q/kT)}{<\sigma_R v>},\qquad(10)$$

where $\mu \simeq m_e$ is the reduced mass of the atomic system, Q is the ionization energy of the system and is always negative, and J_1, J_2 and J_3 are the spins of the nucleus, the electron and the atomic system respectively. The ionization energy Q, for a system $(a-e^-)$ is given by $Q = -Z_a^2\alpha^2 m_e c^2/2$, where α is the fine-structure constant. Using Eqs. (6),(10) we obtain

$$\frac{\epsilon_{a,b}}{n_{a,b}} = Z_{a,b}n_{a,b}\left(\frac{2\pi\hbar^2}{m_e kT}\right)^{3/2}\left[\frac{2J_3+1}{(2J_1+1)(2J_2+1)}\right]_{a,b}\exp\left(\frac{Z_{a,b}^2\alpha^2 m_e c^2}{2kT}\right)\qquad(11)$$

where the spin part is labeled a or b if it belongs to the atomic system $(a-e^-)$ or $(b-e^-)$. From this last equation it is easily seen that the ratio F_2/F_1 depends only on the temperature T, the screening factor f, the densities n_a and n_b and the charges Z_a and Z_b of the nuclei a and b.

We now want to elucidate our arguments with a few numerical results. Let us consider fusion between two hydrogen atoms or between an hydrogen atom and a proton. If we use for the screening factor the well known expression $\exp(\Delta/kT)$, we find that in the solar core, taking $\Delta \simeq 30\,\mathrm{eV}$ (see ref. [1]), the pure screening enhancement f is about 2%. The effective electron density in the solar core is of the order of 10^{26} cm^{-3} and therefore the evaluation of the ratio F_2/F_1 shows an enhancement, due to the atomic recombination, of the order of few percent. This enhancement decreases very rapidly with increasing temperature and decreasing electron density.

The same approach can be used to evaluate the recombination effect in the fusion of helium isotopes. In this case one has to consider the recombination with one and with two electrons and the lifetime of the neutral or singly ionized He atoms. We have found that the enhancement is slightly larger than in the hydrogen isotopes case. We have used the expression $\exp(\Delta/kT)$ for the screening factor, and this may be criticized [3]. In ref.[3] it is argued that a proton gives rise to fusion with a H atom from a distance approximately equal to the Bohr radius and not from infinity, and therefore one should expect $f = 1$ in this case. However, we can show that the neutral systems present in the solar core plasma must have dimensions smaller than the free atoms, for instance, the neutral system $(\alpha + 2p + 4e)$ occupies a spherical volume of radius about one half the Bohr radius. Therefore we argue that one can still use the expression $\exp(\Delta/kT)$ for the screening factor.

In conclusion we would like to remark that, in a high temperature dense plasma, the radiative recombination effect is quite important among the various screening mechanisms that take place and that probably the effect could be even bigger if one was able to consider the collective effects that bring together electrons and ions to form neutral systems [8] (the collective effects could be included by using stationary non-Maxwellian distributions, as discussed in [9]).

It is well known that in the center of the sun the mass density of H is only about 36%, while the mass density of He is about 64%. As a consequence if we fix our attention to the solar core we should consider also the formation of neutral or singly ionized He atoms. In this case one or two electrons can be radiatively captured by the He atoms and we should also consider the fusion of two neutral He atoms or the fusion of singly ionized He with similar systems or with H and protons. In addition we must say that the dimensions of neutral systems in a dense plasma are probably smaller than in normal matter condition, therefore their binding energy is stronger than in the free case and their lifetime are longer than the values we can obtain from Eq.(7). Nevertheless, in this work, we are not interested in a full and complete description of these various processes in the sun interior, rather in calling the attention to the importance of the screening mechanism due to recombination. In this context the preliminary results on the ratio F_2/F_1 we report in this work are probably underestimated and their importance might even be greater than indicated in this contribution.

References

[1] L.Bracci, G.Fiorentini, V.S.Melezhik, G.Mezzorani and P. Quarati, Nucl.Phys. A 513 (1990) 316.

[2] S.Ichimaru, Rev.Mod.Phys. 65 (1993) 255.

[3] A.Dar and G.Shaviv, Phys.Rev.Lett. preprint submitted (1994).

[4] J.Bachall et al. IASSNS-ATS 94/13 Princeton, unpublished (1994).

[5] A.Erdas, G.Mezzorani and P.Quarati, Phys.Rev.A 48, 452 (1993).

[6] C.Rolfs and W.Rondney: *Cauldrons in the Cosmos,* Chicago U.P., Chicago (1988).

[7] G. Kaniadakis and P. Quarati, Physica A 192 (1993) 677.

[8] V.Tsytovich: *Solar neutrino problem,* Edited by.V.Berezinsky and E.Fiorini, Lab.Naz. Gran Sasso (1994) 238.

[9] G. Kaniadakis, A.Erdas, G.Mezzorani and P.Quarati, *Classical distributions for interacting particles,* Proceed. Nuclei in the Cosmos, Gran Sasso, 1994.

QUANTUM STATISTICAL DISTRIBUTIONS FOR INTERACTING PARTICLES

G.Kaniadakis and P.Quarati

Dipartimento di Fisica - Politecnico di Torino , Italy
Istituto Nazionale di Fisica Nucleare - Sezioni di Cagliari e di Torino

Abstract We consider quantum statistical distributions describing systems of particles obeying an exclusion or an inclusion principle together with Maxwell-Boltzmann distributions of classical particles. Particles belonging to any intermediate statistics can be described by a set of distributions, varying a parameter which is the degree of symmetrization or antisymmetrization. Components of astrophysical plasma with partial degeneracy can be usefully described by the proposed distributions.

In many physical fields of research we have often to deal with systems of particles which are with certainty bosons or fermions. We have to deal with classical particles when the thermal wave length is much smaller than the interparticle distance. For instance, in the case of a neutral plasma, the above condition is given by

$$\lambda_{ij} = \frac{\sqrt{2\pi}\hbar}{a_{ij}\sqrt{2\mu_{ij}kT}} \ll 1 \ , \tag{1}$$

where the interparticle distance a_{ij} is

$$a_{ij} = \frac{a_i + a_j}{2} \ , \quad a_i = \left(\frac{3Z_i}{4\pi n_e}\right)^{1/3} \ , \tag{2}$$

Z_i is the charge of the i-th ion, n_e is the electron density and $\mu_{ij} = m_i m_j/(m_i + m_j)$ is the reduced mass. In high energy physics, high temperature superconductivity and in quantum Hall effect we have sometime to describe particles obeying statistics which are different from the pure Bose-Einstein (BE) or Fermi-Dirac (FD) statistics. This task can be accomplished or giving an ad hoc statistical distribution or by mixing with certain weights the two different BE and FD distributions. On the other hand, we have often particles behaving only partially as classical particles, because their distribution must have a quantal component (for instance a partially degenerate electron gas). Also in this case we can search for an ad hoc distribution function or we can mix the classical Maxwell-Boltzmann (MB) with the quantum statistical distributions. This kind of problems are very common in astrophysical dense plasma and particularly in the solar interior [1,2]. In the case of electron components in astrophysical plasma, there is a gradual transition from nondegeneracy toward complete degeneracy as the density rises. An approach to derive the distribution of partially degenerate electrons is well known [1].

© 1995 American Institute of Physics

In this paper we want to show how new distributions, recentely proposed by us [3-5], can be used to completely describe the behavior of systems of particles in the particular situations mentioned above and with a particular interest in the plasma of the solar or stellar interiors.

Let us pose our attention to the function $n(t, \mathbf{v})$ which is the occupational number related to the velocity $\mathbf{v}$. We have recently shown [4] that $n(t, \mathbf{v})$ satisfies the following partial differential equation:

$$\frac{\partial n(t, \mathbf{v})}{\partial t} = \nabla\{[\mathbf{J}(t, \mathbf{v}) + \nabla D(t, \mathbf{v})]n(t, \mathbf{v})[1 + \kappa n(t, \mathbf{v})] + D(t, \mathbf{v})\nabla n(t, \mathbf{v})\} \, , \quad (3)$$

where $\mathbf{J}(t, \mathbf{v})$ is the drift coefficient and $D(t, \mathbf{v})$ is the diffusion coefficient in the velocity space. Eq.(3) is the continuity equation in the velocity space. We have used, to derive this equation, the following expression of the transition probability:

$$\pi(t, \mathbf{v} \to \mathbf{u}) = r(t, \mathbf{v})n(t, \mathbf{v})[1 + \kappa n(t, \mathbf{u})] \quad (4)$$

where $r(t, \mathbf{v})$ is the transition rate.

The quantity in the square brackets is an inhibition factor, when $\kappa = -1$: if the site $\mathbf{u}$ is fully occupied ($n(t, \mathbf{u}) = 1$), then the transition from $\mathbf{v}$ to $\mathbf{u}$ is forbidden. On the other hand, if the site is empty the transition probability is not affected by the occupation of the arrival site. In this case the stationary solution of Eq.(3) is the FD statistical distribution.

When $\kappa = 1$ the statistical distribution describes particles wich are bosons. In this case the quantity in the square bracket in Eq.(4) represents an enhancement factor. The number of particles leaving the site $\mathbf{v}$ increases linearly with the population of the arrival site $\mathbf{u}$.

The two cases describe systems of particles submitted to an exclusion (fermions) or to an inclusion (bosons) principle. Both the principles induce a potential, repulsive or attractive respectively, acting on the particles that, consequently, must be considered as interacting particles.

Let us take κ as a continuous variable with values in the range $-1 \leq \kappa \leq +1$, and solve Eq.(3) in the case of Brownian particles; for $t \to \infty$ we find the following statistical distribution:

$$n(\mathbf{v}) = \frac{1}{\exp[\beta(E - \mu)] - \kappa} \, . \quad (5)$$

with $E = mv^2/2$ and $\beta = 1/kT$. The expression of $n(\mathbf{v})$ reduces to the standard FD($\kappa = -1$) and MB($\kappa = 0$), BE($\kappa = 1$) distribution. We call κ the degree of symmetrization or antisymmetrization, μ is the chemical potential. These two quantities can be determined (as explained below) through normalization conditions, using the known density and pressure of the particle system under consideration.

To make an example let us fix our attention to a one or two component dense plasma with an electron density n_e satisfying the neutrality condition

$$\sum_i Z_i n_i = n_e \ . \qquad (6)$$

Degrees of Fermi degeneracy for electrons at temperature T are measured by the ratio

$$\Theta = \frac{kT}{E_F} \ , \qquad (7)$$

where

$$E_F = m_e c^2 \left(\sqrt{1 + x_F^2} - 1\right) \ , \qquad (8)$$

$$x_F = \frac{\hbar k_F}{m_e c} = \frac{\hbar (3\pi^2 n_e)^{1/3}}{m_e c} \ , \qquad (9)$$

When $\Theta < 0.1$ we have complete degeneracy, when $0.1 \leq \Theta \leq 10$ we have intermediate degeneracy (quantum and classical effects coexist), when $\Theta > 10$ we have not degeneracy and the particles can be considered classical. For instance we have in solar interior (SI) $\Theta = 2.23$, in brown dwarf (BD) $\Theta = 0.01$, in giant planet (GP) $\Theta = 0.023$, in inertial confinement fusion (ICF) $31.8 \leq \Theta \leq 387$[2]. To be more explicit, let us consider the solar interior. The correct normalization conditions are (non relativistic case):

$$8\pi \left(\frac{m_e}{h}\right)^3 \int_0^\infty \frac{v^2 dv}{\exp[\beta(E-\mu)] - \kappa} = n_e \ , \qquad (10)$$

$$8\pi m_e \left(\frac{m_e}{h}\right)^3 \int_0^\infty \frac{v^4 dv}{\exp[\beta(E-\mu)] - \kappa} = P_e \ . \qquad (11)$$

We know temperature, density and pressure in SI, then we may derive from Eq.s (10) and (11) the value of the chemical potential μ and the degree of antisymmetrization $-1 \leq \kappa \leq 0$ [5]. Let us define now the fuction $F_l(x)$

$$F_l(x) = \int_0^\infty \frac{t^l dt}{\exp(t+x) + 1} \ , \qquad (12)$$

from Eq.s (10) and (11) we obtain the following expressions of the density and pressure:

$$n_e = \frac{4\pi}{h^3}(2mkT)^{3/2}\frac{1}{|\kappa|} F_{1/2}\left(\ln\frac{1}{|\kappa|} - \frac{\mu}{kT}\right) \ , \qquad (13)$$

$$P_e = \frac{8\pi kT}{3h^3}(2mkT)^{3/2}\frac{1}{|\kappa|} F_{3/2}\left(\ln\frac{1}{|\kappa|} - \frac{\mu}{kT}\right) \ . \qquad (14)$$

By using Eq.s (13) and (14) if we know the values of density and pressure we can derive the chemical potential μ and the degree of symmetrization κ. If we define the quantity

$$\Lambda_\kappa\left(\frac{\mu}{kT}\right) = \frac{2}{3}F_{3/2}\left(\ln\frac{1}{|\kappa|} - \frac{\mu}{kT}\right) \bigg/ F_{1/2}\left(\ln\frac{1}{|\kappa|} - \frac{\mu}{kT}\right) , \qquad (15)$$

we obtain the following state equation for the partially quantal system of particles:

$$P_e = n_e\, kT\, \Lambda_\kappa\left(\frac{\mu}{kT}\right) \qquad (16)$$

In the case of completely classical particles ($\kappa = 0$) we have:

$$\Lambda_0\left(\frac{\mu}{kT}\right) = 1 , \qquad (17)$$

while in the case of a fermionic system ($\kappa = -1$) from Eq.(15) we get

$$\Lambda_{-1}\left(\frac{\mu}{kT}\right) = \frac{2}{3}F_{3/2}\left(\frac{\mu}{kT}\right) \bigg/ F_{1/2}\left(\frac{\mu}{kT}\right) . \qquad (18)$$

The normalization conditions given by Eq.s (10) and (11) are also true for systems of bosons (in an astrophysical plasma α particles are only partially classical), the bosonic chemical potential and the degree of symmetrization $0 \leq \kappa \leq 1$ can be derived.

If, in addition to the repulsive and attractive potentials due to the exclusion and inclusion principles, we suppose that other internal and external potentials are active among the particles, then the diffusion and the drift coefficients are different from the Brownian case; the expression of the stationary distribution function in three dimensions is given by:

$$n(\mathbf{v}) = \left\{\exp\left[\sum_{i=1}^{3}\int\frac{dv_i}{D(v_i)}\left(J_i(v_i) + \frac{\partial D(v_i)}{\partial v_i}\right) - \beta\mu\right] - \kappa\right\}^{-1} . \qquad (19)$$

References

[1] D.D.Clayton: *Principles of stellar evolution and nucleosynthesis* McGraw-Hill, New York 1968.

[2] S.Ichimaru, Rev.Mod.Phys. 65 (1993) 255.

[3] G.Kaniadakis and P.Quarati Phys.Rev.E 48 (1993) 4263.

[4] G.Kaniadakis and P.Quarati Phys.Rev.E 49 (1994) June.

[5] G.Kaniadakis Phys.Rev.E 49 (1994) June.

The Rapid Hydrogen Burning Process and the Solar Abundances of ^{92}Mo, ^{94}Mo, ^{96}Ru and ^{98}Ru

M. Hencheck(1), R.N. Boyd(1,2), B.S. Meyer(3), M. Hellström(4),
D.J. Morrissey(4,5), M.J. Balbes(1), F.R. Chloupek(1), M. Fauerbach(4),
C.A. Mitchell(1), R. Pfaff(4), C.F. Powell(4), G. Raimann(1),
B.M. Sherrill(4,6), M. Steiner(4), J. Vandegriff(1), and S.J. Yennello(7)

(1) Department of Physics, Ohio State University, Columbus, Ohio 43210
(2) Department of Astronomy, Ohio State University, Columbus, Ohio 43210
(3) Department of Physics and Astronomy, Clemson University, Clems, SC 29634
(4) National Superconducting Cyclotron Laboratory, MSU, East Lansing, MI 48824
(5) Department of Chemistry, Michigan State University, East Lansing, MI 48824
(6) Department of Physics & Astronomy, Michigan State University, East Lansing, MI 48824
(7) Department of Chemistry, Texas A&M University, College Station, TX 77843

Abstract. We explore the possibility of creating the light p-nuclei ^{92}Mo, ^{94}Mo, ^{96}Ru, and ^{98}Ru through a rapid proton capture process. It is shown that for temperatures from 0.95×10^9 K to 1.4×10^9 K, significant abundances of these nuclides can be built up. Thorne-Żytkow objects are suggested as possible sites for the process. The results are sensitive to the location of the proton drip-line. In order to determine which proton rich nuclei in this mass range are proton bound, we analyzed the reaction products of an E/A=60 MeV ^{106}Cd beam using the A1200 projectile fragment separator at the National Superconducting Cyclotron Laboratory. Nine new very neutron-deficient isotopes of Ag, Pd, Rh, and Ru have been identified among them. The resulting mass spectra are presented.

Key words: abundances — nuclear reactions — fragmentation

THE PRODUCTION OF ^{92}MO, ^{94}MO, ^{96}RU, AND ^{98}RU

The p-process nuclei (p-nuclei) are those proton rich nuclei more massive than iron that are shielded from production by both the s- or r-processes. Their nucleosynthesis has been described in a variety of ways, including the photodissociation of heavy nuclides (the γ-process) (1) and the rapid capture of protons by lighter nuclides (the rp-process) (2,3). The γ-process has proven successful in explaining the abundances of the heavy p-process nuclei (4). This is because the r-process and s-process nuclei that serve as γ-process seeds are typically 100 times more abundant than the heavy p-process nuclei. However, some of the lighter p-nuclei, specifically ^{92}Mo, ^{94}Mo, ^{96}Ru, and ^{98}Ru, present a different case. The lighter of these light p-process isotopes have abundances comparable to their r-process and s-process counterparts. Any γ-process on a solar distribution of seed nuclei that accounts for the heavy p-nuclei will underproduce the light p-nuclei by a factor of order 10 – 100. With this in mind, we have investigated the production of these nuclei in an rp-process on an initial solar composition of material.

The rp-process adds protons to the existing nuclei, driving them rapidly toward the proton drip line. After the process ends, the subsequent β^+ decays or electron captures back towards stability ultimately produce the p-process nuclei. With a large enough temperature and proton fluence, N = 46 nuclei from ^{80}Se to ^{84}Sr are swept up the N = 46 chain mostly to ^{92}Pd, and the N = 48 nuclei from ^{84}Kr to ^{86}Sr to ^{94}Pd. If we assume initial solar seeds and neglect weak flows out of N=46 and N=48, we would find overproduction factors $O(^{92}\text{Mo}) = 68$ and $O(^{94}\text{Mo}) = 84$. (The overproduction factor is defined as the abundance produced divided by the solar abundance.) These overproduction factors indicate that if roughly one percent of the Galaxy's mass experienced such conditions, the rp-process might have contributed significantly to the solar system's supply of ^{92}Mo and ^{94}Mo. In a similar manner, the overproduction factors for the ruthenium isotopes would be $O(^{96}\text{Ru}) = 290$ and $O(^{98}\text{Ru}) = 1060$. This indicates that ^{96,98}Ru may be overproduced relative to ^{92,94}Mo. However, if any of the progenitors of the Ru isotopes decay by β^+p decay (as is suggested by their predicted masses) instead of simple β^+ decay, the amount of ^{96,98}Ru produced would be greatly reduced.

Of critical importance to calculations involving the production of ^{92}Mo, ^{94}Mo, ^{96}Ru, and ^{98}Ru through proton capture is the determination of the location of the proton drip-line in the A= 80 – 95 range. In order to learn more about the proton drip-line in this mass range, we conducted a fragmentation experiment at the National Superconducting Cyclotron Laboratory at Michigan State University.

FRAGMENTATION EXPERIMENT AT NSCL

At NSCL, an E/A = 60 MeV beam of the rare isotope ^{106}Cd was focused onto a 44.4 mg/cm^2 natural Ni target with a 9.4 mg/cm^2 ^{9}Be backing at the target position of the A1200. The ^{9}Be backing was used to increase the fraction of reaction products leaving the target in their fully stripped charge state. The reaction products were separated and identified using the A1200 fragment separator (5). The experimental method employed closely follows that of Bazin et al. (6), Mohar et al. (7), and Yennello et al. (8).

Information from two position-sensitive parallel-plate avalanche counters (PPACs) placed at the second dispersive image of the A1200 separator together with NMR measurements of the dipole fields of the separator were used to determine the magnetic rigidity of the product nuclei. Thin plastic scintillators at the first dispersive image and at the final focus were used to measure the time of flight (TOF) through the device and, hence, the velocity of the reaction products. A four element silicon telescope (ΔE_1, ΔE_2, E_1, E_2) placed at the focal plane of the A1200 provided two independent energy loss measurements as well as the total kinetic energy of the particles.

It was observed that the reaction products included nine previously unob-

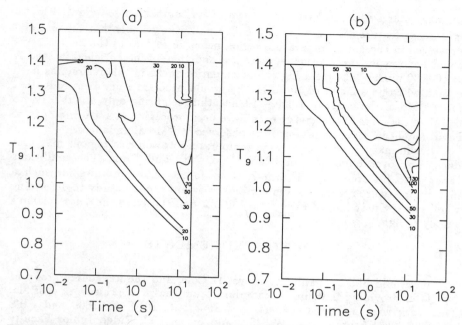

FIGURE 1. Overproduction of ^{92}Mo(a) and ^{94}Mo(b) as a function of Temperature and Processing Time

served nuclei. These nuclei were ^{88}Ru, ^{90,91,92,93}Rh, ^{92,93}Pd, and ^{94,95}Ag. The results represent approximately 19 hours of data collection at a beam current of ≈ 0.2 pnA. The observation of an isotope in this experiment implies that the ion has a lifetime longer than its flight time through the A1200 separator, which is on the order of 150 ns.

RESULTS OF STUDY

The information gained at NSCL was included in a network code written to calculate the change in nuclear abundances in a high temperature proton rich environment. This code includes the nuclei from iron through indium that are either stable or on the proton rich side of the valley of stability. The results of the analysis for ^{92,94}Mo are presented in figure 1. Shown are contour plots of the overproduction of the light p-nuclei for the indicated burning temperatures and times. Notice that overproduction factors on the order of 70 can be achieved for both isotopes.

Production of the molybdenum isotopes in an rp-process requires an environment in which both high temperatures and protons coexist. One such environment is a Thorne-Żytkow object, which forms when a neutron star in a binary star system sinks to the center of its companion red giant (9). An interesting feature associated with the Thorne-Żytkow site is that the material processed in the high temperature region near the surface of the neutron star could have experienced prior s-processing and thereby be enriched in seed A = 60 – 90 nuclei. The resulting enhancement of nuclei in this region has been estimated to be a factor of 50 (10). This would produce a net overproduction for ^{92,94}Mo of the order of 1,000 – 10,000, thus requiring only about 0.1% to 0.01% of the mass of the galaxy need have been processed in a Thorne-Żytkow environment in order to produce the observed ^{92,94}Mo abundances.

In summary, processing in high-temperature proton-rich regions may produce large overabundances of ^{92}Mo, ^{94}Mo, ^{96}Ru and ^{98}Ru. The conditions thought to be present in Thorne-Żytkow objects show promise for such a high temperature rp-process and certainly warrant further study to determine whether they may indeed have contributed significantly to the solar system's supply of light p-isotopes.

ACKNOWLEDGEMENTS

We would like to thank T. Antaya, D. Cole, and the ion source group at NSCL for providing the unusual cadmium beam that made this work possible. This work was supported in part by NSF grants PHY 92-21669 and PHY 92-14992, NASA grant NAGW-3480, and by an Oak Ridge Junior Faculty Enhancement Award.

REFERENCES

1. Woosley, S.E., and Howard, W.M., *Ap. J. Suppl.* **36**, 285–304 (1978).
2. Audouze, J., and Truran, J.W., *Ap. J.* **202**, 204–213 (1975).
3. Cannon, R.C., *Mon.Not.R.Astron.Soc.* **263**, 817–838 (1993).
4. Howard, W.M., Meyer, B.S., Woosley, S.E., *Ap. J.* **373**, L5—L8 (1991).
5. Sherrill, B.M., Morrissey, D.J., Nolen Jr., J.A., and Winger, J.A., *Nucl. Inst. & Methods Phys. Res.* **B56**, 1106–1112 (1991).
6. Bazin, D., Guerreau, D., Anne, R., Guillemaud-Mueller, D., Mueller, A.C., and Saint-Laurent, M.G., *Nuclear Physics* **A515**, 349–364 (1990).
7. Mohar, M.F., Bazin, D., Benenson, W., Morrissey, D.J., Orr, N.A., Sherrill, B.M., Swan, D., and Winger, J.A., *Phys. Rev. Lett.* **66(12)**, 1571–1574 (1991).
8. Yennello, S.J., Winger, J.A., Antaya, T., Benenson, W., Mohar, M.F., Morrissey, D.J., Orr, N.A., and Sherrill, B.M., *Phys. Rev. C* **46(6)**, 2620–2623 (1992).
9. Thorne, K.S., and Żytkow, A.N., *Ap. J.* **199**, L19–L24 (1975).
10. Lamb, S.A., Howard, W.M., Truran, J.W., and Iben, Jr., I., *Ap.J.* **217**, 213–221 (1977).

SECTION III

STELLAR MODELS AND NUCLEOSYNTHESIS

Stellar Evolution with Turbulent Diffusion

Cesare Chiosi(1), Alessandro Bressan(2), Licai Deng(3,4)

(1) Department of Astronomy, University of Padua, Italy
(2) Astronomical Observatory of Padua, Italy
(3) International School for Advanced Studies, Trieste, Italy
(4) International Centre for Theoretical Physics, Trieste, Italy

Abstract. In this paper we briefly report on two recent studies by Deng, Bressan & Chiosi (1994a,b) in which a new formulation of diffusive mixing in stellar interiors is presented. In particular, the analysis is devoted to cast light on the kind of mixing that should take place in the so-called overshoot regions surrounding the inner fully convective cores. Key points of the analysis are the inclusion of intermittence and stirring efficiency and the concept of scale length most effective for mixing, by means of which the diffusion coefficient is formulated. Depending on the value of the diffusion coefficient that holds good in the overshoot region, we go from the case of virtually no mixing (semiconvective like structures) to that of full mixing over there (standard overshoot models). With the aid of this formalism, we find that stellar models of massive and intermediate-mass stars calculated with mild efficiency of mixing in the region of overshoot (in our notation $0.4 \leq P_{dif} \leq 0.8$) possess at the same time evolutionary characteristics that are separately typical of models calculated with different schemes of mixing. In other words, the new models share the same properties of models with standard overshoot, namely a wider main sequence band, higher luminosity, and longer lifetimes than classical models, but they also possess extended loops that are the main signature of the classical (semiconvective) description of convection at the border of the core. These evolutionary models are used to study the HRD of supergiant stars in the Large Magellanic Cloud. The agreement between theory and observations is remarkably good. Finally, we present preliminary results for evolutionary models of massive stars which stand on the above formalism and the concept of *Global Diffusion*. These models are particularly designed to match the properties (surface abundances and location in the HRD) of Wolf-Rayet stars.

Key words: Stars: interiors – Stars : HR Diagram – Stars: supergiants – Convection – Diffusion

1. Introduction

The extension of convective regions and associated mixing processes in stellar interiors are still among the poorly known aspects of stellar structure and evolution. Needless to say that both bear very much on the correct modelization of the stars and the interpretation of the observational data.

In the standard theory of convection, otherwise known as the Mixing Length Theory (MLT), the size of a convective region (either core or envelope) is determined by means of a local stability criterion (either Schwarzschild or Ledoux). This looks for the conditions at which the acceleration imparted to convective elements by the buoyancy force gets zero, thus neglecting the inertia of their motions, and supposes mixing of thermal and chemical properties to be instantaneous, fully efficient, and to take place only at the end of the motion of convective elements. The assumption of instantaneous mixing stems from considering that in most circumstances the time scale of convective motions is much shorter than the nuclear and thermal time scales. In turn, the time scale of convective motions is customarily identified with the lifetime of the largest scale elements before dissolving into the surrounding medium.

The local nature of the MLT immediately leads to the problem of convective overshoot, i.e. whether and how far the convective elements can penetrate into the surrounding, stable radiative regions. In meteorology and laboratory fluid mechanics, convective overshoot is an observed phenomenon whose characteristic extension is comparable to that of the unstable region itself. Unfortunately, the extreme different physical conditions between terrestrial laboratories and stellar interiors prevent us from simply extending to stars what is found to hold in ordinary fluid dynamics. In fact, the extension of laboratory convection (atmosphere, oceans) is often a small fraction of the pressure scale height, whereas in the stars the unstable regions usually extend over a large fraction or even up to several times the local pressure scale height. Although the extent of overshoot and even its existence have been the subject of a vivid debate (see the review papers by Renzini 1987, and Zahn 1991), the problem has been tackled in an impressive number of studies, among which we recall Shaviv & Salpeter (1973), Maeder (1975, 1983, 1990), Cloutman & Whitaker (1980), Roxburgh (1978, 1989), Bressan et al. (1981), Matraka et al. (1982), Xiong (1985, 1986, 1990), Bertelli et al. (1985), Langer (1986), Baker & Kuhfuss (1987), Bressan (1984), Maeder & Maynet (1987, 1988, 1989, 1991), Alongi et al. (1993), Bressan et al. (1993), and Fagotto et al. (1994a,b,c).

There are different formulations for overshoot that vary from author to author sometime with contradictory results. Saslaw & Schwarzschild (1965) found a very small overshoot distance, while Shaviv & Salpeter (1973) argued that this amounts to a significant fraction of the pressure scale height. In most studies overshoot is described by means of the MLT and therefore it contains the scale of mixing as a free parameter (e.g. Maeder 1975, Bressan et al. 1981). On the contrary, the criterion proposed by Roxburgh (1978) is claimed to be parameter free. However, this has been convincingly criticized by Bressan (1984) and Baker & Kuhfub (1987) for its consistency. Nowadays, it is considered to provide the maximum allowed distance of penetration (Roxburg 1989). According to Renzini (1987), the overshoot zone is small if the temperature gradient is radiative there, large if adiabatic. In contrast Xiong (1985, 1986, 1990), using a fully hydrodynamical approach, shows that overshoot is a very complicated process, in which different physical quantities have different distances of penetration, and finds that the overshoot region at the border of the convective core can be very large and radiative at the same time.

As far as mixing itself is concerned, while the complete instantaneous homogeneization sounds resonable in the fully convective regions, this is not straightforward in the overshoot regions. Indeed, it has long been known that a sort of incomplete mixing under the condition of neutrality ($\nabla_R = \nabla_{AD}$)

can take place in suitable circumstances at the border of the convective core, a phenomenon named semiconvection for which many different descriptions have been proposed (cf. Chiosi et al. 1992 for a recent review). Since mixing is a very complex phenomenon requiring the description of a turbulence field at all scales and complete mixing results from the contribution of scales much smaller than those implicit in the MLT, we seek for a phenomenological picture of it easy to incorporate in stellar calculations. Furthermore, we look for a general formalism able to deal with a large variety of possibilities with the minimum number of assumptions.

The problem is split in three parts: determination of the size of the unstable regions, thermodynamical structure of these, and finally formulation of the mixing technique. A great deal of simplification comes from finding that the thermodynamic structure of the overshoot region has little effect on the overall structure of the stars (Deng 1992). Indeed, an adiabatic description of the overshoot region leads to the same results as a purely radiative one, in agreement with the analysis by Xiong (1985).

2. Penetration across a Gradient in Molecular Weight

In order to study the mixing efficiency, we must address the question whether convective elements can travel across an inhomogeneous medium (gradient in molecular weight). This situation can occur either at the border of a growing convective core or at the base of a penetrating envelope.

If the analysis is limited to radial motions along which the restoring force is most effective (Renzini 1977), a gradient in molecular weight can fully inhibit the propagation of convective motions. However, if we consider that in a real situation elements may undergo stretching or deformation, motions in other directions may be produced when the convective elements reach the inhomogeneous region (barrier of molecular wight). In the following, we will argue that convective motions at the bottom of the barrier can generate strong enough shears to trigger turbulence over there.

An elementary analysis shows that the overshoot distance traveled by a rigid cylindrical test element across the gradient is ~ 3 orders of magnitude smaller than the local value of the mixing length λ, so that overshoot is virtually zero (Renzini 1977).

However as the element is not a rigid body, it is hard to conceive that it would immediately stop. Similarly, it is difficult to accept that an element moving with the large speed of the bulk motion, keeps its shape when it matches the barrier. We can picture the real situation imagining that the element will change shape and move toward another direction. Since the total mass of the element must be conserved, the only possible motion is along the tangential direction, along which no buoyancy force, and negligible viscous force are experienced. If the tangential motion turns out to be turbulent, mixing along this direction would be easier than along the vertical direction (e.g. Zahn 1991, Charbonnel & Vauclair 1992).

Because the matter above the transition level is basically at rest, the tangential motion will surely create shears at the interface. The question arises whether these shears are strong enough to trigger an instability. To check this possibility we look at the Richardson number for a stratified fluid (Zahn 1987). The Richardson number can be expressed as

$$J_r = \frac{g}{H_p} \frac{d\ln\mu/d\ln P}{(dv_0/dz)^2} \tag{1}$$

where g is the local gravity, μ is the molecular weight, P is the pressure, and v_0 and H_p are the local values of the radial velocity and pressure scale height, respectively. The Richardson condition says that the region with a gradient is stable if $J_r > \frac{1}{4}$ everywhere.

It can be easily seen that for typical values at the border of an expanding convective core (for instance during the core He-burning phase), one gets $J_r \sim 3.14 \, 10^{-3}$. The region is therefore highly unstable and mixing is likely to occur.

3. The New Mixing Scheme

As already anticipated we split the problem in three steps: first we fix the size of the unstable region (both full convection and overshoot), secondly we calculate the flux of energy carried by convection and determine the thermodynamical structure of the overshoot region, finally we apply the mixing technique to the unstable zones.

3.1. Size of the Unstable Region

The size of the unstable region is obtained following the method of Bressan et al. (1981): i.e. the region of full convection is fixed by the the classical Schwarzschild condition, whereas that of overshoot is calculated from a ballistic approach which determines the layer of the stars at which the velocities of convective elements vanish. Since the method of Bressan et al. (1981) rests on the MLT, it contains the mean free path of convective elements as a free parameter. This is assumed to be proportional to the local value of the pressure scale height ($\lambda = \Lambda H_p$). In the following we adopt $\Lambda = 1.0$. See also Bertelli et al. (1985), Chiosi et al. (1992), Alongi et al. (1993) for more details.

3.2. Energy Transport

The energy transport or equivalently the stratification of the thermodynamical quantities in the unstable region (full convection and overshoot) is less of a problem in our context, because the largest eddies, arising form the instability, are found to carry almost all the energy. At the same time, they excite the smaller eddies which become more and more important for the mixing process. This is also true in the more complicated treatment of stellar convection (Xiong 1985,1986,1990). In the case of an instantaneous mixing in the overshoot region, the very high efficiency of energy transport by the largest eddies makes the region adiabatic (Bressan et al. 1981, Maeder 1975). In contrast, Xiong's (1985) theory shows that the energy transport becomes less efficient and a radiative structure is established in the overshoot region. However, as both adiabatic and radiative structures in the overshoot region yield similar results (Deng 1992), we adopt the adiabatic temperature stratification over there.

3.3. Mixing as a Diffusive Process

Because of the high velocities predicted by the MLT, in almost all stellar model calculations, mixing of chemical elements inside the unstable regions is assumed to be instantaneous. However, the MLT does not specify the detailed structure of the velocity field, but considers only the mean properties of the convective elements, i.e. the large scale motions. It goes without saying that fine mixing cannot be the result of considering only large scales. Specifically, contrary to what happens with other physical quantities, the homogenization of chemical elements requires motions all over the possible scales down to very small eddies. In fluid mechanics, the mixing process is viewed as a sort of stretching of the material in the turbulent field. What we learn from laboratory experiments is that the speed of mixing is not high enough to wipe out inhomogeneities over the time scale of the large scale motions. Indeed fine mixing can be reached only asymptotically. This means that mixing is rather slow, at least much slower than the characteristic lifetime of the largest elements of fluid. Furthermore, laboratory experiments, meteorological studies, and numerical simulations have shown that motion at small scales is slowed down by intermittence, because small elements fill up much smaller volumes than the big ones. Finally, we notice that while the motions at the largest scales can be safely described by the MLT because of the highly anisotropic behaviour of the fluctuations of physical quantities, at the smallest scales the turbulence field becomes isotropic so that the diffusive description can be applied.

In light of the above considerations we suppose that the mixing of chemical elements (as resulting from the contribution of all possible scales) can be derived from solving the diffusion equation

$$\frac{dC}{dt} = \left(\frac{\partial C}{\partial t}\right)_{nucl} + \frac{\partial}{\partial m_r}\left[(4\pi r^2 \rho)^2 D_t \frac{\partial C}{\partial m_r}\right] \qquad (2)$$

where C is the chemical abundance of the element under consideration and D_t is a suitable diffusion coefficient.

In general, the diffusion coefficient D_t can be written as

$$D_t = \frac{1}{3} F_i F_s v_d L \qquad (3)$$

where, v_d is the velocity of the characteristic mean effective scale and L is the dimension of the region interested by diffusion. The factor F_i accounts for intermittence, while the factor F_s accounts for the *stirring efficiency* at the largest scale. Both are described below.

4. Intermittence and Stirring

In this section, we briefly discuss the effects of intermittence and stirring on mixing and derive the factors F_i and F_s. To this aim, we adopt the Frisch (1977) phenomenological description of stellar turbulence, in which the velocity distribution at all scales is derived from the conservation of the kinetic energy flux, and intermittence is obtained from the so-called β-model. As in stars the turbulent regions may be much larger than the typical distance of one pressure scale hight, the effect on mixing induced by the motion of the largest eddies needs to be properly taken into account (so-called stirring). We start introducing the following notation for the various physical quantities in use. They are summarized below

- l_d: Characteristic length scale of mixing;
- v_d: Corresponding velocity of l_d;
- l_0: Dimension of the largest eddy in the region under consideration;
- v_0: Corresponding velocity to l_0 as given by the MLT;
- L_0: Linear dimension of the unstable region;
- L_{ov}: Linear dimension of the overshoot region.

4.1. The Characteristic Velocity of Diffusion

The relation between velocity and linear dimension at any given scale is derived from the conservation of kinetic energy flux

$$\frac{v_0^3}{l_0} = \frac{v_d^3}{l_d} \tag{4}$$

where the velocity v_0 can be calculated from the MLT.

4.2. Intermittence

According to Frisch's (1977) β-model of intermittance, for two successive generations of turbulent elements whose linear dimensions decrease by a factor of two, the intermittence factor is $\beta = N/2^3$, where $N=2^\alpha$ is the real number of off-springs from each parental element, 2^3 is the relative volume filling of the off-springs, and $\alpha \simeq 2.5$ from experimental observations (Frisch 1977). Therefore, the relative intermittence factor is $\beta = 2^{-1/2}$.

We generalize β to a continuous spectrum of scales by replacing the factor '2' with $(\frac{l_i}{l_{i+1}})$. It follows that intermittence from the largest scale l_0 down to the scale l_d is

$$\beta = (\frac{l_d}{l_0})^{1/2}. \tag{5}$$

When intermittence is taken into account, the kinetic energy flux conservation becomes

$$\frac{v_0^3}{l_0} = \beta \frac{v_d^3}{l_d} \tag{6}$$

which provides the current definition of β. This means that the space originally filled by the largest scales is only partially occupied by the small ones so that over the time scale of diffusion ($\tau \sim L_0^2/D$), only a fraction $\beta_{int} = (l_d/l_0)^{\frac{1}{2}}$ of the volume is mixed. In order to have the whole volume statistically filled, a much longer time is required, or equivalently the diffusion coefficient must be decreased by the factor β^n. Mixing turns out to be complete if $n = 3$. Finally, the correction for intermittence to the diffusion coefficient is

$$F_i = (\frac{l_d}{l_0})^{3/2} \tag{7}$$

4.3. Stirring

The largest eddy in a turbulence field works as a rigid stick stirring the material in a mixer and inducing smaller scale motions. However, if the stick (largest eddy) is comparable in size with the dimension of the total volume to be mixed, the net mixing efficiency turns out to be much less. In order to take this into account we correct the diffusion coefficient by the factor

$$F_s = (\frac{L_0 - l_0}{l_0})^3 = (\frac{L_0}{l_0} - 1)^3 \tag{8}$$

This correction turns out to be important only in the convective envelope because it extends over several pressure scale heights.

5. Prescriptions for the Diffusion Coefficient

Given the above premises, we specify the diffusion coefficient for the various unstable regions that might happen to occur in a star, namely the innermost convective core, sometime a fully convective intermediate region, and the external convective envelope. In turn, the fully convective regions extend into their adjacent overshoot regions. In the following we will not consider the case of intermediate fully convective zones. Finally, a great deal of simplification is possible in the case of a convective core with active nuclear burning in it. The case of nuclear burning in a convective shell is not considered here.

The Inner Core with Nuclear Burning. Homogeneization in the core can be obtained independently of details of the mixing process. Indeed the very fast motion at the largest scale of convection guarantees that all the material has the same probability of being exposed to nuclear reactions. Therefore, an homogeneous chemical distribution is obtained over a time scale equal to the lifetime of the convective eddies, $\tau = L_0/v_0$, where L_0 is the dimension of the unstable region and v_0 is the velocity of the largest element in it. Both are derived from the MLT. Therefore, the diffusion coefficient for the inner convective core is simply given by

$$D = \frac{1}{3}v_0 L_0 \tag{9}$$

The Overshoot Region. To determine the diffusion coefficient for the overshoot region we proceed as follows. At any layer X inside this region, there is a natural maximum scale for turbulent elements which is set by the distance l_X between the current position X and the outer border of the overshoot region. Let v_X be its corresponding characteristic velocity.

If these maximum elements had the same volume filling as in any other unstable regions (see above), the diffusion coefficient would be

$$D = \frac{1}{3}(\frac{L_{ov}}{l_0} - 1)^3 (\frac{l_d}{l_0})^{\frac{5}{3}} v_X L_{ov}. \tag{10}$$

where the corrections for stirring and intermittence from the scale L_{ov} down to l_d have already been taken into account. However, the largest eddy at a level X has to be considered as the off-spring of elements coming from the unstable region underneath, so that proper correction for intermittence must be applied to the velocity v_X, and in turn the diffusion coefficient. The velocity v_X can be expressed in terms of L_{ov}, v_0^S, l_X

$$v_X = v_0^S \left(\frac{l_X}{L_{ov}}\right)^{\frac{1}{6}}. \tag{11}$$

where v_0^S is the velocity at the transition layer between the fully unstable and the overshoot region ($\nabla_R = \nabla_{AD}$). The velocity v_0^S and scale length l_0 are derived from the MLT. The final expression for the diffusion coefficient is

$$D = \frac{1}{3}\left(\frac{l_d}{l_0}\right)^{\frac{5}{3}} \left(\frac{l_X}{L_{ov}}\right)^{\frac{5}{3}} (\frac{L_{ov}}{l_0} - 1)^3 v_0^S L_{ov}. \tag{12}$$

The External Envelope without Nuclear Burning. In this region the diffusion coefficient is simply given by

$$D = \frac{1}{3}(L_0/l_0 - 1)^3 (\frac{l_d}{l_0})^{\frac{5}{3}} v_0 L_0 \tag{13}$$

Once again, the velocity v_0 and the scale length l_0 of the largest element together with the extension of the convective envelope are derived from the MLT.

To summarize, the prescriptions for the diffusion coefficient adopted in the various regions are

$$\mathcal{D} = 0 \quad \textit{Radiative Region} \tag{14}$$

$$\mathcal{D} = \frac{1}{3}v_0(r)L_0 \qquad Convective \ Core \qquad (15)$$

$$\mathcal{D} = \frac{1}{3}(L_0/l_0 - 1)^3 \left(\frac{l_d}{l_0}\right)^{\frac{5}{3}} v_0(r)L_0 \qquad Convective \ Envelope \qquad (16)$$

$$\mathcal{D} = \frac{1}{3}(L_{ov}/l_0 - 1)^3 \left(\frac{l_d}{l_0}\right)^{\frac{5}{3}} \left(\frac{l_X}{L_{ov}}\right)^{\frac{5}{3}} v_0^S L_{ov} \qquad Overshoot \ Region \qquad (17)$$

6. Searching the Most Effective Scale l_d

In order to apply the above formalism to stellar models, we must determine the characteristic scale l_d that best drives the mixing process. In a turbulent region, all scales from the maximum one equal to the dimension of the unstable zone itself down to that of the dissipative processes, are possible. The minimum scale is the Kolmogorov micro-scale given by

$$l_K = (\nu^3/\bar{\epsilon})^{1/4} \qquad (18)$$

where ν is the kinematic viscosity, and $\bar{\epsilon}$ is the kinetic energy flux injected into the turbulence field.

Supposing that l_d is equal to the Kolmogorov micro scale l_K essentially no mixing occurs because $l_K \sim 10^2$ cm. In contrast, assuming l_d to be equal to l_0 (largest scale in the turbulence region) almost instantaneous mixing takes place. The effective scale l_d lies in between these two extreme values. Unfortunately, no theory can be invoked to fix the effective scale l_d a priori so that this must be considered as a sort of parameter.

Since the aim of this study is to describe mixing inside the overshoot region as a slow diffusion process, we have performed many test calculations aimed at evaluating the best value for l_d giving incomplete mixing over a time scale comparable with the evolutionary (nuclear) time scale. For a typical 20 $M_\odot$ star the above requirement is met when l_d is

$$l_d = P_{dif} \times 10^{-5} l_0 \qquad (19)$$

where l_0 is expressed in units of H_p and P_{dif} is a fine tuning parameter of the order of unity.

We will show that by slightly changing P_{dif} we are able to recover all existing evolutionary schemes, going from the so-called semiconvective type of models (cf. Langer 1989a,b) to the fully homogenized overshoot models (cf. Bressan et al. 1981, Bertelli et al. 1985, Alongi et al. 1993).

7. Evolution of the Prototype 20$M_\odot$ Star

The above scheme of mixing has been used to follow the evolution from the main sequence until the core He-exhaustion stage of stars with initial mass in the range 5 to 100 $M_\odot$ (Deng et al. 1994a). The input physics (opacity, nuclear reaction rates, neutrino emission rates, mass loss rates by stellar wind)

Table 1: The lifetimes for the 3 sequences of a 20M$_\odot$ star with Z=0.008 and Y=0.25, mass loss by stellar wind and different efficiencies of mixing (P_{dif}). The last column is the final value of the mass.

P_{dif}	τ_H	τ_{He}	τ_{He}/τ_H	M_f
0.1	8.75	1.17	0.13	18.94
0.8	9.79	0.82	0.08	18.91
1.2	10.10	0.74	0.07	18.30

is the same as in the series of papers by Alongi et al. (1993), Bressan et al. (1993), Fagotto et al. (1994a,b,c). For the sake of brevity, no details are given here, they can be found in the papers quoted above. Suffice it to recall that we have adopted (1) the new radiative opacities of the Livermore group (Iglesias et al. 1992) however implemented at low temperatures and densities by the radiative opacities of Cox & Stewart (1970a,b) and the molecular opacities of Bessell et al. (1989, 1991); (2) the mass loss rates of de Jager et al. (1988); (3) the nuclear reaction rates of Caughlan & Fowler (1988) and Caughlan et al. (1985).

Most of the discussion below will be limited to the case of a 20 M$_\odot$ star with chemical composition Z=0.008 and Y=0.25, which is most suited to stars in the Large Magellanic Cloud. No details for the stars with other value of the mass and chemical composition are given here. They can be found in Deng et al. (1994a). Nevertheless, for the sake of consistency a few general remarks are drawn and a quick comparison with other evolutionary schemes is made.

Figure 1 shows the HRD for the 20 M$_\odot$ star calculated with three different values of the parameter P_{dif}. It is soon evident that the evolutionary path in the HRD critically depends on the efficiency of mixing in use. Indeed the three values of P_{dif} are representative of three distinct evolutionary schemes. The smallest value of P$_{dif}$ gives rise to an evolution of the kind named Case B since Chiosi & Summa (1970): the star, after the main sequence phase, ignites central He-burning in the blue and moves slowly toward the red side of the HR diagram. In contrast, the intermediate value of P$_{dif}$ gives rise to the so-called Case A (Chiosi & Summa 1970): in fact the star ignites central He-burning in the red side of the HRD, and performs an extended blue loop afterwards. Finally, the high value of P_{dif} generates models characterized by having the whole core He-burning phase in the red side of the HRD. We name these models Case C. It is worth recalling that Case A is the sort of evolutionary behaviour we get for constant mass models with semiconvection according to the Ledoux criterion, whereas Case B is the same but for the Schwarzschild criterion (cf. Chiosi et al. 1992 for a recent review of this subject). Finally, Case C is typical of the models calculated with large core overshoot and full homogeneization of the overshoot region (cf. Alongi et al. 1993, Bressan et al. 1993). Clearly the above evolutionary behaviour is the result of the different efficiency of the diffusive mixing in the overshoot region. Indeed, the larger the value of the diffusion parameter, the more extended is the region where an efficient mixing occurs.

The lifetimes of the core H- and He-burning phases also depend on the efficiency of mixing. This is shown by the data displayed in Table 1.

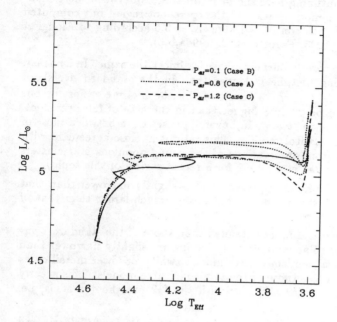

Fig. 1.— The HRD of the 20M$_\odot$ star computed with three different values of P_{dif}, namely 0.1, 0.8 and 1.2.

At increasing P_{dif}, and mixing efficiency in turn, the lifetime of H-burning phase increases, while that of He-burning phase decreases. Furthermore, the lifetimes gradually vary from the values typical of classical (semiconvective) models to those corresponding to the fully homogenized overshoot case. A detailed discussion of the physical structure of 20 M$_\odot$ models with different mixing efficiency and of the main physical quantities driving their evolutionary behaviour is presented in Deng et al. (1994a) to whom the reader should refer.

We would like to call attention on an interesting feature of these stellar models: for small variations of P_{dif} a star can switch from one scheme to another. Indeed within a variation of P_{dif} by as much as 50%, both models with very extended loops and models with no loops at all are possible. If the efficiency of mixing depends on other physical processes such as differential rotation, a wide range of possible evolutionary schemes would be allowed to stars of given mass and chemical composition but different rotational properties with obvious far reaching consequences as far as the interpretation of the HRD is concerned.

As already recalled, two large grids of models with composition Z=0.02 and Y=0.28 (solar vicinity) and Z=0.008 and Y=0.025 (Large Magellanic Cloud) have been calculated by Deng et al. (1994a) adopting intermediate values for P_{dif}, i.e. in the range 0.4 to 0.8. Limiting the discussion to a few key points, it is worth outlining here the main differences between these models and those with full homogenization of the overshoot region but computed with half the mean free path considered here ($\Lambda_c = 0.5$ according to Alongi et al. 1993, Bressan et al. 1993, Fagotto et al. 1994a,b,c).

1. The lifetime of the core H-burning phase is almost the same. In contrast, the core He-burning lifetime is longer than in the standard overshoot models: $\tau_{He}^{dif}/\tau_{He}^{ov} \simeq 1.5$. As a result of this, the lifetime ratios for our models turn out to be 20-50% higher than in the case of full overshoot. The lifetime ratios of the two main burning phases lie in between those of full overshoot and classical models. This ratio is in close agreement with the value inferred from observations of the young populous star clusters of the LMC (see Chiosi et al. 1992 for a recent review of this topic).

2. The extension of the main sequence band is slightly narrower than that of the standard overshoot models, but always much larger than that of the semiconvective ones.

3. The loops are in general more extended than those of the standard overshoot models. For solar metallicity the loops are slightly narrower than those of the classical semiconvective models, while for lower metallicities they are as extended as those of the semiconvective models. This greatly improves upon the major difficulty with the full overshoot models, i.e. their narrow blue loops.

4. These stellar models have been used to study the HRD of Galactic and LMC supergiants reaching an unprecedented level of agreement between theory and observations (Deng et al. 1994a).

8. Are the Wolf-Rayet Stars Undergoing Global Diffusion ?

The above formalism for diffusive mixing has been used to calculate models of massive stars specifically aimed at explaining the properties of Wolf-Rayet (WR) stars, their location in the HRD in particular (Deng et al. 1994b). The goal was achieved introducing the concept of *global diffusion* proposed long ago by Schatzman (1977). Global diffusion stands on the critical Reynolds number and radiative viscosity, and allows a very slow mixing of material to take place between the core and the surface during the whole evolution. The physical process triggering global diffusion is still uncertain even though it could be related to instabilities caused by rotation.

In analogy to the Schatzman (1977) formulation of the mild turbulent diffusion, we make use of the global diffusive mixing governed by the diffusion coefficient

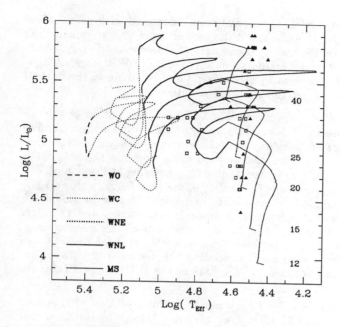

Fig. 2.— Evolutionary tracks with *Global Diffusion*. The composition is Z=0.0080 and Y=0.250. The value of the initial mass (in $M_\odot$) is indicated along each track. Finally, the various phases of WR are shown with different symbols (see the inlet). Superposed are the data for galactic WR stars by Hamann et al. (1993). Solid triangles stand for the WN with detection of surface hydrogen, whereas open squares are for the hydrogen free objects.

$$D_g = \frac{1}{9} R_e \times \nu_{rad}, \quad \nu_{rad} = \frac{4}{15} \frac{aT^4}{\kappa c \rho^2}, \qquad (20)$$

where R_e is the critical Reynolds number for turbulent diffusion and ν_{rad} is the radiative viscosity. Following Schatzman (1977) we adopt $\frac{1}{9}R_e = 166$ and assume that in the radiative regions of the star the material remains close to the physical conditions prevailing at the onset of turbulent instability.

In this scheme, a star is made of a convective core, whose dimensions is defined by a suitable criterion (as in the previous models) and an external stable region, which however is affected by the slow diffusive mixing so that the chemical profile can be slowly changed all the way up to the surface. Although there is no theoretical argument showing that this type of mixing is related to convection, it could result from the exponentially decaying convective overshoot predicted by Xiong (1985, 1986, 1990) if the chemical self-correlation function overshoots in the same manner as the velocity field, i.e. slowly but over an extended region.

We find that stellar models of massive stars calculated with global diffusion offer interesting clues to understand the properties of WR stars and their location in the HRD. In Figure 2 we compare the new evolutionary tracks with the observational data by Hamman et al. (1993) for WR stars. Adopting the current rates of mass loss by stellar wind and referring to the current properties of WR stars, we get the following results:

1. The formation of WR stars is possible at much lower values of the initial mass (therefore much lower luminosity) than with standard models including the effect of mass loss and convective overshoot. Our models show that the evolution, instead of proceeding towards the red, tends to bend to higher effective temperatures thus entering the region of WR stars already during the core H-burning phase (see Fig. 2). No ad hoc assumptions for the rates of mass loss are needed. Mixing by global diffusion does not contain arbitrary, adjustable parameters and therefore the above result depends on the intrinsic efficiency of this phenomenon. We find that a 12 $M_\odot$ star for Z=0.008 and a 9$M_\odot$ star for Z=0.020 can become a WR object, thus much alleviating the discrepancy between expected and observed position of WR stars in the HRD, in particular for the less luminous ones.

2. Global diffusion does not bear very much on the evolution of the most massive stars, say above 30-40 $M_\odot$, as it is overwhelmed by the dominant effect of mass loss.

3. Finally, there is a lower limit below which global diffusion does not alter the classical evolutionary path in the HRD and therefore does not lead to the formation of a WR object.

ACKNOWLEDGEMENTS

L. Deng thanks the ICTP and ISAS of Trieste, Italy for its hospitality and financial support. A. Bressan and C. Chiosi thanks financial support from the Italian Ministry of University, Scientific Research and Technology (MURST) and the Italian Space Agency (ASI).

REFERENCES

Alongi M., Bertelli G., Bressan A., Chiosi C., Fagotto F., Greggio L., Nasi E., 1993, A&AS 97, 851

Baker, N. H., Kuhfuss, R., 1987, A&A 185, 117

Bertelli G., Bressan A., Chiosi C., 1985, A&A 150, 33

Bessell M. S., Brett J. M., Scholz M., Wood P. R., 1989, A&AS 77, 1

Bessell M. S., Brett J. M., Scholz M., Wood P. R., 1991, A&AS 87, 621

Bressan A., 1984, PhD thesis, ISAS, Trieste, Italy

Bressan A., Bertelli G., Chiosi C., 1981, A&A 102, 25

Bressan A., Fagotto F., Bertelli G., Chiosi C., 1993, A&AS 100, 47

Caughlan G. R., Fowler W. A., 1988, Atomic DatA Nuc. Data Tables 40, 283

Caughlan G. R., Fowler W. A., Harris M., Zimmermann B., 1985, Atomic Data Nucl. Data Tables 32, 197

Charbonnel C., Vauclair S., 1992, A&A 265, 55.

Chiosi C., Bertelli G., Bressan A., 1992, ARA&A 30, 305

Chiosi, C., Summa, C., 1970, Astr. Space Sci. 8, 478

Cloutman, L. D., Whitaker, R. W., 1980, ApJ 237, 900

Cox A. N., Stewart J. N., 1970a, ApJS 19, 243

Cox A. N., Stewart J. N., 1970b, ApJS 19, 261

de Jager C., Nieuwenhuijzen H., van der Hucht K. A., 1988, A&AS 72, 295

Deng L., 1992, Master thesis, ISAS, Trieste, Italy

Deng L., Bressan A., Chiosi C., 1994a, A&A submitted

Deng L., Bressan A., Chiosi C., 1994b, A&A submitted

Fagotto F., Bressan A., Bertelli G., Chiosi C., 1994a, A&AS 100, 647

Fagotto F., Bressan A., Bertelli G., Chiosi C., 1994b, A&AS 104, 365

Fagotto F., Bressan A., Bertelli G., Chiosi C., 1994c, A&AS 105, 39

Frisch U., 1977, Lect. Not. Phys., 71, 325.

Hamann W.R., Koesterke L., Wessolowski U., 1993, A&A 274, 397

Iglesias C. A., Rogers F. J., Wilson B. G., 1992, ApJ 397, 717

Langer, N., 1986, A&A 164, 45

Langer N., 1989a, A&A 210, 93

Langer N., 1989b, A&A 220, 135

Maeder, A., 1975, A&A 40, 303

Maeder, A., 1983, A&A 120, 113

Maeder, A., 1990, A&AS 84, 139

Maeder, A., Meynet, G., 1987, A&A 182, 243

Maeder, A., Meynet, G., 1988, A&AS 76, 411

Maeder, A., Meynet, G., 1989, A&A 210, 155

Maeder, A., Meynet, G., 1991, A&AS 89, 451

Matraka, B., Wassermann, C., Weigert, A., 1982, A&A 107, 283

Renzini, A., 1977, in *Advanced Stages of Stellar Evolution*, ed. P. Bouvier & A. Maeder, p. 151, Geneva Observatory

Renzini, A., 1987, A&A 188, 49

Roxburg, I., 1978, A&A 65, 281

Roxburg, I., 1989, A&A 211, 361

Saslow, W. C., Schwarzschild, M., 1965, ApJ 142, 1468

Schatzman E., 1977, A&A 56, 211

Shaviv, G., Salpeter, E. E., 1973, ApJ 184, 191

Xiong D., 1985, A&A 150, 133

Xiong D., 1986, A&A 167, 239

Xiong D., 1990, A&A 232, 31

Zahn J. P., 1987, Proceedings of the Workshop "Instabilities in luminous early type stars", Lunteren, Netherlands, Dordrecht D. Reidel Publishing Co., p. 143

Zahn J. P., 1991 A&A 265, 115

Evolution and Mixing in Low and Intermediate Mass Stars

John C. Lattanzio

*Department of Mathematics, Monash University,
Clayton, Victoria, 3168, Australia;
and
Institute of Astronomy, Madingley Road, Cambridge, U.K.*

Abstract. I present a qualitative picture of our understanding of mixing and evolution in intermediate mass stars, with particular attention to changes in surface composition. The emphasis is on the mechanisms operating rather than a quantitative estimate of their magnitude.

1. INTRODUCTION

In this paper I will present a simple summary of the main phases of evolution of low and intermediate mass stars (roughly, $1-9M_\odot$ on the main sequence), with particular emphasis on nucleosynthesis and mixing, because this is what concerns us most at this conference. To begin I will discuss the pre-Asymptotic Giant Branch (AGB) evolution of typical $1M_\odot$ and $5M_\odot$ stars, introducing terminology and physics as necessary. In section 4 I shall describe the thermally-pulsing AGB phase, with the phenomenon of Hot Bottom Burning (HBB) left to section 5. I leave the observational situation to be summarised by others elsewhere in this volume.

2. SUMMARY OF EVOLUTION AT $\sim 1M_\odot$

We make the usual assumption that a star reaches the zero-age main sequence with a homogeneous chemical composition (for an alternative evolutionary scenario see (1)). Figure 1 shows a schematic HR diagram for this star. Core H-burning occurs radiatively, and the central temperature and density grow in response to the increasing molecular weight (points 1–3). At central H exhaustion (point 4) the H profile is as shown in inset (a) in Figure 1. The star now leaves the main sequence and crosses the Hertzsprung Gap (points 5–7), while the central He core becomes electron degenerate and the nuclear burning is established in a shell surrounding this core. Inset (b) shows the advance of the H-shell during this evolution. Simultaneously, the star is expanding and the outer layers become convective. As the star reaches the Hayashi limit ($\sim$ point 7), convection extends quite deeply inward (in mass) from the surface, and the star ascends the (first) giant branch. The convective envelope

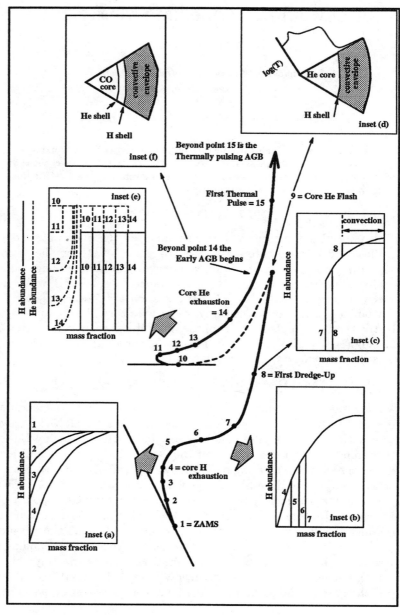

FIGURE 1. Schematic evolution at $\sim 1 M_\odot$

penetrates into the region where partial H-burning has occured earlier in the evolution, as shown in inset (c) of Figure 1. This material is still mostly ^{1}H, but with added ^{4}He together with the products of CN cycling, primarily ^{14}N and ^{13}C. These are now mixed to the surface (point 8) and this phase is known as the "first dredge-up". The most important surface abundance changes are an increase in the ^{4}He mass fraction by about 0.03 (for masses less than about $4M_\odot$), while ^{14}N increases (at the expense of ^{12}C) by around 30%, and the number ratio ^{12}C/^{13}C varies between 18 and 26 (2).

As the star ascends the giant branch the He-core continues to contract and heat. Neutrino energy losses from the centre cause the temperature maximum to move outward, as shown in inset (d) of Figure 1. Eventually triple alpha reactions are ignited at this point of maximum temperature, but with a degenerate equation of state. The temperature and density are decoupled, and the resulting ignition is explosive, being referred to as the "core helium flash" (point 9: see for example (3)). Following this the star quickly moves to the Horizontal Branch where it burns ^{4}He gently in a convective core, and H in a shell (which provides most of the luminosity). This corresponds to points 10–13 in Figure 1. Helium burning increases the mass fraction of ^{12}C and ^{16}O (the latter through ^{12}C$(\alpha,\gamma)^{16}$O) and the outer regions of the convective core become stable to the Schwarzschild convection criterion but unstable to that of Ledoux: a situation referred to as "semiconvection" (space prohibits a discussion of this phenomenon, but an excellent physical description is contained in (4 and 5)). Thus the composition profile in this region adjusts itself to produce convective neutrality, with the profiles as shown in inset (e) of Figure 1.

Following He exhaustion (point 14) the star ascends the giant branch for the second time, and this is known as the Asymptotic Giant Branch or AGB. The C-O core (the proportions of ^{12}C and ^{16}O depend on the uncertain rate for the ^{12}C$(\alpha,\gamma)^{16}$O reaction) becomes degenerate, and the star's energy output is provided by the He-burning shell (which lies immediately above the C-O core) and the H-burning shell. Above both is the deep convective envelope. This structure is shown in inset (f) in figure 1. We will later see that the He-shell is thermally unstable, as witnessed by the recurring "thermal pulses". Thus the AGB is divided into two regions: the early-AGB, prior to (and at lower luminosities than) the first thermal pulse, and the thermally-pulsing AGB beyond this. We will return to this in section 4.

3. SUMMARY OF EVOLUTION AT $\sim 5M_\odot$

A more massive star, say of $\sim 5M_\odot$, begins its life very similarly to the lower mass star discussed above. The main initial difference is that the higher temperature in the core causes CNO cycling to be the main source for H-burning, and the high temperature dependence of these reactions causes a convective core to develop. As H is then burned into He the opacity (due mainly to electron scattering, and hence proportional to the H content) de-

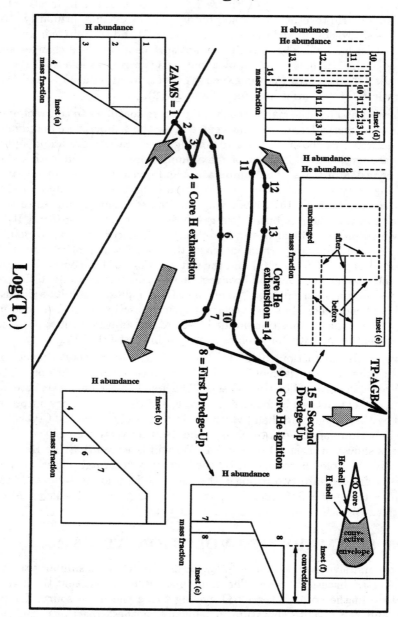

FIGURE 2. Schematic evolution at $\sim 5 M_\odot$.

creases and the extent of the convective core decreases with time. This corresponds to points 1–4 in Figure 2. Following core H exhaustion there is a phase of shell burning as the star crosses the Hertzsprung Gap (points 5–7 and inset (b)), and then ascends the (first) giant branch. Again the inward penetration of the convective envelope (point 8) reaches regions where there has been partial H-burning earlier in the evolution, and thus these products (primarily ^{13}C and ^{14}N, produced at the cost of ^{12}C) are mixed to the surface in the first dredge-up, just as seen at lower masses, and sketched in inset (c) of figure 2.

For these more massive stars the ignition of ^{4}He occurs in the centre and under non-degenerate conditions, and the star settles down to a period of quiescent He-burning in a convective core, together with H-burning in a shell (see inset (d) in figure 2). The competition between these two energy sources determines the occurrence and extent of the subsequent blueward excursion in the HR diagram (e.g. (6)), when the star crosses the instability strip and is observed as a Cepheid variable (points 10–14). Following core He exhaustion the structural re-adjustment to shell He burning results in a strong expansion, and the H-shell is extinguished as the star begins its ascent of the AGB. With this entropy barrier removed, the inner edge of the convective envelope is free to penetrate the erstwhile H-shell. Thus the products of complete H-burning are mixed to the surface in what is called the "second dredge-up" (point 15). This again alters the surface compositions of ^{4}He, ^{12}C, ^{13}C and ^{14}N. This actually reduces the mass of the H-exhausted core, because in the process of mixing ^{4}He outward, of course, we also mix H inward (see inset (e) in figure 2). Note that there is a critical mass (of about $4M_\odot$, but dependent on composition) below which the second dredge-up does not occur. Following this the H-shell is re-ignited and the first thermal pulse occurs soon after: the star has reached the TP-AGB. Note that at this stage the structure is qualitatively similar for all masses.

4. THE THERMALLY-PULSING AGB

This has been reviewed in detail in many places (7, 8, 9, 10) and here we give only a brief summary. The He-burning shell is thermally unstable and experiences periodic outbursts called "shell flashes" or "thermal pulses". The four phases of such a thermal pulse are shown schematically in Figure 3. These are: (a) the off phase, where the structure is basically that of a pre-TP-AGB star. During this phase almost all of the surface lumosity is provided by the H-shell. This phase lasts for 10^4 to 10^5 years, depending on the core-mass; (b) the "on" phase, when the He-shell burns very strongly, producing luminosities up to $\sim 10^8 L_\odot$. The energy deposited by these He-burning reactions is too much for radiation to carry, and a convective shell develops, which extends from the He-shell almost to the H-shell. This convective zone comprises mostly He (about 75%) and ^{12}C (about 22%), and lasts for about 200 years; (c) the "power down" phase, where the He shell begins to die down, and the

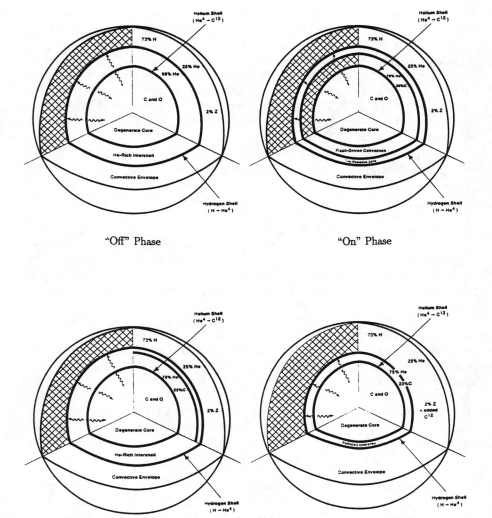

FIGURE 3. The four phases of a thermal pulse cycle.

convection is shut-off. The previously released energy drives a substantial expansion, pushing the H-shell to such low temperatures and densities that it is extinguished; (d) the "dredge-up" phase, where the convective envelope, in response to the cooling of the outer layers, extends inward and, in later pulses, beyond the H/He discontinuity (which previously was the H-shell) and can even penetrate the flash-driven convective zone. This results in the ^{12}C

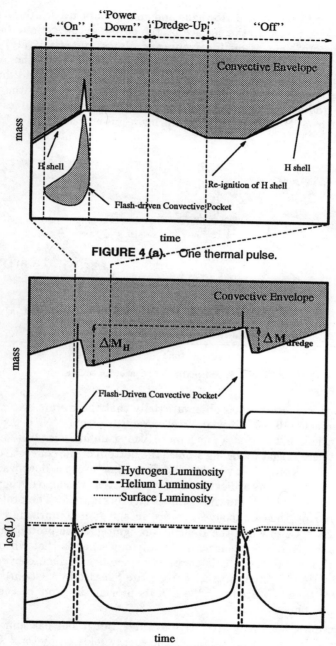

FIGURE 4 (a). One thermal pulse.

FIGURE 4 (b). Two consecutive thermal pulses

which was produced by the He-shell, and mixed outward by the convective shell, now being mixed to the surface by the envelope convection. This is the "third dredge-up", and it qualitatively (and almost quantitatively) accounts for the occurrence of carbon stars at higher luminosities on the AGB (11, 12). Figure 4 shows these four phases during one pulse (top) and during two consecutive pulses (bottom). Also shown in the bottom panel is the variation of the total radiated luminosity and the two nuclear energy sources (i.e. the luminosities from H and He burning) during a pulse cycle.

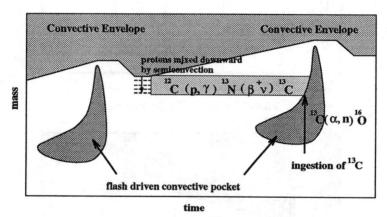

FIGURE 5. Schematic of ^{13}C neutron source.

It is now well established observationally that these stars also produce s-process elements (10, 13). It is also well established that ^{22}Ne is probably not the neutron source, but that a much more likely candidate is ^{13}C (14, 15). The basic result here is that following a pulse the bottom of the convective envelope can become semiconvective which will diffuse some protons downward beyond the formal maximum inward extent of the convective envelope during the third dredge-up phase. This is shown schematically in Figure 5. The protons which are deposited by this semiconvection are in a region comprising about 75% ^{4}He and 22% ^{12}C, so when the H-shell is re-ignited these protons are burned into ^{13}C (and ^{14}N). The scenario as usually envisaged is that when the next thermal pulse occurs this ^{13}C is engulfed by the flash-driven convection, and then in this ^{4}He-rich environment neutrons are released by ^{13}C$(\alpha,n)^{16}$O. These neutrons are then captured by ^{56}Fe and its progeny to produce the observed s-process elements.

This scenario has many attractive features, but it also has some problems (8). In particular, one usually ignores the energy released by the ^{13}C$(\alpha,n)^{16}$O reactions, although this can be substantial, depending on the exact time of ingestion (16, 17). Perhaps a more serious concern is the discovery by Straniero

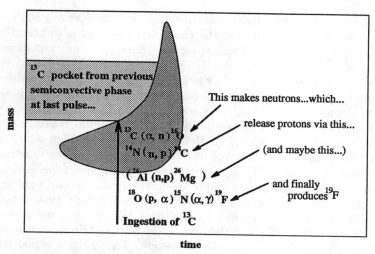

FIGURE 6. Schematic of probable ^{19}F production mechanism.

et al (18) that any ^{13}C formed during re-ignition of the H-shell is burned *radiatively* during the interpulse phase, so that there is no ^{13}C to be ingested at the next pulse! Thus the neutrons are produced, and the s-processing occurs, in a radiative zone. The consequences of this discovery are yet to be examined.

The recent discovery that the ^{19}F/^{16}O ratio in AGB stars increases with the ^{12}C/^{16}O ratio (19) implicates thermal pulses in the origin of this ^{19}F. The only paper to address this situation from the theoretical view is that of Forestini et al (20). They investigated many possible formation scenarios, and the one they judge as most favourable is illustrated in Figure 6. Here, some ^{13}C produces neutrons via the ^{13}C(α,n)^{16}O reaction discussed above. But some of these neutrons are captured by ^{14}N to produce ^{14}C and protons. These protons, plus possibly some from ^{26}Al(n,p)^{26}Mg, are then captured by ^{18}O and the sequence ^{18}O(p,α)^{15}N(α,γ)^{19}F then produces the observed ^{19}F, which is then dredged to the surface in the usual way, following the pulse. For all except those stars with the highest abundances of ^{19}F it appears that the amount of ^{13}C left from the CN cycling H-shell is sufficient. Nevertheless, some extra source of ^{13}C may be required. At present, however, the models overproduce ^{18}O. It remains to be seen if this problem can be overcome, but it may be tied closely to HBB (see below).

5. HOT BOTTOM BURNING

When Wood, Bessell and Fox (21) found bright AGB stars (possessing enhanced abundances of s-process elements) in the Magellanic Clouds it came as some surprise that these were not carbon stars, given the scenario outlined above. These authors speculated that these stars were undergoing HBB, where the temperature at the bottom of the convective envelope is sufficiently high that CN cycling can occur. Thus, some (or perhaps all, or even *more* than all !) of the ^{12}C that is added to the envelope via third dredge-up is then processed into ^{14}N, and the star may never become a carbon star. Subsequent observations by Smith and Lambert (22) showed that these stars possessed extremely strong lithium lines. This was interpreted as further evidence for HBB, as ^{7}Li is believed to be produced by the "beryllium transport mechanism" of Cameron and Fowler (23). Here we require fairly high temperatures to begin the cycle with α-captures on ^{3}He to produce ^{7}Be, as illustrated in the top panel of Figure 7. If this ^{7}Be is to remain at high temperatures, then we would find the PPIII cycle going to completion. Rather, we want the ^{7}Be to β-decay to ^{7}Li. Likewise, this ^{7}Li could capture a proton, thus completing the PPII chain, which we do not want either. We see that the mechanism requires production of ^{7}Be and then transport of that ^{7}Be to cooler regions, as would be expected in a convective envelope. That such a scenario works quantitatively was demonstrated by Boothroyd and Sackmann (24). The lower panel in Figure 7 shows some calculations performed by Cannon et al (25). In this calculation we calculate separately the composition of the upward moving (dashed lines) and downward moving (solid lines) flows. Thus we see clearly the production of ^{7}Be at the bottom of the convective envelope, as well as the reduced abundance of ^{7}Be in the downflows. This is because in the outer envelope much of the ^{7}Be has decayed into ^{7}Li, which is also seen in the enhanced ^{7}Li composition in the downflows (note that the scale is logarithmic). Hence the mechanism works exactly as envisaged. Of course, the duration of the super-lithium-rich phase is limited, for the ^{7}Li produced is mixed through the hot bottomed envelope on each convective cycle, and thus some ^{7}Li is destroyed on each passage. Further, the amount of ^{3}He available for synthesis into ^{7}Li is also limited, and once it is used it is gone. For further details, and a quantitative analysis, see Boothroyd and Sackmann (24).

It was shown by Blöcker and Schönberner (26) that stars undergoing HBB (roughly, those with masses greater than about $5M_\odot$) do not follow the core-mass–luminosity relation (see also (27) and (28)). Further, convective envelopes have been found to reach nearly to 100 million K during the interpulse phase. This is certainly hot enough for the Ne-Na and Mg-Al cycles to operate. No quantitative studies of this phase have been performed to date (but see 25 for some preliminary studies). Such calculations hold the promise of explaining some of the aluminium enhancements seen in SiC and oxide grains (see papers on meteoritic analysis elsewhere in this volume). Indeed, Boothroyd, Sackmann and Wasserburg (29) have applied these models to a quantitative understanding of the oxygen isotopic ratios in meteorites. (See this paper, also, for a discussion of the changes in oxygen isotoic ratios which result from the first and second dredge-up episodes.)

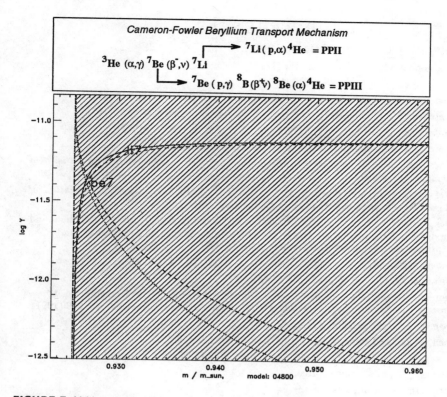

FIGURE 7. Lithium production by HBB in the $5M_\odot$ model of Cannon et al (25). The cross-hatched region denotes convection.

6. CONCLUSIONS

In this paper I have tried to give a qualitative summary of the different nuclear paths and mixing events which occur in the lives of low and intermediate mass stars (roughly, 1–9 $M_\odot$). Although this is an extremely short phase of a star's life (under 10^6 years) the TP-AGB is important for many reasons. It exhibits complex nucleosynthesis (I have barely touched on the subject of s-processing !) combined with normal convective mixing and semi-convection. One must add to this the possibility of overshoot. Hence an understanding of these stars could help us to understand much of the physics operating in their interior. Further, the strong stellar winds which accompany this phase are responsible for depositing the products of the nucleosynthesis into the interstellar medium, where they form the material from which the next stellar generation will form. We are now on the verge of a new quantitative understanding of these processes because of the close interation between theory and observation, including recent meteoritic analysis. The future looks very exciting indeed !

REFERENCES

1. Lattanzio, J. C., *MNRAS*, **207**, 309 (1984).
2. Charbonnel, C., *A. & A.*, **282**, 811 (1994).
3. Deupree, R. G., *Ap. J.*, **287**, 268 (1984).
4. Castellani, V., Giannone, P., and Renzini, A., *Astr. Sp. Sci*, **10**, 340 (1971).
5. Castellani, V., Giannone, P., and Renzini, A., *Astr. Sp. Sci*, **10**, 355 (1971).
6. Lauterborn, D., Refsdal, S., and Weigert, A., *A. & A.*, **10**, 97 (1971).
7. Iben, I., Jr., and Renzini, A., *A. R. A. & A.*, 21, 271 (1983).
8. Lattanzio, J. C., "Evolution and Mixing on the AGB" in *Evolution of Peculiar Red Giants*, 1989, p 161.
9. Sackmann, I.-J., and Boothroyd, A., "On Low Mass AGB Stars" in *Evolution of Stars: The Photospheric Abundance Connection*", 1991, p 275.
10. Iben, I., Jr., "Asymptotic Giant Branch Stars: Thermal Pulses, Carbon Production, and Dredge-Up; Neutron Sources and s-process Nucleosynthesis", in *Evolution of Stars: The Photospheric Abundance Connection*", 1991, p 257.
11. Groenewegen, M. A. T., and de Jong, T., *A. & A.*, **267**, 410 (1994).
12. Lattanzio, J. C., *Ap. J.*, **313**, L15 (1987).
13. Lambert, D. L., "The Chemical Composition of Asymptotic Giant Branch Stars–The s-Process", in *Evolution of Stars: The Photospheric Abundance Connection*", 299 (1991).
14. Iben, I., Jr., and Renzini, A., *Ap. J. Lett.*, **259**, L79 (1982).
15. Iben, I., Jr., and Renzini, A., *Ap. J. Lett.*, **263**, L23 (1982).
16. Bazan, G., and Lattanzio, J. C., *Ap. J.*, **409**, 762 (1993).
17. Mowlavi, N., et al., this volume.
18. Straniero, O., et al., this volume.
19. Jorissen, A., Smith, V. V.,and Lambert, D., L., *A. & A.*, **261**, 164 (1992).
20. Forestini, M., Goriely, S., Jorissen, A., and Arnould, M., *A. & A.*, **261**, 157 (1992).
21. Wood, P. R., Bessell, M. S., and Fox, M. W., *Ap. J.*, **272**, 99 (1983).
22. Smith, V. V., and Lambert, D. L., *Ap. J.*, **345**, L75 (1989).
23. Cameron, A. G. W., and Fowler, W. A., *Ap. J.*, **164**, 111 (1971).
24. Boothroyd, A. I., and Sackmann, I.-J., *Ap. J. Lett.*, **392**, L71 (1992).
25. Cannon, R. C., et al., this volume.
26. Blöcker, T., and Schönberner, D., *A. & A.*, **244**, L43 (1991).
27. Lattanzio, J. C., *Proc. Astron. Soc. Aust.*, **10**, 120 (1992).
28. Boothroyd, A. I., and Sackmann, I.-J., *Ap. J. Lett.*, **393**, L21 (1993).
29. Boothroyd, A. I., Sackmann, I.-J., and Wasserburg, G. J., preprint.

Nucleosynthesis and Supernovae in Massive Stars

S. E. Woosley(1,2), Thomas A. Weaver(2)

(1) Department of Astronomy, University of California, Santa Cruz and (2) General Studies Division, Lawrence Livermore National Laboratory

Abstract. The evolution and explosion of stars heavier than 11 solar masses is explored. Particular emphasis is given to the factors affecting nucleosynthesis and the mass and nature of the collapsed remnant. The neutron star initial mass function will be bimodal with discontinuously larger characteristic masses expected for main sequence stars heavier than 19 $M_\odot$ (the exact value depending upon the rate for $^{12}C(\alpha,\gamma)^{16}O$ and whether carbon burns radiatively or convectively in the core). Above some critical mass, black holes may be produced in otherwise successful supernovae. The ratio of heavy elements to helium ejected in such explosions depends critically upon the explosion energy. For this reason and others, observed values of dY/dZ cannot be used in a straightforward way to imply a limit on the mass of a star that explodes as a Type II supernova.

Key words: Supernovae   nucleosynthesis   neutron star   black hole

1. PRELIMINARIES: STELLAR EVOLUTION

1.1. Basics

The principles governing the evolution of a star are fundamental and well known. Gravity, on the one hand, seeks to contract the star to a point; pressure resists, and a balance, hydrostatic equilibrium, exists. Evolution occurs because radiation and neutrinos carry away energy and entropy. Unless degeneracy intervenes, the star contracts, becomes hotter and burns progressively heavier fuels. At the same time the entropy in the central regions of the star decreases. This is not a hard and fast rule. Nuclear reactions can raise the entropy locally, energy transport can move it around, and of course the second law of thermodynamics says that the total entropy in the universe must *increase*, not decrease. Still the central entropy almost always decreases (Fig. 1). This is, in part, because a negative radial entropy gradient cannot be tolerated in the star (convection occurs) and because of the extreme efficiency of neutrinos in removing entropy during the late stages.

As a consequence of their declining entropy, the cores of all stars have a tendency towards degeneracy as they evolve (Fig. 2). Thus there are well known critical masses for the ignition of each fuel prior to the onset of degeneracy. The critical main sequence mass for igniting carbon non-degenerately, about 8 $M_\odot$ depending upon the choice of initial helium abundance and treatment of convection, defines the lightest stars that are ultimately capable of

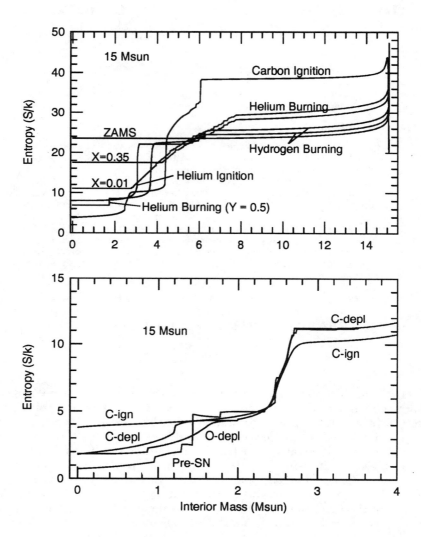

Fig. 1.— Entropy in a 15 $M_\odot$ star at several points during the evolution.

experiencing iron core collapse and making a gravitationally powered supernova. The evolution of stars in the 8 to 11 $M_\odot$ range is quite complicated and will not be reviewed here (Miyaji et al. 1980; Nomoto 1984, 1987; Woosley & Weaver 1980; Weaver & Woosley 1994). Possibilities include, in order of increasing mass, the production of AGB stars with neon-oxygen white dwarfs at their cores; neon-oxygen deflagration leading to core collapse; and violent off center shell flashes leading to envelope ejection followed by a Type II supernovae.

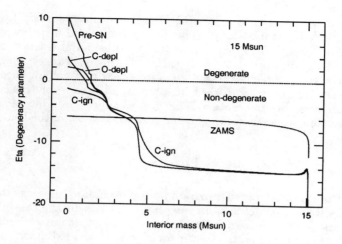

Fig. 2.— Degeneracy parameter, η, in a 15 M$_\odot$ star at various points in its evolution.

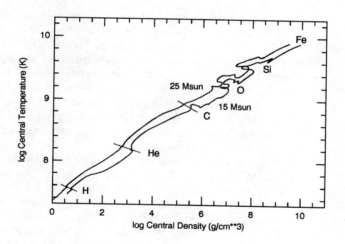

Fig. 3.— Evolution of central density and temperature in 15 and 25 solar mass model stars.

The heaviest stars experience increasingly large amounts of mass loss which must be included in their study. This is especially true for stars of solar metallicity over about 40 $M_\odot$ as well as for the substantial fraction of stars (up to 1/3; Podsiadlowski, Joss, & Hsu 1992; Tutukov, Yungelson, & Iben 1992) that lose their envelopes in binary systems early in helium burning. For other stars, the ones considered in some detail here, evolution at constant mass is a reasonable assumption. This is especially appropriate if all that concerns us are the properties of the helium core where most of the heavy elements are made.

1.2. Input Physics

Aside from the straightforward programming of standard physics, the results one obtains from a model of a massive star (and the differing results obtained by the various groups) are a consequence of the assumed nuclear cross sections, the starting composition, the treatment of convection in all its forms, the prescription for mass loss, the inclusion (or usually, neglect) of rotation, and how the supernova explosion is calculated (or parametrized).

Numerous cross sections influence the production of specific isotopes, but the structural evolution of a massive star is most sensitive to the rate employed for $^{12}C(\alpha,\gamma)^{16}O$ and, to a lesser extent, the uncertain weak interaction rates employed during oxygen and silicon burning and core collapse. As discussed at the meeting, there has been substantial recent progress in determining $12C(\alpha,\gamma)^{16}O$. The nucleosynthetic optimum, $S(300\text{ keV}) = 170 \pm 50$ keV barns (Weaver & Woosley 1993; Woosley, Timmes, & Weaver 1993) is consistent with current experimental measurements and estimates - namely 79 ± 21 or 82 ± 26 keV barns (R- and K-matrix fits respectively) for the E1 part of this rate (Azuma et al 1994) and the theoretical expectation that E2 be comparable to E1 (Mohr et al., these proceedings). Work on the weak interaction rates is needed, especially for beta decay and electron capture rates for masses between 60 (the Fuller, Fowler, and Newman 1980, 1982 cut off) and 80, and is in progress (e.g., Aufderheide et al 1994).

The treatment of convection in all its manifestations is the greatest source of uncertainty and diversity in models for massive stars. We employ a time dependent mixing length theory (Weaver, Zimmerman & Woosley 1978; Woosley & Weaver 1988). In our standard treatment of semiconvective regions, a diffusion coefficient is employed that is related to the radiation diffusion coefficient by a parameter $\alpha \sim 0.1$. Elsewhere in these proceedings, El Eid discusses this prescription and its success compared to various observational constraints such as the ratio of red and blue supergiants and the nucleosynthesis of ^{23}Na. He concludes that a good vale for α is ~ 0.03. For the time being we regard this as excellent agreement.

Convective overshoot mixing is discussed elsewhere in these proceedings by Chiosi. Our treatment is simple by comparison, almost certainly too simple. Each convective region is bounded, top and bottom, by a single semiconvective zone. This allows for the slow growth of convective regions that might otherwise be inhibited, but lacks any physical basis. There are at least two places where this may cause problems. One is the boundary of the helium convective core in quite massive stars. Some of our earlier calculations (Weaver & Woosley 1993) showed, for example, the production of primary nitrogen in very metal deficient stars of 35 $M_\odot$ and more. This occurred as the convective helium core reached into the envelope and mixed hydrogen down into an active burning. More recent calculations (Weaver & Woosley 1994) that use finer zoning (and therefore artificially reduce the overshoot) do not show this nitrogen production. Another place where overshoot mixing is certainly important, at least in some masses, is during oxygen shell burning just prior to the collapse of the iron core (Arnett 1994). In our most recent 25 $M_\odot$ model, (S25A in Table 1 later) the temperature at the base of the oxygen shell rises from 2.4 billion to 3.1 billion K during the last 500 s of the star's life as the contraction of the iron core and silicon shell cause the oxygen burning shell to be superheated to nearly explosive conditions.

This oxygen burning shell is separated by a narrow entropy barrier from an extended convective region farther out where both carbon and neon shells have already merged. In a realistic multi-dimensional calculation these convective regions will all merge (and may have combined a long time before). Indeed the most likely outcome in this situation is a merging of all three shells (oxygen, neon, carbon, and maybe silicon as well) into one extended region reaching almost to the helium shell. Time dependence in the convection will keep the various fuels burning at different altitudes, carbon farthest out, neon next in, and so on. Furthermore the time available for mixing decreases rapidly as these shells are superheated, so some nucleosynthetic segregation will persist, but interesting variations may occur (e.g., for ^{26}Al?). The calculation may be complicated by the differential rotation of the various shells.

The effects of mass loss can be important, especially for the supernova light curve and, if more than the hydrogen envelope is lost, for the nucleosynthesis of heavy elements. Of course the production of some light isotopes - ^{14}N, ^{17}O, ^{23}Na, ^{26}Al, and others - will always be affected by mass loss, but heavier elements are only sensitive to the mass of the helium core until that core itself begins to shrink. For nucleosynthesis in various mass losing models see Woosley, Langer, & Weaver (1993, 1994) and Woosley et al (1994).

Just how the star explodes can also greatly affect the nucleosynthesis, especially of those isotopes heavier than calcium (and the r- and p-processes). Considerable progress has occurred in recent years in successfully modelling the explosion. Convection is central to the mechanism (e.g., Bethe 1990; Miller, Wilson, & Mayle 1993; Herant, Benz, & Colgate 1992; Herant et al 1994; Burrows & Fryxell 1992, 1993; Janka & Müller 1993, 1994) and how the explosion develops will very likely affect the nucleosynthesis of the iron group and of ^{44}Ti. Presently these models have severe nucleosynthetic problems, namely a large overproduction of neutron-rich species in the layer where neutrinos drive convection. There are also technical difficulties in coupling the 2D calculations to those in 1D that preclude a complete realistic simulation of explosive nucleosynthesis. These problems are under study and likely to be resolved soon.

For now however, the results to be discussed are based on explosions simulated using a piston.

2. RECENT MODELS AND RESULTS

Weaver & Woosley (1994) have considered the evolution of a set of massive stars characterized by variable initial metallicity and mass (all evolved at constant mass). These include 46 models of solar metallicity in the range 11 to 25 $M_\odot$ plus stars of 12, 13, 15, 18, 20, 22, 25, 30, 35, and 40 $M_\odot$ and metallicities 0, 10^{-4}, 0.01, and 0.1 $Z_\odot$. An additional series of models is being calculated for 0.5 and 2 $Z_\odot$. In each model approximately 1000 zones are carried with a 200 isotope network for nucleosynthesis studies updated in every zone every cycle. Convective and semiconvective coupling are included in all zones where appropriate as the star is evolved from the main sequence to iron core collapse. In addition, 53 of these presupernova stars have been exploded using a piston situated near the edge of the iron core (more precisely where the electron mole number, Y_e suddenly decreases below 0.50) and the explosive nucleosynthesis (including the ν-process) determined (Woosley & Weaver 1994). In many cases, especially the stars above 25 $M_\odot$, several explosions have been calculated characterized by variable kinetic energies for the ejecta at infinity. Some characteristic values are given for solar metallicity stars in Table 1.

The amount of nucleosynthetic information generated in these calculations is far too much to present here and the Astrophysical Journal papers should be available approximately contemporaneously with these proceedings. The yields of all these models (solar and other metallicities) have been incorporated into a model for Galactic chemical evolution and the time evolution determined for all isotopes from helium to zinc (excepting those made by cosmic rays or in novae). The results are reported in Timmes, Woosley, and Weaver (1994). Certain aspects of the work are discussed briefly by Timmes in these proceedings. Production of the gamma-ray emitting isotopes - ^{26}Al, ^{44}Ti, and ^{60}Fe - has also been discussed by Hoffman et al. (1994).

While the comparison that Timmes finds with respect to solar abundances is very good (Fig. 4), it is noteworthy that no single supernova produces a solar set. Fig. 5 shows the ratio of the mass fraction of the most abundant heavy elements in the ejecta of supernovae of masses 11, 15, 20, 25, and 35 $M_\odot$ compared to the sun. Very little nucleosynthesis occurs in the 11 $M_\odot$ star (at least for elements from carbon to iron) because the heavy element shells are thin. The presupernova star is almost purely an iron core plus helium mantle at the time of collapse. Heavier stars produce more - note the monotonic rise of oxygen, the most abundant heavy element ejected. The iron group yields are sensitive to explosion energy and mass cut (Table 1). The yields of intermediate mass elements vary significantly with mass and in order to get solar ratios of magnesium, neon, and oxygen to silicon, for example, it is useful to have the contribution from the 35 $M_\odot$ star.

TABLE 1: Explosion Characteristics of $Z_\odot$ Models

Mass Model	Fe Core ($M_\odot$)	Piston ($M_\odot$)	M_9 ($M_\odot$)	BE_9 (10^{50} erg)	Rem. ($M_\odot$)	KE_∞ (10^{51} erg)	M_{56} ($M_\odot$)
S11A	1.32	1.32	1.52	0.18	1.32	1.29	0.069
S12A	1.32	1.32	1.53	0.37	1.32	1.17	0.043
S13A	1.41	1.41	1.86	0.56	1.46	1.31	0.133
S15A	1.32	1.29	1.99	1.48	1.43	1.22	0.115
S18A	1.46	1.42	2.34	2.84	1.76	1.17	0.066
S19A	1.66	1.66	2.86	4.14	1.98	1.19	0.100
S20A	1.74	1.74	2.93	5.16	2.06	1.17	0.088
S22A	1.82	1.82	3.10	7.12	2.02	1.47	0.205
S25A	1.78	1.78	3.14	9.78	2.07	1.18	0.129
S30A	1.83	1.83	3.13	10.7	4.24	1.13	0
S30B	1.83	1.83	3.13	10.7	1.94	2.01	0.440
S35A	2.03	2.03	3.63	15.9	7.38	1.23	0
S35B	2.03	2.03	3.63	15.9	3.86	1.88	0
S35C	2.03	2.03	3.63	15.9	2.03	2.22	0.568
S40A	1.98	1.98	3.90	20.1	10.34	1.19	0
S40B	1.98	1.98	3.90	20.1	5.45	1.93	0
S40C	1.98	1.98	3.90	20.1	1.98	2.57	0.691

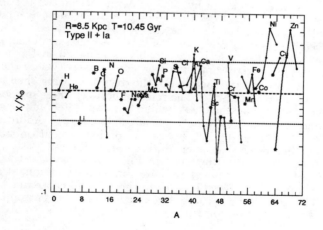

Fig. 4.— Production of intermediate mass elements compared to the sun evaluated at a time and radius in our galaxy when the sun was born (Timmes, Woosley, & Weaver, 1994). A small contribution ($\sim 1/3$) to the iron group from SN Ia is included.

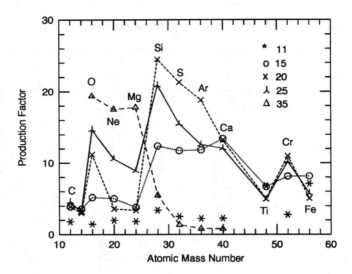

Fig. 5.— Production factors for heavy elements in supernovae of various masses.

Fig. 5 shows that the particular 35 M$_\odot$ star sampled (Model S35B; Table 1) makes essentially no elements heavier than magnesium. Different results would be obtained for Model S35A (less oxygen and magnesium) or S35C (much more intermediate mass elements and iron group). The reason is a highly variable amount of reimplosion that occurs for the different energies. Because the actual energy developed by the explosion mechanism is still difficult to determine with any accuracy, this adds an element of uncertainty to all such nucleosynthesis calculations (see next section).

Another interesting result of our survey of massive stellar evolution is the systematics of iron core mass, which presumably is reflected in the masses of bound remnants. Fig. 6 gives both the iron core masses (where the iron mass fraction rises to 1%) and the location of the oxygen burning shell for 46 presupernova stars of solar metallicity. The relation between these masses and that of the neutron star or black hole left behind is complicated and sensitive to the hydrodynamics of the explosion (not calculated here). Additionally the mass should be reduced by 15 20% before comparing to neutron stars to account for the neutrino losses that occur during the explosion.

Several characteristics are noteworthy in Fig. 6. First, the iron core masses (after the neutrino adjustment) are lower limits to the remnant mass. Ejection of the neutron rich material interior to there would cause nucleosynthetic problems. The oxygen shell usually has a large associated entropy jump which might be natural location for a mass separation, but this is only speculation which must be backed up by detailed calculations. A mass separation there would give (gravitational) masses around 1.3 $M_\odot$ for the most abundant supernovae (12 to 20 $M_\odot$), close to observations but perhaps a little on the light side. The remnant mass is not monotonic owing to variation in the location where the many different convective shell burning episodes ignite in each star. Recalculation of models of a given mass (with symmetry broken by time step or zoning criteria) typically give the same result to 0.01 $M_\odot$.

Superimposed upon the overall tendency for monotonic increase of the core mass (caused by the rise in final entropy of the presupernova star), is an abrupt jump around 19 $M_\odot$. This mass marks the transition between stars that burn carbon convectively in the core and those that burn radiatively. Below 19 $M_\odot$ the star typically experiences convective carbon core burning and three episodes of convective carbon shell burning. Neon core burning is also convective in such stars. Above 19 $M_\odot$ however, for our choice of $^{12}C(\alpha,\gamma)^{16}O$, the abundance of carbon is too small to provide sufficient power (above the neutrino losses) to drive convection in the core. It burns radiatively as does neon (there are episodes of convective carbon and neon *shell* burning however). Lacking the time that it would have spent during these phases, the star jumps almost immediately from helium depletion to oxygen ignition. The entropy remains high in the core and thus the adjusted Chandrasekhar mass and iron core mass, larger.

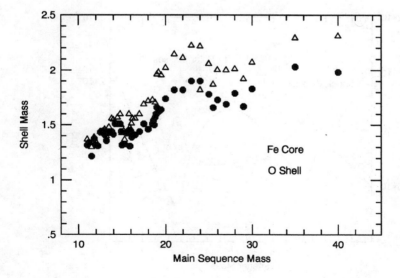

Fig. 6.— Iron core and oxygen shell masses for 46 presupernova stars of solar metallicity.

This bimodality in the iron core mass may be reflected in the neutron star mass birth function or in the bifurcation between black hole and neutron star remnants.

3. FALL BACK AND BLACK HOLE FORMATION

Qualitatively, the reason for increased fall back in the higher mass is the larger binding energy of the mantle (see the quantity BE_9, the gravitational binding energy exterior to 10^9 cm in the presupernova star in Table 1). Unless the explosion mechanism naturally provides a larger energy for bigger stars, there will come a critical mass where the mantle cannot all be ejected. There is some hope that this increase might actually occur because the additional ram pressure of the collapsing mantle might bottle up the neutrino powered energy deposition until a successful explosion is assured, but for now we don't know.

Hydrodynamically the fall back occurs as the shock passes through regions of increasing ρr^3 (Bethe 1990; Herant & Woosley 1994; Woosley & Weaver 1994). Figs. 7 and 8 show this effect for Models 15A and 35B. Where ρr^3 increases, the shock slows; conversely the shock accelerates where ρr^3 decreases. The slowing of the shock is communicated back to the core, first by sound waves and later by a "reverse shock" that forms when the outgoing shock

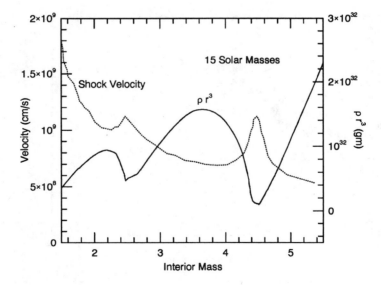

Fig. 7.— The quantity ρr^3 and the velocity of material just behind the shock as a function of mass in Model S15A.

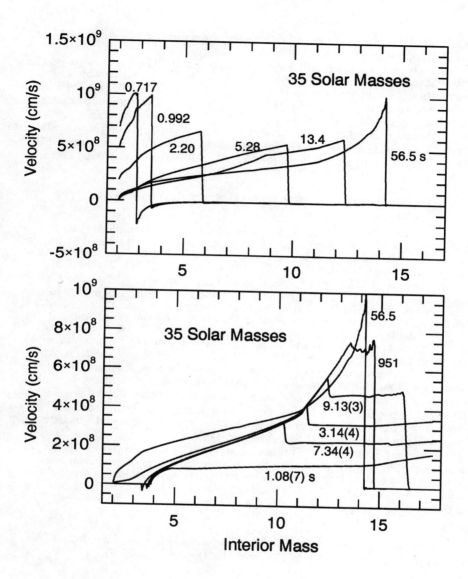

Fig. 8.— Propagation of the shock through Model S35B. Note the fall back of over 3.5 $M_\odot$ by 9000 seconds, well before the reverse shock has reached the center of the supernova.

exits the helium core and runs into the hydrogen envelope (located at 14 $M_\odot$ in the 35 $M_\odot$ star).

Clearly a black hole will form in Model S35B. The interaction of the material falling back into this hole, both of which have considerable angular momentum is an interesting topic for future study (Woosley 1993), but for now we note obvious implications for the nucleosynthesis. The amount of material that falls back is very sensitive to the energy of the outgoing shock. A 20% increase causes the ejection of all material external to the piston (Model S35C, Table 1) including a great deal of iron group elements; a 30% decrease and the mass of the remnant almost doubles to 7.4 $M_\odot$ with an accompanying decrease in oxygen synthesis.

4. HELIUM AND HEAVY ELEMENT PRODUCTION - DY/DZ

Because of this fall back, the relative contribution of massive stars to the Galactic abundances of helium and heavy elements is uncertain and so too are any arguments relating a "cut off" mass for nucleosynthesis to the nature of the remnant (e.g., Maeder 1992, 1993; Brown & Bethe 1994). Model S35A has a very large dY/dZ, Model S35C, a small one. The large dY/dZ from S35A might allow some other massive star to leave a neutron star while ejecting more heavy elements. The explosion energy may not be uniquely determined by the star's main sequence mass and thus there is some randomness in the final remnant mass.

Moreover, all this is based on arguments which do not include mass loss. Woosley, Langer, & Weaver (1993, 1994) have studied the evolution to supernova of massive stars that lose all their hydrogen envelope, either to a stellar wind or a close companion, and then enter a phase of rapid "mass dependent mass loss" (Langer 1989). These stars may be characterized by large values of dY/dZ ranging from 2 to 6 and more. Thus when mass loss is included, precisely those stars which others have suggested must collapse into black holes may contribute appreciably to *increasing* dY/dZ.

For what it is worth, Timmes et al (1994) have computed the dY/dZ resulting from our stellar models for an assumed range of explosion energies but neglecting mass loss. Under these restrictive assumptions, they find that a cut off around 25 to 30 $M_\odot$ is needed in order to give dY/dZ near 3 to 4 as is observed (see Pagel these proceedings).

This work has been supported by the National Science Foundation (AST 91-15367), NASA (NAGW-2525), and the Department of Energy (W-7405-ENG-48).

REFERENCES

Arnett, W. D. 1994, ApJ, 427, 932

Aufderheide, M., Fushiki, I., Woosley, S. E., & Hartmann, D. H. 1994, ApJS, 91, 389

Azuma, R. E., Buchman, L., Barker, F. C., Barnes, C. A., and 13 others 1994, Phys. Rev. C, 50, 000.

Bethe, H. A. 1990, Rev. Mod. Phys., 62, 801

Brown, G. E., & Bethe, H. A. 1994, ApJ, 423, 659
Burrows, A. & Fryxell, B. A. 1992, Science, 258, 430
Burrows, A. & Fryxell, B. A. 1993, ApJ, 418, L33
Fuller, G. M., Fowler, W. A., & Newman, M. 1980, ApJS, 42, 447
Fuller, G. M., Fowler, W. A., & Newman, M. 1982, ApJS, 48, 279
Hashimoto, M., Iwamoto, K., & Nomoto, K. 1993, ApJ, 414, L105
Herant, M., Benz, W., & Colgate, S. 1992, ApJ, 395, 642
Herant, M., Benz, W., Hix, W. R., Fryer, C. L., & Colgate, S. 1994, ApJ, in press
Herant, M., & Woosley, S. E., ApJ, 425, 814
Hoffman, R. D., Woosley, S. E., Weaver, T. A., Timmes, F. X., Eastman, R. G., and Hartmann, D. H. 1994, in The Gamma-Ray Sky with CGRO and Sigma, ed. M. Signore, P. Salati, & G. Vedrenne, (Kluwer Acad. Pub.: Dordrect), in press
Janka, H.-Th., & Müller, E. 1993, in Supernovae and Supernova Remnants - Proc. IAU 145, eds. R. McCray and Wang Zhenru, Cambridge Univ. Press, in press
Janka, H.-Th., & Müller, E. 1994, A&A, in press
Langer, N. 1989, A&A, 220, 135
Maeder, A. 1992, A&A, 264, 105
Maeder, A. 1993, A&A, 268, 833
Miller, D. S., Wilson, J. R., & Mayle, R. W. 1993, ApJ, 415, 278
Miyaji, S., Nomoto, K., Yokoi, K., & Sugimoto, D. 1980, PASJ, 32, 303
Nomoto, K. 1984, ApJ, 277, 291
Nomoto, K. 1987, ApJ, 322, 206
Podsiadlowski, , Ph., Joss, P. C., & Hsu, J. J. L. 1992, ApJ, 391, 246
Timmes, F. X., Woosley, S. E., & Weaver, T. A. 1994, ApJ, submitted
Tutukov, A. V., Yungelson, L. R., & Iben, I 1992, ApJ, 386, 197
Weaver, T. A., Zimmerman, G., & Woosley, S. E. 1978, ApJ, 225, 1021
Weaver, T. A. & Woosley, S. E. 1993, Phys. Rept, 227, 65
Weaver, T. A. & Woosley, S. E. 1994, ApJ, in preparation
Woosley, S. E. 1993, ApJ, 405, 273
Woosley, S. E., Weaver, T. A., & Taam, R. E. 1980, Proceedings of the Austin Conference on Type II Supernovae, J. C. Wheeler, Austin: U. Texas Press, 1980, 96
Woosley, S. E., & Weaver, T. A. 1988 Phys. Rept, 163, 79
Woosley, S. E., Timmes, F. X. & Weaver, T. A. 1992, Nuclei in the Cosmos, F. Käppeler and K. Wisshak, Institute of Physics: Bristol, 1993, 531
Woosley, S. E., Langer, N., & Weaver, T. A. 1993, ApJ, 411, 823
Woosley, S. E., Eastman, R. E., & Weaver, T. A., & Pinto, P. A. 1994, ApJ, 429, 300
Woosley, S. E., Langer, N., & Weaver, T. A. 1994, ApJ, submitted
Woosley, S. E. & Weaver, T. A. 1994, ApJ, in preparation

Nucleosynthesis in Core Collapse Supernovae

F.-K. Thielemann[1], Ken'ichi Nomoto[2], Masaaki Hashimoto[3]

[1] *Institut für theoretische Physik, Universität Basel, Switzerland*
[2] *Department of Astronomy, School of Science, University of Tokyo, Japan*
[3] *Department of Physics, Faculty of Science, Kyushu University, Japan*

We performed nucleosynthesis calculations for 13, 15, 20 and 25$M_\odot$ stars, based on induced supernova explosions. The calculations made use of mass cuts between the central neutron star and the ejected envelope by requiring ejected ^{56}Ni-masses in agreement with supernova light curve observations. Specific emphasis was put on the treatment of the innermost layers, which are the source of ^{56}Ni, the Fe-group composition in general, and some intermediate-mass alpha-elements like Ti. The predictions are compared with abundances from specific supernova observations (e.g. SN 1987A, 1993J) or supernova remnants (e.g. G292.0+1.8, N132D). The amount of detected ^{16}O and ^{12}C or products from carbon and explosive oxygen burning can constrain our knowledge of the *effective* ^{12}C$(\alpha, \gamma)^{16}$O rate in He-burning. The ^{57}Ni/^{56}Ni ratio can give constraints on Y_e in the innermost ejected zones. This helps to estimate the necessary delay time between collapse and the neutrino-driven explosion. Provided that the stellar pre-collapse models are reliable, this allows additional insight into the exact working of the supernova explosion mechanism.

I. INTRODUCTION

Except for type Ia supernovae, which are explained by exploding white dwarfs in binary systems, all other supernova types seem to be linked to massive stars with main sequence masses M>8$M_\odot$ (e.g. Hashimoto, Iwamoto & Nomoto 1993). All stars in that mass range produce a collapsing core after the end of their hydrostatic evolution, which proceeds to nuclear densities (for a review see e.g. Bethe 1990). The total energy released, 2-3$\times 10^{53}$erg, equals the gravitational binding energy of a neutron star. Because neutrinos are the particles with the longest mean free path, they are able to carry away that energy in the fastest fashion. This was proven by the neutrino emission of supernova 1987A, detected in the Kamiokande and IMB experiments (see Burrows 1990 for an overview).

The most promising supernova mechanism is the delayed explosion mechanism, caused by neutrino heating (neutrino and anti-neutrino captures on neutrons and protons) on a time scale of seconds or less after the collapse.

The exact delay time t_{de} depends on the question whether neutrinos diffuse out from the core (>0.5s), weak convection occurs due to composition gradients ("saltfinger convection"), or convective turnover due to entropy gradients shortens this escape time substantially (e.g. Burrows & Fryxell 1992; Janka & Müller 1993; Wilson & Mayle 1993; Herant et al. 1994). The behavior of t_{de} as a function of stellar mass is still an open question and quantitative results of self-consistent calculations still have uncertainties. Therefore, we make here still use of the fact that typical kinetic energies of 10^{51} erg are observed and light curve as well as explosive nucleosynthesis calculations can be performed by introducing a shock of appropriate energy in the precollapse stellar model (see e.g. Woosley & Weaver 1986; Shigeyama, Nomoto & Hashimoto 1988; Weaver and Woosley 1993). Uncertainties are expected from the missing knowledge of the exact core structure at the time when the shock wave starts propagating outward. The present set of calculations is meant to explore these uncertainties in t_{de}, being guided by comparison to observations (e.g. SN1987A, a 20M$_\odot$ during the main sequence stage – see e.g. Arnett et al. 1989, McCray 1993; SN 1993J, a 14±1M$_\odot$ star during main sequence – see e.g. Nomoto et al. 1993; type Ib and Ic supernova light curves like SN 1994I, which due to the lack of a large H-envelope and their early X-ray and gamma-ray losses are steeper than those of SNe II, but are also core collapse events – see e.g. Shigeyama et al. 1990, Nomoto et al. 1994; the ^{57}Ni/^{56}Ni ratio deduced from γ-rays of the ^{56,57}Co decay or spectral features changing during the decay time – see e.g. Clayton et al. 1992, Kumagai et al. 1993; or supernova remnants like G292.0+1.8, N132D – Hughes and Singh 1994, Blair et al. 1994; and comparison with abundances in low metallicity stars, which reflect the average SNe II composition).

We concentrate here on the composition of the ejecta from such core collapse supernovae as an extension to earlier work (Hashimoto, Nomoto & Shigeyama 1989; Thielemann, Hashimoto & Nomoto 1990; Thielemann, Nomoto & Hashimoto 1993, 1994b; Hashimoto et al. 1993b). A more detailed account is given in Thielemann, Nomoto & Hashimoto (1994c). We ignore with our approach the very small but important amount of matter in the high entropy bubble, where neutrino heating and neutronization can cause the production of r-process nuclei as well as possibly the neutron-rich isotopes of ^{48}Ca, ^{50}Ti, ^{54}Cr, ^{58}Fe, and ^{64}Zn. A recent analysis of the solar r-process abundance pattern revealed that it can only be reproduced in detail by a superposition of several steady flow components with neutron number densities $n_n > 10^{20}$cm^{-3} and temperatures $T > 10^9$K (Thielemann et al. 1993, 1994a; Kratz et al. 1993). The conditions in the *high-entropy bubble* seem to agree reasonably well with these requirements (see e.g. Woosley & Hoffman 1992, Meyer et al. 1992, Witti et al. 1994, Takahashi et al. 1994, Woosley et al. 1994). The question still remains, whether for all supernova progenitor masses these conditions can be met (Cowan, Thielemann & Truran 1991; and Mathews, Bazan & Cowan 1992).

One of the still uncertain parameters in stellar evolution, and thus for the pre-supernova models, is the $^{12}C(\alpha,\gamma)^{16}O$ reaction (see Filippone, Humblet & Langanke 1989; Caughlan et al. 1985; Caughlan & Fowler 1988; Barker & Kajino 1991, Buchmann et al. 1993, Zhao et al. 1993, Azuma et al. 1994). The amount of ^{12}C and ^{16}O produced by a star, depends on the nuclear rate *and* the treatment of convection, which governs the mixing of fresh He-fuel into the He-burning core. This *combined* dependence can hopefully be disentangled in the future. The calculations presented here are based on stellar models which employ the rate of Caughlan et al. (1985), the Schwarzschild criterion of convection, and no overshooting, described in detail in Nomoto and Hashimoto (1988) and Hashimoto et al. (1993b). We will show that it predicts results in very good agreement with observations.

II. EXPLOSIVE NUCLEOSYNTHESIS AND THE MASS CUT

A. Basic Features

The calculations were performed by depositing a total thermal energy of the order $E = 10^{51}$ erg + the gravitational binding energy of the ejected envelope into several mass zones of the stellar Fe-core (for details see Shigeyama et al. 1988, Thielemann et al. 1990, and for the nuclear physics input Thielemann et al. 1994a).

A first overview of results from the explosion calculations is displayed in Table 1 for elemental abundances. They can be characterized by the following behavior: the amount of ejected mass from the unaltered (essentially only hydrostatically processed) C-core and from explosive Ne/C-burning (C, O, Ne, Mg) varies strongly over the progenitor mass range, while the amount of mass from explosive O- and Si-burning (S, Ar, and Ca) is almost the same for all massive stars. Si has some contribution from hydrostatic burning and varies by a factor of 2-3. The amount of Fe-group nuclei ejected depends directly on the explosion mechanism. The values listed for the $20M_\odot$ star have been chosen to reproduce the $0.07M_\odot$ of ^{56}Ni deduced from light curve observations of SN 1987A. The choice for the other progenitor masses is also based on supernova light curve observations, but their uncertain nature should be underlined.

Thus, we have essentially three types of elements, which test different aspects of supernovae, when comparing with individual observations. (i) The first set (C, O, Ne, Mg) tests the stellar progenitor models, (ii) the second (Si, S, Ar, Ca) the progenitor models and the explosion energy in the shock wave, while (iii) the Fe-group (beyond Ti) probes clearly in addition the actual supernova mechanism. Only when all three aspects of the predicted abundance yields can be verified with individual observational checks, it will be reasonably secure to utilize these results in chemical evolution calculations of galaxies. In general we should keep in mind, that as long as the explosion mechanism is not completely and quantitatively understood yet, one has to

TABLE 1. MAJOR NUCLEOSYNTHESIS YIELDS

Element	13M$_\odot$	15M$_\odot$	20M$_\odot$	25M$_\odot$
C	0.060	0.083	0.115	0.148
O	0.218	0.433	1.480	3.000
Ne	0.028	0.039	0.257	0.631
Mg	0.012	0.046	0.182	0.219
Si	0.047	0.071	0.095	0.116
S	0.026	0.023	0.025	0.040
Ar	0.0055	0.0040	0.0045	0.0072
Ca	0.0053	0.0033	0.0037	0.0062
Fe	0.150*	0.130*	0.075	0.050*

assume a position of the mass cut, and dependent on that position, which is a function of the delay time between collapse and final explosion, the ejected mass zones will have a different neutron excess or $Y_e = <Z/A>$ of the nuclear composition. We will discuss this in more detail in the following subsections.

B. Ni(Fe)-Ejecta and the Explosion Mechanism

Figures 1a and 1b (both presenting a 13M$_\odot$ star) make clear how strongly a Y_e change can affect the resulting composition. Figure 1a makes use of a constant $Y_e=0.4989$ in the inner ejecta, experiencing incomplete and complete Si-burning. Figure 1b makes use of the original Y_e resulting from the pre-collapse burning phases. Here Y_e drops to 0.4915 for mass zones below $M(r)=1.5$M$_\odot$. Huge changes in the Fe-group composition can be noticed. The change in Y_e from 0.4989 to 0.4915 causes a tremendous change in the isotopic composition of the Fe-group for the affected mass regions (<1.5M$_\odot$). In the latter case the abundances of ^{58}Ni and ^{56}Ni become comparable. All neutron-rich isotopes increase (^{57}Ni, ^{58}Ni, ^{59}Cu, ^{61}Zn, and ^{62}Zn), the even-mass isotopes (^{58}Ni and ^{62}Zn) show the strongest effect. In Figure 1a one can recognize the increase of ^{40}Ca, ^{44}Ti, ^{48}Cr, and ^{52}Fe with an increasing remaining He mass fraction. These are direct consequences of a so-called alpha-rich freeze-out with increasing entropy.

While these calculations were performed by depositing energy at a specific radius inside the Fe-core and letting the shock wave propagate outward, in reality stellar models *at the time t_{de}, when the successful shock wave is initiated,* have to be utilized. Instead they were taken at the onset of core collapse, which corresponds to a prompt explosion. Aufderheide et al. (1991) performed a calculation with a model at 0.29s after core collapse for a 20M$_\odot$ star, when the prompt shock had failed, and found an increase of the mass cut by roughly 0.02M$_\odot$. A delayed explosion would set in after a delay of up to 1s, with the exact time being somewhat uncertain and dependent on the details of neutrino transport.

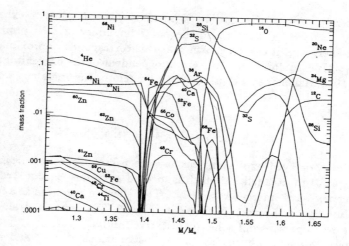

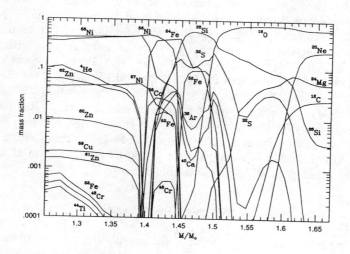

FIG. 1. Isotopic composition of the ejecta for a core collapse supernovae from a 13M$_\odot$ star (3.3M$_\odot$ He-core). Only the dominant abundances of intermediate mass nuclei are plotted, while the Fe-group composition is presented in full detail. The exact mass cut in $M(r)$ between neutron star and ejecta depends on the details of the delayed explosion mechanism. Figures 1a and 1b show how strongly a Y_e-change can affect the resulting composition. Figure 1a makes use of a constant $Y_e=0.4989$ in the inner ejecta, Figure 1b makes use of the original Y_e resulting from the pre-collapse burning phases, which drops to 0.4915.

The outer boundary of explosive Si-burning with complete Si-exhaustion is given by $T=5\times 10^9$K and is also the outer boundary of ^{56}Ni production. From pure energetics it can be shown that this corresponds approximately to a radius $r_5=3700$ km for $E_{SN}\approx 10^{51}$ erg, independent of the progenitor models (Woosley 1988, Thielemann et al. 1990). Therefore, the mass cut would be at

$$M_{cut} = M(r_5) - M_{ej}(^{56}\text{Ni}). \qquad (1)$$

In case of a delayed explosion, we have to ask the question from which radius $r_{0.5}(t=0)$ matter fell in, which is located at radius $r_5(t=t_{de})=3700$km when the shock wave emerges at time t_{de}. This effect of accretion as a function of delay time t_{de} has been studied in detail by us (Thielemann et al. 1994c). Here we want to present only the quantitative results.

In Figures 2a and b we display the Y_e-distributions of a 13 and a 20M$_\odot$ star and the position of the outer boundary of explosive Si-burning with complete Si exhaustion, M_{ex-Si}, as a function of the delay time t_{de}. We consider for each star delay times of 0, 0.3, 0.5, 1, and 2s, resulting in $r_{0.5}$=3700, 4042, 4412, 5410, and 7348km. Inside this boundary ^{56}Ni is produced as the dominant nucleus and the mass cuts would have to be positioned at $M_{cut}=M(r_{ex-Si}) - M(^{56}\text{Ni})=M(r_{0.5}(0)) - M(^{56}\text{Ni})$. When Ni-ejecta of 0.15, 0.13, 0.07, and 0.05M$_\odot$ are used for 13, 15, 20, and 25M$_\odot$ stars, the mass cuts M_{cut} of Table 2 result for a vanishing delay time. For t_{de}=0.3, 0.5, 1, and 2s the masses $\Delta M_{acc,i}$, i=1, 2, 3, and 4 have to be added to M_{cut}. It is recognizable that especially for the 13M$_\odot$ star the Y_e's encountered for these different delay times vary strongly, and differences of the Fe-group composition can be expected. On the other extreme, the Y_e in the innermost ejecta of a 25M$_\odot$ star are not affected at all by the available choices.

C. Neutron Star Masses

When taking into account the results of the previous section, we obtain baryonic neutron star masses M_b for the sequence of 13, 15, 20, and 25 M$_\odot$ SN II progenitors as noted in Table 2. Column 1 indicates the progenitor mass, column 2 gives the original Fe-core mass, the absolute lower limit for M_b. The third column lists the outer boundary of explosive Si-burning with Si-exhaustion M_{ex-Si}, which represents the outer edge of ^{56}Ni production. The fourth column indicates the location of the mass cut under the assumption that the amount of ^{56}Ni is ejected as required from light curve and/or chemical evolution arguments, using Eq.(1), i.e. assuming a prompt explosion. Probably the largest source of uncertainty is the still existing lack of a complete understanding of the SN II explosion mechanism. In case of a delayed explosion, accretion onto the proto-neutron star will occur until finally after a delay period t_{de} a shock wave is formed, leading to the ejection of the outer

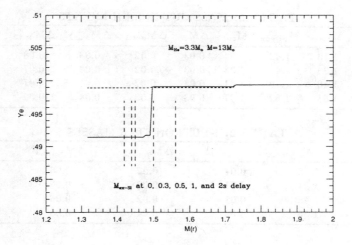

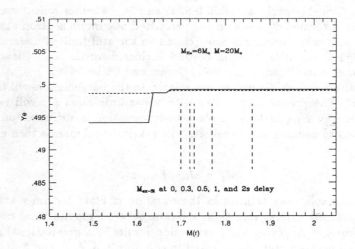

FIG. 2. Figures 2a and b present the Y_e-distributions of a 13 and 20$M_\odot$ star and the position of the outer boundary of explosive Si-burning with complete Si-exhaustion, M_{ex-Si}, as a function of the delay/accretion period t_{de}. For each star delay times of 0, 0.3, 0.5, 1, and 2s are considered, resulting in $r_{0.5}$=3700, 4042, 4412, 5410, and 7348km. ^{56}Ni is produced inside this boundary $r_{0.5}$ as the dominant nucleus. For a given amount of Ni-ejecta, mass cuts would have to be positioned at $M_{cut}=M(r_{ex-Si})-M(^{56}\text{Ni})=M(r_{0.5}(0))-M(^{56}\text{Ni})$. The delay times t_{de} and the required $M(^{56}\text{Ni})$ determine Y_e in the ejected material (solid=original, dashed=experienced for sufficiently large t_{de}, when low Y_e-matter is accreted onto the neutron star.

TABLE 2. MASS CUT IN SN II EVENTS

$M/M_\odot$	M_{core}	M_{Si-ex}	M_{cut}	ΔM_E	$\Delta M_{acc,1}$	$\Delta M_{acc,2}$	$\Delta M_{acc,3}$	$\Delta M_{acc,4}$
13	1.18	1.42	1.27	0.03	0.02	0.03	0.08	0.14
15	1.28	1.46	1.33	0.03	0.02	0.03	0.07	0.15
20	1.40	1.70	1.61	0.03	0.02	0.03	0.07	0.16
25	1.61	1.82	1.77	0.03	0.03	0.04	0.09	0.19

TABLE 3. NEUTRON STAR MASSES

$M/M_\odot$	M_g	$\Delta M_{g,acc,1}$	$\Delta M_{g,acc,2}$	$\Delta M_{g,acc,3}$	$\Delta M_{acc,4}$
13	1.16	0.01	0.02	0.06	0.11
15	1.21	0.01	0.02	0.06	0.12
20	1.45	0.01	0.02	0.06	0.12
25	1.56	0.03	0.04	0.08	0.19

layers. The neutron star boundary would have to be moved outward, accordingly. We give ΔM_{acc} in columns 6-9, the growth of the proto-neutron star by accretion for delay periods of 0.3, 0.5, 1, and 2s. Whether a neutron star or black hole is formed depends on the permitted maximum neutron star mass, which is somewhat uncertain and related to the still limited understanding of the nuclear equation of state beyond nuclear densities (e.g. Glendenning 1991, Weber and Glendenning 1991, Brown and Bethe 1994).

All of the previous discussion was related to the baryonic mass of the neutron star. The proto-neutron star with a baryonic mass M_b will release a binding energy E_{bin} in form of black body radiation in neutrinos during its contraction to neutron star densities. The gravitational mass is then given by

$$M_g = M_b - E_{bin}/c^2. \qquad (2)$$

For reasonable uncertainties in the equation of state Lattimer and Yahil (1989) obtained a relatively tight relation between gravitational mass and binding energy. Applying their expression results in a gravitational mass of the formed neutron star M_g as listed in columns 2, 3, 4, 5, and 6 of Table 3 for the corresponding progenitor masses and baryon masses of Table 2. An error of roughly ±15% for the difference $M_b - M_g$ applies. ΔM_{acc}, due to the uncertainty of the accretion period or delay time, and the choice of $M(^{56}\text{Ni})$ from Table 1 which determines M_{cut} in Table Table 2, dominate the error in M_g. A delay time of about 1s is expected to be an upper bound for the delayed explosions. This is close to a pure neutrino diffusion time scale without any convective turnover.

The results indicate a clear spread of neutron star masses. This spread would be preserved in real supernova events, unless a conspiracy in the combination of proto-neutron star masses, delay times, and explosion energetics leads to a unique neutron star mass. Such a spread is also found in neutron star masses from observations (e.g. Nagase 1989, Page and Baron 1990, van Paradijs 1991), but it is not clear whether it is just due to the large observational errors. We do not know whether the spread predicted in Table 3 already includes the uncertain upper mass limit of neutron stars due to the nuclear equation of state (Baym 1991, Weber and Glendenning 1991). If it does, we would expect for these cases the formation of a central black hole during the delay period. Thus, no supernova explosion would occur and no yields be ejected. Different maximum stable masses for the initially hot and a cold neutron star (see e.g. Brown and Bethe 1994) could result in a supernova explosion and a central black hole.

III. OBSERVATIONAL CONSTRAINTS AND CONCLUSIONS

A. Stellar models

There exist a number of quantitative comparisons for SN1987A (a 20M$_\odot$ star during its main sequence evolution) between nucleosynthesis predictions and observations [see e.g. Table 2 in Danziger et al. (1990), section IVb in Thielemann et al. (1990) or McCray (1993)], which show reasonable agreement for C, O, Si, Cl, Ar, Co, and Ni (or Fe) between observation and theory. We want to concentrate here on a crucial aspect, the O abundance.

The amount of ^{16}O is closely linked to the "effective" ^{12}C$(\alpha,\gamma)^{16}$O rate during core He-burning. This effective rate is determined by three factors: (1) the actual nuclear rate, (2) the amount of overshooting, mixing fresh He-fuel into the core at late phases of He-burning, when the temperatures are relatively high and favor alpha-captures on ^{12}C, and (3) the stellar mass or He-core size, which determine the central temperature during He-burning.

The nuclear rate is still not fully determined. We performed these investigations with the rate by Caughlan et al. (1985), which is one choice within the uncertainties left by experiments (see e.g. Filippone et al. 1989, Humblet et al. 1991, Barker and Kajino 1991). The rate is based on an astrophysical S-factor of $S_{tot}(0.3\text{MeV})=0.24$MeV barn, which is within the error bars of the evaluation undertaken by Barker and Kajino (1991). The resonances of interest are $J^\pi = 1^-$ and 2^+ states, which emit electric dipole (E1) and quadrupole (E2) radiation when decaying to the 0^+ ground state of ^{16}O. The S-factor quoted above is composed of an E1 component in the range 0.08-0.40 and an E2-component of 0.06-0.19 MeV barn (Barker and Kajino 1991). Their result was obtained from best fits to the then available data on alpha capture of ^{12}C, ^{12}C$+\alpha$ elastic-scattering phase shifts, and the delayed alpha spectrum from ^{16}N-decay.

More recent experiments by a Triumf and a Yale collaboration (Buchmann et al. 1993, Zhao et al. 1993, Azuma et al. 1994) are also based on the beta-delayed alpha emission of ^{16}N. While the quoted results still differ somewhat, S_{E1}=0.044-0.070 MeV b from a T-matrix analysis (Buchmann et al.) and

0.095±0.006 MeV b from an R-matrix analysis (Zhao et al.), a subsequent R-matrix analysis of both experiments yields a value close to 0.080 MeV b. Unfortunately the final result for the E2-component is still open. With an expected contribution of the same size as the E1-component (0.080 MeV b) but uncertain by roughly a factor of 2, we expect a total S-factor at 300 keV of roughly 0.120-0.240 MeV b, which corresponds to the Caughlan and Fowler (1988) rate multiplied by a factor 1.3-2.4 and includes the value of the Caughlan et al. (1985) rate. As the effect on the astrophysical outcome is dramatic, this issue still requires a final solution.

As the rate by Caughlan et al. (1985) seems to be close to the barely permitted upper limit, it is crucial to check the observations for individual stellar models, in order to normalize the O-production correctly. The model calculations for a 20 $M_\odot$ star predict 1.48$M_\odot$ of ejected ^{16}O. This is within the early observational constraints of 0.3-3.0$M_\odot$ (see Table 2 in Danziger el. al. 1990) but somewhat unsatisfying. The improved analysis of observations for SN1987A by Spyromilio and Pinto (1991) helped to put tighter constraints on the pre-collapse models by increasing the lower limit to 0.7$M_\odot$. Major improvements were possible by modeling of the late nebular spectra. Franson, Houck & Kozma (1993) found a value of about 1.5$M_\odot$. Chugai (1994) determined 1.2-1.5$M_\odot$. Our value lies well in the remaining uncertainty range, which seems not to ask for a smaller ^{12}C$(\alpha,\gamma)^{16}$O rate than Caughlan et al. (1985). A total S-factor of 0.150MeV b would reduce the ^{16}O mass – within our treatment of convection – below 1$M_\odot$, a value which seems now excluded. It should, however, be clear that these observations test only the combined effect of nuclear rate and convection treatment (here Schwarzschild without overshooting). Similar results were found by Werner et al. (1994) when analyzing spectra of young white dwarfs with models of D'Antona and Mazzitelli (1992).

The first results from O-determinations for SN 1993J are also available now. Houck and Fransson (1994) find a value of ≈0.4$M_\odot$. Our prediction of 0.423$M_\odot$ for a 15$M_\odot$ main sequence star agrees fairly well, SN 1993J was determined to be a 14±1$M_\odot$ star (see discussion in section I). This leads to the conclusion that the Caughlan et al. (1995) rate, used in conjunction with the Schwarzschild criterion for convection and no overshooting, give a very good agreement with observations for individual supernovae. This is the best possible check and preferable over methods which make use of integrals over stellar populations.

Recently other diagnostics became available for abundance determinations in supernova remnants. In that case the progenitor mass is not known, but the relative abundance ratios between different elements can be tested for consistency with abundance predictions for a variety of progenitor masses. Hughes and Singh (1994) made use of X-ray spectra of the supernova remnant G292.0+1.8 and found remarkable agreement for all element ratios from O through Ar with our 25$M_\odot$ calculations (15% rms deviation). This tests implicitely the ^{12}C$(\alpha,\gamma)^{16}$O-rate, as it is also reflected in the ratios between

C-burning products like Ne and Mg and explosive O-burning products like Ar and S. Comparisons with model predictions, which made use of smaller $^{12}C(\alpha,\gamma)^{16}O$-rates, did not pass that consistency check. UV and optical observations of supernova remnant N132D by Blair, Raymond & Long (1994) give very good agreement with our element predictions for a 20$M_\odot$ star, with slight deviations for Mg. Thus, we have direct observations of supernovae and supernova remnants ranging from 15 over 20 to 25$M_\odot$, which agree well with our model predictions and indicate that their application for other purposes should be quite reliable.

B. Y_e at the Mass Cut and Clues for the Explosion Mechanism

The formation of the nuclei ^{58,61,62}Ni, which are produced in form of the neutron-rich species ^{58}Ni and ^{61,62}Zn, is strongly dependent on Y_e and varies therefore with the position of the mass cut between ejected matter and the remaining neutron star (see the discussion in Thielemann et al. 1990 and Kumagai et al. 1991, 1993). Especially for the Ni-abundances the position of the mass cut is crucial. The ^{57}Ni/^{56}Ni ratio is correlated with the abundances of stable Ni isotopes, predominantly ^{58}Ni, i.e. with ^{58}Ni/^{56}Ni. Light curve observations of SN1987A (Elias et al. 1991, Bouchet et al. 1991) could be interpreted with a high 57/56 ratio of 4 times solar, but this would also require too large stable Ni abundances not substantiated from observations (Witteborn et al. 1989, Wooden et al. 1993). In order to meet the stable Ni constraints of 3-5$\times 10^{-3} M_\odot$ (Danziger et al. 1990, Witteborn et al. 1989, and Wooden et al. 1993) only an upper limit of 1.4-1.7 times solar is permitted for the 57/56 ratio from our results, given in detail in Thielemann et al. (1994c). This also agrees well with the observations by Varani et al. (1990) and γ-ray line observations by GRO (Kurfess et al. 1992, Clayton et al. 1992). The apparent discrepancy was solved by correct light curve and spectra modeling with a non-equilibrium treatment of the involved ionization stages at late times (Fransson et al. 1994). This gives a consistent picture for observations of stable Ni, light curve observations which are sensitive to ^{56}Co and ^{57}Co decay, and the γ-ray lines emitted from both decays.

This corresponds to a Y_e at the mass cut of 0.497 within the little nitch in Figure 2b. A mass cut at deeper layers, where Y_e decreases to 0.494, would imply 57/56 ratios larger than 2.5 times solar. A mass cut further out, implying a Y_e of 0.4989 results in a 57/56 ratio of the order of 1 times solar. This means that in order to meet the Y_e-constraint with an ejection of 0.075$M_\odot$ of ^{56}Ni, we have a required delay time of 0.3-0.5s. Keeping all uncertainties of the model in mind, this can be taken as a strong support that SN 1987A did not explode via a prompt explosion, and did not have to wait either for a delayed explosion with a long delay time $t_{de} > 0.5$s. The latter would correspond more to a pure neutrino diffusion case, while this result supports the current understanding that entropy gradients drive the convective turnover and cause a faster neutrino transport. The multidimensional calculations by Janka and Müller (1993) and Herant et al. (1994) predict delay times of ≈ 0.3s, in agreement with this finding.

Unfortunately, we do not have yet similar observational and computational results for other supernovae. This would be a strong test for the explosion mechanism as a function of progenitor mass. It is feasable that progenitors exist, where the entropy gradients do not cause strong Rayleigh-Taylor instabilities. Wilson and Mayle (1988) showed clearly that such instabilities can depend on the equation of state used, it would accordingly also depend on the progenitor structure. Therefore, it is important to explore the whole progenitor mass range with multidimensional explosion calculations in order to find out what Y_e self-consistent calculations would predict for the inner ejecta.

A further test for the correct behavior of the ejecta composition as a function of progenitor mass is the comparison with abundances in low metallicity stars. These reflect the average SNe II composition, integrated over an initial mass function of progenitor stars. First individual tests have been done in Thielemann et al. (1994c), a complete analysis is in work. A verification of SNe II-ejecta in such a way permits a correct application in chemical evolution calculations together with SNe Ia- and PN-ejecta.

We want to thank our collaborators T. Shigeyama and T. Tsujimoto, who contributed to material presented here, and R. Azuma, G. Brown, C. Fransson, P. Höflich, J. Houck, J. Hughes, K. Langanke, N. Langer, and B. Schmidt for helpful discussions and providing results before publication. This research was supported in part by NSF grant AST 89-13799, the Grants-in-Aid for Scientific Research of the Ministry of Education, Science, and Culture in Japan (05242102, 06233101), and the Swiss National Fonds.

References

Arnett, W.D., Bahcall, J.N., Kirshner, R.P., Woosley, S.E. 1989, *Ann. Rev. Astron. Astrophys.* **27**, 629
Aufderheide, M., Baron, E., Thielemann, F.-K. 1991, *Ap. J.* **370**, 630
Azuma, R. et al. 1994, *Phys. Rev. C*, submitted
Barker, F.C., Kajino, T. 1991, *Aust. J. Phys.* **44**, 369
Baym, G. 1991, in *Neutron Stars, Theory and Observations*, eds., J. Ventura and D. Pines, Kluwer Acad. Publ., Dordrecht, p.37
Bethe, H.A. 1990, *Rev. Mod. Phys.* **62**, 801
Blair, W.P., Raymond, J.C., Long, K.S. 1994, *Ap. J.*, in press
Bouchet, P., Danziger, I.J. Lucy, L. 1991, in *SN1987A and Other Supernovae*, ed. I.J. Danziger (ESO, Garching), p. 217
Brown, G., Bethe, H.A. 1994, *Ap. J.* **423**, 659
Buchmann, L. et al. 1993, *Phys. Rev. Lett.* **70**, 726
Burrows, A. 1990, *Ann. Rev. Nucl. Part. Sci.* **40**, 181
Burrows, A., Fryxell, B. 1992, *Science* **258**, 430
Caughlan, G.R., Fowler, W.A., Harris, M.J, Zimmerman, G.E. 1985, *At. Data Nucl. Data Tables* **32**, 197

Caughlan, G.R., Fowler, W.A. 1988, *At. Nucl. Data Tables* **40**, 283
Chugai, N.N. 1994, *Ap. J. Letters*, in press
Clayton, D.D., Leising, M.D., The, L.-S., Johnson, W.N., Kurfess, J.D. 1992, *Ap. J. Letters* **399**, L14
Cowan, J.J., Thielemann, F.-K., Truran, J. W. 1991, *Phys. Rep.* **208**, 267
d'Antona, Mazzitelli, I. 1991, in IAU Symposium 145 *Evolution of Stars: The Photospheric Abundance Connection*, eds. G. Michaud, A. Tutukov, Kluwer Acad. Publ., Dordrecht, p.399
Danziger, I.J., Bouchet, P., Gouiffes, C., Lucy, L. 1990, in *Supernovae*, ed S.E. Woosley, (Springer-Verlag, New York), p.69
Elias, J.H., Depoy, D.L., Gregory, B., Suntzeff, N.B. 1991, in *SN1987A and Other Supernovae*, ed. I.J. Danziger (ESO, Garching), p. 293
Filippone, B.W., Humblet, J., Langanke, K. 1989, *Phys. Rev.* **C40**, 515
Fransson, C., Kozma, C. 1993, *Ap. J. Lett.* **408**, L25
Fransson, C., Houck, J., Kozma, C. 1994, in IAU Coll. 145 *Supernovae and SN Remnants*, ed. R. McCray, Cambridge Univ. Press, in press
Glendenning, N.K. 1991, *Nucl. Phys.* **B24B**, 110
Hashimoto, M., Iwamoto, K., Nomoto, K., 1993, *Ap.J.* **414**, L105
Hashimoto, M., Nomoto, K., Shigeyama, T. 1989, *Astron. Astrophys.* **210**, L5
Hashimoto, M., Nomoto, K., Tsujimoto, T., Thielemann, F.-K. 1993, in *Nuclei in the Cosmos II*, ed. F. Käppeler (IOP Bristol), p.587
Herant, M., Benz, W., Hix, W. R., Fryer, C.L., Colgate, S. A. 1994, *Ap. J.*, in press
Houck, J.C., Fransson, C. 1994, in preparation
Hughes, J.P., Singh, K.P. 1994, *Ap.J.* **422**, 126
Humblet, J. Filippone, B.W., Koonin, S.E. 1991, *Phys. Rev.* **C44**, 2530
Janka, H.-T., Müller, E. 1993, in *Frontiers of Neutrino Astrophysics*, eds. Y. Suzuki, K. Nakamura, Universal Academy Press, Tokyo, p.203
Kratz, K.-L., Bitouzet, J.-P., Thielemann, F.-K., Möller, P., Pfeiffer, B. 1993, *Ap. J.* **402**, 216
Kumagai, S., Shigeyama, T., Hashimoto, M., Nomoto, K. 1991, *Astron. Astrophys.* **243**, L13
Kumagai, S., Nomoto, K., Shigeyama, T., Hashimoto, M., Itoh, M. 1993, *Astron. Astrophys.* **273**, 153
Kurfess, J. D. 1992, *Ap. J. Lett.* **399**, L137
Lattimer, J.M., Yahil, A. 1989, *Ap. J.* **340**, 426
Mathews, G.J., Bazan, G., Cowan, J.J. 1992, *Ap. J.* **391**, 719
McCray, R. 1993, *Ann. Rev. Astron. Astrophys.* **31**, 175
Meyer, B.S., Mathews, G.J., Howard, W.M., Woosley, S.E., Hoffman, R.D. 1992, *Ap. J* **399**, 656
Nagase, F. 1989, *Publ. Astron. Soc. Japan* **41**, 1
Nomoto, K., Hashimoto, M. 1988, *Phys. Rep.* **163**, 13
Nomoto, K., Suzuki, T., Shigeyama, T., Kumigai, S., Yamaoka, H., Saio, H. 1993, *Nature* **364**, 507

Nomoto, K., Yamaoka, H., Pols, O. R., van den Heuvel, E. P. J., Iwamoto, K., Kumagai, S., Shigeyama, T. 1994, *Nature* **371**, 227
Page, D., Baron, E. 1990, *Ap. J. Lett.*, **354**, L17
Shigeyama, T., Nomoto, K., Hashimoto, M. 1988, *Astron. Astrophys.* **196**, 141
Shigeyama, T., Nomoto, K., Tsujimoto. T.,Hashimoto, M. 1990, *Ap. J. Lett.* **361**, L23
Spyromilio, J., Pinto, P.A. 1991, in *SN1987A and Other Supernovae*, ed I.J. Danziger, K. Kjär, ESO/EIPC, p.423
Takahashi, K., Witti, J., Janka, H.-T. 1994, *Astron. Astrophys.* **286**, 857
Thielemann, F.-K., Hashimoto, M., Nomoto, K. 1990, *Ap. J.* **349**, 222
Thielemann, F.-K., Kratz, K.-L., Pfeiffer, B., Rauscher, T., van Wormer, L., Wiescher, M. 1994, Nucl. Phys. A **A570**, 329c
Thielemann, F.-K. Nomoto, K., Hashimoto, M., 1993 in *Origin and Evolution of the Elements*, ed. N. Prantzos, E. Vangoni-Flam, M. Cassé, (Cambridge Univ. Press), p.297
Thielemann, F.-K., Nomoto, K., Hashimoto, M. 1994, in *Supernovae, Les Houches, Session LIV*, eds. S. Bludman, R. Mochkovitch, J. Zinn-Justin (Elsevier, Amsterdam), p. 629
Thielemann, F.-K., Nomoto, K., Hashimoto, M. 1994, *Ap. J.*, submitted
van Paradijs, J.A. 1991, in *Neutron Stars, Theory and Observations*, eds., J. Ventura and D. Pines, Kluwer Acad. Publ., Dordrecht, p.289
Varani, G.-F., Meikle, W.P.S., Spyromilio, J., Allen, D.A. 1990, *MNRAS* **245**, 570
Weaver, T.A., Woosley, S.E. 1993, *Phys. Rep.* **227**, 65
Weber, F., Glendenning, N.K. 1991, *Phys. Lett.* **B265**, 1
Werner K., Dreizler S., Heber U., Rauch T. 1994, in *Nuclei in the Cosmos III*, ed. C. Raiteri, AIP Publ., in press
Wilson, J.R., Mayle, R.W. 1988, *Phys. Rep.* **163**, 63
Wilson, J.R., Mayle, R.W. 1993, *Phys. Rep.* **227**, 97
Witteborn, F.C., Bregman, J.D., Wooden, D.H., Pinto, P.A., Rank, D.M., Woosley, S.E., Cohen, M. 1989, *Ap. J. Letters* **338**, L9
Witti, J., Janka, H.-T., Takahashi, K. 1994, *Astron. Astron.* **286**, 841
Wooden, D.H., Rank, D.M., Bregman, J.D., Witteborn, E.C., Cohen, M., Pinto, P.A., Axelrod, T.S. 1993, *Ap. J.Suppl.* **88**, 477
Woosley, S.E. 1988, *Ap. J.* **330**, 218
Woosley, S.E., Hoffman, R. 1992, *Ap. J.* **395**, 202
Woosley, S.E., Weaver, T.A. 1986, *Ann. Rev. Astron. Astrophys.* **24**, 205
Woosley, S.E., Weaver, T.A. 1994, in *Les Houches, Session LIV, Supernovae*, eds. S.R. Bludman, R. Mochkovitch, J. Zinn-Justin (Elsevier Science Publ.), p. 63
Woosley, S.E., Wilson, J.R., Mathews, G.J., Hoffman, R.D., Meyer, B.S. 1994, *Ap. J.*, in press
Zhao, Z., France, R.H., Lai, K.S., Rugari, S.L., Gai, M., Wilds, E.L., Kryger, R.A., Winger, J.A., Beard, K.B. 1993, *Phys. Rev.*, **C48**, 429

Sodium Enrichment and the Evolution of A–F Type Supergiants

Mounib F. El Eid

Universitäts-Sternwarte Göttingen, FRG and Dept. of Physics, American University of Beirut (AUB), Beirut, Lebanon

Abstract. We have investigated the sodium (^{23}Na) nucleosynthesis in stars of masses $M = 5$ to $19\,M_\odot$ having solar-like initial chemical composition. The values obtained for the Na–excess after the first dredge–up phase are in close agreement with recent observations suggesting a moderate Na–excess in F–type supergiants. We also found a positive correlation between the overabundance factors [N/H] and [Na/H] which seems indicate that Na–enrichment originates from the Ne–Na cycle operating simultaneously with the CNO tri-cycle in these stars. We emphasize that our results were obtained on the basis of standard physical assumptions in the stellar model calculations.

Key words: Abundance anomalies in A–F type supergiants

1. INTRODUCTION

It is still under debate whether the observed Na–excess in A–F type supergiants stems from the products of Ne–Na cycle, which may be mixed to the surface after the first dredge–up phase. A determination of the Na–excess in nine F–type and four A–type supergiants, based on the non–LTE analysis of the Na I lines (including the strong line NaI $\lambda 8195$), has recently been provided by Takeda & Takada-Hidai (1994, hereafter TT94). Their results indicate that the F Supergiants exhibit a moderate Na–excess of [Na/H]$\approx$ +0.07 to +0.50 ([X]=log X_{star}− log $X_\odot$). The observed values for [Na/H] seem to increase with stellar mass and show a positive correlation with [N/H]. For the A–type supergiants, TT94 found rather high values of [Na/H]$\approx$ +0.80 to +1.0, which were not correlated with the stellar mass, in contrast to the F-type supergiants.

According to TT94, the data for the Na–excess derived by Boyarchuck et al. (1988) were overestimated for many F–type supergiants due to an improper treatment of the non–LTE effects. Therefore, previous attempts to explain such observations have invoked ad–hoc assumptions such as increasing the initial abundances of ^{22}Ne (Denissenkov 1989), or by artificially enhancing the rate of the ^{20}Ne(p,γ) reaction (Prantzos et al. 1991; El Eid & Prantzos 1993).

In the following section, we present our predictions for the CNO isotopes and Na–excess obtained on the basis of a comprehensive grid for stellar masses of $M = 5$ to $19\,M_\odot$. It will be shown that the values of [Na/H] obtained by TT94 for the F-type supergiants can be generally explained on the basis of standard physical assumptions in the stellar models and by adopting the view that many of these stars are post red–supergiants (see also Luck & Lambert 1985) performing blue loops in the H–R diagram.

© 1995 American Institute of Physics

2. STELLAR MODELS AND SODIUM EXCESS

Our evolutionary computations were based on the numerical code and physical assumptions as described by El Eid (1994a). The evolutionary sequences presented here (see Fig. 1) have been described and analyzed in a separate paper (El Eid 1994b). For completeness, we just mention that the nucleosynthesis results (see Table 1) were obtained by using a network of thermonuclear reactions including 29 nuclei linked by 73 reactions. For hydrogen burning, we included besides the pp–chain, the CNO tri–cycle, Ne–Na cycle and Mg–Al cycle. The reaction rates were basically taken from the compilation by Caughlan & Fowler (1988, hereafter CF88). However, we used the lowest rates obtained by Landré et al. (1990) for the reactions $^{17}O(p,\alpha)$ and $^{17}O(p,\gamma)$. These rates are still unsettled experimentally (for a discussion, see El Eid 1994a). For the nuclear reactions involved in the Ne–Na cycle, we have used the rates up–dated by Champagne (1994). The most uncertain reaction in this cycle is the $^{22}Ne(p,\gamma)$, whose rate is larger higher by a factor in excess of 1,000 in the range T_8=0.20–0.80 as compared with that tabulated by CF88. For He–burning, we used the principal reactions responsible for the energy production. In particular, for the $^{12}C(\alpha,\gamma)$ reaction we adopted the rate of CF88, but multiplied by a factor of 1.70 following the suggestion by Weaver & Woosley (1993).

In the present stellar models, the initial chemical composition was assumed to be solar–like, where the chemical abundances were chosen according to Anders & Grevesse (1989). In particular, the abundances of ^{22}Ne and ^{23}Na (relative to Si=10^6) are 2.34×10^5 and 5.74×10^4, respectively, i.e. with a ratio of 4.08 for $^{22}Ne/^{23}Na$.

The evolutionary sequences, plotted in Fig. 1, were calculated from the ZAMS up to the stage of He–shell burning. As described in details by El Eid (1994b), no red–blue loops were obtained for stellar masses above 13 $M_\odot$, when the Schwarzschild criterion for convection was used. However, such loops were recovered by adopting the Ledoux criterion for convection but combined with semiconvective mixing in the layers where the molecular weight gradient varies radially. This type of mixing has been described as a diffusion process with a diffusion coefficient, $D_{SC} = \beta D_r$, where β is an efficiency parameter and $D_r = 4\sigma T^3/3\kappa\varrho^2 C_P$ is the radiation diffusion coefficient (κ is the opacity, and C_P is the specific heat at constant pressure). This treatment is similar to that used by Woosley & Weaver (1988).

Unfortunately, it is very difficult to obtain the parameter β from the first principle. Referring to our detailed calculations (El Eid 1994b), we found that blue loops are obtained up to $M = 19 M_\odot$ (see Fig. 1), when the variation of β is confined to the range $0.01 < \beta \leq 0.05$. The tracks shown in Fig. 1 for the stellar masses above 13 $M_\odot$ were obtained with $\beta = 0.05$. Notice that the extension of the blue loops was found to be insensitive to the variation of β in the above range. However, the time at which the loops are triggered during core He–burning is rather sensitive to this variation.

Table 1: Predicted surface abundance ratios (by numbers) for the CNO isotopes and the sodium overproduction factors, [Na/H], [Na/Fe], and nitrogen excess [N/H]. At each initial stellar mass (in $M_\odot$), the first or second row gives the results after first dredge–up phase or second dredge–up phase (see text)

M	$^{12}C/^{13}C$	$^{16}O/^{17}O$	$^{16}O/^{18}O$	$^{14}N/^{15}N$	$^{12}C/^{14}N$	[Na/Fe]	[Na/H]	[N/H]
5	20.29	446	634	1439	0.812	0.180	0.188	0.407
	19.70	412	643	1523	0.777	0.190	0.199	0.417
7	19.77	508	641	1572	0.756	0.202	0.212	0.434
	18.92	507	645	1645	0.740	0.202	0.212	0.437
9	19.23	576	647	1698	0.710	0.215	0.227	0.457
	18.08	540	663	1864	0.670	0.226	0.238	0.468
11	18.61	628	646	1893	0.644	0.237	0.256	0.497
	17.29	585	665	2113	0.603	0.249	0.268	0.509
12	18.14	715	644	2095	0.585	0.259	0.286	0.537
	16.50	661	665	2393	0.545	0.271	0.297	0.549
13	17.84	790	640	2181	0.566	0.270	0.299	0.553
	13.45	714	683	2658	0.489	0.282	0.312	0.574
15	17.51	805	638	2273	0.550	0.274	0.305	0.565
	16.58	606	801	3121	0.391	0.298	0.329	0.615
18	16.93	949	623	2460	0.519	0.293	0.332	0.597
	15.94	494	864	3789	0.323	0.383	0.423	0.663
19	16.73	1023	617	2539	0.506	0.300	0.343	0.611
	15.66	524	867	3970	0.310	0.400	0.443	0.679

In Table 1, we summarized our predicted values for the surface abundance ratios of the CNO isotopes and for the overproduction factors [Na/H], [Na/Fe] and [N/H]. These results show the expected modification of the CNO abundances after the first and second dredge-up phases, and also a clear enrichments of N and Na. The predicted values for [Na/H] are consistent with those derived by TT94 for the F-type supergiants (see sect. 1). In addition, they increase with stellar mass and show a positive correlation with with [N/H] in agreement with the trend suggested by the observed data.

It remains to see whether our evolutionary tracks (Fig. 1) are consistent with the model parameters (log g, log $T_{\rm eff}$) adopted in the abundance analysis. These parameters are displayed in Fig. 2 for the supergiants investigated by TT94, and are compared with our evolutionary tracks. It can be seen that only three F supergiants (ρ Cas, γ Cyg, δ CMa) lie outside the range of our tracks and appear to be more massive than 20 $M_\odot$.

The masses of the stars shown in Fig. 2 have been determined by TT94 by using the relation:

$$\log(M/M_\odot) = \frac{1}{3}\left[4 \log (T_{\rm eff}/T_{\rm eff}^\odot) - \log (g/g^\odot)\right].$$

This expression has been obtained on the basis of a mass–luminosity relation $L/L_\odot = (M/M_\odot)^4$ assumed to be valid in the range log $T_{\rm eff} \approx 3.8-4.2$ K. Our evolutionary tracks (Fig. 1) in this range lead to $(L/L_\odot) \approx 1.23(M/M_\odot)^4$ during the evolution of the stars upon their blue loops, where the luminosity is nearly constant. With this relation, the stellar masses are smaller by ~ 0.03 dex as compared with those obtained by TT94. Clearly, these mass estimates rely on the log g and log T_{eff} values adopted in the model atmospheres which are not accurately determined (for discussion see Venn 1993).

In Fig. 3, our predictions for [Na/H] and [Na/Fe] are plotted versus stellar mass and are compared with the observed values for several stars. Notice that the difference between these quantities increases with increasing stellar mass due to the effect of increasing mass loss with increasing stellar masse. The comparison of our predictions with the observed data in Fig. 3 reveals that we can explain the observed [Na/H] values for four F supergiants including the Cepheid ζ Gem. It is not clear why the Cepheid η Aql and the F star α UMi have lower Na–excess than found for ζ Gem, despite the similarity of their atmospheric parameters (see TT94). The relatively high [Na/H] values proposed for the stars ρ Cas and δ CMa are clearly out of the range of our predictions. The value for γ Cyg is within this range despite its proposed high mass. The conclusion from this comparison is that the moderate Na–excess obtained by TT94 for the F–Type supergiants can be consistently explained on the basis of standard physical assumptions in the stellar model calculations. The situation concerning the A–type supergiants is not clear. Except for the star η Leo, the observed Na–excess is too high to be explained in a standard way. More observations and further theoretical work are needed in order to to understand their evolutionary status and abundance anomalies. In any case, we argue that the interpretation of the abundance anomalies in A and F supergiants represents a very useful tool for testing the mixing processes in massive stars.

REFERENCES

Anders, E., and Grevesse, N. 1989, Geochem. Et Cosmochem. Acta, 53, 197

Boyarchuck, A.A., Gubney, I., Kubat, I., Lyubimkov, L.S., and Sakhibullin, N.A. 1988, Afz 28, 335, (1989, Astrophysics, 29, 197)

Champagne, A. 1994, private communication

Denissenkov P.A. 1990, Astrophys., 31, 588

El Eid M.F. 1994a, A&A, 285, 915

El Eid M.F. 1994b, Effect of convective mixing on the blue loops in the H–R diagram, submitted to MNRAS

El Eid, M.F., and Prantzos, N. 1993, in: Origin and Evolution of the Elements, eds. N. Prantzos, E. Vangioni–Flam and M. Cassé r, Cambridge:Cambridge University Press), p. 279

Caughlan, G.R., Fowler, W.A., Harris, M.J., and Zimmerman, B.A. 1985 Atomic Data and Nuclear Data Tables, 32, 197

Caughlan, G.R., and Fowler, W.A. 1988 Atomic Data and Nuclear Data Tables, 40, 283, (CF88)

Iglesias, C.A., Rogers, F.J., and Wilson B.G. 1992, ApJ, 397, 717

Landré, V., Prantzos, N., Auger, P., Bogaert, G., Levebre, A., and Thibaud K.P. 1990, A&A 240, 85

Luck, R.E., and Lambert, D.L. 1985, APJ, 298, 782

Prantzos, N., Coc, A., and Thibaud, J.P., 1991, APJ, 379, 729

Takeda, Y., and Takada-Hidai, M. 1994, On the Abundance of Sodium in A-F Supergiants, preprint, (TT94)

Venn, K. 1993, APJ, 414, 316

Weaver, T.A., and Woosley, S.E. 1993 Phys. Rep., 227, 65

Woosley, S.E., and Weaver, T.A. 1988, Phys. Rep., 163, 79

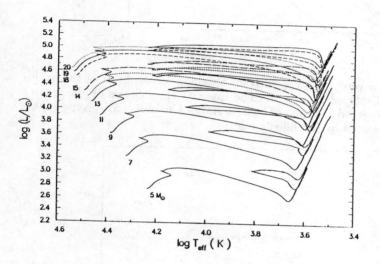

Fig. 1.— Theoretical H-R diagram showing the present evolutionary tracks for initial masses in the range 5 and 19 $M_\odot$, (see text)

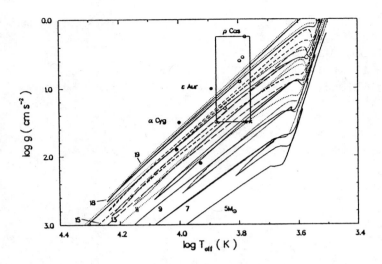

Fig. 2.— A log g versus log $T_{\rm eff}$ diagram obtained on the basis of the evolutionary tracks displayed in Fig. 1. The inserted open circles or the crosses correspond to observed F–type or Cepheid stars. The observed A–type supergiants are represented by the filled circles

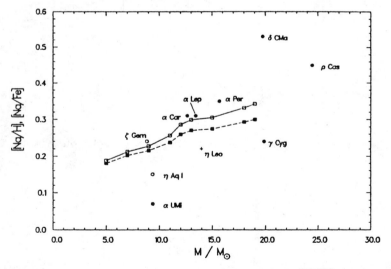

Fig. 3.— The predicted surface sodium excess, [N/H] (open squares) and [Na/Fe] (filled squares) versus stellar mass compared with the values derived from recent observations (see text). The inserted stars are F–type supergiants, except the A–type supergiant η Leo

Structure and Evolution of AGB Stars

Thomas Blöcker and Detlef Schönberner

Astrophysikalisches Institut Potsdam, Telegrafenberg, D-14473 Potsdam, Germany

Abstract. We present stellar evolution calculations for initial masses between 1 and $7M_\odot$ covering the whole evolution from the main sequence towards the regime of white dwarfs. Mass loss was considered since the base of the RGB. Along the AGB we applied a mass-loss law derived from dynamical calculations for Mira-like stars. Our results are consistent with empirical initial-final mass relationships. Some thermal-pulse properties as well as hot bottom burning models are briefly discussed. We note that mass-loss has to be taken into account for the evolution and nucleosynthesis of bottom burning models.

Key words: mass loss — thermal pulses — hot bottom burning

1. INTRODUCTION

We have calculated the evolution of 1, 3, 4, 5 and $7M_\odot$ stars from the main sequence through the AGB towards the stage of white dwarfs. The initial composition was $(Y, Z) = (0.24, 0.021)$, convection was treated within the mixing length theory (Böhm-Vitense 1958) with $\alpha = 1.5$ and 2, resp.
Concerning the description of mass loss we applied the Reimers (1975) rate $\dot{M}_R$ along the RGB and during central helium-burning. During the AGB evolution we used an adaption to the numerical results of Bowen (1988). Taking his standard model with the inclusion of dust for atmospheres of Mira-like stars we got: $\dot{M}_B = 4.8 \cdot 10^{-9} \dot{M}_R \cdot L^{2.7}/M_{ZAMS}^{2.1}$. For $M_{ZAMS} = 3, 4, 5$ and $7M_\odot$ this rate allowed 20, 15, 9 and 15 thermal pulses, resp., to occur until the AGB evolution terminated. Fig. 1a shows that our adapted mass-loss formula $\dot{M}_B$ yields final masses which are consistent with the empirical initial-final mass relationship of Weidemann (1987). For comparison, the calculations of Vassiliadis & Wood (1993) are also given. Their final masses are somewhat larger than our ones and are up to $\approx 0.1 M_\odot$ above the empirical relation. Additional calculations with the mass-loss formula of Baud & Habing (1983), $\dot{M}_{BH}$, led to final masses which are considerably too high. For instance, for $M_{ZAMS} = 3M_\odot$ the final mass was $0.814 M_\odot$ corresponding to 71 thermal pulses.
The whole mass-loss evolution along the AGB of a $3 M_\odot$ star calculated with $\dot{M}_B$ is shown in Fig. 1b. The occurence of thermal pulses leads to strong mass-loss modulations. More details concerning the treatment of mass loss are given in Blöcker (1994a).

2. THERMAL PULSE PROPERTIES

During the recurrent helium shell flashes sufficiently large neutron exposures for the production of heavy s-nuclei can be provided in the convective

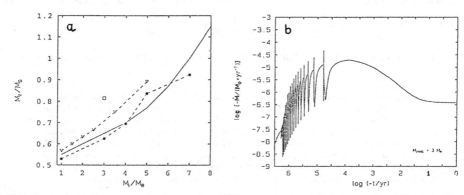

Fig. 1.— a Initial mass vs. final mass for different mass-loss histories on the AGB. Circles refer to $\dot{M}_B$, the square to a $3M_\odot$ sequence calculated with $\dot{M}_{BH}$, and the triangles to the results of Vassiliadis & Wood (1993). The thick line is the initial-final mass relation of Weidemann (1987). — b Temporal evolution of the mass-loss rate $\dot{M}_B$ along the AGB for a $3M_\odot$ sequence (20 thermal pulses). The point $t = 0$ refers to the zero point of the central star evolution corresponding to a pulsational period of $P_0 = 50$ d.

unstable He shell (e.g. Truran & Iben 1977, Iben & Renzini 1982, see also Busso et al. 1994 and references therein). Concerning nucleosynthesis calculations the overlap parameter r and the maximum temperature of the convevtive He shell, $T_{\rm conv}^{\rm max}$, are crucial quantities. The overlap is defined as the mass fraction of the convective He shell at maximum size which is incorporated in the convective shell of the next pulse, giving for a particluar mass zone a measure for multiple neutron exposures. Fig. 2 shows $T_{\rm conv}^{\rm max}$ and r vs. the mass of the hydrogen exhausted core, M_H, for different sequences calculated with low mass-loss rates (e.g. $\dot{M}_{BH}$). Please note, that the number of thermal pulses is noticeably smaller for sequences with larger (and more realistic) mass-loss rates, like $\dot{M}_B$. More thermal pulse properties will be given in Blöcker (1994b).

3. HOT BOTTOM BURNING

Besides of mass loss and thermal pulses the occurence of hot bottom burning is another important feature of the AGB evolution. While the model climbs up the AGB the envelope convection gets more and more extended downwards due to the increasing radiation pressure. For more massive models, i.e. $M_{\rm ZAMS} > 5M_\odot$ and $M_H > 0.8M_\odot$, it can even reach and deeply cut into the hydrogen burning shell leading to very high temperatures at the base of the convective envelope (e.g. Iben 1975, Scalo et al. 1975). The base temperature as a function of M_H is shown in Fig. 3a. During hot bottom burning ^{7}Li can be produced and mixed to the surface (e.g. Scalo et al. 1975). Indeed, model calculations predict super-rich Li AGB-stars just in the observed luminosity range (cf. Blöcker & Schönberner 1991, Sackmann & Boothroyd 1992).

Furthermore, hot bottom burning models become very luminous and do *not* follow the classical core-mass luminosity relation of Paczyński (1970) any-

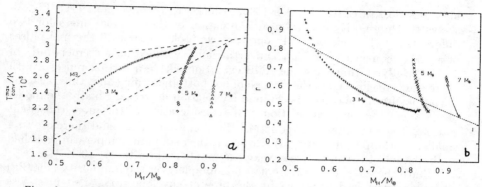

Fig. 2.— a Maximum temperature of the convective unstable helium shell vs. M_H for different initial masses ($M_{ZAMS} = 3, 5, 7M_\odot$; $N_{TP} = 91, 43, 38$). The dashed lines refer to relations of Iben (1977) [I] and Malaney & Boothroyd (1987) [MB]. — b Overlap parameter r vs. M_H for different initial masses. The dashed line refers to the relation of Iben (1977).

more (Blöcker & Schönberner 1991, Lattanzio 1992, Boothroyd & Sackmann 1992). Instead, they evolve very quickly to higher luminosities: Our $7M_\odot$ model needs only 61000 yrs for $M_{bol} = -6 \rightarrow -7$ instead of $\approx 10^6$ yrs according to Paczyński (1970). This leads to a very low detection probability for luminous C stars. In fact, such stars are only scarcely observed in this luminosity range (Frogel et al. 1990). Moreover, C stars can be turned into N-rich stars during this phase (Iben 1975, Renzini & Voli 1981, Boothroyd et al. 1993). Finally, we want to discuss the role of mass loss. It has already been demonstrated that if hot bottom burning is efficient it cannot be prevented even by

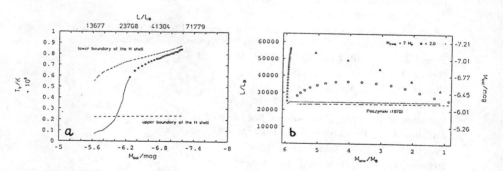

Fig. 3.— a Base temperature of the convective envelope vs. M_{bol} for a $7M_\odot$ AGB model ($\alpha = 2$). Rhombs refer to our models prior pulse No. 1-38. The solid line shows the evolution on the E-AGB since the re-ignition of the hydrogen shell. — b Luminosity vs. envelope mass M_{env} for a $7M_\odot$ AGB model ($\alpha = 2$). Triangles refer to a sequence with $\dot{M}_{BH}$ ($N_{TP} = 26$) and $\dot{M} = 4 \cdot 10^{-4} M_\odot$/yr ($N_{TP} = 5$), circles to one with $\dot{M}_B$ ($N_{TP} = 15$). The solid and dashed line indicate what is to be expected according to Paczyński (1970).

high mass-loss rates (cf. Blöcker 1994a, Blöcker & Schönberner 1994). Fig. 3b shows the luminosity evolution as a function of the envelope mass for two different mass-loss histories. In one case we took the first 26 thermal pulses of a $7M_\odot$ sequence with $\dot{M}_{BH}$ and artificially switched on a "superwind" of $4 \cdot 10^{-4} M_\odot/\text{yr}$ terminating the AGB evolution within 5 further pulses. With decreasing envelope mass the strengh of hot bottom burning and, thus, also the deviations from the core-mass luminosity relation decrease. Finally the model obeys again the classical relation. The consideration of the more consistent rate $\dot{M}_B$ leads to an earlier reduction of the envelope mass and therefore to a smaller overluminosity (Fig. 3b). However, the maximum temperature reached at the base of the convective envelope amounts still up to $75 \cdot 10^6$K.

Since the base temperature depends uniquely on the luminosity (Fig. 3a) which, in turn, is determined by the present envelope mass, i.e. mass-loss history (Fig. 3b), there exists for a given envelope mass and mass-loss law a maximum luminosity with a concomitant maximum of the base temperature. Consequently, nucleosynthesis and mass loss are closely connected for hot bottom burning models.

REFERENCES

Baud, B., Habing, H.J.: 1983, A&A 127, 73
Blöcker, T.: 1994a, A&A, submitted
Blöcker, T.: 1994b, in preparation
Blöcker, T., Schönberner, D., 1991, A&A 244, L43
Blöcker, T., Schönberner, D., 1994, in Proc. European Workshop on *Heavy Element Nucleosynthesis*, eds. E. Somorjai, Z. Fülöp, p. 5
Böhm-Vitense, E., 1958, Z. Astrophys. 46, 108
Boothroyd, A.D., Sackmann, I.-J., 1992, ApJ 393, L21
Boothroyd, A.D., Sackmann, I.-J., Ahern, S.C., 1993, ApJ 416, 762
Bowen, G.H., 1988, ApJ 329, 299
Busso, M., Gallino, R., Raiteri, C.M., Straniero, O., 1994, in Proc. European Workshop on *Heavy Element Nucleosynthesis*, eds. E. Somorjai, Z. Fülöp, p. 26
Frogel, J.A., Mould, J.R., Blanco, V.M, 1990, ApJ 352, 96
Iben, I. Jr., 1975, ApJ 196, 525
Iben, I. Jr., 1977, ApJ 217, 788
Iben, I. Jr., Renzini, A., 1982, ApJ 299, L79
Lattanzio, J.C., 1992, Proc. Astron. Soc. Austr., 10, 120
Malaney, R.A., Boothroyd, A.D., 1987, ApJ 320, 866
Paczyński, B., 1970, Acta Astron. 20, 47
Reimers, D., 1975, Mem. Soc. Sci. Liege 8, 369
Renzini, A., Voli, M., 1981, A&A 94, 175
Sackmann, I.-J., Boothroyd, A.D., 1992, ApJ 392, L71
Scalo, J.M., Despain, K.H., Ulrich, R.K., 1975, ApJ 196, 805
Truran, J.W., Iben, I. Jr., 1977, ApJ 216, 797
Vassiliadis, E., Wood, P.R., 1993, ApJ 413, 641
Weidemann, V.: 1987, A&A 188, 74

S-process calculations in thermal pulses: a word of caution

N. Mowlavi[*,1], A. Jorissen[*,2], M. Forestini[**,3], M. Arnould[*,2]

*Institut d'Astronomie et d'Astrophysique, Université Libre de Bruxelles, C.P. 226, Bd du Triomphe, B-1050 Bruxelles, Belgium
**Laboratoire d'Astrophysique, Observatoire de Grenoble, BP53X, F-38041 Grenoble Cedex, France

Abstract. The s-process nucleosynthesis resulting from the ingestion of ^{13}C in a thermal pulse of a solar-metallicity 2.5 $M_\odot$ star is computed with due consideration of the feedback of the s-process energetics on the stellar model. It is shown that the pulse structure along with the s-process predictions are dramatically affected by this feedback. The common practice of neglecting this effect may thus lead to erroneous results, and must be abandoned.

Key words: Stars: evolution, AGB – nucleosynthesis – Solar system: meteorites, isotopic anomalies

1. Introduction

AGB stars are generally considered as the most efficient galactic contributors to the so-called "main" s-process component involving species with mass number $A > 90$ (e.g. Käppeler et al. 1989). They have also been viewed as the progenitors of isotopically anomalous SiC grains of supposedly circumstellar origin identified recently in primitive meteorites, and exhibiting, among other characteristics, an enrichment of s-process nuclides (e.g. Ott 1993, Anders & Zinner 1993).

Calculations by Hollowell & Iben (1988; HI88) of a low-mass low-metallicity AGB star indicate that enough ^{13}C can be produced during the interpulse phase for allowing, when ingested in a next pulse, the development of a s–process thanks to the neutrons liberated by ^{13}C$(\alpha,n)^{16}$O. However, the production of ^{13}C in this scenario depends upon complex mixing processes, and other calculations have not been able to replicate these results.

Given the repeated failures of self-consistent models to predict a s–process in solar-metallicity low-mass stars, a phenomenological approach has been widely used. It just parametrizes the available amount of neutrons in order to fit the observed abundances in S and C-type stars (Busso et al. 1992), the bulk solar system composition (Käppeler et al. 1990), or the isotopic patterns of anomalous SiC grains (Gallino et al. 1990).

---

[1] Boursier I.R.S.I.A.

[2] Research Associate, National Fund for Scientific Research, Belgium

[3] Research Assistant, National Fund for Scientific Research, Belgium

The (sometimes far reaching) conclusions drawn from the parametrized model mentioned above have to be taken with care and caution. This situation relates in particular to the complete neglect by that simple model of the feedback on the stellar structure of the energy produced by the s–process neutron captures and β-decays.

We report here on a large re-adjustment of the pulse structure in a low-mass *solar*-metallicity star, consecutive to ^{13}C injection that has a strong impact on the s-process nucleosynthesis (see Bazan & Lattanzio 1993 for a discussion of the pulse re-adjustment in a *low*-metallicity star).

2. Results

The consequences of ^{13}C injection on the structure of a pulse have been analyzed on the 12th pulse of a solar-metallicity 2.5 M$_\odot$ star whose evolution has been followed from the T-Tauri phase up to the thermally-pulsing AGB stage (see Mowlavi & Forestini 1994 and Forestini 1994 for a description of the stellar evolution code). Details about the numerical method used to handle the energy feedback on the pulse structure can be found in Mowlavi et al. (1994).

A first set of models of pulse 12, the "standard model" (ST12), are calculated using the abundance profiles resulting from pulse 11, whereas a second set of models (EN12) are calculated by increasing *artificially* the amount of ^{13}C ingested by the 12th pulse. A rectangular profile is adopted for ^{13}C and ^{14}N in the region formerly at the tip of pulse 11 (Fig. 1). The peak ^{13}C and ^{14}N mass fractions of 10^{-2} are determined from the requirement that a few neutrons be captured per pulse and per initial ^{56}Fe seed nucleus ($n_c < {}^{13}\text{C}/^{56}\text{Fe} = 4$ in the pulse convective tongue), as is usually the case in the phenomenological models. The total mass of ^{13}C ingested by the pulse amounts to $3.5\ 10^{-5}$ M$_\odot$. It should be noticed that the injection of $5\ 10^{-6}$ M$_\odot$ of ^{13}C, like in HI88 or Bazan & Lattanzio (1993), would lead in our *solar*-metallicity star to only $n_c < 0.5$ per pulse according to the above expression, which would yield a rather weak s-process.

Figure 1 clearly shows that the structural evolution of pulse 12 is substantially different in the ST12 and EN12 cases. In particular, the growth rate of the pulse suddenly increases when ^{13}C starts to be ingested. This difference relates directly to the change in the energy production rate (Despain & Scalo 1976), being equal to $3.6\ 10^8$ erg g^{-1} s^{-1} at the base of the EN12 pulse, as compared to $1.5\ 10^7$ erg g^{-1} s^{-1} in the standard case ST12. The EN12 energy production is dominated by ^{13}C$(\alpha,\text{n})^{16}$O (21.6%) and the associated s-process (60.4%) at the time of ^{13}C ingestion. The pulse resumes its normal evolution after ^{13}C has been burned.

The impact of the different pulse shapes EN12 and ST12 on the s-process nucleosynthesis has then been evaluated in the following way. In a first calculation (**NOFBK**), the pulse shape corresponding to ST12 is adopted to derive the temperature and density profiles, as well as the ^{13}C ingestion rate, but the amount of ^{13}C ingested is as in EN12. This situation corresponds to the usual

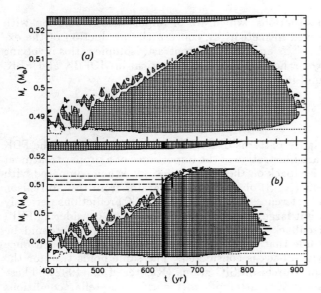

Fig. 1.— Structural evolution of pulse 12 in the ST12 (a) and EN12 (b) cases. The dotted lines represent the location of the maximum energy production in the H- and He-burning layers. The dashed and dot-dashed lines delineate the regions where the ^{13}C and ^{14}N mass fractions have been set to 10^{-2}, respectively. Each vertical line corresponds to a model

phenomenological approach, where the energy feedback from the s-process on the pulse structure is neglected. These predictions are then compared to those (referred to as **FBK** in the following) from the s-process nucleosynthesis performed with the pulse shape EN12 taking into account the energy feedback.

Even if the same amount of ^{13}C is ingested by **NOFBK** and by **FBK**, **FBK** produces much higher neutron concentrations ($n_n = 5.8 \, 10^{10}$ cm^{-3}, corresponding to the peak value averaged over the convective zone) than **NOFBK** ($n_n = 10^9$ cm^{-3}). This relates directly to the much faster growth rate of EN12. In contrast, the s-process efficiency is slightly higher in **NOFBK** ($n_c = 2.5$ per pulse) than in **FBK** ($n_c = 1.8$), the corresponding neutron irradiations (at 30 keV) amounting to 0.14 and 0.11 mb^{-1}, respectively. This is due to the fact that the light neutron poisons become comparatively more important at higher n_n, because the s-process path then involves neutron-rich isotopes that have smaller neutron-capture cross sections on the average than isotopes closer to the valley of stability.

The different neutron-density histories of **NOFBK** and **FBK** imprint a clear signature on r/s abundance ratios involving nuclei close to a branching developing at a critical neutron density of the order of 10^{10} cm^{-3}. This is the case for example for ^{134}Xe, ^{136}Xe and ^{148}Nd. In the **FBK** asymptotic regime (i.e. after 25 identical EN12 pulses with a 68% overlap), they are significantly produced, with $\delta \sim -250, -750$ and 600, respectively [$\delta_i = ((N_i/N_{ref})/(N_i/N_{ref})_\odot - 1) \times 1000$, ^{130}Xe and ^{144}Nd being the reference isotopes], while they are not produced in the **NOFBK** conditions. Other differences between the two situations, which are of importance in relation with isotopic anomalies observed in meteorites, concern the production of the extinct radio-nuclides ^{60}Fe, ^{135}Cs and ^{182}Hf. In contrast, isotopic ratios involving nuclides not affected by branching points are identical in **NOFBK** and **FBK**. More details can be found in Mowlavi et al. (1994).

3. Conclusions

The different abundance patterns of the Xe isotopes obtained in the **FBK** and **NOFBK** situations are illustrative of the importance of a proper treatment of the s-process energy feedback on the pulse structure. When compared with the pure-s Xe component (Xe-S) observed in SiC grains from the Murchison meteorite (Ott 1993 and references therein), the **NOFBK** predictions perfectly match the SiC pattern, but large discrepancies appear when considering **FBK**. In absence of energy feedback, the 12th pulse of our $2.5\,M_\odot$ star could be a viable site of Xe-S, while that conclusion must clearly be abandoned once the effect of the energy feedback is considered. The same holds true for the s-process Nd observed in Murchison SiC grains (Richter et al. 1992). These few examples clearly show that attempts to identify the stellar conditions responsible for the synthesis of the various anomalous components observed in meteorites should rely on a fully self-consistent approach including the effect on the pulse structure of the energy liberated by the s-process. An approach simply based on the phenomenological model may be very misleading.

REFERENCES

Anders E., Zinner E., 1993, Meteoritics 28, 490

Bazan G., Lattanzio J.C., 1993, ApJ 409, 762 (BL93)

Busso M., Gallino R., Lambert D.L., Raiteri C.M., Smith V.V., 1992, ApJ 399, 218

Despain K.H., Scalo J.M., 1976, ApJ 208, 789

Forestini M., 1994, A&A, in press

Gallino R., Busso M., Picchio G., Raiteri C.M., 1990, Nature 348, 298

Hollowell D., Iben I.Jr., 1988, ApJ 333, L25 (HI88)

Käppeler F., Beer H., Wisshak K., 1989, Rep. Prog. Phys. 52, 945

Käppeler F., Gallino R., Busso M., Picchio G., Raiteri C.M., 1990, ApJ 354, 630

Mowlavi N., Forestini M., 1994, A&A 282, 843

Mowlavi N., Jorissen A., Forestini M., Arnould M., 1994, A&A, submitted

Ott U., 1993, Nature 364, 25

Richter S., Ott U., Begemann F., 1992, Lun. Planet. Sci. XXIII, 1147

Radiative ^{13}C Burning in AGB Stars and s-Processing

O. Straniero(1), R. Gallino(2), M. Busso(3), A. Chieffi(4), C.M. Raiteri(3), M. Salaris(1), and M. Limongi(5)

(1) Osservatorio di Collurania - Teramo (Italy)
(2) Istituto di Fisica Generale dell'Università - Torino (Italy)
(3) Osservatorio Astronomico di Torino (Italy)
(4) Istituto di Astrofisica Spaziale del CNR - Frascati (Italy)
(5) Osservatorio Astronomico di Roma (Italy)

Abstract. We present new evolutionary calculations for a 3 $M_\odot$ star, up to the 24th thermal pulse of the Asymptotic Giant Branch. The third dredge-up is naturally produced by the model starting from the 14th pulse. We show that, if some ^{13}C is formed below the convective border due to partial proton mixing, it is subsequently burnt completely through (α,n) reactions due to the heating of the region between the H and the He shells, before the next pulse develops. As a consequence, s-elements are produced in a radiative environment, at a neutron density of at most a few 10^7 n/cm^{-3}. These s-elements are then ingested by the convective pulse, where a second (minor) neutron burst occurs, driven by the ^{22}Ne$(\alpha,n)^{25}$Mg reaction that is marginally activated in the last stages of the He-shell instability. Despite the differences of the present model with respect to past calculations, we predict s-process distributions very close to what was previously computed by allowing ^{13}C to burn in convective pulses: ^{13}C burning and the ensuing s-processing is confirmed to account for the main observational and experimental constraints. The reasons for this are discussed. No ^{13}C survives the interpulse phase: hence we conclude that no modifications of the pulse shape due to the energy feedback from the ^{13}C$(\alpha,n)^{16}$O reaction occur.

Key words: Stars: evolution of — Nucleosynthesis — s-process

1. THE CALCULATIONS OF AGB EVOLUTION

As part of a recent set of evolutionary models for stars with masses between 1.5 and 7 $M_\odot$, starting from the Main Sequence and prosecuted up to advanced Thermally Pulsing Asymptotic Giant Branch (TP-AGB) phases (see e.g. [1]) we present here calculations referring to a 3 $M_\odot$ star of solar metallicity. A very fine grid both in mass coordinate ($\geq$ 2000 mesh points) and in time was used. The model adopts opacities from [2] and [3]. With these opacities, from the fit to the solar structure we get a value for the ratio of the mixing length to the pressure scale height of $\alpha_p = 1.6$. During the TP phase, up to the 14th pulse, the structure of our model star reproduces well the typical characteristics found in the literature for the same stages (see e.g. [4-7]). In particular, the mass of the convective He shell decreases monotonically from $\sim 3\ 10^{-2}\ M_\odot$ at the first pulse down to $1\ 10^{-2}\ M_\odot$; meanwhile, the interpulse duration declines from $\sim 10^5$ yr to $\sim 4\ 10^4$ yr and the overlapping factor between adjacent pulses from ~ 0.8 to 0.5. After the 14th pulse, the third dredge-up is found to occur in a self-consistent way; the envelope convection penetrates into the intershell region swept by the previous convective instability by an increasing amount, starting from $\sim 10^{-4}$ $M_\odot$ at the 14th pulse and reaching $\sim 10^{-3}$ $M_\odot$ at the 24th. The onset of dredge-up has the effect of stabilyzing the intershell structure, so that the mass of the thermally unstable region does not change any longer, being fixed to around 10^{-2} $M_\odot$. So far, a proper accounting of convective or semiconvective dredge-up in TP-AGB stars of population I with low initial mass ($M \leq 3$ $M_\odot$) and low core mass ($M_H \leq 0.8$ $M_\odot$) was generally very difficult to obtain in the models (see an exception in [8]). Dredge-up was instead more easily found for higher core masses [9].

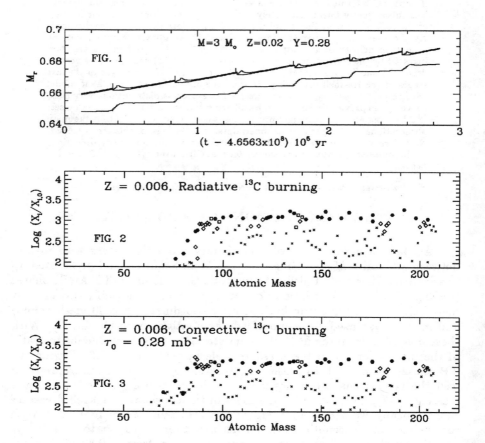

Some relevant features of our model are shown in Figure 1, where the position of the convective envelope border, of the H-burning shell, and of the He-burning shell are represented as a function of time for the thermal instabilities from 15th to 20th. The occurrence of the pulses is clearly indicated by the places where the He-shell suddenly moves upward. Taking into account the effect of the third dredge-up and of the shrinkage of the envelope mass due to both the advance of the H-burning shell and the continuos mass-loss by stellar winds, the surface C/O ratio would reach the typical carbon star value (C/O $\sim$ 1.0) some pulses after the point where we stopped the calculations, near the 30th pulse.

2. ON THE RADIATIVE ^{13}C BURNING

We have used our stellar model to investigate the effect of a possible mixing of protons into a thin zone at the top of the carbon-rich region during each dredge-up episode; when the H-burning shell is reactivated, such mixing would lead to the formation of a tiny region enriched in primary ^{13}C (and ^{14}N), and hence to s-process nucleosynthesis.

A mechanism suitable to mix a small amount of protons from the envelope into the intershell layers was presented by [10,11] and by [12] in the framework of population II stellar models. It is related to the extra opacity of recombining ^{12}C, during the expansion and cooling phase that follows a pulse, driving semiconvective regions at the intershell zone.

This partial mixing mechanism was not found to work [13] for population I stars like the peculiar red giants of classes MS, S and C, showing enrichments of ^{12}C and of s-process elements in their spectra [14]. However, these stars are of low mass, with relatively low He-shell temperatures, suitable to ignite efficiently only the ^{13}C neutron source, and only marginally the alternative ^{22}Ne$(\alpha,n)^{25}$Mg reaction, which has a much slower reaction rate. The required downward penetration of protons, to produce ^{13}C, might in this case be ascribed to phenomena like shear effects, Rayleigh-Taylor instabilities or chemical diffusion: the possible occurrence of them still requires careful scrutiny.

The nucleosynthesis ensuing from the activation of the ^{13}C$(\alpha,n)^{16}$O reaction in TP-AGB stars has been in recent years the object of several studies (see e.g. [15-19]). In these studies, a certain amount of ^{13}C (taken as a free parameter) was assumed to be ingested by each thermal instability, and the ensuing s-process nucleosynthesis was followed (see e.g. [20] for details). Some caution was recently advanced on this procedure by [21] suggesting that the energy liberated in the thermal pulse by the ^{13}C$(\alpha,n)^{16}$O reaction could modify the upper profile of the convective zone, thus affecting the process of neutron capture nucleosynthesis.

The stellar model we present here, accounting for dredge-up since the 14th pulse, gives us the possibility of testing which is the *real* fate of the extra ^{13}C. Indeed, by knowing to which extent the bottom of the envelope convection has penetrated, pulse after pulse, into the intershell layers, we can follow from then on the actual conditions of density and temperature in these layers, during the interpulse phase and up to their engulfment into the next convective pulse.

In the interpulse phase (that lasts for about 40,000 years in late pulses), following the advance in mass of the H-burning shell, the temperature and density of the zone where the ^{13}C-pocket is to be expected increase from about $T_6 \sim 10$ (T_6 being the temperature in units of 10^6 K) and $\rho \sim 1$ g/cm^{-3} to $T_6 \sim 100$ and $\rho \sim 10^4$ g/cm^{-3}. After one third of the interpulse duration, the temperature has already reached values higher than $T_6 = 90$, so that the α-capture lifetime of ^{13}C becomes short enough to easily consume all ^{13}C well before the end of the interpulse phase. Hence, the s-process nucleosynthesis occurs in radiative conditions. No ^{13}C survives the radiative interpulse phase and can ever be ingested into a pulse. The energy liberated by ^{13}C at low temperature ($\sim$8 keV), in a stage where the H shell is the major nuclear source in the star, is so small that no effect at all is seen on the stellar structure. Moreover, the resulting neutron density remains very low. Actually, it is now more difficult than before to define a proper *average* neutron density (and a proper *mean* neutron exposure), because each radiative layer is characterized by different conditions, strictly related to the distribution of primary ^{13}C and ^{14}N (this last being the dominant neutron poison). However, in the most efficient layers the neutron density never exceeds 2×10^7 n/cm^{-3}.

Once the s-process nuclei are produced in a radiative environment, they are engulfed and diluted by the growing convective profile of the next thermal instability, in the last stages of which marginal activation of the ^{22}Ne source gives rise to a second burst of neutrons, as already known [17].

3. COMPARISON WITH PREVIOUS RESULTS

In order to verify the consequences of the new model for ^{13}C burning, we have made several calculations, adopting the same input parameters (^{13}C abundance, stratification of ^{14}N in the ^{13}C-rich layers, metallicity of the star) as in past studies performed in the convective scenario. Let us give here only the most significant results.

Our previous analysis pointed out [20] that the *main* s-process component of the solar system could be reproduced by stars of relatively low metallicity ($Z = 1/3\ Z_\odot$), burning (4-5)$\times 10^{-6}\ M_\odot$ of ^{13}C. Running our new radiative code with the above parameters we get an asymptotic distribution of s-element abundances earlier than before (10 cycles are sufficient, against 15-20 before). Apart from this, differences of the resulting new distribution (Figure 2) with respect to the previous best-fit to the solar abundances of the main component (Figure 3) are minimal, limited to some neutron-density dependent branchings on the s-chain. These are due to the much lower neutron density of the radiative model, whose effects are however moderated by the second neutron burst driven by the ^{22}Ne-source, which remains the same as before with a peak average neutron density of a few 10^8 n/cm^{-3}. In practice, most predictions already made in the convective scenario can now be confirmed. An exception is the production of nuclei lying on the neutron-rich side of branchings, such as ^{86}Kr, ^{87}Rb, ^{96}Zr, which are now considerably reduced due to the lower neutron density, and hence allow for a higher r-process contribution. Another difference is that, for $A < 90$, there is a stronger drop in the s-process contribution by TP-AGB stars. These problems, together with a detailed analysis of branching points, need further investigation.

It must not be surprising that the new scenario confirms the predictions of the old one: indeed, it is well known that the s-process distribution is essentially a function of three parameters (τ_0, n_n, T_{max}: see [17]), of which the first plays the dominant role. When, in the same model, i) the total number of neutrons released (hence τ_0) is not changed, ii) the temperature of the hotter part of the pulse, where the ^{22}Ne source is marginally activated, remains the same, and iii) the maximum neutron density is varied, but not by orders of magnitude, also the s-process distributions have to be very similar.

A further result of our work is that, since no ^{13}C is ingested by the convective pulses, the structural modifications of them suggested by [21] do not occur in real conditions.

REFERENCES

[1]. Busso, M., Chieffi, A., Gallino, R., Limongi, M., Raiteri, C.M., and Straniero, O. "New Calculations of Thermal Pulses and s-Process Nucleosynthesis in AGB Stars" in *Planetary Nebulae*, IAU Symp. N. 155, ed. R. Weinberger and A. Acker, (Dordrecht: Kluwer), 361 (1993).
[2]. Huebner, W.F., Merts, A.L., Magee, N.H., Jr., and Argo, M.F., Los Alamos Sci. Lab., LA-6760-M, (1977).
[3]. Cox, A.N., and Tabor, J.E., *Astrophys. Journal Suppl.* **31**, 371, (1976).
[4]. Schönberner, D. 1979, *Astron. and Astrophys.* **79**, 108, (1979).
[5]. Iben, I.Jr., and Renzini, A., *Ann. Rev. Astron. and Astrophys.* **21**, 271, (1983).
[6]. Boothroyd, A. I., and Sackmann, I.-J., *Astrophys. Journal* **328**, 653 (1988a).
[7]. Boothroyd, A. I., and Sackmann, I.-J., *Astrophys. Journal* **328**, 671 (1988b).
[8]. Lattanzio, J.C., *Astrophys. Journal* **344**, L25, (1989).
[9]. Iben, I.Jr., *Astrophys. Journal* **196**, 525, (1975).
[10]. Iben, I.Jr., and Renzini, A., *Astrophys. Journal* **259**, L79, (1982a).
[11]. Iben, I.Jr., and Renzini, A., *Astrophys. Journal* **263**, L23, (1982b).
[12]. Hollowell, D.E., and Iben, I.Jr., *Astrophys. Journal* **333**, L25, (1988).
[13]. Iben, I.Jr., *Astrophys. Journal* **275**, L65, (1983).
[14]. Smith, V.V., and Lambert, D.L., *Astrophys. Journal Suppl.* **72**, 387, (1990).
[15]. Gallino, R., Busso, M., Picchio, G., Raiteri, C.M., and Renzini, A., *Astrophys. Journal* **334**, L45, (1988).
[16]. Busso, M., Gallino, R., Lambert, D.L., Raiteri, C.M., and Smith, V.V., *Astrophys. Journal* **399**, 218, (1992).
[17]. Käppeler, F., Gallino, R., Busso, M., Picchio, G., and Raiteri, C.M., *Astrophys. Journal* **354**, 630, (1990).
[18]. Gallino, R., Busso, M., Picchio, G., and Raiteri, C.M., *Nature* **348**, 298, (1990).
[19]. Hollowell, D.E., and Iben, I.Jr., *Astrophys. Journal* **349**, 208, (1990).
[20]. Gallino, R., Raiteri, C.M., and Busso, M., *Astrophys. Journal* **410**, 400, (1993).
[21]. Bazan, G., and Lattanzio, J.C., *Astrophys. Journal* **409**, 762, (1993).

Production of CNO Isotopes and 26Al in Massive Single and Binary Stars

Norbert Langer[†] and Christian Henkel[‡]

[†] *MPI für Astrophysik, 85740 Garching, FRG, and*
[‡] *MPI für Radioastronomie, 53121 Bonn, FRG*

Abstract. In this paper we present CNO isotopic yields for massive mass loosing stars in the initial mass range $15 M_\odot \leq M_i \leq 50 M_\odot$. We investigate their dependence on assumptions about mass loss rates, internal mixing processes, and metallicity. Furthermore, we compute the production of ^{26}Al in hydrogen burning zones of massive and very massive stars.

INTRODUCTION

The CNO isotopes are of particular relevance for the chemical evolution problem, since the observational data are of particularly high quality, but on the other side our understanding of those observations is still rather limited (1), (2). In the present paper, we discuss for the first time the CNO isotopic yields of massive mass loosing stars, and investigate the dependence of these yields on various physical parameters.

For this purpose we have calculated a set of stellar models in the mass range $15 M_\odot \leq M_i \leq 50 M_\odot$ from the ZAMS up to silicon ignition, which was found to be sufficient in order to obtain reliable final yields for the CNO isotopes. Those are only subject to minor changes due to explosive processing during the supernova outburst (3). The nuclear reaction rates used are due to (4), except for the ^{12}C$(\alpha, \gamma)^{16}$O-rate which we choose like in (3), and the ^{25}Mg$(p, \gamma)^{26}$Al-rate (5). The remnant mass is computed as function of the final C/O-core mass (cf. Table 1) as $M_{\rm rem} = 1.18 M_\odot + 0.112 M_{C/O}$, which is a linear fit to the results of (6) and (7). Note that the errors in the CNO yields due to inaccuracies of this fit are very small and may only affect ^{12}C and ^{16}O (as long as the stars result in supernovae at all, and not in black holes). For convection we have used the model described in (8), which involves a semiconvection parameter $\alpha_{\rm sc}$; note that $\alpha_{\rm sc} = 0$ is equivalent to using the Ledoux criterion for convection, while $\alpha_{\rm sc} = \infty$ implies the Schwarzschild criterion (cf. Table 1).

Production of CNO Isotopes

Table 1: Nucleosynthesis results for 15 sequences computed up to silicon ignition. The symbols have the following meanings: M_i is the initial mass, Z the metallicity, α_{sc} the semiconvective mixing parameter (see text), M_f is the final stellar mass, $M_{C/O}$ the final C/O-core mass, and M_{rem} the approximated remnant mass. M_C and M_O are the total mass of carbon and oxygen ejected by stellar wind mass loss and by the supernova explosion (initially present amounts are *not* subtracted). $f_{13} \ldots f_{18}$ are the overproduction factors of ^{13}C, ^{14}N, ^{17}O, and ^{18}O, and $\Delta Y/\Delta Z$ is the ratio of the **net** yields of helium and metals.

#	M_i $M_\odot$	Z %	α_{sc}	M_f $M_\odot$	$M_{C/O}$ $M_\odot$	M_{rem} $M_\odot$	M_C $M_\odot$	M_O $M_\odot$	f_{13}	f_{14}	f_{17}	f_{18}	$\Delta Y/\Delta Z$
1	15	2	0.04	14.7	1.8	1.38	0.11	0.34	3.0	3.3	17	32	2.9
2	20	2	0.04	16.5	2.4	1.45	0.45	0.89	4.2	4.0	12	36	1.5
3	25	2	0.04	18.3	2.7	1.49	0.93	1.4	2.5	3.5	12	13	1.2
4	30	2	0.04	18.3	3.7	1.60	1.1	2.1	2.4	3.4	12	22	0.9
5	40	2	0.04	9.3	4.6	1.69	1.8	3.2	2.0	5.1	11	3.1	1.2
6	50	2	0.04	6.3	4.8	1.72	3.1	3.5	1.6	5.8	12	5.6	1.3
7	20	2	∞	15.9	4.2	1.65	0.31	2.3	3.3	4.0	18	2.5	0.3
8	25	2	∞	13.2	5.2	1.76	0.55	3.5	1.8	3.4	15	0.4	0.1
9	30	2	∞	13.6	8.4	2.10	1.1	4.8	1.4	3.8	17	0.4	0.1
7	20	2	∞	15.9	4.2	1.65	0.31	2.3	3.3	4.0	18	2.5	0.3
2	20	2	0.04	16.5	2.4	1.45	0.45	0.89	4.2	4.0	12	36	1.5
10	20	2	0.01	16.5	2.3	1.43	0.43	0.70	4.2	3.9	13	53	1.8
11	20	2	0.0	16.0	2.2	1.43	0.38	0.68	4.1	3.6	13	57	2.1
12	20	0.2	0.04	19.1	2.2	1.43	0.43	0.75	2.3	4.3	7.1	40	1.8
13	25	0.2	0.04	23.3	2.9	1.51	0.78	1.1	2.3	5.3	5.9	21	1.2
14	40	0.2	0.04	36.8	6.9	1.96	1.4	4.1	2.0	4.2	3.6	9.8	0.7
15	25	4	0.04	13.0	2.8	1.50	0.90	1.8	2.6	4.3	23	3.7	0.9

Table 2: Reaction of ejected total carbon ($M(^{12}C)$) and oxygen ($M(^{16}O)$) mass and of the production factors for ^{13}C, ^{14}N, ^{17}O, and ^{18}O, on an increase of the parameters initial mass (M_i), mass loss rate (meaning binary or WR type mass loss), semiconvective diffusion coefficient D_{sc} (i.e. mixing speed), and metallicity Z. Vertical up- or down-arrows mean sensitive reactions, inclined arrows indicate moderate trends, and the equal sign designates insensitivity.

	$M_i \uparrow$	$\dot M \uparrow$	$D_{sc} \uparrow$	$Z \uparrow$
$M(^{12}C)$	$\uparrow$	$\uparrow$	$\searrow$	=
$M(^{16}O)$	$\uparrow$	$\downarrow$	$\uparrow$	=
$f(^{13}C)$	$\searrow$	$\searrow$	=	=
$f(^{14}N)$	=	$\nearrow$	=	=
$f(^{17}O)$	=	=	=	$\uparrow$
$f(^{18}O)$	$\downarrow$	$\nearrow$	$\downarrow$	=
$\Delta Y/\Delta Z$	$\downarrow$	$\uparrow$	$\downarrow$	=

RESULTS

^{12}C and ^{16}O: Both are primary isotopes and their yields depend sensitively on the still uncertain ^{12}C$(\alpha,\gamma)^{16}$O nuclear reaction rate. Table 1 gives the total amounts ejected by the considered stars (i.e. *not* the net production). These values increase with increasing initial mass, since the He- and C/O-core masses are steeper functions of M_i than the remnant mass. However, for very massive stars (of not too small metallicity) the ejected oxygen mass may be decreasing for increasing initial mass due to mass loss effects (9). For stars with $M_i < 30 M_\odot$ we find generally good agreement between our models with $\alpha_{sc} = 0.04$ and the models in (3) with "restricted semiconvection", and also between our models with $\alpha_{sc} = \infty$ and models of (3) with "nominal semiconvection". For larger initial masses effects of mass loss become important, which was not considered in (3). Mass loss (i.e. WR-type mass loss or Roche lobe overflow in a binary) tends to increase the ^{12}C yields but to decrease the ^{16}O production (cf. also (9) and (1)). While the ^{16}O yield was found to depend much on the semiconvection parameter due to its effect on the C/O-core mass (cf. Table 1, sequences # 7,2,10,11) a weaker trend for the ejected ^{12}C mass is found in the opposite direction. Due to the fact that the He- and C/O-core masses depend only very weakly on Z, there is no direct dependence of M_C and M_O on Z (though there may be indirect effects, e.g. via $\dot{M}(Z)$).

^{13}C, ^{14}N, ^{15}N, ^{17}O: The isotope ^{15}N is effectively destroyed in our models, with production factors (defined as in (3)) of 0.3 ... 0.5, basically independent of the parameters discussed in Table 2 (cf. also (3)). The other three isotopes are secondary products of CNO burning. Semiconvection during hydrogen burning does practically not affect their nucleosynthesis (cf. Table 1). While for ^{13}C a slight decrease of the production factor with increasing initial mass is found, ^{14}N and ^{17}O are independent of M_i. Mass loss affects only slightly ^{13}C and ^{14}N, in opposite directions (see Tables 1 and 2). Only for ^{17}O we find a metallicity dependence of the production factor. This may be due to the fact that CNO-burning occurs at higher temperature for smaller Z, and the equilibrium abundance of ^{17}O varies sensitively with temperature (10). Note that the ^{17}O production factor depends on uncertainties in the proton capture reaction on ^{17}O (cf. (10)).

^{18}O: This isotope is produced by α-capture on ^{14}N, and is also destroyed by α-capture (the ^{18}O(α,γ) cross section is very uncertain; cf. (11)). Thus, the ^{18}O yield depends sensitively on the conditions below the hydrogen burning shell (cf. (3)). We find drastic changes of the ^{18}O production factor during the evolution up to core oxygen burning. As already noted in (3), the ^{18}O yield depends greatly on the semiconvection parameter (cf. Table 1, sequences # 7,2,10,11). We also find a pronounced trend with M_i (cf. Tables 1 and 2). Strong mass loss appears to somewhat increase the ^{18}O yield (cf. also

Table 3: ^{26}Al yields of various stellar sequences (cf. Table 1). All numbers refer to ^{26}Al produced during hydrogen burning only, i.e. contributions from neon burning and due to explosive processing during the SN explosion are ignored! M_{tot} represents the ^{26}Al mass corresponding to the total number of 1.8 MeV photons produced in the circumstellar medium during the hydrostatic evolution and after the SN explosion. $M_{wind,max}$ is the maximum circumstellar ^{26}Al mass occurring before the SN explosion. $M_{*,SN}$ is the amount of ^{26}Al present in the star at the time of core collapse, and $M_{wind,SN}$ is the amount present in the circumstellar medium at that time.

#	M_i $M_\odot$	Z %	α_{sc}	M_{tot} $M_\odot$	$M_{wind,max}$ $M_\odot$	$M_{*,SN}$ $M_\odot$	$M_{wind,SN}$ $M_\odot$
1	15	2	0.04	3.0(-6)	0.0	3.0(-6)	0.0
3	25	2	0.04	1.1(-5)	5.1(-7)	1.1(-5)	5.1(-7)
4	30	2	0.04	1.7(-5)	3.0(-6)	1.3(-5)	3.0(-6)
5	40	2	0.04	6.0(-5)	5.0(-5)	0.0	4.9(-5)
6	50	2	0.04	1.4(-4)	1.2(-4)	0.0	1.0(-4)
16	140	2	0.04	1.0(-3)	8.0(-4)	0.0	8.0(-4)
8	25	2	∞	2.1(-5)	2.0(-6)	1.9(-5)	2.0(-6)
9	30	2	∞	4.1(-5)	9.0(-6)	3.1(-5)	9.0(-6)
12	20	0.2	0.04	6.2(-8)	0.0	6.2(-8)	0.0
13	25	0.2	0.04	1.3(-7)	0.0	1.3(-7)	0.0
14	40	0.2	0.04	1.1(-6)	0.0	1.1(-6)	0.0
15	25	4	0.04	1.7(-5)	3.1(-6)	1.3(-5)	3.1(-6)
17	50	2	over.	4.3(-5)	3.4(-5)	0.0	3.2(-5)

(7)). Finally, note that large surface overabundances of ^{18}O are found for the WN/WC-transition Wolf-Rayet phase if $\alpha_{sc} < 0.1$ (cf. (12)).

$\Delta Y/\Delta Z$: Due to its effect on the C/O-core mass, semiconvection affects greatly the total net production of metals, in the sense that $\Delta Y/\Delta Z$ is much smaller for models computed with the Schwarzschild criterion compared to models with intermediate semiconvection parameters (Table 1). Also the variations of $\Delta Y/\Delta Z$ with initial mass and mass loss rate can be understood in terms of the C/O-core mass.

^{26}Al: Our results for the production of ^{26}Al due to the NeNa-cycle during hydrogen burning (cf. (10)) are summarized in Table 3. Sequence # 17 is computed with an input physics very similar to that used in ref. (13), for the purpose of comparison. M_{tot} in Table 3 may be used to obtain time integrated γ-ray fluxes per stellar sequence, while the maximum of $M_{wind,max}$ and $M_{*,SN} + M_{wind,SN}$ yields the upper limit for the 1.8 MeV photon flux from a given star.

ACKNOWLEDGEMENT

This work has been supported by the Deutsche Forschungsgemeinschaft through grant No. La 587/8-1.

REFERENCES

1. Prantzos N., Vangioni-Flam E., Chauveau S., A&A 285, 132 (1994).
2. Henkel C., Wilson T.L., Langer N., Chin Y.-N., Mauersberger R., in *The Structure and Content of Molecular Clouds*, T.L. Wilson, ed., Springer, in press (1994).
3. Weaver T.A., Woosley S.E., Phys. Rep. 227, 65 (1993).
4. Caughlan G.A., Fowler W.A., Atomic Data and Nuclear Data Tabels 40, 238 (1988).
5. Iliadis Ch., et al., Nuc. Phys. A 512, 509 (1990).
6. Woosley S.E., Langer N., Weaver T.A., ApJ 411, 823 (1993).
7. Woosley S.E., Langer N., Weaver T.A., ApJ, submitted (1994).
8. Langer N., El Eid M.F., Fricke K.J., A&A 145, 179 (1985).
9. Maeder A., A&A 264, 105 (1992).
10. Arnould M., Mowlavi N., in *Inside The Stars*, Proc. IAU-Colloq. No. 137, ASP Conf. Ser. Vol. 40, W.W. Weiss et al., eds., p. 310 (1993).
11. Käppeler F., et al., preprint (1994).
12. Langer N., A&A 248, 531 (1991).
13. Prantzos N., Doom C., Arnould M., de Loore C., ApJ 304, 695 (1986).

Presupernova Evolution and Explosive Nucleosynthesis in Massive Stars

Masa-aki Hashimoto(1), Ken'ichi Nomoto(2) and F.-K. Thielemann(3)

(1) Department of Physics, Kyushu University, (2) Department of Astronomy, University of Tokyo, and (3) Institut für theor. Physik, , Universität Basel

Abstract.
We investigate the presupernova evolution and the explosive nucleosynthesis in massive stars from 13 to 70 $M_\odot$. The explosion energy is assumed to be $1-1.5 \times 10^{51}$ erg. We found that the amount of ^{56}Ni produced ranges from 0.07 to 0.15 $M_\odot$. The ratio between the produced and solar abundance agrees with each other within factor 2-3 for $A \leq 32$. For the $^{12}C(\alpha,\gamma)^{16}O$ rate, our results indicate that the higher rate in 1985 will be better than the lower one.

Subject headings: Stellar Evolution, Nucleosynthesis — Supernovae

1. PRESUPERNOVA EVOLUTION AND HYDROSTATIC NUCLEOSYNTHESIS

Presupernova models are obtained evolving helium stars with masses $M_\alpha = 3.3\ M_\odot$, 4 $M_\odot$, 5 $M_\odot$, 6 $M_\odot$, 8 $M_\odot$ (model 8A and 8B), 16 $M_\odot$, and 32 $M_\odot$ (Nomoto & Hashimoto 1988; Hashimoto et al. 1993). The smallest star corresponds to the case that neon ignites in the center. The largest star corresponds to the upper bound of the supernova explosion beyond which no explosion occurs (Tsujimoto et al. 1994). We have used the Schwarzschild criterion for convection and adopted the time dependent mixing theory (Unno 1967) and neglected both overshooting and semiconvection. With this choice, these helium star masses approximately correspond to the main-sequence masses of $M_{\rm ms} \sim 13, 15, 18, 20, 25, 40$, and 70 $M_\odot$, respectively. These helium stars are evolved from helium burning to the onset of the iron core collapse. The initial compositions are: $X(^4\text{He}) = 0.9879$ and $X(^{14}\text{N}) = 0.0121$, where all the original CNO elements are assumed to be converted into ^{14}N during core hydrogen burning. During the hydrostatic evolution, many secondary nuclei are synthesized (Weaver & Woosley 1993). However, density structure of the presupernova model is not affected by the secondary nuclei. Nuclear reaction rates are mostly taken from Caughlan & Fowler (1988), but for the uncertain rate of $^{12}C(\alpha,\gamma)^{16}O$, we have mainly used the rate by Caughlan et al. (1985, CFHZ85), which is larger than the rate by Caughlan & Fowler (1988, CF88) by a factor of ~ 2.4. Fig. 1 shows the evolutionary track of the central density and temperature of massive stars. Fig. 2 shows the pre-collpase models of 25 $M_\odot$ star. The upper figure (model 8A) shows the result with use of CFHZ85 and the lower is with CF88. We can see that the subsequent nucleosynthesis after core helium burning is affected by the $^{12}C(\alpha,\gamma)^{16}O$ rate significantly.

2. Explosive Nucleosynthesis of 25 $M_\odot$

Since 25 $M_\odot$ star gives the largest contribution to the oxygen production, [O/Fe] exceeds zero. For typical hydrostatic burning products below $A < 27$, however, the relative abundance ratios coming from massive stars are in good agreement with the solar ratios as is seen in Fig. 4. The good agreement suggests that the $^{12}C(\alpha,\gamma)^{16}O$ reaction rate of CFHZ85, which is a factor of $\sim$ 2.4 larger than that of CF88, is more consistent with the observations than the latter, if no convective overshooting or no semiconvection is included. We can recognize this situation from Fig. 3 which shows the explosive nucleosynthesis inside the core and from Fig. 4 which shows the integrated abundances of the ejecta relative to the solar values. Obviously, Ne, Na, Al are overproduced compared with the solar values. This is due to the extra carbon burning compared with the case of CFHZ85. Since 25 $M_\odot$ is the typical massive star, the rate of CF88 is clearly too low. If overshooting at core helium burning would reduce the C/O ratio, a smaller $^{12}C(\alpha,\gamma)^{16}O$ rate would be a better choice. Weaver & Woosley (1993) have included 'nominal' semiconvection, and they concluded that 1.7 ± 0.5 times that given by CF88 is the best value. In the case of their 'restricted' semiconvection, they got 1.6 ± 0.7 times that given by CF88. Since we haven't included both overshooting and semiconvection, our results will not contradict with those of Weaver & Woosley (1993).

3. DISCUSSION

Fig. 5 shows the isotopic abundances relative to their solar values (Anders & Grevesse 1989) after integrating over the mass range from 13 to 70 $M_\odot$ with the initial mass function $\propto M^{-1.35}$. If we combine the contributions from both Type Ia and Type II supernovae with an appropriate ratio, the supernova abundances agree with the solar abundance ratios within a factor of 2-3 for the above typical species (see Tsujimoto et al. 1992, 1993, Thielemann et al. 1994, Hashimoto et al. 1994).

The effect of the initial metallicity on the yields from hydrostatic and explosive nuclear burning needs further quantitative studies. In Fig. 6, shown is the pre-collpase model of 20 $M_\odot$ star (helium core of 6 $M_\odot$) which we have evolved starting from pure helium core (X(^{4}He)=1.0). Evolution of the pure helium core simulates the evolution of the zero-metallicity star. Though the structure of the inner-core is essentially the same as that with use of the solar metallicity initially (compare with the Fig. 1 in Hashimoto et. al 1989), not a few abundances except for the alpha particle nuclei should be different drastically. In the subsequent studies, we will show what kind of products appear from this kind of low melallicity stars.

REFERENCES

Caughlan, G.R., Fowler, W.A., Harris, M.J., & Zimmerman, B.A. 1985, Atomic Data & Nuclear Data Tables, 32, 197

Caughlan, G.R., and Fowler, W.A. 1988, Atomic Data & Nuclear Data Tables, 40, 283

Hashimoto, M., Nomoto, K., Shigeyama, T., 1989, A&A, 210, L5

Hashimoto, M., Nomoto, K., and Tsujimoto, T., and Thielemann F-K. 1993, Proceedings of Nuclei in the Cosmos, 587

Hashimoto, M., Nomoto, K., and Tsujimoto, T., and Thielemann F-K. 1994, in preparation

Nomoto, K., and Hashimoto, M. 1988, Phys. Rep., 163, 13

Thielemann, F.-K., Hashimoto, M., and Nomoto, K. 1991, ApJ, 349, 222

Thielemann, F.-K., Nomoto, K., and Hashimoto, M. 1994, Les Houches, Session LIV, 629

Thielemann, F.-K., Nomoto, K., and Hashimoto, M. 1994, in preparation

Tsujimoto, T., Nomoto, K., Hashimoto, M., S. Yanagida, and Thielemann, F.-K. 1994, submitted to MNR

Tsujimoto, T., Nomoto, K., Hashimoto, M., and Thielemann, F.-K. 1994, submitted to ApJ

Weaver, T. A., and Woosley S.E. 1993, Phys. Rep., 227, 65

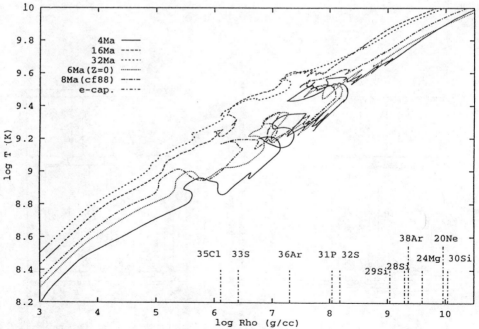

Fig. 1.— The evolution of the central density and temperature through the onset of iron core collpase.

422 Presupernova Evolution

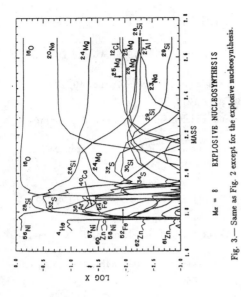

Fig. 3.— Same as Fig. 2 except for the explosive nucleosynthesis.

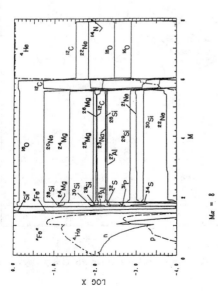

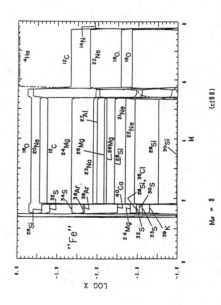

Fig. 2.— The presupernova abundances for 25 $M_\odot$ (8 $M_\odot$ helium) star with use of the CFHZ85 reaction rate (upper) and CF88 (lower).

M. Hashimoto et al. 423

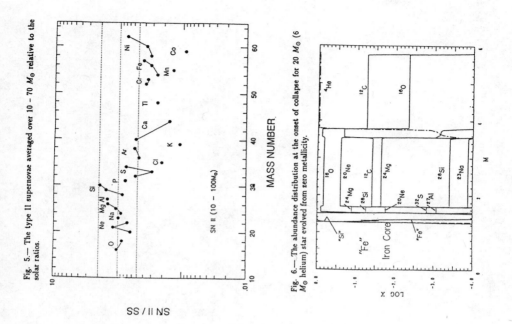

Fig. 5.— The type II supernovae averaged over 10 – 70 $M_\odot$ relative to the solar ratios.

Fig. 6.— The abundance distribution at the onset of collapse for 20 $M_\odot$ (6 $M_\odot$ helium) star evolved from zero metallicity.

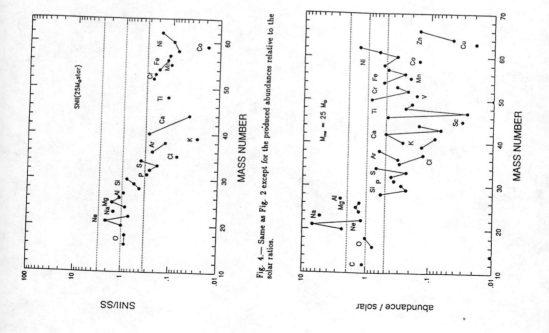

Fig. 4.— Same as Fig. 2 except for the produced abundances relative to the solar ratios.

Effect of the equation of state on the rapidly rotating neutron star

Masa-aki Hashimoto(1), Kazuhiro Oyamatsu(2), Yoshiharu Eriguchi(3) and Masataka Kan(1)

(1) Department of Physics, Kyushu University, (2) Department of Energy Engineering and Science, Nagoya University, and (3)Department of Earth Science and Astronomy, University of Tokyo

Abstract. The maximum angular velocity of uniformly rotating neutron stars is investigated. If we consider a hot neutron star stage after the birth from supernova explosions, an upper limit of the angular velocity for uniformly rotating neutron stars becomes a little smaller than that obtained by the equation of state for cold neutron matters.

Key words: Rotating Neutron Stars — equation of state

1. INTRODUCTION

Rapidly rotating neutron stars can be used to determine equations of state (EOS) for high density matters (see e.g. Friedman et al. 1989; Eriguchi et al. 1994). This can be done because the maximum angular velocity of rotating neutron stars depends crucially on the EOS if the angular velocity becomes very large. The angular velocity can be larger for the softer EOS because the softer EOS gives models with a smaller equatorial radius. In this respect if we could find very short period pulsars, we would be able to exclude stiffer EOS's by which we could not construct rotating models with the observed period.

At the present time, however, even the fastest pulsar, PSR1937+21, is rather slowly rotating compared with the fastest models which can be constructed numerically by using various EOS's (see e.g. Friedman et al. 1989; Eriguchi et al. 1994). This may be because we have not found faster pulsars yet, though they exist, or there may be other physical reasons why there are no pulsars which rotate faster than PSR1937+21. In this paper we will study the second possibility from the neutron star formation scenario.

We will expect many neutron stars are born through supernova explosions. During formation stages the structure of neutron stars will change in a very small time scale of order of a second. If we assume neutron stars to be spherical, proto-neutron stars ($T \sim 50 - 100$ Mev) will contract rapidly and cool down to hot neutron stars ($T \sim 10$ Mev) in 0.1 sec to 1 sec after the bounce.

The thermal effect on the EOS of neutron star matter becomes important when temperature of the star exceeds some MeV ($\sim 10^{10}$ K). The equation of state of high temperatures is different from that of cold states. In general high temperatures will make the radius of the hot neutron stars larger. Neutron stars with a larger equatorial radius cannot rotate rapidly. This possibility will set limit to rotation rates of neutron stars.

© 1995 American Institute of Physics

In the present paper, we will study an upper limit of the angular velocity of uniformly rotating hot neutron stars by computing their critical configurations due to mass shedding.

2. ASSUMPTIONS AND RESULTS

2.1. Assumptions

Since thermal structures of hot neutron stars have not been fully established due to the uncertainty of the EOS, we will assume that the temperature distribution in the most part of neutron stars is isothermal. However, temperature of the crust of the hot neutron star must be much cooler than the inner hotter region. Thus the temperature of regions where the density is below a certain value is assumed to be zero. Here, for the low density region ($\rho < 10^{10}$ gcm^{-3}), we adopt the EOS by Baym et al. (1971).

As for the rotation of neutron stars, it is expected that the rotation will become uniform before the star cools down to $T \sim 0$ K. Thus we assume that the temperature at which uniform rotation is established is $T \sim 10$ MeV. Therefore we treat uniformly rotating isothermal hot neutron stars with the temperature of ≤ 10 MeV.

2.2. Spherical Hot Neutron Stars

Before we discuss the effect of the rotation on structures of hot neutron stars, we will summarize structures of spherical neutron stars with high temperatures. For the equation of state at high temperatures we will adopt three different EOS's: 1) the EOS of Friedman & Pandharipande (1981, FP hereafter), 2) Model I EOS by Oyamatsu (1993) is modified by taking into account the finite temperatures, and 3) the stiff EOS for finite temperatures by Sumiyoshi & Toki (1994, RMF hereafter). It should be noted that Model I and FP differ only for very high density regions. The slope of FP is steeper than that of Model I.

In Fig. 1 (upper), gravitational masses of neutron stars are shown against the central density. The maximum masses for each EOS are 1.55 (Model I), 1.95 (FP) and 2.9 (RMF) $M_\odot$, respectively. The masses of neutron stars are governed mainly by the EOS for the regions with $\rho > 5 \times 10^{14}$ gcm^{-3}. As far as the temperature remains below 10 MeV, masses of neutron stars do not depend on temperatures so significantly due to the degeneracy of the neutron gas.

2.3. Rapidly Rotating Hot Neutron Stars

We solve structures of rapidly rotating neutron stars with the Model I EOS by Oyamatsu (1993) which will set limit to the angular velocity.

The maximum angular velocity for stable configurations is determined by the following conditions: 1) the gravitational mass of the model is extreme against the central energy density on the equilibrium sequence with the constant total angular momentum, i.e.

$$\partial M/\partial \rho_c \mid_{J=const} = 0, \qquad (1)$$

where M, ρ_c and J are the gravitational mass, the central energy density and the total angular momentum, respectively, and 2) the model is in a critical state, i.e. at the mass shedding state, beyond which no equilibrium states are allowed due to the centrifugal force. In Fig. 1 (lower), we have shown the result for Model I.

As seen from Table 1, if the neutron stars are settled down to a uniform rotation with the temperature of 10 MeV, it will evolve to a rotating neutron star with the rotational period of 0.9 msec. In Table 1 the critical state for the temperature of 0.75 MeV and the central energy density $\rho = 2.5 \times 10^{14}$ g cm^{-3} is also tabulated. Thus the neutron stars which experience very hot stage will rotate with the angular velocity of 2/3 of the critical one or the Kepler angular velocity.

3. Conclusions

It is concluded that most single pulsars rotate with periods of larger than roughly 1 msec (Hshimoto et al. 1994). Though our senario will not change, for some special EOS's, our result could be modified (Oyamatsu et al. 1994)

EOS	T(MeV)	$\rho_c(g cm^{-3})$	$M_G/M_\odot$	$\Omega(s^{-1})$	P(ms)
Model I	9.9	4.0E15	1.61	8.1E3	0.8
	0.75	2.5E15	1.59	7.0E3	0.9
Critical Model	0.75	2.5E15	1.75	1.0E4	0.6

Table 1: The physical quantities for several models.

REFERENCES

Baym, G., Pethick, C.J., Sutherland, P., 1971, ApJ, 170, 299

Eriguchi, Y., Hachisu, I., and Nomoto, K., 1994, MNRAS, 266, 179

Friedman, B., Pandharipande, V.R., 1981, Nucl. Phys., A361, 502

Friedman, J.L., Ipser, J. R., and Parker, L., 1989 Phys. Rev. Lett., 62, 3015

Hashimoto, M., Oyamatsu, K., and Eriguchi, Y., 1994 ApJ, in press

Oyamatsu, K., 1993, Nucl. Phys., A561, 431

Oyamatsu, K., Hashimoto, M., Kan, M., and Eriguchi, Y., 1994, in preparation

Sumiyoshi, K., and Toki, H., 1994, ApJ, 422, 700

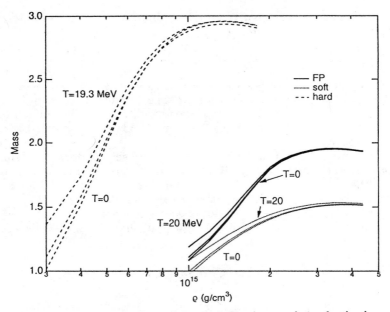

Fig. 1.— The energy density and the gravitational mass relation for the three different EOS's (upper) and for Model I with rotation (lower).

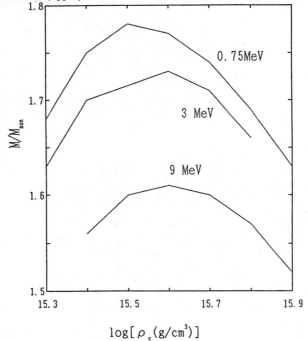

Explosion of a C-O Accreting White Dwarf

Eduardo Bravo[1,2], Amedeo Tornambé[3], Jordi Isern[2]

(1) Departament de Física i Enginyeria Nuclear, Universitat Politècnica de Catalunya, Spain, (2) Centre d'Estudis Avancats de Blanes, CSIC, Spain, and (3) Osservatorio Astronomico di Roma, Italy

Abstract. The evolution of binary systems leading to a supernova explosion and the nature of the thermonuclear flame propagating through the white dwarf are the two most uncertain aspects of the standard model of Type Ia Supernovae. Both aspects are connected by the transition from the accretion phase to the thermonuclear runaway. In this communication a scenario is investigated from the onset of the accretion phase up to the explosion and disruption of the white dwarf. Starting from a $0.8\ M_\odot$ CO white dwarf accreting C+O at a constant rate of $5\,10^{-7}\ M_\odot/\mathrm{yr}$, explosive ignition of carbon occurs in the center at a rather low density, $1.8\,10^9\ \mathrm{g/cm^3}$. The thermal conditions set up by the accretion process, nuclear energy generation, and convection turn out to be critical for determining the runaway properties. A detonation or a deflagration result for slightly different treatments of the transition from the hydrostatic evolution to the hydrodynamic one. The nucleosynthesis has been computed for both solutions.

1. INTRODUCTION, DISCUSSION AND CONCLUSIONS

Type Ia Supernovae (SNIa) are thought to be responsible for about 30-50% of the Fe-group elements production during the history of the Milky Way (Bravo et al 1993a), while gravitational supernovae would account for the remaining 50-70%. The exact percentage by which each element is produced in each type of supernovae varies from one element to another. For instance, there is observational evidence that Mn is mainly due to SNIa. In contrast, many models of SNII predict ^{62}Ni being synthesized in excess independently of the values of the unknown parameters of the explosion (explosion energy, mass cut, and the precise mechanism of the explosion itself).

The nucleosynthetic composition of the ejecta of SNIa depends strongly on the type of thermonuclear wave that propagates through the white dwarf: detonation, deflagration or a combination of both (late detonation, delayed detonation). The most popular model rely on the properties of a thermonuclear deflagration that, starting from the center, propagates all the way subsonically throughout the star (Nomoto et al 1984). This model reproduces quite well the main properties of SNIa (light curves and spectra) but still poses some problems due to the overproduction of neutron rich isotopes of the Fe-peak elements (see e.g. Wheeler & Harkness 1990). Another popular class of models are the so-called delayed detonations, whose main point is the transition from a deflagration to a detonation at some point inside the white dwarf, at a density lower than $\sim 3\,10^7$ g/cm^3 (Khokhlov 1991, Woosley & Weaver 1994). All

Table 1: Type of thermonuclear flame resulting for different conditions at central explosive ignition

T_c (10^8 K)	∇T	$M_{inc}(M_\odot)^a$	
8.0	ad	$2\,10^{-4}$	slow deflagration
8.5	ad	$2\,10^{-4}$	fast deflagration
9.0	ad	$2\,10^{-4}$	detonation
10.0	ad	$< 10^{-2}$	detonation
10.0	ad	$> 2\,10^{-2}$	slow deflagration
10.0	$2\times$ad	$< 2\,10^{-3}$	detonation
10.0	$2\times$ad	$> 5\,10^{-3}$	slow deflagration

^{a}amount of mass instantaneously incinerated at the center

those models have in common a slow velocity of the flame near to the center, afterwards accelerated as a result of mixing by the Rayleigh-Taylor instability through the thermonuclear front.

The purpose of this paper is to study the conditions set up in the center of the white dwarf by the accretion process, and its consequences on the initial stages of the propagation of the flame through the star. To do that, the evolution of the white dwarf has been followed with a hydrostatic code (FRANEC, see e.g. Chieffi & Straniero 1989) from the start of the accretion up to the onset of the hydrodynamic instability and, from that point on, the star has been evolved with the same hydrodynamic code used by Bravo et al (1993b). The initial model was a 0.8 $M_\odot$ CO white dwarf, that has been followed under accretion, at a rate of $\dot M = 5\,10^{-7}$ $M_\odot$/yr. The white dwarf center reached the ^{12}C ignition curve at a density of $2.67\,10^9$ g/cm^3. Shortly afterwards, a convective core developed, which allowed the white dwarf to evolve up to the hydrodynamic instability in a time of $\sim 4\,10^3$ yr. At that moment the central density had decreased to $1.8\,10^9$ g/cm^3.

Due to the operation of convection, the thermal gradient was nearly the adiabatic one and the chemical composition was homogeneous in the inner 90% of the star, the mass fraction of ^{12}C being 0.38. When the temperature is as high as $T = 0.85\,10^9$ K the burning time is comparable to the white dwarf sound crossing time (for the ignition density of $1.8\,10^9$ g/cm^3). From that point on the star can not be in hydrostatic equilibrium. As the temperature rises, the burning time goes down, and the convection becomes inefficient to maintain the adiabatic gradient except for an increasingly small central zone. However, for an extended region (the central ~ 1 $M_\odot$) the temperature was sensitively higher than that corresponding to the ignition curve, so that one could speak of a 'hot' white dwarf explosion. To set the bounds of the problem

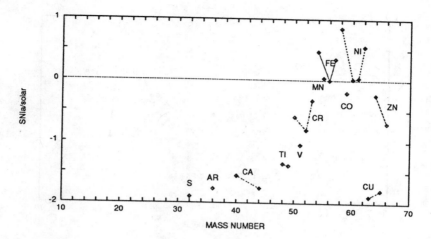

Fig. 1.— Log of the abundance of the ejected elements normalized to ^{56}Fe, for the detonation model.

of the propagation of the thermonuclear flame through the white dwarf, we followed the hydrodynamic evolution of the star for two different assumptions: a) explosion at a central temperature near to the explosive ignition temperature, with an adiabatic thermal profile (which overestimates the effectiveness of convection); b) instantaneous incineration of the central mass-zone ($2\,10^{-4}\ M_\odot$) without changing the temperature of the surrounding layers (which underestimates convection). The results of those experiments are presented in Table 1.

Our results show that a slight change in the assumptions about the effectiveness of convection when the center is reaching the explosive ignition condition results in different modes of propagation of the flame. A high central temperature and an adiabatic thermal gradient both favour the formation of a detonation. The mass resolution at the center is also a critical factor to determine if a detonation will form. The models with $T_c = 10^{10}$ K were developing a detonation if the central mass resolution was better than $10^{-2}\ M_\odot$ and the thermal gradient was the adiabatic, or if the central mass resolution was better than $2\,10^{-3}\ M_\odot$ for a thermal gradient two times the adiabatic one.

It is interesting to note that for a central temperature of $0.85\,10^9$ K the flame propagates as a 'fast' deflagration in the central regions. In this case, the detonation fails to form, but the temperature is still close enough to the explosive ignition temperature, so that the passage of the pressure waves sent by the incineration of the central layers heat the material, and appreciably shortens its burning time. As a result, the velocity of the deflagration in the central $\sim 0.02\ M_\odot$ is as high as $0.2 v_{\text{sound}}$. Behind that point, the flame speed slows down to the conductive velocity, before being again accelerated by the mixing induced by the Rayleigh-Taylor instability.

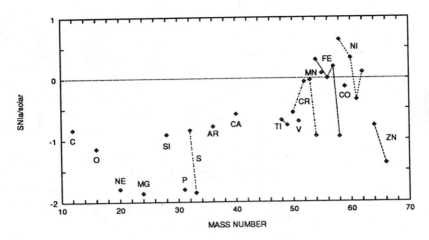

Fig. 2.— Log of the abundance of the ejected elements normalized to ^{56}Fe, for the deflagration model.

The nucleosynthesis output is very different for both solutions, the detonation and the deflagration. It has been computed and the results are presented in Figs. 1 and 2. The mass of ^{56}Fe ejected in each model is: 0.87 $M_\odot$ in the detonation, and 0.54 $M_\odot$ in the deflagration.

ACKNOWLEDGEMENTS

This work has been supported in part by the DGICYT grants PB91-0060 and PB90-0912, by the CESCA project "Evolution of galaxies", by the PICS 114 of the CSIC and by an international cooperation grant between Spain (CSIC) and Italy (CNR). One of us (ID) is very indebted for a CIRIT grant.

REFERENCES

Bravo E., Isern J., Canal R., 1993, A& A 270, 288

Bravo E., Domínguez I., Isern J., Canal R., Hofflich P., Labay J., 1993, A& A 269, 187

Chieffi A., Straniero O., 1989, ApJS 71, 47

Khokhlov A., 1992, A& A 245, 114

Nomoto K., Thielemann F.K., Yokoi K., 1984, ApJ 286, 644

Wheeler J.C., Harkness R.P., 1990, Rep.Prog.Phys. 53, 1467

Woosley S.E., Weaver T.A., 1994, ApJ 423, 371

CNO anomalies on the red giant branch : Tests for classical evolutionary models of low mass stars

Corinne Charbonnel

Observatoire Midi-Pyrénées, Toulouse, France

Abstract.

Observations of chemical anomalies in different points of the Hertzsprung-Russel diagram suggest that some "non-standard" processes mix the stellar interior. We show how we can use the observations of the $^{12}C/^{13}C$ ratios in evolved low-mass stars to identify and describe the mixing processes which should be part of stellar modelling.

Key words: Stellar models — Low mass stars — Non-standard mixing processes

1. THEORETICAL AND OBSERVATIONAL FIRST DREDGE-UP

The first dredge-up phase offers a unique opportunity to test the physical and chemical description of the stellar interior. During this phase, the deepening convective envelope mixes the outer layers of the red giant with internal matter which has been CN-processed while the star was on the main sequence. This particularly induces a change of the atmospheric abundances of ^{12}C and ^{13}C (which were unaltered until this phase) which respectively decrease and increase by amounts that depend on the internal chemical structure at the end of the main sequence and on the maximum extend of the external convective zone during the ascent along the red giant branch.

Figure 1 shows the standard theoretical post dredge-up $^{12}C/^{13}C$ ratio as a function of the stellar mass as it is given by the evolutionary tracks at solar metallicity computed by the Geneva group (Schaller et al. 1992). These predictions are compared with observations in subgiants and giants of open galactic clusters (Gilroy 1989; Gilroy & Brown 1991) with turn-off masses lower than 2 $M_\odot$. The scatter of the observed $^{12}C/^{13}C$ ratio may be due to a dispersion of the evolutionary status of the studied objects. The domain of observations in globular cluster giants is also indicated.

In a large fraction of low-mass evolved stars, the observed $^{12}C/^{13}C$ ratio is substantially lower than the post-dredge up value predicted in the framework of standard stellar theory. Moreover, the observations reveal a decrease of the $^{12}C/^{13}C$ ratio as the stellar mass decreases (this tendency is accentuated by the observations in globular giants), whereas standard theory predicts the exact opposite trend. These discrepancies between real and model stars reveal the existence of an extra-mixing process which additionally lowers the post-dredge up carbon isotopic ratio in stars with masses lower than 2 $M_\odot$.

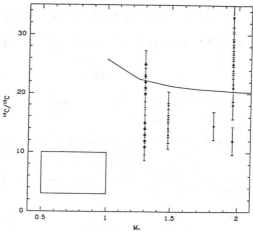

Fig. 1.— Standard theoretical $^{12}C/^{13}C$ ratio at the end of the dredge-up as a function of the stellar mass (solid line) compared to observations in subgiants and giants of galactic clusters (Gilroy 1989; Gilroy & Brown 1991). The rectangle indicates the domain of observations in globular cluster giants.

2. EXTRA-MIXING ON THE RED GIANT BRANCH

In order to determine at what evolutionary step the extra-mixing does occur, let us focus on M67 evolved stars. In figure 2 we compare the theoretical evolution of the $^{12}C/^{13}C$ ratio along the isochrone of M67 with the isotopic ratios observed in subgiants and giants of this cluster (Gilroy & Brown 1991). Three points show up from this comparison.

1. The theoretical and observational first dredge-up are in complete agreement : Predictions perfectly reproduce the observations for the stars which are ascending the giant branch up to luminosities lower than 100 times the solar luminosity. This fact indicates that the standard theoretical main sequence profiles of ^{12}C and ^{13}C match the real chemical profiles. The extra-mixing process which is necessary to explain low observed $^{12}C/^{13}C$ ratios is thus effective only on the giant branch after the completion of the first dredge-up.

2. The helium flash does not modify the surface value of the $^{12}C/^{13}C$ ratio. Indeed, clump giants and giants with Log $L/L_\odot > 2$ exhibit the same low carbon isotopic ratios.

3. In M67, the disagreement between theory and observation appears above log $L/L_\odot=2$. On the evolutionary track of the 1.25 $M_\odot$ model (approximatively the turn-off mass of M67), this luminosity is slightly higher than the luminosity at which the hydrogen-burning shell reaches the chemical discontinuity created by the convective envelope at its maximum extent. Above this evolutionary point, hydrogen burning occurs in a region of constant molecular weight. Below this point, the molecular weight gradient certainly probably acts as a barrier to the development of the new mixing.

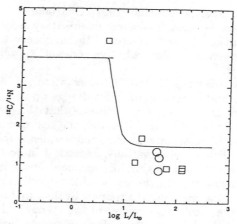

Fig. 2.— Theoretical evolution of the $^{12}C/^{13}C$ ratio along the isochrone of M67. Observational values of $^{12}C/^{13}C$ in giants of M67 from Gilroy & Brown (1991) (squares for pre-helium flash stars, circles for clump stars)

Let us stress a crucial point. In more massive stars ($M_* \geq 2M_\odot$), central helium burning is ignited before the core becomes degenerate, and the surrounding hydrogen-burning shell never reaches the region of constant molecular weight that has been homogenized by the convective envelope. Thus the extra-mixing is not expected to occur in these stars, as it is confirmed by the observations in galactic clusters with turnoff masses higher than 2 $M_\odot$.

3. SPECULATIONS ON THE NATURE OF THE EXTRA-MIXING PROCESS

Zahn (1992) proposed a consistent picture of the interaction between meridional circulation and turbulence induced by rotation in stars. The resulting mixing depends mainly of the loss of angular momentum via a stellar wind. Moreover, some mixing can take place wherever the rotation profile presents steep vertical gradients, and in shell burning regions. All the conditions for the presence of such a mixing are fulfilled during the evolution on the red giant branch. This leads us to hold Zahn's process responsible for the required extra-mixing on the red giant branch of low mass stars.

4. CONSTRAINTS ON THE AMOUNT OF EXTRA-MIXING IN DEEP SOLAR INTERIOR

Up to now, the study of lithium anomalies at the surface of stars in various evolutionary stages provided the strongest constraints to the physical description of the non-standard mixing process described above (Charbonnel et al. 1992, Charbonnel & Vauclair 1992, Charbonnel et al. 1994). The results presented here suggest to use complementary informations from the CNO elements evolution. Indeed, whereas lithium observations help to sound the most external layers of stellar radiative zones, CNO observations help to sound deeper regions.

As we have seen previously, $^{12}C/^{13}C$ observations in M67 cluster suggest that the theoretical and observational first dredge-up are in complete agreement. This point is crucial, since it indicates that the standard theoretical main sequence profiles of ^{12}C and ^{13}C match the real chemical profiles. This result gives additional constraints on the main sequence mixing processes invoked to explain the lithium observations. Indeed, it indicates that the associated diffusion coefficient decreases rapidly as one deepens inside the star, and this is in perfect agreement with the shape of the effective diffusion coefficient given by Zahn for rotation-induced mixing. Such a mixing should thus not affect significantly the most inner region of the stars. It does not lead to an important change in the solar neutrino flux, as can be seen in table 1 where we compare the total capture rates for the chlorine and gallium experiments for different solar models. These models where computed with the Geneva evolutionary code with the same input microphysics (eos of Mihalas et al. (1988) and Iglesias et al. (1992) opacities), but with different prescriptions for the macrophysics : Model 1 is a "standard" model, model 2 was computed with microscopic diffusion of helium and metals, model 3 with both microscopic diffusion and rotation-induced mixing by Gaige (1994). Model 3 gives better agreement with helioseismological informations (depth of the base of the convective zone and P/ρ profile in the radiative zone). The inclusion of rotation-induced mixing leads to a small reduction for the predicted neutrino capture rates compared to the case of pure microscopic diffusion.

Table 1: Predicted neutrino capture rates for the chlorine and gallium experiments

	Φ_{37Cl} (SNU)	Φ_{71Ga} (SNU)
Model 1	7.3	126.3
Model 2	7.9	129.5
Model 3	7.5	127.2

REFERENCES

Charbonnel C., Vauclair S., Zahn J.P., 1992, A&A 255, 191
Charbonnel C., Vauclair S., 1992, A&A 265, 55
Charbonnel C., Vauclair S., Maeder A., Meynet G., Schaller G., 1994, A&A 283, 155
Gaigé Y., 1994, PhD Thesis
Gilroy K.K., Brown J.A., 1991, ApJ 371, 578
Iglesias C.A., Rogers F.J., Wilson B.G., 1992, ApJ 397, 717
Mihalas D., Hummer D.G., Däppen W., 1988, ApJ 331, 815
Schaller G., Schaerer D., Meynet G., Maeder A., 1992, A&AS 96, 269
Zahn J.P., 1992, A&A 265, 115

Simulation of a White Dwarf Explosion with a 3D Particle Hydrocode

Eduardo Bravo[1,3], Domingo García[1,2], Nuria Serichol[1]

(1) Departament de Física i Enginyeria Nuclear, Universitat Politècnica de Catalunya, Spain, (2) Lick Observatory, University of California at Santa Cruz, USA, and (3) Centre d'Estudis Avancats de Blanes, CSIC, Spain

Abstract. The hydrodynamic evolution of a exploding white dwarf has been followed with a 3D particle hydrocode. The fractal dimension of the thermonuclear front surface is analysed for various cases.

1. INTRODUCTION, DISCUSSION AND CONCLUSIONS

Numerical simulations of thermonuclear stellar explosions favour the propagation of combustion flames as deflagration waves, specially in the early phases of combustion near the center of the star. These simulations are mainly one dimensional calculations that assume spherical symmetry. From theoretical arguments one expects that turbulence will appear as the front advances through the star. One dimensional calculations are, by nature, unable to follow the motion of the fluid in a turbulence, that is intrinsically three dimensional. In recent years, some steps have been given towards the proper resolution of the hydrodynamics of turbulent deflagrations, and nowadays extensive multi-dimensional calculations with a realistic physics seem to be reliable.

Timmes & Woosley (1992) solved the structure of a nuclear laminar front in 1D, at the densities relevant for a SNIa explosion, and found the flame speed normal to the surface. Khokhlov (1993), and Livne & Arnett (1993) succeeded in tracking the evolution of the turbulent flame in 2D by using a PPM code with a flame capturing technique. The flame front was resolved till scales where the flame is not affected by the Rayleigh-Taylor instability. This critical scale was found to be of order $5\,10^4 - 5\,10^5$ cm, at a front radius $3\,10^7 - 10^8$ cm. Above these scalelengths, the geometry of the front surface is complicated due to the simultaneous evolution and interaction of structures of different sizes, with the result that the surface of the front increases and, consequently, the heat transfer efficiency increases too. As time goes on, the large structures absorb the small ones and, although the effective flame speed increases, it remains always smaller than $0.1 v_{sound}$, a value too small to unbind the whole star.

The only 3D calculation reported in the literature is that of Khokhlov (1994), in which the front structure was solved within a box of size $(2\,10^6)$ x $(2\,10^6)$ x $(6\,10^6)$ cm. Its main result is that the properties of the flame in 2D and 3D are remarkably different, with the result that turbulence accelerates more efficiently the flame in 3D calculations.

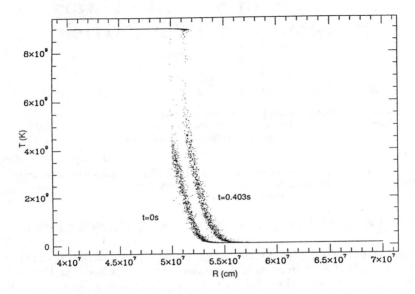

Fig. 1.— Tracking of the flame front with SPH. The temperature is shown as a function of the particles position, for two times.

Table 1: Evolution of the fractal dimension of the flame front

Model	ρ_c[a]	D_c	ρ_c	D_c	ρ_c	D_c	$r_{\rm inc}/r_{\rm wd}$
L0M0[b,d]	$6\,10^9$	2.05	10^8	2.04	$2\,10^7$	2.05	0.30
L7M0[b,d]	$6\,10^9$	2.09	$9\,10^7$	2.12	$2\,10^6$	2.34	0.30
L7M0B[b,d]	$6\,10^9$	2.09	$5\,10^8$	2.06	$3\,10^6$	2.11	0.15
L7M0C[b,e]	$6\,10^9$	2.09	$8\,10^7$	2.06	$4\,10^8$	2.06	0.30
L7M5[b,e]	$6\,10^9$	2.07	$8\,10^8$	2.08	$2\,10^7$	2.11	0.30
RAND[c,e]	$6\,10^9$	2.06	$4\,10^8$	2.09	$2\,10^7$	2.10	0.30

[a] Initial model in hydrostatic equilibrium, immediately after incineration
[b] Explosion induced by the instantaneous incineration of the particles inside a sphere of radius $r_{\rm inc}$, whose surface was perturbed by a spherical harmonic: $Y_l^m = 0.05 P_l^m(\cos\theta)\cos(m\phi)$, with l, m as given in the model name
[c] In this case, the incinerated sphere was perturbed randomly, with an amplitude $0.05 r_{\rm inc}$
[d] No flame propagation was allowed, the temperature in NSE was kept constant during the expansion, to simulate the high heat capacity of incinerated matter
[e] No flame propagation was allowed, the overpressure behind the flame front was limited to 20%

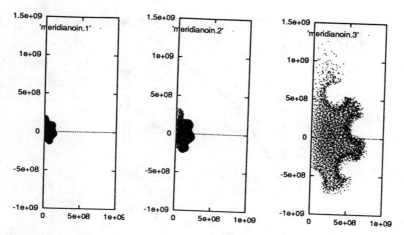

Fig. 2.— Evolution of the particles in the incinerated core in model L7M0. The snapshots correspond to central densities of $9\,10^7$, $3\,10^7$, and $2\,10^6$ g/cm^3.

The aim of this communication is to show that the structure of the deflagration can be reasonably handled in 3D simultaneously with the hydrodynamic evolution of the whole star using a realistic physics. The final goal will be to establish the temporal evolution of the scaling properties of the flame front (unstable wavelengths, fractal dimension). Up to the time of the preparation of this manuscript, some steps have been given in that direction:

1) Construction of an initial model for SPH (Smooth Particle Hydrocode) in hydrostatic equilibrium, at near the Chandrasekhar Mass. The electronic EOS that has been used is valid for any arbitrary degree of degeneracy, for relativistic as well as for non-relativistic electrons, and includes pair creation. An ideal gas law is adopted for the ions, and the diffusion approximation for the radiation field.

2) Development of a tracking method for the flame front, in the context of SPH, that has probed to be very efficient. Nuclear Statistical Equilibrium (NSE) is assumed behind the flame front. The flame advance is a result of the competence of two processes: thermal diffusion, and nuclear energy generation. For the thermal conductivity and the specific heat approximate analytical expressions have been adopted that are valid for the physical conditions in the thermonuclear front (Woosley 1986). The nuclear energy generation has been assumed to be dominated by the ^{12}C fusion reaction. Adjusting the flame thickness to the resolution of the code ($2h$, where h is the interpolation distance for SPH), the front is resolved, and the flame speed given by Timmes & Woosley (1992) is recovered (Figure 1).

3) Simulation of the explosion of a White Dwarf without flame front propagation. The explosion is triggered by the artificial instantaneous incineration of a region near the center, for which some perturbation of the spherical symmetry is introduced. In this case, the changes in the geometry of the flame surface are purely due to hydrodynamic effects. The results are given in Table

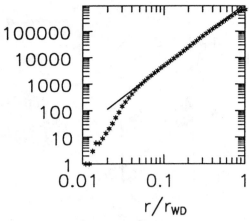

Fig. 3.— Determination of the fractal dimension of the front surface for model L7M0, at a central density of $2\,10^6$ g/cm^3. The vertical axis is the number of particles at the front that are inside a sphere of radius r, and centered at some point inside the front. At low radii the numbers are meaningless due to poor resolution, while at large radii there can be seen a saturation effect.

1.

4) Simulation of the evolution of the flame within an orange-section like portion of the star. In this case, which has a higher spatial resolution ($6\,10^5$ cm at the center), the flame tracking algorithm is used.

In Table 1 there is given the evolution of the fractal dimension of the flame front, D_c, for different numerical models, and at different times during the expansion. The results, that refer to the simulations with the flame front 'frozen', show no clear increase in the fractal dimension until the density has decreased below $\sim 10^7$ g/cm^3. Only for one of the computed models (L7M0, Figure 2) has the fractal dimension grown appreciably, but only up to $D_c = 2.34$ (Figure 3), a value much lower than the one that is usually adopted in current SNIa models.

ACKNOWLEDGEMENTS

This work has been supported in part by the DGICYT grant PB90-0912, and the CESCA-UPC project "Hydrodynamic Evolution of Compact Stars". One of us (DG) wants to thank the hospitality of Lick Observatory, where part of the calculations have been done

REFERENCES

Khokhlov A., 1993, ApJ 419, L77
Khokhlov A., 1994, ApJ 424, L115
Livne E., Arnett W.D., 1993, ApJ 415, L107
Timmes F.X., Woosley S.E., 1992, ApJ 396, 649

Neutrino Spallation Reactions on ^{4}He and the r-Process

B. S. Meyer(1), S. E. Woosley(2), R. D. Hoffman(2), G. J. Mathews(3), J. R. Wilson(3)

(1) Department of Physics and Astronomy, Clemson University, Clemson, SC 29634-1911, USA; (2) University of California Observatories/Lick Observatory, Board of Studies in Astronomy and Astrophysics, University of California at Santa Cruz, Santa Cruz, CA 95064, USA; (3) Lawrence Livermore National Laboratory, Livermore, CA 94550, USA

Abstract. We study the effects of neutrino spallation of neutrons and protons from ^{4}He during the r-process. These reactions tend to hinder the r-process by allowing the assembly of too many heavy seed nuclei. In order to make sufficient quantities of high-mass nuclei, the current r-process models seem to require temperatures for the mu and tau neutrinos of less than or roughly 6 MeV, although this is a very preliminary conclusion. More study of this interesting effect is definitely required. Nonetheless, we can at present conclude that these reactions should be included in calculations of the r-process in neutrino-driven winds and that we may be able to obtain from nucleosynthesis interesting constraints on the neutrino cooling of neutron stars.

Key words: Nucleosynthesis, supernovae, neutrinos

1. Introduction

Our current picture is that the r-process occurs in the winds blowing from newly-born neutron stars (Woosley and Hoffman 1992; Meyer et al. 1992; Takahashi et al. 1994; Woosley et al. 1994). Neutrinos from the cooling neutron star play a crucial role in this scenario. First they deposit the energy that allows mass to escape from the strong gravitational well of the star. The resulting net heating from the neutrinos raises the entropy per baryon in the wind to the high values (~400k) necessary for the r-process. The second crucial role the neutrinos play is that they set the neutron richness of the wind material (Qian et al. 1993). Electron anti-neutrinos come from deeper in the neutron star where it is hotter than do the electron neutrinos. The anti-neutrinos thus have a higher energy and interact more readily with wind material than do the neutrinos. Since anti-neutrinos capture on protons to produce neutrons, the wind material becomes increasingly neutron rich as the anti-neutrinosphere moves further inward. In these two ways neutrinos establish the conditions necessary for the r-process to occur.

Neutrino effects do not stop at the onset of nucleosynthesis, however. Neutrinos will continue to interact with nucleons and nuclei even as the r-process occurs. In general, one would expect that such reactions would not significantly affect r-process yields because neutrino-nucleus scattering cross sections are small. One set of interactions can have a large effect, however. These are neutrino spallations of nucleons from ^{4}He nuclei. Such reactions can dramatically affect the nucleosynthesis because of the large abundance of alpha particles in the high-entropy wind. Since the efficacy of these reactions depends strongly on the neutrino temperatures and luminosities, there exists the possibility that the r-process may provide important constraints on the neutrino cooling of a nascent neutron star.

2. The Calculations

We followed the nucleosynthesis in mass elements calculated in a realistic supernova model (Woosley et al. 1994). This supernova model is fully relativistic and follows detailed neutrino transport. Extremely fine mass zoning allowed us to resolve the last 0.03 $M_\odot$ of material that escapes the neutron star in a wind. We have already shown that this model produces r-process yields in excellent agreement with the solar r-process curve, although there is a serious overproduction of N=50 nuclei in some of the first mass trajectories to leave the neutron star (Woosley et al. 1994). The limitation of the supernova model to one dimension possibly accounts for this overproduction. In the first few seconds following core bounce, large-scale convection probably dominates mass transport (Herant et al. 1992). The conditions in the convective regime are quite different from those in the wind, thus the wind results are likely not applicable until several seconds post-bounce when the convection dies down and the winds take over.

In the calculations of Woosley et al. (1994) we did not treat the effect of neutrino-nucleus interactions during nucleosynthesis. In this paper we study neutrino-spallation reactions on abundant ^{4}He. The nucleosynthesis code we use for the present calculations is a fully implicit single network code that includes over 3000 isotopes. This code can follow the assembly of nucleons into ^{4}He, the combination of alpha particles into heavier nuclei, and the subsequent r-process. Nuclear reaction rates are from Cauglin and Fowler (1988) and Thielemann et al. (1987) where available and from Woosley et al (1975) where they are not. Neutron capture rates above krypton are from Cowan et al. (1992) and beta-decay rates for neutron-rich nuclei are from Klapdor et al. (1984).

We have included neutrino-spallation of neutrons and protons from ^{4}He. To do this, we assume the neutron star emits a total energy in neutrinos of 3×10^{53} ergs. We assume this energy is equally shared among the six types of neutrinos: $\nu_e, \bar{\nu}_e, \nu_\mu, \bar{\nu}_\mu, \nu_\tau, \bar{\nu}_\tau$. We also assume the neutrino luminosity falls exponentially on a three second timescale. We used the neutrino-alpha interaction cross section in Woosley et al. (1991) with the neutron and proton branching ratios appropriate for a neutrino temperature of 8 MeV.

We made four runs. Run 1 was a standard without the effects of neutrino-alpha interactions. Run 2 was the same as the first except that we included the neutrino-alpha interactions with neutrino temperatures $T_{\nu_e} = 4$ MeV, $T_{\bar{\nu}_e} = 6$ MeV, and $T_{\nu_x} = 8$ MeV. Here ν_x refers to the other four types of neutrinos. These neutrino temperatures are fairly realistic in light of detailed supernova models (Wilson and Mayle 1990; Woosley et al. 1994). Run 3 was identical to the second except that $T_{\nu_x} = 6$ MeV. Run 4 was identical to run 2 except that $T_{\nu_x} = 10$ MeV. For each run we computed the nucleosynthesis in 20 trajectories–specifically trajectories 4, 6, 8, 10, 12, 14, 16, 18, 20, 22, 24, 26, 28, 30, 32, 34, 36, 38, 39, and 40 of Woosley et al. (1994). We then mass-averaged over the results of these trajectories according to their contribution to the mass lost. The results were then scaled to solar at ^{129}Xe.

3. Results

The results of run 1 are shown in figure 1. We see that there is excellent agreement between the calculations and the solar r-process abundances from Käppeler et al. (1989), shown as points with error bars, above mass number A=120. We see the overproduction for N=50 nuclei (A≈90), as discussed previously.

Figure 1 also shows the results of run 2. For this run, the A=130 peak is well reproduced but the A=195 peak is reduced by roughly a factor of three relative to solar. This is at first a somewhat surprising result since neutrino-alpha spallations give off neutrons which should drive the r-process. Epstein et al. (1988) proposed that the r-process resulted from this mechanism well out in the helium shell, although this model has not proven successful. In the present case, because there are so many neutrons present throughout much of the r-process, the ^{3}He nucleus resulting from a neutrino spallation reaction immediately captures back a neutron. There is thus no effect on the r-process until the neutron abundance is so low that the r-process has essentially ceased.

The reason the neutrinos in fact hinder the r-process is that neutrinos also spall protons off of alpha particles. The resulting tritium nucleus immediately captures the proton back when the proton abundance is high. As the temperature falls to around $T_9 = 2$ in the r-process, protons become increasingly locked up into ^{4}He and heavier nuclei. This means that the free proton abundance decreases below 1/10,000 the abundance of the alpha particles. Once this occurs a tritium nucleus most likely captures one of the abundant ^{4}He nuclei to become ^{7}Li. ^{7}Li then captures another ^{4}He nucleus to become ^{11}B which can further capture alpha particles. In this way, neutrino spallation of protons from ^{4}He nuclei leads to additional assembly of alpha particles into heavy nuclei beyond the usual routes of $\alpha + \alpha + \alpha \rightarrow ^{12}C$ and $\alpha + \alpha + n \rightarrow ^9$Be followed by ^{9}Be$(\alpha, n)^{12}$C. More seed nuclei means there are fewer free neutrons per seed nucleus; thus, less r-process material moves to high nuclear mass number.

We note that the neutrino-alpha interactions do not solve the overproduction problem associated with N=50 nuclei. It is the early, lower entropy wind trajectories that overproduce the N=50 nuclei. Because of the low entropies, the abundance of ^{4}He nuclei at $T_9 = 2$ is already low, and the N=50 nuclei have already been made. Only for higher entropy trajectories, for which the ^{4}He abundance will be large near $T_9 = 2$, will the neutrino spallation from alpha particles be important. These are the trajectories that make the A=195 peak material.

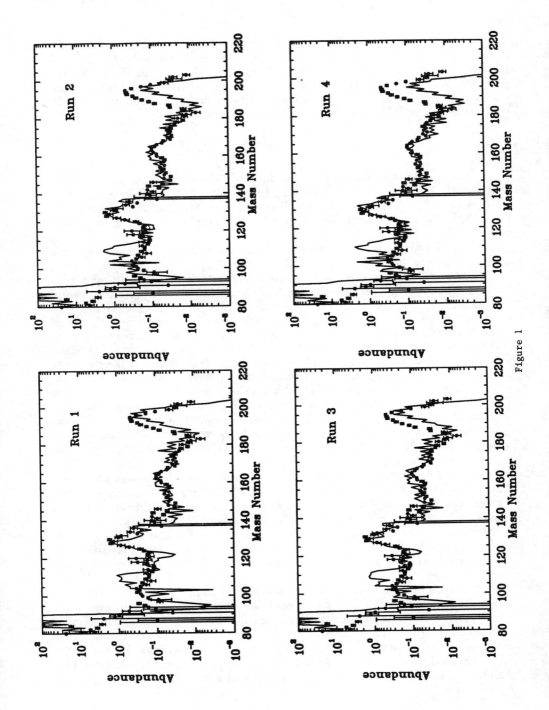

Figure 1

The spallation effect depends on the neutrino temperatures and luminosities because the neutrino-alpha cross section increases with neutrino temperature. Figure 1 shows the results for run 3, with $T_{\nu_x} = 6$ MeV, and for run 4 with $T_{\nu_x} = 10$ MeV. As expected, the effect of neutrino-alpha spallations is less dramatic for the lower temperature and more dramatic for the higher ν_x temperature. Of course it is not realistic simply to alter the neutrino temperatures because such a change will feed back into the wind dynamics and modify the thermodynamic conditions for the r-process. Nevertheless, runs 3 and 4 do give some indication of the effect of changing the neutrino temperatures.

4. Conclusions

The r-process in neutrino-driven winds from nascent neutron stars is rather sensitive to the temperature of the neutrinos. The present r-process models require $T_{\nu_x} \lesssim 6$ MeV in order to satisfactorily reproduce the A=195 r-process abundance peak. These conclusions will also be dependent on the neutrino luminosity and the wind dynamics in addition to the neutrino temperatures, and these will all be coupled together. We thus must continue to explore realistic r-process models that include neutrino interactions. For now we can say that neutrino-alpha spallation interactions need to be included in calculations of the r-process in neutrino-driven winds and that the r-process may give us important constraints on the temperatures of neutrinos from cooling nascent neutron stars and the velocity of the neutrino-driven wind.

REFERENCES

Caughlin G R and Fowler W A 1988 *Atomic Data and Nucl. Data Tables* **40** 283

Cowan J J, Thielemann F-K, and Truran J W 1991 *Phys. Rep.* **208** 267

Epstein R I, Colgate S A, and Haxton W C 1988 *Phys. Rev. Lett.* **61** 2038

Herant M, Benz W, and Colgate S 1992 *Ap. J.* **395** 642

Käppeler F, Beer H, and Wisshak K 1989 *Rep. Prog. Phys.* **52** 945

Klapdor, H V, Metzinger J, and Oda T 1984 *Atomic Data and Nuclear Data Tables* **31** 81

Meyer B S, Mathews G J, Howard W M, Woosley S E, and Hoffman R 1992 *Ap. J.* **399** 656

Qian Y-Z, Fuller G M, Mathews G J, Mayle R W, Wilson J R, and Woosley S E 1993 em Phys. Rev. Lett. **71** 1965

Takahashi K, Witti J, Janka H-Th 1994 *Astron. Astrophys* in press

Thielemann F-K, Arnould M, and Truran J W 1986 Advances in Nuclear Astrophysics (Gif-sur-Yvette: Editions Frontieres) p 525

Woosley S E, Fowler W A, Holmes J A, and Zimmerman B A 1975 Cal. Inst. of Tech., W. K. Kellogg Rad. Lab. Preprint No. OAP-422

Woosley S E, Hartmann D H, Hoffman R D, and Haxton W C 1990 *Ap. J.* **356** 272

Woosley S E and Hoffman R D 1992 *Ap. J.* **395** 202

Woosley S E, Wilson J R, Mathews G J, Hoffman R D, and Meyer B S 1994 *Ap. J.* in press

Zero Age Main Sequence Stars: Structure and Neutrino Emission

D. Hartmann, B. Meyer, D. Clayton, N. Luo, & T. Krishnan

Department of Physics & Astronomy
Clemson University, Clemson, SC 29634, USA

Clayton (1986) introduced an analytical ZAMS model that provided a good fit to the interior of the present Sun, despite assuming chemical homogeneity. We extend this solar ZAMS model to stars over the full mass range of the main sequence, $0.08 - 100$ $M_\odot$ and show that the difference between pp-chain and cno cycle causes two distinct branches of the interior main sequence.

We calculate the neutrino spectrum as a function of stellar mass and integrate over the mass spectrum to estimate the neutrino background due to hydrogen burning in the Milky Way and in external galaxies. Although many orders of magnitude below the solar flux, the Galactic stellar background exceeds the supernova background for neutrino energies below ~ 15 MeV. The extragalactic background is comparable or larger than the supernova background for neutrino energies below ~ 8 MeV.

STRUCTURE ON THE ZERO AGE MAIN SEQUENCE

The theory of stellar structure and evolution constitutes one of the foundations of modern astrophysics. Observationally, we are directly interested in the surface properties of stars, but important aspects of stellar astrophysics, such as energy generation, neutrino emission, and nucleosynthesis, depend on thermodynamic conditions in stellar interiors. To determine the internal structure of stars in hydrostatic equilibrium one solves a set of differential equations for radius, pressure, luminosity, and temperature. We consider only "static models" where all time derivatives are neglected. Boundary conditions have to be specified and one must provide the equation of state, $P(\rho, T, C)$, the nuclear energy generation rate, $\epsilon(\rho, T, C)$, and the opacity, $\kappa(\rho, T, C)$. All of these equations depend on the composition of stellar material, $C = (X,Y,Z)$. Technical procedures and details of the input physics for computer models are described in many textbooks. We restrict analysis to chemically homogeneous ZAMS stars. For $r \to 0$, hydrostatic equilibrium requires

$$\partial_r P(r) = -g(r)\rho(r) = -\frac{GM(r)}{r^2}\rho(r) \to -\frac{4\pi}{3} G \rho_c^2 r, \tag{1}$$

© 1995 American Institute of Physics

448 Zero Age Main Sequence Stars

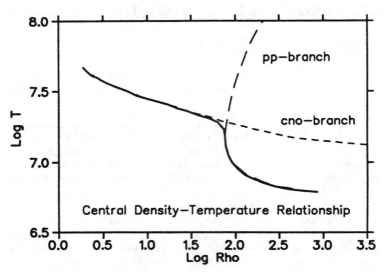

FIG. 1. Sequence of central temperature (K) versus central density (g cm^{-3}) for stellar masses between 0.08 and 100 M$_\odot$. The general trend is increasing temperature and decreasing density with increasing stellar mass. A break occurs near 1 M$_\odot$. The IMS has two branches, determined from either pp or cno burning.

where ρ_c is the central density. Both the pressure force and gravity approach zero at the center, as they must. Stability requires that the density monotonically declines with increasing radius, so that this linear law for the pressure gradient can not hold for all radii. Clayton (1986) made an exponential ansatz

$$\partial_r P(r) = -\frac{4\pi}{3} G \rho_c^2 \, r \, \exp\left\{-(r/a)^2\right\}, \quad (2)$$

where a is the single free parameter of the model. This ansatz leads to analytic forms for the pressure and density profiles. The temperature profile is then determined by specifying the equation of state, which here includes ideal gas pressure and radiation pressure.

The structure of ZAMS stars in hydrostatic equilibrium is fully determined by the above equations, except for the parameter $\eta = R/a$, which can be determined from

$$L = \int_0^M dm \, \epsilon = (4\pi R^3) \, \eta^3 \int_0^\eta dx \, x^2 \, \rho(x) \, \epsilon(x). \quad (3)$$

We do not use the radiation transport equation in constructing models, but global information on this aspect of stellar structure is preserved through the requirement that the model must yield the observed luminosity. Stellar luminosities and radii are fixed through observations, and we use a standard solar mix $C_\odot = (0.70, 0.28, 0.02)$ to calculate a mass sequence from 0.08 M$_\odot$

to 100 $M_\odot$. Figure 1 shows the interior main seqence (IMS), i.e., a plot of central temperature versus density. A break occurs near 1 $M_\odot$. To show that the IMS consists of two branches (pp and cno), we display solutions obtained from separately using the energy generation prescriptions for either pp or cno. The transition between upper and lower main sequence is essentially the result of the two distinct hydrogen burning modes. Note that the model equations do not contain information on the mode of energy transport.

THE ELECTRON NEUTRINO BACKGROUND

In neutrino light, as in the optical band, the Sun dominates over all terrestrial and cosmic background sources. The total solar neutrino flux at Earth is about 10^{11} cm^{-2} s^{-1} (e.g., Bahcall 1989) and Figure (2) shows the complete spectrum. The Sun has competition at high energies where the integrated neutrino flux from cosmic supernovae is estimated to yield a flux of ~ 1 cm^{-2} s^{-1} (e.g., Woosley, Wilson, & Mayle 1986). The temperature of supernova (electron) neutrinos is ~ 5 MeV which implies typical neutrino energies of ~ 15 MeV with substantial emission exceeding the solar cut-off, but cosmological expansion shifts the emission spectrum to lower energies. Krauss, Glashow, & Schramm (1984) and Woosley, Wilson, & Mayle (1986) consider the neutrino background from Type II and Ib supernovae. The rate per galaxy is $r_0 \sim 10^{-2}$ r_2 yr^{-1} and the local galaxy density is $n_0 \sim 10^7$ n_7 Gpc^{-3}. The differential neutrino flux from cosmological core collapses is given by

$$\frac{\partial F_\nu}{\partial E_\nu} = L_H\ n_0\ r_0 \int_0^{z_{max}} dz\ E(z)^{-1}\ \eta(z)\ \Phi(E_\nu(1+z))\ (\text{MeV cm}^2\ \text{s})^{-1}, \quad (4)$$

where z is the cosmological redshift, $L_H = c/H_0$ is the Hubble length, Φ is the time integrated neutrino spectrum of a typical core collapse supernova, and $E(z)$ is given by

$$E(z) = (1+z)\ (1+\Omega z)^{1/2}. \quad (5)$$

We assume a universe with zero cosmological constant and constant comoving supernova rate. Woosley, Wilson, & Mayle (1986) calculate antineutrino spectra for several stars ranging in mass from 10 $M_\odot$ to 5 10^5 $M_\odot$. If we consider stars less massive than $\sim 10^2$ $M_\odot$ their calculations show that the spectra do not depend sensitively on the initial main sequence mass. We approximate all core collapse spectra by a single blackbody (BB) spectrum. The total binding energy of a neutron star, $E = 10^{53}$ E_{53} ergs, is predominantly liberated through emission of neutrinos. The BB assumption implies electron neutrino temperatures of $T_\nu \sim 4-5$ MeV. The time integrated neutrino spectrum is

$$\Phi(E_\nu) = 1.6\ 10^{57}\ E_{53}\ T_\nu^{-4}\ E_\nu^2\ [\exp(E_\nu/T_\nu) - 1]^{-1}\ \text{neutrinos/MeV}. \quad (6)$$

The background spectrum due to supernovae is then

$$\frac{\partial F_\nu}{\partial E_\nu} \sim 1.6\, h_{100}^{-1}\, n_7\, r_2\, E_{53}\, T_\nu^{-4}\, E_\nu^2\, G\left(z_{\max}, E_\nu, T_\nu\right), \tag{7}$$

where h_{100} is the Hubble constant in units of 10^2 km s^{-1} Mpc^{-1} and the cosmological effects are contained in

$$G = \int_0^{z_{\max}} dz\, E(z)^{-1}\, (1+z)^2\, [\exp\left(E_\nu(1+z)/T_\nu\right) - 1]^{-1}. \tag{8}$$

The resulting background spectrum for $z_{\max} = 2$ and $\Omega = 1$ is shown in Figure 2. The total integrated flux for these parameters is ~ 1 cm^{-2} s^{-1}, very close to the value given by Woosley, Wilson, & Mayle (1986). Larger galaxy densities, supernova rates, and increased early galactic evolution could boost the flux (Krauss, Glashow, & Schramm 1984). Supernovae dominate over the solar spectrum only at energies higher than 15 MeV, but the total flux is many orders of magnitude below the solar peak value.

The simple stellar models discussed above allow us to calculate the electron neutrino spectrum as a function of stellar mass. We assume solar composition for all stars and include all neutrino producing reactions of the pp chain and the cno cycle (e.g. Bahcall 1989). Thermonuclear reaction rates are from Caughlan & Fowler (1988). The stellar spectrum follows from integrating neutrino production over the density-, temperature profile of the simple structure models. To determine the total Galactic background we assume a constant star formation rate and a universal IMF, $\zeta(m)$. The galactic neutrino luminosity is then

$$L_g(E_\nu) = 10^{11}\, N_{11} \int dm\, \zeta(m)\, \tau'(m)\, L_\nu(m), \tag{9}$$

where we adopt an arbitrary normalization of 10^{11} typical stars. Not all of the gas that went into stars contributes to the neutrino background at the present time. Only the fraction of stars ever formed that are still on the main sequence will in fact do so. The factor $\tau' = \min\{\tau(m), h_{100}^{-1}\}$ assures that only these stars are included. Stellar lifetimes, $\tau(m)$, are in units of 10^{10} yrs. While the stellar neutrino luminosity increases rapidly with stellar mass, the steep decrease with mass of IMF and stellar lifetime reduces the contribution of the upper main sequence. The neutrino flux at Earth then follows from a model of the stellar density in the Galaxy (we assume a uniform, thin disk with radius $R_g = 10$ kpc).

Supernovae produce large numbers ($\sim 10^{58}$) of electron neutrinos, but the event rate is only $\sim 10^{-2}$ yr^{-1} so that the total galactic neutrino production in a Hubble time is $\sim 10^{66}$. The solar neutrino luminosity is about 10^{38} s^{-1}, so that $\sim 10^{11}$ suns would also produce $\sim 10^{66}$ neutrinos on the same time scale. For a constant star formation rate and universal IMF, $\zeta(m)$, a typical galactic neutrino luminosity is

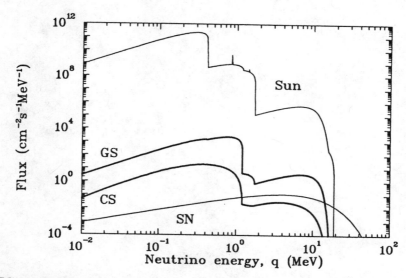

FIG. 2. Differential neutrino background fluxes from the Sun (top curve), supernovae (bottom curve), and stellar hydrogen burning in stars (middle curves). The Galactic stellar background exceeds that from the integrated cosmological stellar backgrond, and both exceed the supernova contribution for most energies.

$$L_g(z, E_\nu) = 10^{11} \, N_{11} \int dm \, \zeta(m) \, \tau'(m,z) \, L_\nu(m) \qquad (10)$$

As before, the factor τ' takes into account that only stars on the main sequence contribute to the neutrino luminosity, but the redshift-dependent lookback time is now included. The total cosmological background flux (Figure 2) then follows from an expression similar to equation (4). We integrate over galaxies within redshift $z_{max} = 2$. The background from stars in our Galaxy and in external galaxies is much smaller than the Solar flux, but exceeds the background from supernovae for a wide range of neutrino energies.

REFERENCES

1. Bahcall, J. N. 1989, *Neutrino Astrophysics*, (Cambridge Univ. Press)
2. Caughlan, G. R., & Fowler, W. A. 1988, *Atomic Data Nucl. Data Tables*, 40, 283
3. Clayton, D. D. 1986, Am J. Phys., 54, 354
4. Hartmann, D. H., Krishnan, T. D., & Clayton, D. D., 1994, ApJ, in preparation
5. Hartmann, D. H., Meyer, B. S., & Luo, N., 1994, ApJL, in preparation
6. Krauss, L. M., Glashow, S. L., & Schramm, D. N. 1984, Nature, 310, 191
7. Luo, N., Hartmann, D. H., & Meyer, B. S. 1994, in preparation
8. Woosley, S. E., Wilson, J. R., & Mayle, R. 1986, ApJ, 302, 19

ONeMg Novae :
Nuclear Uncertainties
on the ^{26}Al and ^{22}Na Yields

Alain Coc(1), Robert Mochkovitch(2), Yvette Oberto(2),
Jean-Pierre Thibaud(1) and Elisabeth Vangioni-Flam(2)

(1) CSNSM, F-91905 Orsay Campus (France) and
(2) IAP, 75 Bd. Arago, F-75014 Paris (France)

> The influence of the nuclear uncertainties on the reaction rates involved in ^{26}Al and ^{22}Na production in ONeMg novae are discussed. A semi-analytical model of novae envelope is used with an improved calculation of the time evolution. Comparison is made with the results of hydrodynamical models.

INTRODUCTION

Recent hydrodynamical calculations (1) show that a fraction of the ^{26}Al observed in the galaxy by its γ-ray emission may be produced by ONeMg novae and that γ's from ^{22}Na could be observed in the case of a nearby nova. However, some of the reaction rates involved in the formation and destruction of these isotopes are not well known at low energy. The uncertainty can reach several orders of magnitude for proton capture on radioactive isotopes. We studied these rates, estimated their uncertainties and calculated their influence on ^{26}Al and ^{22}Na yields (2).

THE MODEL

Our simulation of novae outburst is based on McDonald's semi-analytical model (3). Spherical symmetry is assumed for the inert white dwarf core onto which lies an expanding envelope of hydrogen rich material. The model provides temperature and density profile for successive steps of the outburst and the time evolution is given by energy conservation. We improved the calculation of time evolution by coupling the model with a network of nuclear reactions so that energy generation is obtained by a detailed nucleosynthesis calculation. For comparison, we used the initial parameters of the three hydrodynamical models of Starrfield et al. (4) and Politano et al. (1), representing ONeMg models of 1., 1.25 and 1.35 M$_\odot$; the initial abundances are 50% solar, and 50% ^{16}O,^{20}Ne, ^{24}Mg in the ratio 3:5:2. The temperature profiles and with the improved energetics the time scale of the outburst (2) show a reasonable agreement with the hydrodynamical calculations.

The nuclear network we used extends up to potassium. The rates are derived from the most recent nuclear data when available, otherwise we adopted the rates from Caughlam & Fowler (5) compilation (CF88) or for the heavier elements the SMOKER theoretical calculations. The rates are averaged over the envelope and the abundances are treated as uniform (fast convection). The main nuclear uncertainties come from reactions involving radioactive nuclei; in particular below 10^8 K, ^{26}Al(p,γ)^{27}Si and ^{22}Na(p,γ)^{23}Mg suffers from large uncertainties. (The peak temperature at the base of the envelope is much higher, but convection bring matter to cooler regions.) Even at higher temperatures, ^{25}Al(p,γ)^{26}Si remains poorly known. For these rates, we obtained upper and lower limit from the available experimental data or calculated theoretical limits (2).

^{26}Al YIELD

Since CF88, new measurements (6) (7) have improved the spectroscopic data for ^{27}Si states between the ^{26}Al + p threshold and the known resonance at E_p=0.277 MeV. From these data, we deduced a rate which differs from CF88 by the contributions of the resonances at E_p=68 keV, 93 keV, 188 keV and 277 keV. For the 68 keV and 93 keV ones, we incorporated factors '0 to 1' to their resonant contribution to account for the unknown spectroscopic factors, resulting in a High and Low limit for the rate. Table 1 represents the ^{26}Al yield for the three novae models calculated with different reaction rates. When CF88 rate for ^{26}Al(p,γ)^{27}Si is used, our results agree well with those obtained from hydrodynamical calculations (1).

TABLE 1. ^{26}Al yields for different reaction rates

Rate / Mass	1.0 M$_\odot$	1.25 M$_\odot$	1.35 M$_\odot$
Taken from ref. (1)	1.96×10^{-2}	9.39×10^{-3}	7.39×10^{-3}
^{26}Al(p,γ)=CF88	2.44×10^{-2}	9.37×10^{-3}	1.24×10^{-2}
^{26}Al(p,γ)=High from ref. (2)	1.07×10^{-2}	5.38×10^{-3}	8.11×10^{-3}
^{26}Al(p,γ)=Low from ref. (2)	1.11×10^{-2}	5.51×10^{-3}	8.2×10^{-3}
ω_γ(188 keV)$\times$3	3.46×10^{-3}	2.0×10^{-3}	3.43×10^{-3}
ω_γ(188 keV)$\times$1/3	2.08×10^{-2}	9.09×10^{-3}	1.26×10^{-2}
^{25}Al(p,γ) from ref. (8) $\times10^2$	6.01×10^{-3}	1.95×10^{-3}	3.52×10^{-3}

However, when using the updated rates (High or Low limit), the ^{26}Al yields are reduced by the same factor $\approx$2. This means that the 68 keV and 93 keV resonances uncertainties have no effect on ^{26}Al yield. The difference with the results obtained with CF88 can be traced to the introduction of the 188 keV resonance whose strength is taken from the only available (but yet unpublished) measurement (7). The sensitivity of the ^{26}Al yield to the strength of this resonance can be seen in the table when arbitrarily scaling its value by a factor of three.

The link ^{24}Mg$(p,\gamma)^{25}$Al$(p,\gamma)^{26}$Si$(\beta)^{26m}$Al$(\beta)^{26}$Mg goes through the ^{26}Al isomeric state, bypasses the long-lived $^{26g.s.}$Al and therefore does not contribute to ^{26}Al production. As a consequence of the scarcity of spectroscopic informations on ^{26}Si, they have to be supplemented by data on the mirror nucleus ^{26}Mg to obtain the ^{25}Al$(p,\gamma)^{26}$Si rate. Even the location of some levels in ^{26}Si is not known, thus leading to large uncertainties over a wide temperature interval for this rate. For instance, the effect of multiplying by a factor 100 the rate proposed by Wiescher et al. (8) reduces the ^{26}Al yield by a factor ≈ 2 as shown in the table, but the use of a lower rate lower does not increase the yield.

^{22}Na YIELD

Even though, large uncertainties remain for $T_8 \lesssim 2$, the data on ^{22}Na$(p,\gamma)^{23}$Mg at low energy has been improved by a new measurement (9). The results are summarized in figure 1 where the left part represents ^{22}Na yields as a function of white dwarf mass from ref. (1) (circles) and the ones we obtained (triangles) with CF88 and the upper and lower limits from ref. (9).

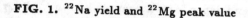

FIG. 1. ^{22}Na yield and ^{22}Mg peak value

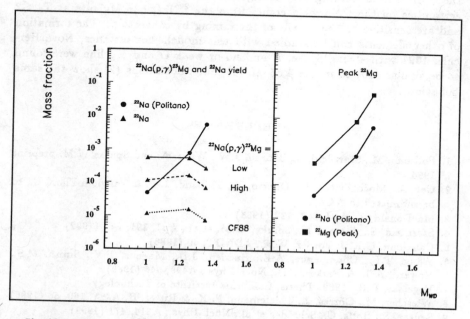

Significantly more ^{22}Na is produced with the new rate than with CF88

reflecting the fact that it is at least one order of magnitude lower than CF88 for $T_8 \gtrsim 1$. However, the ^{22}Na yields we obtained are low are approximately constant with respect to mass in contrast to hydrodynamical results. This difference can be explained if a fraction of the ^{22}Mg produced during the peak of the outburst is saved in the outer and cooler regions of the envelope. Due to the low Q value of the ^{22}Mg$(p,\gamma)^{23}$Al reaction, rapid photodisintegration of ^{23}Al block the nuclear flow resulting in a large ^{22}Mg abundance around the climax of the outburst. The right part of the figure shows the ^{22}Mg peak mass fraction (squares) we obtained, compared with the ^{22}Na yields from ref (1) (circles). They show the same evolution as a function of mass and thus their high ^{22}Na yields can be reproduced if we assume that 10% of the ^{22}Mg is saved when convection retreats and subsequently decays to ^{22}Na.

Unlike ^{26}Al, ^{22}Na can be produced by two ways starting from ^{20}Ne via ^{21}Ne$(p,\gamma)^{22}$Na or via ^{22}Mg$(\beta)^{22}$Na. Consequently, the still uncertain ^{21}Ne$(p,\gamma)^{22}$Na rate has little influence on the final ^{22}Na yield.

CONCLUSION

McDonald's model coupled with a nuclear network can be used to investigate the influence of nuclear reactions on novae yields. The ^{26}Al yield of ONeMg novae is strongly sensitive to the strength of the $E_p=188$ keV resonance in ^{26}Al+p and depends on the poorly known ^{25}Al$(p,\gamma)^{26}$Si rate. Hence, the ^{26}Al yield in ONeMg may be significantly lower than expected. The uncertainties on the ^{22}Na yield comes from the ^{22}Na$(p,\gamma)^{23}$Mg rate at $T_8 \lesssim 2$ and are sensitive to the details of the mixing by convection. The formation of other elements can be studied with this model. For instance, Nova Herculis 1991 with strong S, Ne, N and no or weak O and Mg line were found to be similar (10) to model 3 of ref. (1); our calculations (2) show the same behaviour.

REFERENCES

1. Politano, M., Starrfield, S., Truran, J.W., Weiss, A. and Sparks W.M. preprint 1994
2. Coc, A., Mochkovitch, R., Oberto, Y., Thibaud, J.-P. & Vangioni-Flam, E., to be submitted to A&A
3. MacDonald, J., ApJ, 267, 732 (1983)
4. Starrfield, S., Shore, S.N., Sparks, W.M. et al., ApJ, 391, L71 (1992)
5. Caughlan, G.R. & Fowler, W.A., ADNDT, 40 (1988)
6. Wang, T.F., Champagne, A.E., Hadden, J.D., Magnus, P.V., Smith, M.S., Howard, A.J., & Parker, P.D., Nucl. Phys., A499, 546 (1989)
7. Vogelaar, R.B., 1989, Thesis, California Institute of Technology
8. Wiescher, M., Görres, J., Thielemann, F.-K. & Ritter, H. A&A, 160, 56 (1986)
9. Seuthe, S., Rolfs, C., Schröder et al., Nucl. Phys., A514, 471 (1993)
10. Matheson, T., Filippenko, A.V. & Ho, L.C., ApJ, 418, L29 (1993)

^{26}Al and ^{22}Na Production in Neon Novae

Shinya Wanajoh[1], Masa-aki Hashimoto[2] and Ken'ichi Nomoto[1]

(1) *Department of Astronomy, School of Science, University of Tokyo*
7-3-1 Hongo, Bunkyo-ku, Tokyo 113
(2) *Department of Physics, College of General Education, Kyushu University*
4-2-1 Ropponmatsu, Chyuo-ku, Fukuoka 810

Abstract. We have calculated nucleosynthesis during neon nova explosions over wide ranges of the white dwarfs and envelope masses. As the results, we found considerably high mass ratio of ^{26}Al can be produced in some nova events with higher masses of the white dwarfs and the envelopes. In addition, such novae produce substantial amount of ^{22}Na, which has been expected to be detected by *CGRO*, as well as the clue of the Ne-E problem.

1 Introduction

It is commonly accepted that classical nova outbursts are triggered by thermonuclear runaways in accreted hydrogen rich shells on white dwarf components of close binary systems. Recent observations have indicated that about a quarter of events are neon novae, explosions on the surface of ONeMg white dwarfs. Neon novae have been suggested to be promising site of ^{26}Al and ^{22}Na production because it is easy to synthesize them in the hydrogen rich gas enhanced with neon and magnesium.

Radioactive isotope ^{26}Al, decaying to ^{26}Mg with gamma-ray line emissions in $\sim$ a million years, was found to have existed as much as $3M_\odot$ in our Galaxy by recent gamma-ray observations (Mahoney *et al.* 1984). Though neon novae, supernovae, Wolf-Rayet stars, and AGB stars have been suggested to be the origins of ^{26}Al (Starrfield *et al.* 1993), no consensus has been reached.

© 1995 American Institute of Physics

Another radioactive isotope ^{22}Na, decaying to ^{22}Ne with gamma ray line emissions in ~ 3.75 years, has been considered as the origin of the neon isotopic anomalies in some meteorites, so-called Ne-E problem (^{20}Ne/^{22}Ne ≈ 0.01, as low as a thousand times of the terrestrial ratio!!). However, there has been no clear evidence of the production of enough ^{22}Na in theoretical models or observations.

The aim of this paper is to report briefly the results of our study of the production of ^{26}Al and ^{22}Na in neon novae. These calculations have been performed using a nuclear reaction network coupled to a quasi-analytical nova model.

2 Numerical Results

We have calculated the nuclear abundance evolution in neon novae, applying the quasi-analytical model of hydrogen shell flashes on white dwarfs (Fujimoto 1982). The model predicts temperature and density histories of the hydrogen-rich envelope during a nova outburst for only two parameters, the mass of the white dwarf (M) and the envelope (ΔM). We selected the range of the parameters as $M = 1.1$–$1.35\ M_\odot$, $\Delta M = 1 \times 10^{-6} - 1 \times 10^{-4} M_\odot$, predicted in previous works (Fujimoto 1982, Nomoto 1984). The reaction network we used here contains 471 isotopes from hydrogen to krypton, enough to study nucleosynthesis in neon novae. The initial abundance is enriched with oxygen, neon, and magnesium to a mass fraction 0.25, assuming that these elements are mixed from the surface of an ONeMg white dwarf into the envelope of solar abundance.

The results show that the novae involving the most massive ONeMg white dwarfs and envelopes can produce significant amount of ^{26}Al and ^{22}Na (Table 1,2). That can be explained as follows. Below 3×10^8 K, ^{20}Ne is fairly stable, thus ^{22}Na can not produced at all. However, ^{24}Mg burns easily over 1×10^8 K, hence ^{26}Al is produced in the Mg-Al cycle by the reaction ^{25}Mg$(p,\gamma)^{26}$Al. On the contrary, ^{20}Ne starts burning finally when temperature exceeds 3×10^8 K. Most of ^{20}Ne is converted to ^{22}Mg by two proton capture reactions. At the same time ^{25}Al becomes to burn to ^{26}Si quickly. This neon burning releases a large quantity of nuclear energy so that progress of the explosion get faster. The gas reaches to the maximum temperature in several seconds after the explosive hydrogen burning is driven. Furthermore, it exhausts most of ^{20}Ne immediately. Before most of ^{22}Mg and ^{26}Si decay to ^{22}Na and ^{26}Al, therefore, the main reactions get frozen. Thus large amount of ^{22}Na and ^{26}Al can remain without decaying by proton capture reactions in novae with high temperature gases, in other words, with massive white dwarfs and envelopes.

3 Discussions and Conclusions

The calculations presented here imply that neon novae contribute substantially to the production of ^{26}Al observed in our Galaxy. We estimated the amount of ^{26}Al from neon novae as $M(^{26}\text{Al}) = \tau(^{26}\text{Al})X(^{26}\text{Al})\Delta M R_{nova} f_{\text{ONeMg}}$, where we apply the mean life time of ^{26}Al $\tau(^{26}\text{Al}) \approx 10^6$ yr, a mass fraction of ^{26}Al $X(^{26}\text{Al})$, frequency of nova events in our Galaxy $R_{nova} \approx 100$/yr, the ratio of the appearances of neon novae against all events $f_{\text{ONeMg}} \approx 0.25$, assuming that all nova systems have the same masses of white dwarfs and envelopes.

We found some results are in good agreement with the observations (Table 3). However, it is difficult to show that all ^{26}Al have been produced in novae, taking into account that most of neon nova systems involve white dwarfs less massive than $1.2 M_\odot$. Furthermore, gamma-ray flux of decaying ^{26}Al from a single neon nova is $\sim 10^{-8}$. That indicates the nova must exist in less than 100 pc even in the best case. It implies that novae should be excluded as the origins of clumpy distributed ^{26}Al around the center of our Galaxy.

We also estimated the gamma-ray line flux of decaying ^{22}Na from neon novae. For the neon novae involving white dwarfs as massive as $1.3 M_\odot$, almost of all ^{20}Ne contained initially burns into ^{22}Na, because of high temperature in the gases. Hence the mass fraction of ^{22}Na in the ejected gases reaches to $0.01-0.1$. If the neon novae occur within a few kiloparsec, they can produce gamma-ray line flux detectable by $CGRO$.

At the same time, such kind of novae can erupt the gases with the ratio of ^{20}Ne/^{22}Na ≈ 0.01 (Table 4). That imply the neon novae involving massive white dwarfs can be the origins of the Ne-E problem if the gases from them form dust immediately, within a few years before ^{22}Na decays to ^{22}Ne.

The mass of the white dwarf in Nova Cygni 1992 is estimated $\sim 1.3 M_\odot$, and the ejected mass $\sim 1 \times 10^{-4} M_\odot$, according to the report by Quirrenbach et al.(1993). Further observations and analyses of the nova are desired to get more information.

References

Fujimoto, M.Y. 1982, ApJ, 257, 752

Mahoney, W.A., Ling, J.C., Wheaton, W.A., & Jacobson, A.S. 1984, ApJ, 286, 578

Nomoto, K. 1984, ApJ, 277, 291

Quirrenbach, A., *et al.* 1993, AJ, 106, 1118

Starrfield, S., Truran, J.W., Politano, M., Sparks, W.M., Nofar, I., & Shaviv, G., 1993, Phys.Rep., 227, 223

TABLE 1. Mass fraction of ^{26}Al

$\Delta M/M_\odot$	$M/M_\odot$					
	1.1	1.15	1.2	1.25	1.3	1.35
1(-6)	2.7(-4)	4.8(-4)	7.2(-4)	9.5(-4)	2.5(-3)	7.8(-4)
5(-6)	7.9(-4)	1.2(-3)	3.0(-3)	1.3(-3)	2.8(-4)	4.5(-4)
1(-5)	1.3(-3)	3.1(-3)	1.5(-3)	2.4(-4)	3.8(-4)	1.7(-3)
5(-5)	2.5(-4)	2.3(-4)	3.0(-4)	1.9(-3)	5.7(-3)	2.1(-2)
1(-4)	2.1(-4)	2.7(-4)	1.3(-3)	5.5(-3)	1.5(-2)	2.4(-2)

TABLE 2. Mass fraction of ^{22}Na

$\Delta M/M_\odot$	$M/M_\odot$					
	1.1	1.15	1.2	1.25	1.3	1.35
1(-6)	8.4(-4)	9.1(-4)	9.9(-4)	1.1(-3)	1.1(-3)	5.0(-4)
5(-6)	1.0(-3)	1.1(-3)	1.1(-3)	5.5(-4)	5.0(-4)	3.7(-3)
1(-5)	1.1(-3)	1.1(-3)	5.8(-4)	5.0(-4)	2.3(-3)	2.0(-2)
5(-5)	5.0(-4)	8.2(-4)	4.0(-3)	1.7(-2)	5.1(-2)	6.8(-2)
1(-4)	1.0(-3)	4.6(-3)	1.5(-2)	4.1(-2)	7.3(-2)	6.4(-2)

TABLE 3. Total mass of ^{26}Al from neon novae in our Galaxy

$\Delta M/M_\odot$	$M/M_\odot$					
	1.1	1.15	1.2	1.25	1.3	1.35
1(-6)	6.7(-3)	1.2(-2)	1.8(-2)	2.4(-2)	6.3(-2)	2.0(-2)
5(-6)	9.9(-2)	0.15	0.38	0.16	3.5(-2)	5.6(-2)
1(-5)	0.33	0.79	0.38	6.0(-2)	9.6(-2)	0.44
5(-5)	0.31	0.29	0.38	2.3	7.2	27
1(-4)	0.52	0.67	3.2	14	38	60

TABLE 4. Ratio of ^{20}Ne/^{22}Na

$\Delta M/M_\odot$	$M/M_\odot$					
	1.1	1.15	1.2	1.25	1.3	1.35
1(-6)	100	93	86	80	81	172
5(-6)	83	78	80	156	169	19
1(-5)	78	81	15	170	34	1.4
5(-5)	173	101	19	3.8	2.0	4.8(-4)
1(-4)	80	17	0.29	0.97	1.1(-2)	7.6(-5)

Rare Neutron-rich Nucleosynthesis in Type IA Supernovae

S. E. Woosley(1,2), Thomas A. Weaver(2), and R. D. Hoffman(1)

(1) Department of Astronomy, University of California, Santa Cruz and (2) General Studies Division, Lawrence Livermore National Laboratory

Abstract. We find nucleosynthetic evidence for at least the occasional explosion of white dwarfs near the Chadrasekhar mass. This evidence is the solar abundances of at least four rare isotopes that seemingly are produced nowhere else in nature except the high density, low entropy regions characteristic of the cores of Type Ia deflagration models. However, because the abundances of these isotopes, ^{48}Ca, ^{50}Ti, ^{54}Cr, and ^{66}Zn, is very low in nature, the frequency of such explosions is severely restricted, perhaps no more than 10% of the observed Type Ia supernova rate. Amounts of ^{60}Fe of interest to gamma-ray astronomy are also synthesized in these same explosions.

Key words: Supernovae — nucleosynthesis

1. TYPE Ia SUPERNOVAE

1.1. The Model

We assume here a standard model of a carbon-oxygen white dwarf accreting at the proper rate for a sufficiently long time to grow to nearly the Chandrasekhar mass (1.39 M$_\odot$), at which point it ignites carbon burning at or near its center. Most relevant is the central density at the time a convective carbon runaway develops, typically 3×10^9 g cm^{-3} (e.g., Woosley 1990). We assume, as do many, that the flame becomes localized to near the center of the white dwarf, and commences as if ignited at a single point.

Once born, the flame initially propagates as a conductive burning front of infinitesimal thickness and uncertain surface geometry. The laminar flame speed is now known following the work of Timmes & Woosley (1992). The flame is Rayleigh-Taylor unstable and becomes deformed. Its effective velocity in a spherical remap of the flame sheet structure is given by $v_{eff} = (Actual\ Area/4\pi r_b^2)v_{cond}$ where r_b is the approximate radius of the burned out region (a spherical volume equivalent to the actual deformed, and maybe even disjoint, burned out region(s)). This "actual area" can also be expressed (Woosley 1990; Timmes & Woosley 1992) as $4\pi r_b^2 (\lambda_{max}/\lambda_{min})^{D-2}$ where λ_{min} and λ_{max} are the minimum and maximum possible wavelengths for the Rayleigh-Taylor instability and D is the *fractal dimension* of the surface.

Timmes (1994) suggests, on physical grounds, a time dependent prescription for the fractal dimension. Timmes, for example, gives

$$D(t) = \frac{3D_{max}(1 - (v_{cond}/v_{eff})) + 2(v_{cond}/v_{eff})}{3 - 2(v_{cond}/v_{eff})}$$

and suggests a maximum value $D_{max} = 8/3$, though this is very uncertain. Garcia uses $D_{max} = 2.5$ and Peters & Frank (1990) suggest 7/3. If nothing else this equation has the desired effect of taking D initially equal to 2 and raising it smoothly to an asymptotic value.

The minimum wavelength is taken to be its standard value

$$\lambda_{min} = \frac{4\pi f_\lambda v_{cond}^2}{g_{eff}}$$

with $g_{eff} = g(r)(\Delta\rho/\rho) \approx (4/3)\pi G r \rho (\Delta\rho/\rho)$ and the maximum wavelength, equal to the radius of the equivalent burned out volume. A dimensionless factor f_λ has been inserted in the definition of λ_{min} here and will be varied in the study. Recent numerical experiments by Khokhlov (1993) suggest $f_\lambda \gg 1$.

When the range of unstable wavelengths is reduced (f_λ greater than unity) for a given value of fractal dimension (D = 2.66) much more intermediate mass elements can be made. The results are given in Table 1 and Fig. 1. The flame moves more slowly at first (for a given D_{max}) and then accelerates rapidly, provoking a detonation in some cases well after the density has gone down a great deal. It is interesting that the values that give both substantial intermediate mass elements and ^{56}Ni have $f_\lambda \sim 50 - 100$, about the same as inferred in the numerical experiments of Khokhlov (1993).

We have explored in detail the nucleosynthesis for one representative case in Fig. 1, $f_\lambda = 75$ and varied the central density. For comparison one model with $f_\lambda = 5$ was also computed.

TABLE 1: Variation of λ_{min} At D = 2.66

f_λ	Ni Mass ($M_\odot$)	Si Mass ($M_\odot$)	KE_∞ (10^{51} erg)
1	0.96	-	1.74
5	1.09	0.01	1.56
25	1.09	0.05	1.53
50	0.64	0.30	1.31
62.5	0.40	0.34	0.95
75	0.34	0.17	0.53
100	0.28	0.12	0.34

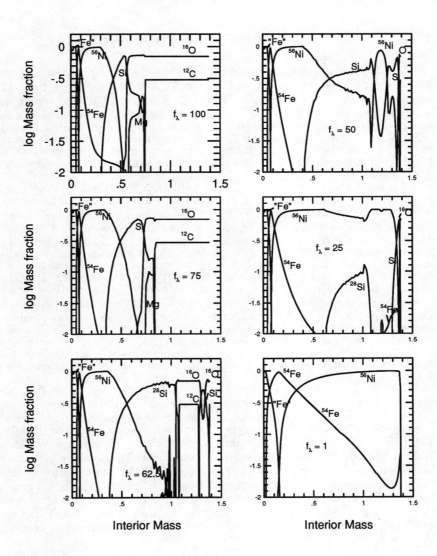

Fig. 1.— Final composition of models that used a maximum fractal dimension D = 2.66 and multiplied the minimum wavelength scale by the indicated factor (see also Table 1).

1.2. Neutron-Rich Nucleosynthesis

The flame in the standard model is born at high density, $\sim 3 \times 10^9$ g cm^{-3}, perhaps more if the chilling effect of the URCA process is included (Iben, 1982, found $\gtrsim 4 \times 10^9$ g cm^{-3}, see also Barkat & Wheeler, 1990) and moves quite slowly at first. It may take a second or so before a few hundredths of a solar mass is burned and the white dwarf expands appreciably. During that time there is a lot of electron capture and the nucleosynthesis that emerges from nuclear statistical equilibrium in the inner regions must be very neutron rich. It is here that we will find a potential site for the synthesis of the rare neutron-rich isotopes ^{48}Ca, ^{50}Ti, and ^{54}Cr.

For example, the model in Table 1 with D = 2.666 and $f_\lambda = 75$ reaches a central value of Y_e of 0.423. In a similar model in which ignition occurs at 4.8×10^9 g cm^{-3} instead of 3.1×10^9 g cm^{-3} (caused here by simply lowering the accretion rate of carbon and oxygen), the central value of Y_e is farther decreased to 0.413. However, the final value of Y_e in the central regions of the two models was 0.434 and 0.424 respectively showing that Y_e actually *increased* as the explosion developed. Beta-equilibrium is approached in the inner regions of these supernovae and the shift to lower densities allows, to a small extent, a shift back towards neutron-proton equality. A third run also used ignition density of 4.8×10^9 g cm^{-3} but a smaller value of $f_\lambda = 5$ (see Fig. 1 and Table 1) in order to cause a faster flame speed. This resulted in the production of more iron, but interestingly a *smaller* central Y_e than in the model having $f_\lambda = 75$. The lowest value reached, shortly after ignition was 0.410; the final value 0.413. The faster expansion allowed less time for neutron rich material to shift back towards neutron-proton equality as the density went down.

The final abundances of some key isotopes are given in Table 2 for these three runs and are fascinating. First, the aficionado will recognize that ^{48}Ca, ^{50}Ti, and ^{54}Cr are three species whose nucleosynthesis has hitherto been mysterious. They are not made in Type II supernovae or in standard models for Type Ia (e.g., Nomoto, Thielemann, & Yokoi 1984; though see Woosley & Weaver 1986b). The need for neutron-rich nuclear statistical equilibrium has been recognized (Hartmann, Woosley, & El Eid 1985), but the site not determined. It seems clear that the site is the cores of (some) Type Ia supernovae. Incidentally, in order to track this nucleosynthesis very fine central zoning (5×10^{-5} M$_\odot$) and a large nuclear reaction network (236 isotopes) had to be employed. The network included all the necessary weak and strong couplings, at least for nuclei up to mass A = 60.

So it is a triumph to make these nuclei, especially in solar proportions in the higher density models (might beta equilibrium lead all models that ignite above 4×10^9 g cm^{-3} to converge on these abundances?). But it may be too much of a good thing.

These supernovae all make about 30 - 100 times too much ^{54}Cr, for example, compared to ^{56}Fe. That is they could only make a few percent of solar iron. Modern chemical evolution studies (eg., Timmes, Woosley, & Weaver 1994) suggest that Type Ia supernovae need to make about one-half to one-third of the solar iron. This suggests that the common Type Ia event may not be a Chandrasekhar mass white dwarf igniting a subsonic flame. Maybe only a few percent are and the rest are sub-Chandrasekhar mass models. Similar

TABLE 2: Nucleosynthesis of Rare Isotopes

Nucleus	$\rho_{c9} = 3.1$ $f_\lambda = 75$ ($M_\odot$)	Rel. Fe56	$\rho_{c9} = 4.7$ $f_\lambda = 75$ ($M_\odot$)	Rel. Fe56	$\rho_{c9} = 4.7$ $f_\lambda = 5$ ($M_\odot$)	Rel. Fe56
^{48}Ca	2.5(-4)	5.2	4.6(-3)	93	4.3(-3)	37
^{50}Ti	4.6(-3)	85	1.1(-2)	190	5.8(-3)	41
^{51}V	1.5(-4)	1.4	3.7(-4)	3.2	3.2(-4)	1.1
^{54}Cr	9.6(-3)	70	1.4(-2)	97	1.1(-2)	31
^{54}Fe	2.5(-2)	1.0	2.7(-2)	1.1	8.7(-2)	1.4
^{56}Fe	3.9(-1)	1.0	4.1(-1)	1.0	1.02	1.0
^{58}Fe	1.0(-2)	6.9	1.2(-2)	7.8	1.5(-2)	3.9
^{60}Fe	2.2(-3)	-	6.1(-3)	-	3.2(-3)	-
^{58}Ni	2.5(-2)	1.6	2.6(-2)	1.6	1.1(-1)	2.7
^{66}Zn	1.0(-3)	5.1	6.5(-3)	32	4.0(-3)	7.8

restrictions result from considering the mass of ^{60}Fe in Table 2 and the fact that gamma-ray astronomers have not yet seen a signal from this long lived radioactivity. We note, however, that ^{60}Fe is co-produced with ^{26}Al in Type II supernovae (Hoffman et al. 1994), and therefore detection of ^{60}Fe without an accompanying signal from the more abundant (in Type II's) ^{26}Al could provide a convincing diagnostic of the Type Ia event we describe. The abundance of ^{60}Fe in the interstellar medium, replenished about every 2 million years, cannot exceed about 5 solar masses.

At a minimum these results suggest that nucleosynthesis must be considered very carefully when crafting the "standard" Type Ia model. Perhaps a way can be found to make the ignition density substantially lower than 3×10^9 g cm^{-3} (e.g., Aparicio & Isern 1993); cause the ignition to occur substantially off center; or make the flame move much faster through the first few hundredths of a solar mass (multiple ignition points?).

This work has been supported by the National Science Foundation (AST 91-15367), NASA (NAGW-2525), the Department of Energy (W-7405-ENG-48), and the Institute for Geophysics and Planetary Physics at Lawrence Livermore National Laboratory (IGPP 92-94).

REFERENCES

Aparicio, J. M., & Isern, J. 1993, A&A, 5, 280

Barkat, Z., & Wheeler, J. C. 1990, ApJ, 355, 602

Hartmann, D. H., Woosley, S. E. & El Eid, M. 1985, ApJ, 297, 837

Hoffman, R. D., Woosley, S. E., Weaver, T. A., Timmes, F. X., Eastman, R. G., and Hartmann, D. H. 1994, in The Gamma-Ray Sky with CGRO and Sigma, ed. M. Signore, P. Salati, & G. Vedrenne, (Kluwer Acad. Pub.: Dordrect), in press

Iben, I., Jr. 1982, ApJ, 253, 248

Khokhlov, A. 1993 ApJ, 419, 77

Nomoto, K., Thielemann, F. K., & Yokoi, K. 1984, ApJ, 286, 644

Peters, N. & Frank, Ch. 1990 Dissipative Structures in Transport Processes and Combustion, D. Meinkohn ed., Berlin, Springer Verlag, 1990, 40

Timmes, F. X. 1994, ApJ, in press

Timmes, F. X., & Woosley, S. E. 1992, ApJ, 396, 649

Timmes, F. X., Woosley, S. E., & Weaver, T. A. 1994, ApJS, in press

Woosley, S. E. 1990, Supernovae, A. G. Petschek ed., Springer Verlag, New York, 1990, 182

Woosley, S. E., & Weaver, T. A. 1986, Radiation Hydrodynamics in Stars and Compact Objects, D. Mihalas and K-H. A. Winkler eds., Springer Verlag, Berlin, 1986, 91

Hot Bottom Burning in Red Giants

Robert C. Cannon[2,3], Cheryl A. Frost[1,3], John C. Lattanzio[1,3], and Peter R. Wood[4]

[1] *Department of Mathematics, Monash University,
Clayton, Victoria 3168, Australia*
[2] *Observatoire de Paris Meudon, 92195, Meudon France*
[3] *Institute of Astronomy, Madingley Road, Cambridge, UK*
[4] *Mount Stromlo and Siding Spring Observatories,
The Australian National University,
Private Bag, Weston Creek P.O., A.C.T.2611, Australia*

Abstract. We present results of hot-bottom burning calculations for a Population II $5M_\odot$ star using OPAL opacities, and a post-processing code to compute detailed nucleosynthesis with time-dependent convective mixing. In particular we look at the abundances of Li, Al and CNO isotopes.

INTRODUCTION

Over the past few years, evidence has been accumulating for hot-bottom burning (HBB) in red-giants. A number of AGB stars have been identified in the LMC and SMC (1,2) which are enhanced in both lithium and s-process elements, the latter indicating third dredge-up. None were carbon stars. Further observations of four Li-rich SMC stars (3) have shown that these have a low (C^{12}/C^{13}) ratio of about 7±3. It has been suggested that the low ratio is due to C^{12} depletion as a result of HBB, and the high Li abundance ($\log\epsilon(Li^7)\sim 3$ for the SMC), to the beryllium-transport mechanism proposed by Cameron and Fowler (4). This requires temperatures in excess of 50×10^6 K.

Other clues to nuclear processes in AGB stars come from isotopic abundances in meteorites. Zinner et. al. (5) found (Al^{26}/Al^{27}) ratios of up to 0·06 in graphite and 0·2 in SiC grains, and Amari et. al. (6) found that SiC grains with lower (C^{12}/C^{13}) ratios ($3<(C^{12}/C^{13})<20$) tend to have higher (Al^{26}/Al^{27}) ratios. As AGB stars are considered to be a possible source of graphite and SiC grains, these ratios may be an important guide to the production of Al^{26} and the Mg-Al cycle in these stars. Again, these processes require high temperatures.

To investigate the synthesis of these elements, we intend to evolve a series of models with a range of masses and compositions from the pre-Main Sequence through to the TP-AGB. We have begun with a $5M_\odot$ star with initial $Z=0.003$ and $Y=0.25$, and followed its evolution through 20 thermal pulses.

COMPUTATIONAL METHODS

The Mount Stromlo Stellar Structure Code was used to evolve the star from pre-Main Sequence through to the TP-AGB. OPAL opacities (7), including tables enhanced in C and O (8), were used. Nuclear reaction rates were taken from CF88 (9), and mass loss was not included. A value of 1.71 was assigned to the mixing-length parameter α after using this to construct a standard solar model.

A post-processing code used the structure obtained from the evolution code to calculate detailed nucleosynthesis with time-dependent convective mixing (10). A network of 39 nuclear species ranging from H^1 to Si^{30} was included, with reaction rates taken from Thielemann et. al. (11). Reactions involving neutrons were not included.

RESULTS
CNO Ratios

The changes of the surface (C^{12}/O^{16}) and (C^{12}/C^{13}) ratios with time are shown in Figure 1. The (C^{12}/O^{16}) ratio begins to increase with the sixth pulse (for which the dredge-up parameter, λ, is about 0.7), reaching a peak of ~ 0.36 after the eighth pulse. During each subsequent interpulse phase, $T_{bce} \geq 72 \times 10^6$K, and CN cycling destroys the C^{12}, thus decreasing the (C^{12}/O^{16}) ratio until the dredge-up associated with the next pulse. A minimum of ~ 0.098 is reached just prior to the thirteenth pulse, after which (C^{12}/O^{16}) increases again. This is because the amount of carbon destroyed by HBB is slightly smaller than the amount dredged-up in the previous pulse. Carbon is also being replenished by CNO-cycling, which gradually destroys O^{16}. The total result is an increase in the (C^{12}/O^{16}) ratio.

On the other hand, the (C^{12}/C^{13}) ratio decreases during the interpulse phase until the tenth pulse, after which the interpulse value becomes nearly constant at ~ 3.2, indicating CN equilibrium and agreeing with the minimum

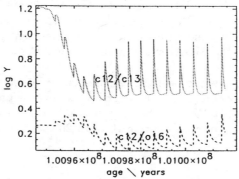

FIGURE 1. Plot of the surface abundance ratios $(Y(C^{12})/Y(O^{16}))$ and $\log(Y(C^{12})/Y(C^{13}))$ versus time, where Y is the number density.

of the observed values (4-10) in the SMC (1,2). The abundance of C^{13} peaks after the ninth pulse, declining afterwards as it is transformed into N^{14}. A minimum is reached prior to the thirteenth pulse, corresponding to the minimum in C^{12}, and thereafter C^{13} increases again.

Lithium

The surface abundance of Li^7 as a function of time is shown in Figure 2. The abundance increases during each interpulse period, reaching a maximum after the seventh pulse, and then decreases slowly. This behaviour was noted by Sackmann and Boothroyd (12), who ascribed it to the destruction of large amounts of Be^7 at the base of the envelope before it could be mixed to lower temperatures where it would decay to Li^7. The maximum value of $\log\epsilon(Li^7) \sim 4.2$ agrees well with (12), who found $\log\epsilon(Li^7) \sim 4.5$ across a range of masses and compositions.

Aluminium

The production site of Al^{26} in AGB stars has been debated. Lattanzio (13) suggested that Al^{26} might be produced by HBB. However, Forestini et al (14) found that for a $3M_\odot$ star, Al^{26} was synthesised in the hydrogen shell during the interpulse phase and mixed to the surface during third dredge-up. Our model indicates that both processes are occurring. Figure 2 shows the surface abundance of Al^{26}. Initially the change in abundance is anti-correlated with Li^7, increasing rapidly with dredge-up and only very slowly between pulses. This is because the majority of Al^{26} is produced in the H-shell. However, the scenario changes with later pulses as Al^{26} *decreases* during dredge-up, despite having a higher abundance in the shell than in the envelope. The reason is that during the expansion following a pulse, the Al^{26} decays prior to dredge-up. When the envelope penetrates into the H-processed region, it now adds material of lower Al^{26} abundance, causing a decrease at dredge-up. However, the abundance increases overall because of HBB.

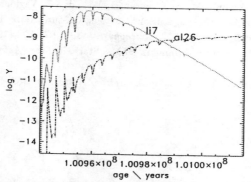

FIGURE 2. Plot of the log(Y) versus time at the surface for Li^7 and Al^{26}.

The (Al^{26}/Al^{27}) ratio approaches a constant value of $\sim$0·001. This is considerably smaller than the ratios found in meteorites (5,6). However, our results for aluminium are uncertain because in these preliminary calculations we have not treated the Al^{26} ground and meta-stable states separately. The high temperature dependance of aluminium decay will make it very sensitive to pulse-induced expansion; this is the subject of further study.

DISCUSSION

The preliminary results presented here show good agreement with observations and with the theoretical research of others. Much work will be done in expanding the study to include more nuclear species, especially neutrons to investigate the production of fluorine and s-process elements.

ACKNOWLEDGEMENTS

We would like to thank Forrest Rogers and Carlos Iglesias for providing OPAL opacity tables and interpolation routines before publication, and Robin Humble for technical assistance. One of us (C.A.F.) would like to thank the British Council for support given towards this project.

REFERENCES

1. Smith, V. V., and Lambert, D. L., *Astrophys. J. Lett.* **345**, L75-L78 (1989)
2. Smith, V. V., and Lambert, D. L., *Astrophys. J. Lett.* **361**, L69-L72 (1990)
3. Plez, B., Smith, V. V., and Lambert, D. L., *Astrophys. J.* **418**, 812-831 (1993)
4. Cameron, A. C., and Fowler, W. A., *Astrophys. J.* **164**, 111-114 (1971)
5. Zinner, E., Amari, S., Anders, E., and Lewis, R., *Nature.* **349**, 51-54 (1991)
6. Amari, S., Hoppe, P., Zinner, E., Lewis, R. S., *Nature.* **365**, 806-809 (1993)
7. Rogers, F. J., and Iglesias, C. A., *Astrophys. J. Suppl. Ser.* **79**, 507-568 (1992)
8. Iglesias, C. A., and Rogers, F. J., *Astrophys. J.* **412**, 752-760 (1993)
9. Caughlan, G. R., and Fowler, W. A., *Atomic Data Nucl. Data Tables* **40**, 283-334 (1988)
10. Cannon, R. C., *Mon. Not. R. Astr. Soc.* **263**, 817-838 (1993)
11. Thielemann, F.-K., Arnould M., and Truran, J. W., eds. E. Vagioni-flam et. al. in *Advances in Nuclear Astrophysics*, Editions frontiers, Gif-sur-Yvette, 1987, p525
12. Sackmann, I.-J., and Boothroyd, A. I., *Astrophys. J.* **392**, L71-L74 (1992)
13. Lattanzio, J. C., "Hot Bottom Burning in a $5M_\odot$ Star", in *Proc. Astron. Soc. Austr.*, 1992, pp.120-121
14. Forestini, M., Paulus, G., and Arnould, M., *Astron. Astrophys.* **252**, 597-604 (1991)

Advanced Evolution of Massive Stars: Central Carbon Burning

O. Aubert(1), I. Baraffe(2), N. Prantzos(1)

(1) Institut d'Astrophysique de Paris and (2) Ecole Normale Supérieure de Lyon

Abstract. We present preliminary results of calculations of the advanced evolutionary stages of massive stars. These results concern the central carbon burning phase of a 20 $M_\odot$ star with an initial metallicity Z=0.02.

Key words: models — stellar evolution

1. INTRODUCTION

The advanced evolution and nucleosynthesis of massive stars has been extensively investigated in recent years, particularly in view of the important role those stars play in the chemical evolution of galaxies. Important uncertainties remain, however, associated to the role of e.g. mass loss, semi-convection, overshooting and nuclear reaction rates, like the $^{12}C(\alpha,\gamma)^{16}O$ reaction. Further investigations are certainly needed to clarify those aspects of stellar evolution.

In the following we present preliminary results for the carbon burning stage in a star of 20 $M_\odot$ with solar metallicity.

2. MODELS

In our calculations the Schwarzschild criterion for convection is used, without semiconvection or overshooting. The adopted mixing length parameter is $\alpha = 1.5$. The adopted opacities are from Rogers et al. 1992. With these opacities the transition between blue giant and red giant is made when $X_{He} \cong .40$ in the center (Fig. 1), earlier than with Los-Alamos opacities.

The nuclear network includes 39 nuclei (up to ^{35}Cl) and is solved with the Gear method. We are currently implementing a 231 nuclei network (up to ^{77}Ge) for detailed nucleosynthesis calculations. The most recent results for thermonuclear reaction rates are adopted. Neutron capture cross-sections are from Beer et al. 1992, and weak interaction rates from Fuller et al. 1982. For the $^{12}C(\alpha,\gamma)^{16}O$ rate, which plays a determinant role in the evolution of the star after cenral He-burning, we take twice the value given by Caughlan and Fowler (1988). The high value adopted for this reaction leads to a low abundance of ^{12}C at the end of He-burning.

3. RESULTS

We compare our preliminary results concerning the masses of He and C-O cores with those of some previous works (Barkat et al. 1990, Nomoto et al. 1986, Woosley et al. 1988). As can be seen from Figs. 2 and 3 there is an excellent agreement between all those works: at the end of core H-burning, the mass of the He-core is $M_\alpha = 6\ M_\odot$, while at the end of the core He-burning, the mass of the C-O core is $M_{C-O} = 4\ M_\odot$. At the end of He-core burning the central amount of ^{12}C is only 0.07 by mass fraction . Because of this low abundance of ^{12}C, the nuclear energy can not compete with the neutrino energy losses, and the central C-burning takes place in a radiative core . The central abundances of the most important nuclei evolve during central C-burning as shown in Fig. 4. The chemical profile of the star in the central region $(M < M_\alpha)$ at C exhaustion shows a large amount of ^{22}Ne and ^{12}C outside the C-exhausted core because C-burning was radiative . (Fig. 5). Strong shell C-burning may occur in such conditions, leading to the development of an important carbon convective shell and to some neutron capture nucleosynthesis (the neutron source being $^{22}Ne(\alpha,n)^{25}Mg$). This process will probably modify the results of the previous neutron capture nucleosynthesis in the He-burning core of the star, as suggested by post-processing calculations in (Raiteri et al. 1991) . This possibility deserves to be further investigated, since core He-burning in massive stars is currently thought to be at the origin of the so-called "weak component" of the s-process.

4. CONCLUSION

In this work preliminary results of the central C-burning phase in a 20 $M_\odot$ star of solar metallicity are presented. They are compared to, and found to be in excellent agreement with, previous works (Barkat et al. 1990, Nomoto et al. 1986, Woosley et al. 1988). Those results constitute a first step towards a full calculation of the advanced stages in the evolution of massive stars, up to the pre-supernova stage.

REFERENCES

Barkat, Z. and A. Marom 1990. in supernovae, Jerusalem Winter School, Eds Wheeler, J.C., Piran, T. and Weinberg, S., Vol 6., p 95
Beer, H. , Voss, F. and Winters, R. R. 1992, Ap. J. Suppl. 80, p 403
Caughlan G. and Fowler W. 1988, At. Data Nucl. Data Tables, 40-283
Fuller, G.M., Fowler, W.A. and Newman, M. 1982, Ap. J. Suppl. 48, p 279
Nomoto K. and Hashimoto K. 1986, Prog. Part. Nuc. Phys. 17, p. 267
Raiteri C., Busso M., Gallino R. Picchio G 1991, Ap. J. 371, 665
Rogers, F. J. and Iglesias, C. A. 1992, Ap. J. Suppl. 79, p 507
Woosley S. E. and Weaver T. 1988, Phys. Rep. 163, p. 79

FIGURE CAPTIONS

Fig. 1 Evolution of the star in the H-R diagram

Fig. 2 Mass of He-core M_α as a function of main sequence mass from calculations by Nomoto and Hashimoto (1986;black dots), Woosley and Weaver (1988;crosses). The result of this work is noticed with an hexagon.

Fig. 3 Mass of M_{CO} as a function of mass M_α from calculations by Nomoto and Hashimoto (1986;black dots), Woosley and Weaver (1988,crosses) and Barkat and Marom (1990; squares). The result of this work is noticed with an hexagon.

Fig. 4 Evolution of the central abundances of several nuclei during core C-burning (t_f is the time of central C-exhaustion).

Fig. 5 Chemical profile of the most important nuclei inside M_α at central C-exhaustion.

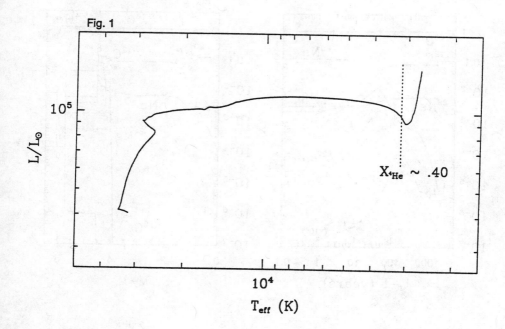

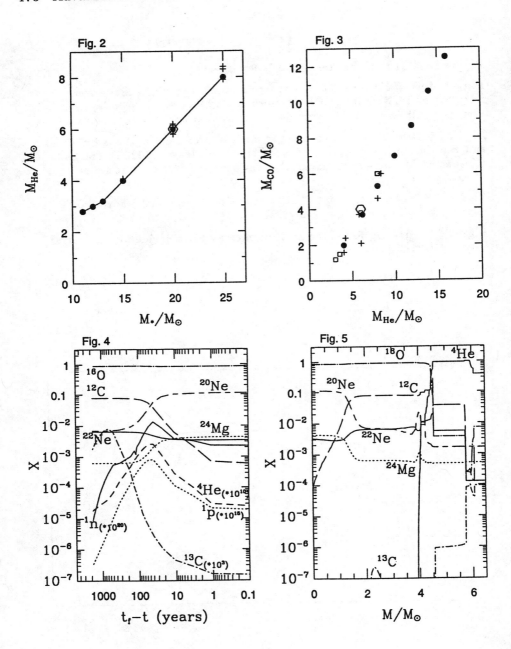

Nucleosynthesis in TP-AGB stars

Paola Marigo

Department of Astronomy, Vicolo dell'Osservatorio 5, 35122 Padova, Italy

Abstract. This study deals with the chemical enrichment at the surface of TP-AGB stars and the luminosity function of C-stars in the LMC.

1. Basic assumptions of the TP-AGB models

This study deals with the chemical enrichment at the surface of TP-AGB stars and the luminosity function of C-stars in the LMC. To this aim, TP-AGB models are calculated to follow the changes in the surface abundances caused by (1) the intershell nucleosynthesis and convective dredge-up; (2) CNO nuclear burning in the deepest layers of the convective envelope, and (3) mass loss by stellar winds. The evolution of the surface abundances of 13 different elements (H, ^3He, ^4He, ^{12}C, ^{13}C, ^{14}N, ^{15}N, ^{16}O, ^{17}O, ^{18}O, ^{20}Ne, ^{22}Ne, ^{25}Mg) is followed in detail. To follow the evolution along the TP-AGB phase for low and intermediate mass stars a semi-analytical method is adopted (see Renzini & Voli 1981 and Groenewegen & de Jong 1993). It stands on:

- **The initial conditions** at the first thermal pulse of the He-burning shell. They are derived from the evolutionary tracks of Bressan et al. (1993) for $Z = 0.02$ and Fagotto et al. (1994) for $Z = 0.008$.

- **The Core Mass - Luminosity relationship** ($M_c - L$). We use the equation derived by Boothroyd & Sackmann (1988b) in the range $0.5 \leq M_c/M_\odot \leq 0.66$ and the equation by Groenewegen & de Jong (1993) for $0.95 \leq M_c/M_\odot \leq 1.36$ (see also Iben & Truran 1978). A linear interpolation in between. For the first pulses suitable corrections are applied.

- **The evolutionary rate along the TP-AGB phase.** This is expressed by the growth rate of the core mass as a function of the luminosity from the H-burning shell (Groenewegen & de Jong 1993).

- **The Core Mass - Interpulse Period relationship** ($M_c - T_{ip}$). We adopt two equations from Boothroyd & Sackmann (1988 c) corresponding to $Z = 0.02$ and $Z = 0.008$. Linear interpolations are used for other values of metallicity. For the first pulses suitable corrections are also applied.

- **The rate of mass loss** by stellar wind. This is taken from Vassiliadis & Wood (1993) and stems from the theory according to which the mechanism powering the wind resides in the pulsational instability of the stars.

- **The dredge-up algorithm.** The variation of the surface abundances caused by the III dredge-up are analytically calculated following Renzini & Voli (1981). The analytical treatment of the III dredge up is based on the prescriptions for two basic parameters:
  - λ expressing the fraction of the increment in core mass during the interpulse phase which is dredged up to the surface in the next pulse. The adopted value is $\lambda = 3/4$ (Groenewegen & de Jong 1993)
  - M_c^{min} giving the minimum core mass for the dredge-up mechanism to operate. The adopted value is $M_c^{min} = 0.58\ M_\odot$ (Groenewegen & de Jong 1993).
- **An algorithm to derive the effective temperature of the current model.** This is obtained by using an envelope model and imposing that the envelope convection extend down to the top of the C-O core. The adopted mixing length parameter of the envelope convection is $\alpha = 1.63$.
- **The CNO nuclear burning** at the bottom of the convective envelope and the mixing of the nuclear products throughout the envelope. The nuclear reaction rates are from Caughlan & Fowler (1988). For each element, the equilibrium condition is established by comparing the nuclear lifetime against a proton capture and the convective lifetime relative to every envelope layer.

2. The TP-AGB phase: Results

The most interesting results are briefly described in a few figures which show some physical features of the TP-AGB phase (fig. 1 - 2) and the effect on the surface chemical composition due to nucleosynthetic and mixing processes (fig 3).

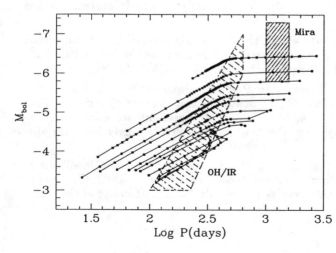

Fig. 1.— The bolometric magnitudes of the TP-AGB models as a function of the pulsational period P compared with the data for LPVs in LMC (Vassiliadis & Wood, 1993), namely the Mira stars and the OH/IR stars (dashed areas).

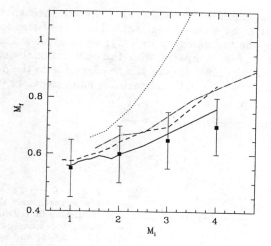

Fig. 2.— The initial-final mass relationship from our models (solid line $Z = 0.02$, dashed line $Z = 0.008$) compared with the one of Vassiliadis & Wood (1993) for $Z = 0.008$ (dot-dashed line), of Bertelli et al. (1994) for standard mass loss rates and $Z = 0.008$ (dotted line), and finally (filled squares) the semi-empirical one of Weidemann (1987).

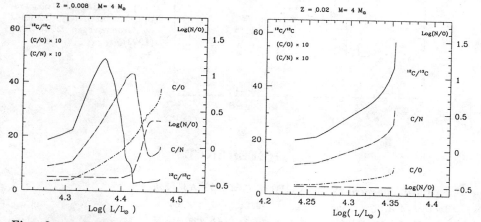

Fig. 3.— The variation of some surface C-N-O abundance ratios during the TP-AGB phase for the 4 $M_\odot$ model in both cases $Z = 0.008$ and $Z = 0.02$. Note the effect of envelope burning which is efficient in the former model ($Z = 0.008$) and almost absent in the latter one ($Z = 0.02$).

3. The luminosity function of C-stars

With the aid of our models corresponding to $Z = 0.008$ we derive the theoretical luminosity function (TLF) of TP-AGB stars, with particular attention to the transition from the M-phase to the C-phase. We assume the Salpeter law for the IMF, the age of the LMC equal to 15 Gyr, and the same kind of star formation suggested by Bertelli et al. (1992) in their study of the CMDs of field stars in selected areas of the LMC. The rate of star formation has been moderate from the beginning up to 8 Gyr ago and since then a factor of ten stronger.The TLF is compared with the observational data taken from Groenewegen & de Jong (1993) (figure 4).The effect due to the post-flash luminosity dip is also included. Note the remarkable agreement.

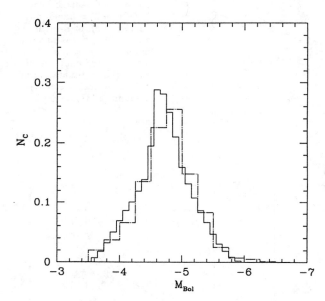

Fig. 4.— The comparison between the theoretical luminosity function of C stars calculated from our models ($Z = 0.008$) and normalized to unity (solid line), and that derived for LMC (dot-dashed line) by using the observational data already discussed by Groenewegen & de Jong (1993).

REFERENCES

Bertelli G., Mateo M., Chiosi C., Wood P.R. 1992, ApJ, 388, 400

Boothroyd A.I., and Sackmann I.-J., 1988b, ApJ, 328, 641

Boothroyd A.I., and Sackmann I.-J., 1988c, ApJ, 328, 653

Bressan A., Fagotto F., Bertelli G., Chiosi C., 1993, A&A, 100, 674

Caughlan G.R., Fowler W.A., 1988, Atomic Data Nuc. Data Tables, 40, 283

Fagotto F., Bressan A., Bertelli G., Chiosi C., 1994, A&A, 1994, 104, 365

Groenewegen M.A.T., and de Jong T., 1993, A&A, 267, 410

Iben I., Truran J.W., 1978, ApJ, 220, 980

Renzini A., Voli M., 1981, A&A, 94, 175

Vassiliadis E., Wood P.R., 1993, ApJ, 413, 641

Weidemann V. 1987, A&A, 188, 74

Stellar Models and Microstructural Investigations of Stardust

Anja C. Andersen[1], Marie-Louise Andersen[1],
Kristian Glejbøl[2], Uffe Gråe Jørgensen[1]

[1] *Niels Bohr Institute, Blegdamsvej 17, DK-2100 Copenhagen, Denmark*
[2] *Physics Institute, Technical University of Denmark, DK-2800 Lyngby, Denmark*

Abstract. We have extracted extra-solar diamonds from the Allende meteorite and measured their optical properties. The optical data are used to incorporate effects of the extra-solar diamonds in carbon star model atmospheres and in synthetic stellar spectra. Eventually such computed spectra will be compared with observed spectral features of AGB stars.

Introduction

Atoms and ions dominate the opacity of most astrophysical plasmas at temperatures above approximately 5000 K. In the range between 5000 K and about 1500 K, molecular processes are important contributors to the opacity. Depending upon the composition and pressure of the plasma, grains will condense at temperatures below 1500 K. For the conditions of interest here, the grains which form are generally small (radius $\leq 0.25 \mu m$, Alexander & Ferguson, 1994) compared to the wavelength of light near the peak of the Planck curve. Because their absorption and scattering cross sections are large compared to the corresponding values for atoms and molecules, grains dominate the mean opacity whenever they are present and must therefore be taken into account whenever such temperatures are encountered. Alexander & Ferguson (1994) found that a failure to do so can lead to errors as large as five orders of magnitude in the mean opacity in static model atmospheres.

We have begun a detailed study of the effect diamond and SiC grains may have on carbon star atmospheres. We do this by studying carbonaceous chondrites, as it has been known since 1987 (Lewis et al. 1987, Zinner et al. 1987) that carbonaceous chondrites contain extra-solar diamond and SiC grains. These grains (and/or noble gas inclusions in the grains) have isotopic compositions markedly different from the rest of the solar system. Jørgensen (1988) has suggested that the identified meteoritic diamond and SiC grains are condensates from the atmospheres of carbon stars, and we will test this theory by calculating synthetic spectra for carbon star atmospheres and comparing them to observations.

Carbon and silicon carbide condense when there is more carbon than oxygen present ($C/O \geq 1$). Carbon grains present a particularly difficult problem because they can take many different physical forms, sometimes with significantly different optical properties. For high temperature grains, the question is whether the grains form as amorphous carbon or in some kind of crystal structure, such as diamond or C_{60}. For low temperature (circumstellar) grains, polycyclic aromatic hydrocarbons (PAHs) probably play an important role in nucleation (Omont, 1994).

Experiments

We have dissolved a piece of the Allende meteorite by alternating treatments with $10M$ HF and $1M$ HCl, using the methode described by Tang & Anders (1987).

The grains were retrieved on Ti grids covered with carbon film, and studied using a Philips EM 430 transmission electron microscope (TEM) operated at 300 keV. With this instrument we did conventional imaging, electron diffraction and energy-dispersive X-ray spectroscopy (EDS). Diamonds were a major constituent of all our samples and we have been able to confirm most of the properties reported by Fraundorf et al. (1989).

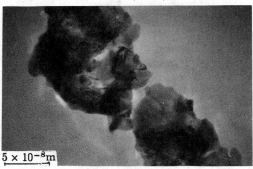

Figure 1. Cluster of extra-solar nm-sized diamond crystalizes.

We have performed several comparisons between extra-solar diamonds and industrial diamonds made by chemical vapor decomposition (CVD), and we have completed measurements of the optical properties of the diamonds in the wavelength region 190nm–820nm. We plan to extend our measurements to include the near infrared region as well. These data are needed in order to incorporate the extra-solar diamond opacities in carbon star model atmosphere and in stellar spectrum computations.

Particles in an astrophysical environment will be present in a variety of different sizes. This size distribution should be explicitly considered when computing the grain opacity, because the extinction efficiency of a grain depends upon its size.

Stellar Atmosphere Models

As stars evolve through the red giant stage, their atmospheres become more extended due to the increasing mass of the degenerate core. At the same time convection generates a variety of acoustic waves, shocks, and magnetic activity, which heat the chromosphere. Stellar pulsations extend the stars, grains may form and mass loss becomes appreciable. The growth regime of the grains extends from the nucleation centre outward to a distance where either the condensable material is totally consumed by grains or where the material becomes too diluted to have significant collision rates. The upper atmosphere of a cool carbon star is found, (e.g., Jørgensen & Johnson 1991), to be highly inhomogenous, with typically 90% of the surface covered by cool photospheric material and 10% covered by hot chromospheric material. It is in this cool photospheric material that the first nucleation is expected (see Fig. 2).

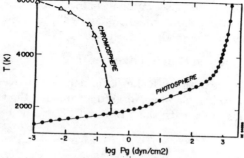

Figure 2. Temperature vs. the logarithmic gas pressure of the photospheric (filled circles) and chromospheric (triangles) model atmospheres. The upper atmosphere of a cool carbon star is found to be highly inhomogenous, with typically 90% of the surface covered by cool photospheric material and 10% covered by hot chromospheric material. The photospheric model has been shown to fit the C_2, CN and HCN spectrum of TX Psc, whereas the chromospheric model is known to fit the IUE observations of the MgII and CaII emission lines of TX Psc (models are from Jørgensen & Johnson, 1991).

The model photospheres we are presently computing are based on an improved version (Jørgensen et al. 1992) of the MARCS code (Gustafsson et al. 1975). It assumes hydrostatic equilibrium and lokal thermodynamic equilibrium (LTE), but includes effects of sphericity and opacity sampling (OS) treatment of molecular opacities from approximately 60 million spectral lines (Jørgensen 1994). Empirical, inhomogeneous chromospheres are included in the spectrum computations. Such models have proven to reproduce well the observed spectral features of carbon stars (Jørgensen & Johnson 1991, Jørgensen 1989, Lambert et al. 1986). The models are, nevertheless, most uncertain in the more shallow (upper) layers where grains are expected, and improvements in the physics and the chemistry included in these layers is desired.

A first estimate (or upper limit) of the expected effect of the diamonds on the structure and the spectra of the various classes of carbon stars, however, can be made by assuming that all the available carbon not bound in CO molecules, is tied up momentarily in carbon grains when the temperature drops below the nucleation temperature. Much more advanced dynamical models of the grain formation region have been constructed for example by Sedlmayr (1994) and by Höfner & Dorfi (1992). Such models are, nevertheless, still based on rather crude assumptions about the underlying photosphere, and no self-consistent model exists of a "unified stellar atmosphere" including both the photosphere-chromosphere and the dynamical grain formation region above it. It is our hope in the future to be able to combine the grain formation models computed by other groups with our photospheric-chromospheric models, and such collaborational efforts are presently in progress. Also collaborational efforts to construct improved fully self-consistent photospheric-chromospheric models for carbon stars based on recent HST observations (Johnson et al. 1994) are in progress. Possible future observational confirmation, of meteoritic extra-solar diamonds in the atmospheres of such stars will put important limits on the future unified carbon star model atmospheres.

Acknowledgements

The authors would like to thank R. Lewis for detailed advices on the extraction of the extra-solar grains, Carsten Nørgård for making the CVD diamonds, Allan Henschel for assistance with the TEM and Peter Thejll for fruitful discussions.

References

Alexander D.R., Ferguson J.W., 1994, in U.G.Jørgensen (ed.), Lecture Notes in Physics, vol.428, *Molecules in the Stellar Environment*, (Berlin:Springer), 149–162.
Fraundorf P., Fraundorf G., Bernatowitz T., Lewis R., Tang M., 1989, *Ultramicroscopy* **27**, 401–412.
Gustafsson B., Bell R.A., Eriksson K., Nordlund Å, 1975, *Astron. Astrophys.* **42**, 407.
Höfner S., Dorfi E.A., 1992, *Astron. Astrophys.* **265**, 207–215.
Johnson H.R. et al., 1994, submitted to *Astrophys. J.*
Jørgensen U.G., 1988, *Nature* **332**, 702–705.
Jørgensen U.G., 1989, *Astrophys. J.* **344**, 901–906.
Jørgensen U.G., 1994 in U.G.Jørgensen (ed.), Lecture Notes in Physics, vol.428, *Molecules in the Stellar Environment*, (Berlin:Springer), 29–48.
Jørgensen U.G., Johnson H.R., 1991, *Astron. Astrophys.* **244**, 462–466.
Jørgensen U.G., Johnson H.R., Nordlund Å, 1992, *Astron. Astrophys.* **261**, 263–273.
Lambert D.L., Gustafsson B., Eriksson K., Hinkle K.H., *Astrophys. J. Suppl.* **62**, 373.
Lewis R.S., Ming T., Wacker J.F., Anders E., Steel E., 1987 *Nature* **326**, 293–298.
Lewis R.S., Anders E., Draine T., 1989, *Nature* **339**, 117–121.
Omont A., 1994 in U.G.Jørgensen (ed.), Lecture Notes in Physics, vol.428, *Molecules in the Stellar Environment*, (Berlin:Springer), 134–148.
Sedlmayr E., 1994 in U.G.Jørgensen (ed.), Lecture Notes in Physics, vol.428, *Molecules in the Stellar Environment*, (Berlin:Springer), 163–185.
Tang M., Anders E., 1988 *Geochimica Cosmoch. Acta* **52**, 1235–1244.
Zinner E., Ming T., Anders E., 1987, *Nature* **330**, 730.

Properties of Strange-Matter Stars

F. Weber(1), Ch. Kettner(1), and N. K. Glendenning(2)

(1) Institute of Theoretical Physics, University of Munich, Theresienstr. 37/III, 80333 Munich, Germany, and (2) Nuclear Science Division, Lawrence Berkeley Laboratory, Berkeley, CA 94720, USA

Abstract. This paper deals with an investigation of the properties of hypothetical strange-matter stars, which are composed of u, d, s quark matter whose energy per baryon number lies below the one of ^{56}Fe (Witten's strange matter hypothesis). Observable quantities which allow to distinguish such objects from their "conventional" counterparts, neutron stars and white dwarfs, are pointed out.

Key words: Strange Matter — Strange Stars – Pulsars

1. INTRODUCTION

The hypothesis that strange quark matter may be the absolute ground state of the strong interaction (not ^{56}Fe) has been raised by Witten in 1984 (Witten 1984). If the hypothesis is true, then a separate class of compact stars could exist, which are called *strange stars*. They form a distinct and disconnected branch of compact stars and are not part of the continuum of equilibrium configurations that include white dwarfs and neutron stars. In principle both strange and neutron stars could exist. However if strange stars exist, the galaxy is likely to be contaminated by strange quark nuggets which would convert all neutron stars that they come into contact with to strange stars (Glendenning 1990, Madsen and Olesen 1991, Caldwell and Friedman 1991). This in turn means that the objects known to astronomers as pulsars are probably rotating strange matter stars, *not* neutron matter stars as is usually assumed. The properties of (rotating) strange stars are discussed in this paper.

2. NEUTRON STARS AND WHITE DWARFS VERSUS STRANGE-MATTER STARS

2.1. Minimum Rotational Periods of Neutron Stars

From our extensive investigation performed somewhere else (Weber and Glendenning 1993a), it is know that the gravitational-radiation reaction driven instability sets a lower limit on the stable rotational period of a rotating neutron star of mass $\sim 1.45\,M_\odot$ of about 1 msec. (For the purpose of comparison, the most rapidly rotating pulsar known to date rotates at 1.56 msec.) This indicates that the nature of any pulsar that is found to have a considerably shorter period than ~ 1 msec must be different from the one of a neutron star.

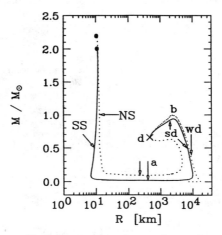

Fig. 1.— Mass versus radius of strange-star configurations with nuclear crust (solid curve) and gravitationally bound stars (dotted curve). The following abbreviations are used: NS=neutron star, SS=strange star, wd=white dwarf, sd=strange dwarf.

As pointed out in the literature, rapidly rotating strange stars arise as natural candidates for such objects (Glendenning 1990). Being bound by the strong force rather than gravity, strange stars possess a mass-radius relationship (cf. Sect. 3.) that allows them to withstand such rapid rotation.

3. Mass-Radius Relationships

The mass-radius relationship of non-rotating strange stars with nuclear crust, whose inner density is equal to neutron drip (maximum possible density, Alcock et al. 1986) is shown in Fig. 1 (Weber and Glendenning 1993b, Weber and Glendenning 1993c). A value for the bag constant of $B^{1/4} = 145$ MeV has been chosen. This choice represents strongly bound strange matter with an energy per baryon ~ 830 MeV. The solid dots denote the maximum-mass stars of the neutron (NS) and strange quark star (SS) sequences. The arrows indicate the minimum-mass star of each sequence ('a': strange star, 'b': neutron star). White dwarf-like strange star configurations ('sd': strange dwarfs) terminate at the crossed point labeled 'd'. The symbol 'wd' indicates the region of ordinary white dwarfs. The sequence of strange stars has a minimum mass of $\sim 0.015\, M_\odot$ (radius of ~ 400 km) or about 15 Jupiter masses, which is smaller than that of the neutron star sequence, about $0.1\, M_\odot$ (Baym et al. 1971). These low-mass strange stars may be of considerable importance since they may be difficult to detect and therefore may effectively *hide baryonic matter*. It is striking that the bulk properties of neutron and strange stars of masses that are typical for neutron stars, $1.1 \lesssim M/M_\odot \lesssim 1.8$, are relatively similar and therefore do not allow the distinction between the two different species of stars. The situation changes as regards the possibility of *fast rotation* of

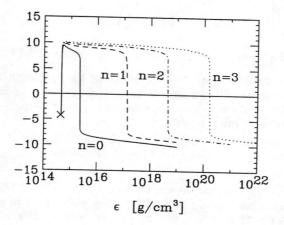

Fig. 2.— Pulsation frequencies, ω_n^2, of the lowest four ($n = 0, 1, 2$, and 3) normal radial modes of strange stars as a function of central star density. Instead of ω_n^2 itself, the quantity $\text{sign}(a) \log(1 + \text{abs}(a))$, where $a \equiv (\omega_n/\text{sec}^{-1})^2$, is plotted on the y-axis. The cross refers to the termination point of the strange dwarf sequence (cf. Fig. 1).

strange stars. This has its origin in the different mass-radius relationships of neutron stars and strange quark stars (Glendenning 1990). As a consequence of this the *entire* family of strange stars can rotate rapidly, at Kepler periods in the range $\sim (0.4 - 0.8)$ msec (depending on mass and bag constant), not just those near the limit of gravitational collapse to a black hole as is the case for neutron stars (Glendenning and Weber 1992).

4. STABILITY OF STRANGE-MATTER STARS AGAINST RADIAL OSCILLATIONS

The normal modes of vibration of the strange-matter stars are computed from the Sturm-Liouville eigenequation, possessing the form (for details, see Glendenning et al. 1993),

$$\frac{d}{dr}\left(\Pi(r)\frac{du_n(r)}{dr}\right) + \left(Q(r) + \omega_n^2 W(r)\right) u_n(r) = 0 \ .$$

The four lowest-lying eigenfrequencies of massive strange stars and strange dwarfs, whose inner crust density is equal to neutron drip, are shown in Fig. 2. We find that the $n = 0$ mode becomes zero for the maximum- and minimum-mass star configurations labeled 'a' and 'b' in Fig. 1 (Glendenning et al. 1993). The $n = 0$ mode passes through zero at 'b' and remains negative (i.e., $\omega_n^2 < 0$) for all densities down to the central density of the strange dwarf star at the termination point (cross). Since $\omega_0^2 < 0$ is associated with an exponentially growing mode of oscillation, all these strange dwarfs are *unstable* against radial oscillations. Thus, no stable strange dwarfs located between 'b' and the crossed termination point can exist in nature!

5. SUMMARY

The main issues of this contribution can be stated as follows:

- Strange pulsars of masses $M \sim 1.45 M_\odot$ can have rotational periods that are clearly smaller than 1 millisecond. Because this range seems to be excluded for neutron stars, submillisecond pulsar periods may serve as **signatures** for strange stars.

- Strange stars can possess nuclear **crusts** of thickness $\sim (1-10^3)$ km, depending on central density. This will be of great importance for their cooling behavior.

- Strange stars possess **masses** in the range $\sim (2-10^{-4}) M_\odot$ and **radii** $\sim 10^3$ km. Since masses and radii of $10^{-4} M_\odot$ and $\sim 10^3$ km are completely excluded for both neutron stars as well as white dwarfs, they may serve as additional signatures for hypothetical strange stars.

- The Oppenheimer-Volkoff equations lead to strange-star configurations whose masses and radii are similar to those of ordinary white dwarfs. We thus call such objects **strange dwarfs**. Those strange dwarfs carrying nuclear crusts whose density at the base is equal to neutron drip are unstable against radial oscillations.

REFERENCES

Witten, E. 1984, Phys. Rev. D, 30, 272

Glendenning, N. K. 1990, Mod. Phys. Lett., A5, 2197

Madsen, J., and Olesen, M. L. 1991, Phys. Rev. D, 43, 1069; ibid., 44, 4150 (erratum)

Caldwell, R. R., and Friedman, J. L. 1991, Phys. Lett., 264B, 143

Weber, F., and Glendenning, N. K. 1993a, Astrophysics and Neutrino Physics, Eds. D. H. Feng, G. Z. He, and X. Q. Li, World Scientific, 63–183

Alcock, C, Farhi, E., and Olinto, A. V. 1986, ApJ, 310, 261

Weber, F., and Glendenning, N. K. 1993b, Proceedings of the 2nd International Conference on Physics and Astrophysics of Quark-Gluon Plasma, January 19-23, 1993, Calcutta, India, to be published by World Scientific, (LBL-33771)

Weber, F., and Glendenning, N. K. 1993c, Proceedings of the NATO Advanced Study Institute *Hot and Dense Nuclear Matter*, Bodrum/Turkey, 26.09.-9.10.1993, ed. by W. Greiner and H. Stöcker, to be published by Plenum Press, (LBL-34783)

Baym, G., Pethick, C, and Sutherland, P. 1971, ApJ, 170 299

Glendenning, N. K., and Weber, F. 1992, ApJ, 400, 647

Glendenning, N. K., Kettner, Ch., and Weber, F. 1993, *Properties of Strange Stars and Strange Dwarfs*, in preparation, (LBL-34869)

SECTION IV

CHEMICAL EVOLUTION AND COSMOCHRONOLOGY

Light Elements: from Big Bang to Stars

B.E.J. Pagel

NORDITA, Blegdamsvej 17, Dk-2100 Copenhagen Ø, Denmark

Abstract. Primordial abundances inferred by extrapolation of astrophysical and cosmochemical measurements continue to support the so-called "Standard" Big Bang nucleosynthesis model, but raise many unanswered questions concerning galactic chemical evolution.

Key words: beryllium, Big Bang, boron, deuterium, helium-3, helium-4, lithium-6, lithium-7, nucleosynthesis

1. INTRODUCTION

The light elements up to ^{11}B are of special interest because they are created in whole or in part by different processes than the others, namely Big Bang and cosmic-ray spallation or $\alpha - \alpha$ fusion (see, e.g., Reeves 1994a). Intensive activity in their study in the past decade has been largely stimulated by cosmology, but results from the last few years have also raised interesting questions about galactic chemical evolution (GCE), e.g. why is there so much beryllium and boron in metal-weak stars and how to explain the ^{3}He abundance (or lack of it) and the ^{6}Li/^{7}Li ratio in the interstellar medium (ISM)? "Standard" Big Bang nucleosynthesis theory (SBBN) still seems to be in good shape, but when we consider the growth of any one of the light elements due to subsequent GCE, we open a whole can of worms. In this review, I shall briefly summarize the inferences made a few years ago by comparing observational data with SBBN theory, and then comment on more recent developments.

2. EXISTING INFERENCES FROM SBBN

Fig 1 shows what has been inferred over the last few years about baryonic density from a comparison of SBBN predictions with primordial abundances deduced by extrapolating astrophysical and cosmochemical measurements, and the resulting implications for baryonic and non-baryonic dark matter. The most restrictive upper limit to baryonic density comes from ^{4}He and the most restrictive lower limit from D+^{3}He, resulting in a narrow region of concordance. Recent developments concerning each of the relevant elements are presented in the succeeding sections.

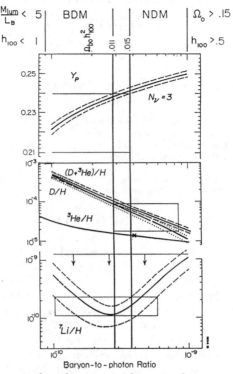

Fig. 1.— Light-element abundances *vs.* baryon-photon ratio for ^{4}He (top); D+^{3}He, D and ^{3}He (middle) and ^{7}Li (bottom) predicted as a function of baryon:photon ratio and with lower and upper limits deduced from various observations, after Walker *et al.* (1991) and M.S. Smith *et al.* (1993). The cross shows an estimate of primordial ^{3}He by Balser *et al.* (1994). Implications for existence of baryonic and non-baryonic dark matter are shown at the top.

3. DEUTERIUM AND HELIUM-3

Protosolar (D+^{3}He)/H $\equiv y_{23\odot}$, ^{3}He/H $\equiv y_{3\odot}$ and D/H$\equiv y_{2\odot}$ are deduced from meteoritic and solar wind measurements to be 4.1±1.0, 1.5±0.3 and 2.6± 1.0 respectively, all in units of 10^{-5} (Geiss 1993), from which by the well known arguments about survival of ^{3}He (Yang *et al.* 1984; Steigman & Tosi 1992; Vangioni-Flam *et al.* 1994) one infers $y_{23P} < 10^{-4}$ and $y_{2P} < 8 \times 10^{-5}$. HST measurements of interstellar deuterium towards Capella give $10^5 y_2 = 1.65 \pm 0.15$ (Linsky *et al.* 1993; Linsky 1994), which may be representative of the results of older *COPERNICUS* observations along lines of sight towards more distant OB stars as well (McCullough 1992), but Lemoine (1994a) suspects some variations in local clouds towards a white dwarf observed with HST.

Much more sensational is the observation of a putative D Ly-α line with a red-shift of 3.32 in a Lyman forest cloud towards the QSO Q0014+813 giving a nominal D/H ratio of $(2.5\pm0.6)\times 10^{-4}$ (Songaila et al. 1994; Carswell et al. 1994). If this is a correct identification, the arguments about ^{3}He must be invalid for some reason (cf. Vangioni-Flam & Cassé 1994), evidence for baryonic and non-baryonic dark matter becomes weaker and stronger, respectively, and the upper limit to primordial helium abundance allows one additional neutrino type. However, as both groups of authors point out, one cannot exclude the possibility of interference from a rogue hydrogen cloud, especially as Carswell et al. find that quite a number of clouds with different velocities are needed anyway to get a satisfactory fit to the data and express some scepticism as to whether the result is more than just an upper limit. More data on other systems will be needed before a firm conclusion can be drawn.

At the same time, there are also problems with ^{3}He. Balser et al. (1994) are making considerable progress in their long-term programme to measure ^{3}He/H in Galactic HII regions from the hfs line at 3.46 cm and (assuming homogeneous HII region models) they find values ranging from 0.7 to 6×10^{-5} with no clear correlation with galactocentric distance that might give a clue to the chemical evolution. Their suggestion is that there is an underlying ^{3}He abundance of about 1.1×10^{-5}, unchanged since the Big Bang (cf. Fig. 1), on which are superposed local variations due to pollution by winds from massive stars, for which there is some evidence in the form of helium variations across W3 (Roelfsema et al. 1992) and some optical extragalactic HII regions (see below). That explanation fits in quite nicely with the old arguments about D+^{3}He, but only on the assumption that there has been little net production of ^{3}He in the Galaxy by stellar processes, whereas substantial production has long been predicted and is, furthermore, supported by their observation of a planetary nebula with ^{3}He/H $\simeq 10^{-3}$ (Rood et al. 1992). Since AGB stars with typically an order-of-magnitude enhancement of s-process elements are plausibly sufficient to account for the presence of those elements in the Solar System, it is implausible that the presence of an enhancement of ^{3}He by nearly 2 orders of magnitude should not have led to a substantial increase in the course of GCE. Thus the arguments about D+^{3}He look somewhat unsafe, whatever one thinks about the observations at high red-shift.

4. HELIUM-4

The position about 10 years ago is well described in the ESO Workshop on "Primordial Helium" (Shaver et al. 1983; cf. also Boesgaard & Steigman 1985 and Pagel 1993, 1994a). Helium is robust and abundant and accordingly there is an abundance of data from various sources; but the accuracy of a few per cent or better that is needed in order to constrain cosmology and astrophysics significantly is difficult to achieve. By the mid-70's it had become clear that it has a universal "floor", as required by SBBN, both from stellar observations (including members of globular clusters and other metal-deficient stars) and from emission lines in ionized nebulae, which are capable of greater accuracy. After the seminal observations of the two Zwicky blue compact galaxies IZw18 and IIZw40 by Searle & Sargent (1972), which showed them to be dominated by giant HII regions with low oxygen but near-normal helium abundance (IZw18 still holds the record among HII regions with about 1/50 solar oxygen abundance), Peimbert & Torres-Peimbert (1974) initiated a programme to determine the primordial mass fraction Y_P and the astrophysically important quantity dY/dZ from observations of the Magellanic Clouds and other irregular and blue compact galaxies by plotting regressions of helium against oxygen abundance, culminating in the papers by Lequeux et al. (1979) who found $Y_P = 0.230$ and $dY/dZ = 3.5$, and by Kunth & Sargent (1983) who with generally better data (but some bias due to high weighting of $\lambda 5876$ in IIZw40 which is absorbed by Galactic sodium), detected no dY/dZ slope and could only conclude that $Y_P < 0.243$.

The spectrophotometric survey of HII galaxies by Roberto Terlevich and colleagues (Campbell, Terlevich & Melnick 1986; Terlevich et al. 1991) seemed to me to provide a good opportunity to improve the situation by combining new observations with a careful re-evaluation of existing data. Davidson & Kinman (1985) had given a formidable survey of the difficulties involved, but I thought I had answers to most of them. Pagel, Terlevich & Melnick (1986) noted that there seemed to be real scatter in helium at a given oxygen abundance; we attributed this to local pollution by winds of hydrogen-burned material from embedded Wolf-Rayet stars that sometimes show up in the form of broad HeII $\lambda 4686$ emission and accordingly introduced the idea of plotting a regression against nitrogen as well as oxygen, in the vain hope that this effect would cancel out to first order. Later we added new data taken with the INT and AAT, some (not enough) of unprecedentedly good quality; the results are described in Simonson (1990) and Pagel et al. (1992) and they are illustrated in Fig. 2.

The main problems in our work are remaining imperfections in the data and certain problems of interpretation, notably

(i) Theoretical recombination coefficients. As usual, we used those of Brocklehurst (1972), but noted that new calculations by Smits (1991ab) gave slightly smaller values for $\lambda 5876$ and $\lambda 6678$. Improved calculations (Smits 1994) confirm Brocklehurst's values for these lines, but substantially change the one for $\lambda 7065$ which is sensitive to collisional and radiative transfer effects.

(ii) Collisional excitation, which we estimated using formulae by Clegg (1987) with electron densities deduced from the [SII] line ratio. Since these came out very small, uncertainties in the theory did not seem important.

(iii) Cosmic scatter in the He,O or He,N relation. In NGC 5253, Campbell et al. found a central peak in Y associated with a highly localized WR feature, and subsequent observations revealed variations across the extended HII region in N/O and probably He/H as well, while in other objects there is just marginal evidence that high He and N abundances may be associated with the WR feature. To eliminate systematic effects from the existence of such a correlation (which may or may not be real) without introducing a statistical bias, we decided to exclude from our regression lines all objects in which the WR feature had been definitely identified; these are shown by filled circles in Fig. 2.

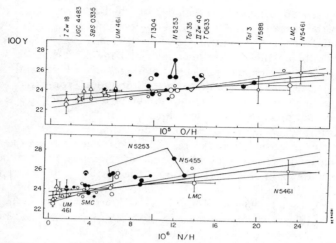

Fig. 2.— Regressions of helium against oxygen and nitrogen in HII galaxies (Pagel et al. 1992). Filled circles show objects with definite WR features, open circles others. Maximum likelihood regression lines are shown with equivalent $\pm 1\sigma$ limits; the short line in the lower panel is the regression finally preferred for nitrogen. Triangles show later results from Skillman & Kennicutt (1993), Skillman et al. (1994a) and Izotov et al. (1994); stars are from a rediscussion of observations of SBS 0355-052 (Melnick et al. 1992).

The results from this were $Y_P = 0.228 \pm 0.005$ (s.e.) or, assuming a systematic error of up to 0.005, $Y_P < 0.242$ with 95 per cent confidence, and $dY/dZ = 4 \pm 1$ for $O/H \leq 2.4 \times 10^{-4}$ or 0.3 solar oxygen abundance, not significantly different from the results of Lequeux et al. (1979). We derived the same Y_P from the regression against nitrogen, taking into account the contribution of "secondary" nitrogen production, which led to some discussion but no serious subsequent change (Pagel & Kazlauskas 1993; Balbes et al. 1993). The large dY/dZ, already implicit in earlier work (Faulkner 1967; Perrin et al. 1977; Lequeux et al. 1979), and supported by studies of planetary nebulae (Chiappini & Maciel 1994), has led to much discussion, since it is about twice the value expected from conventional stellar nucleosynthesis, and basically three different hypotheses have been put forward:-

(i) A low upper limit ($\simeq 25 M_\odot$) to the initial mass of stars undergoing supernova explosions rather than going into black holes (Maeder 1992, 1993; Brown & Bethe 1994). Apart from questions about the initial mass function and resulting absolute yields assumed by Maeder, objections have been raised by Prantzos (1994) on grounds of overproduction of carbon.

(ii) Preferential loss of oxygen and other components of Z by selective galactic winds following bursts of star formation in dwarf galaxies (Pilyugin 1993; Marconi et al. 1994). Pilyugin also includes temporary increases in both

Y and Z by self-pollution in HII regions, for which there is some independent evidence in the case of IZw18 from HST observations giving a still much lower oxygen abundance in the neutral gas (Kunth et al. 1994). Such differences could cast doubt on the whole principle of plotting regression lines, but they cannot be a major factor in objects like the Magellanic Clouds and the Galaxy, where no HII regions are found to be overabundant compared to stars or neutral gas, and the scatter in helium abundance that they are supposed to explain only marginally exceeds experimental errors.

(iii) Our oxygen abundances are underestimated because we neglected fluctuations in electron temperature. This is a controversial question, but there is some evidence now that fluctuations are significant and could result in our having underestimated O/H by up to a factor 2 in some cases (González Delgado et al. 1995). For Galactic HII regions, with temperature fluctuations included, Peimbert (1993) has derived $\Delta Y / \Delta Z$ values close to 3. An upper IMF slope of 1.7 and an upper mass limit of $50 M_\odot$ for SN progenitors gives $\Delta Y / \Delta Z = 3$ (Maeder 1992), so this may be the solution.

At present, efforts are being made towards still further refinements of helium abundance by discovering more low-metallicity systems (which reside in low-luminosity galaxies), better signal:noise and resolution using various telescopes and data reduction systems and improvements in atomic data. Skillman & Kennicutt (1993) have re-observed IZw18 and Skillman et al. (1994a) have studied UGC 4483 with results shown in Fig. 2. Further results from this group (Skillman et al. 1994ab) have been exhaustively analysed together with our data (Pagel et al. 1992) by Olive & Steigman (1994); from various subsets of the data, mostly including the definite WR objects which we had rejected, they find $Y_P = 0.232 \pm 0.003$ and $Y_P^{2\sigma} + \sigma_{syst} \leq 0.243$. Izotov et al. (1994) have observed ten HII galaxies from the Second Byurakan Survey including UGC 4483 and analyzed them with various combinations of Brocklehurst and Smits recombination coefficients. The right combination is presumably Smits (1994) giving the same dY/dZ as Pagel et al. (1992), but a higher $Y_P = 0.239 \pm 0.007$. The Y_P value given by Olive & Steigman looks like the best compromise, but radiative transfer effects not previously taken into account could raise the upper limit, perhaps by as much as 0.01 (Sasselov & Goldwirth 1994).

5. LITHIUM, BERYLLIUM AND BORON

The discovery of lithium in metal-deficient stars (Spite & Spite 1982) was a confirmation of SBBN, whereas the discoveries of beryllium (Gilmore et al. 1991) and boron (Duncan et al. 1992), though stimulated by ideas on inhomogeneous Big-Bang nucleosynthesis, are now thought to have little or no cosmological significance, although they raise plenty of problems about GCE.

A useful compilation of Li/H vs. [Fe/H] by Rebolo, Molaro & Beckman (1988) shows the main outline of the trends observed: Below [Fe/H] = -1.5 or so (and for stars that are not too cool or evolved; cf. Pilachowski et al. 1993), one has the Spite plateau with $12 + \log(\mathrm{Li/H}) \simeq 2.2$, while at higher metallicities there is a vast scatter due to depletion in stellar atmospheres, but with a fairly well defined upper envelope rising to somewhere near the meteoritic abundance of 3.3 at solar and higher metallicity. This is usually ascribed to a contribution from some sort of stellar production, and the upper envelope is fairly well fitted by adding to the primordial abundance another component that evolves *pari passu* with iron (Pagel 1991), or maybe a little more slowly (D'Antona & Matteucci 1991). There is a vast literature on processes leading to the depletion of lithium in stellar atmospheres, which are not well understood, although it seems that with new opacities and other developments some progress is being made (Swenson et al. 1994). The best evidence that depletion is not very great along the lithium plateau is the detection of ^{6}Li in the warm subdwarf HD 84937 (V. Smith et al. 1993).

Despite this, some evidence is emerging that the lithium plateau is not a true plateau, but has a gentle rise as a function of metallicity. This comes out from careful analyses of subdwarf lithium abundances as a function of effective temperature, showing a small but significant decrease with decreasing temperature consistent with the predictions of Pinsonneault et al. (1992) for non-rotating stellar models (Thorburn 1994; Norris et al. 1994). Analysis of the residual scatter around this temperature dependence leads to a noticeable increase of lithium (corrected to a standard effective temperature) with metallicity, implying a primordial lithium abundance possibly as low as $12 + \log(\text{Li/H}) = 2.0$ and an increase by maybe 0.1 of solar-system Li for only 0.01 of solar-system iron. This suggests significant ^{7}Li production by cosmic-ray $\alpha - \alpha$ fusion in the early Galaxy, which is not entirely unexpected in view of the abundance of ^{6}Li and those of beryllium and boron, discussed below.

Further evidence against the usual view that ^{7}Li comes mainly from stellar production on top of the Big Bang comes from measurements of the ^{7}Li/^{6}Li ratio in the interstellar medium which give about 12 (like the Solar System) or lower, and variation between clouds (Meyer et al. 1993; Lemoine et al. 1993), in one case being as low as 2 (Lemoine 1994b). An increase in ^{6}Li/^{7}Li since formation of the Solar System is not easy to understand if ^{7}Li comes from stellar production and ^{6}Li from cosmic-ray spallation and fusion (*cf.* Steigman 1993), and so there are arguments for reinvestigating the role of a low-energy component of cosmic rays (Beckman 1994)—as previously postulated to explain the ^{11}B/^{10}B ratio (Reeves 1994b)—particularly in view of recent evidence for a high flux of low-energy cosmic rays in Orion (Bloemen et al. 1994).

There is now quite a lot of data on beryllium in low-metallicity stars (Ryan et al. 1992; Gilmore et al. 1992; Boesgaard & King 1993) which indicates that there is no plateau at the abundance levels observed, but that beryllium evolves with a nearly constant ratio to iron or oxygen, which is unexpected in view of the need to have CNO elements present for spallation so that (for unchanged cosmic-ray flux relative to supernova rates and oxygen production) one expects it to behave like a "secondary" element rising approximately as the square of primary elements. At the same time, HST observations of HD 140283, with [Fe/H] $\simeq -2.6$, analyzed with the aid of non-LTE model atmosphere calculations, indicate a B/Be ratio of about 17 (Edvardsson et al. 1994), similar to the prediction for conventional cosmic-ray spallation (Walker et al. 1985), although ^{11}B could also have a primary source from neutrino processes in supernovae (Olive et al. 1994).

Several ideas have been put forward to explain the mysterious "primary" behaviour of beryllium and boron at low metallicities. Such behaviour is natural if they result from the accelerated component and if ancient cosmic rays had the same CNO abundance as now (Duncan et al. 1992) or if spallation took place mainly in the near-supernova environment before the dilution of synthesis products (Feltzing & Gustafsson 1994). Ryan et al. pointed out that, if the Galactic halo lost most of its material in an intense Galactic wind (Hartwick 1976), the resulting reduction in star formation rate would lead to a flattening of the initially quadratic dependence of Be on Fe or O, giving a crude fit to the data, and Prantzos et al. (1993) provided physical motivation for such a model by postulating a strengthening of the cosmic ray flux and hardening of its spectrum by a large grammage due to the great extent of the early halo. Spectral hardening had the desirable effect of reducing any lithium production from $\alpha - \alpha$ fusion associated with the large CR flux, as well as explaining a low B/Be ratio that was thought to hold at the time, two aspects that deserve re-examination in the light of the results reported above for both lithium and boron. Pagel (1994b) considered an inhomogeneous "chemo-dynamical" toy model and showed that in this case the cosmic-ray boost required could be considerably reduced, but he was no more successful than Ryan et al. or Prantzos et al. in producing a linear Be, O relation at the lowest metallicities observed, and the fit to the data is crude rather than convincing. However, the assumptions made in all these papers about cosmic rays are very simplistic, and the answer could lie there (cf. Fields et al. 1994).

REFERENCES

Balbes, M.J., Boyd, R.N. & Mathews, G.J. 1993, ApJ, 418, 229

Balser, D.S., Bania, T.M., Brockway, C.J., Rood, R.T. & Wilson, T.L. 1994, ApJ, in press

Beckman, J.E. 1994, The Light Element Abundances, P. Crane, Springer-Verlag

Bloemen, H. et al. 1994, AA, 281, L5

Boesgaard, A.M. & King. J.R. 1993, AJ, 106, 2309

Boesgaard, A.M. & Steigman, G. 1985, Ann.Rev.Astr.Ap., 23, 319

Brocklehurst, M. 1972, MNRAS, 157, 221

Brown, G.E. & Bethe, H.A. 1994, ApJ, 423, 659

Campbell, A., Terlevich, R.J. & Melnick, J. 1986, MNRAS, 223, 811

Carswell, R.F., Rauch, M., Weymann, R.J., Cooke, A.J. & Webb, J.K. 1994, MNRAS, 268, L1

Chiappini, C. & Maciel, W.J. 1994, AA, submitted

Clegg, R.E.S. 1987, MNRAS, 229, 31P

D'Antona, F. & Matteucci, F. 1991, AA, 248, 62

Davidson, K. & Kinman, T.D. 1985, ApJ Suppl., 58, 521

Duncan, D., Lambert, D.L. & Lemke, D. 1992, ApJ, 401, 584

Edvardsson, B., Gustafsson, B., Johansson, S.G., Kiselman, D., Lambert, D.L., Nissen, P.E. & Gilmore, G. 1994, AA, in press

Faulkner, J. 1967, ApJ, 147, 617

Feltzing, S. & Gustafsson, B. 1994, ApJ, 423, 68

Fields, B.D., Olive, K.A. & Schramm, D.N. 1994, FERMILAB-Pub-94/010-A, subm. to ApJ

Geiss, J. 1993, Origin and Evolution of the Elements, N. Prantzos, E. Vangioni-Flam & M. Cassé, Cambridge UP, p. 89

Gilmore, G., Edvardsson, B. & Nissen, P.E. 1991, ApJ, 378, 17

Gilmore, G., Gustafsson, B., Edvardsson, B. & Nissen, P.E. 1992, Nature, 357, 379

González Delgado, R.M. et al. 1995, ApJ, in press

Hartwick, F.D.A. 1976, ApJ, 209, 418

Izotov, Y.I., Thuan, T.X. & Lipovetsky, V.A. 1994, ApJ, in press

Kunth, D., Lequeux, J., Sargent, W.L.W. & Viallefond, F. 1994, AA, 282, 709

Kunth, D. & Sargent, W.L.W. 1983, ApJ, 273, 81

Lemoine, M. 1994ab, The Light Element Abundances, P. Crane, Springer-Verlag

Lemoine, M., Ferlet, R., Vidal-Madjar, A., Emerich, C. & Bertin, P. 1993, AA, 269, 469

Lequeux, J., Peimbert, M., Rayo, J.F., Serrano, A. & Torres-Peimbert, S. 1979, AA, 80, 155

Linsky, J.L. 1994, The Light Element Abundances, P. Crane, Springer-Verlag

Linsky, J.L. et al. 1993, ApJ, 402, 604

McCullogh, P.R. 1992, ApJ, 390, 213

Maeder, A. 1992, AA, 264, 105; 1993, AA, 268, 833

Marconi, G., Matteucci, F. & Tosi, M. 1994, MNRAS, in press

Melnick, J., Haydari-Malayeri, M. & Leisy, P. 1992, AA, 253, 16

Meyer, D.M., Hawkins, I., & Wright, E.L. 1993, ApJ Lett., 409, L61

Norris, J., Ryan, S.G. & Stringfellow, G.S. 1994, ApJ, 423, 386

Olive, K.A., Prantzos, N., Scully, S. & Vangioni-Flam, E. 1994, ApJ, 424, 666

Olive, K.A. & Steigman, G. 1994, Preprint UMN-TH-1230/94; OSU-TA-6/94

Pagel, B.E.J. 1991, Phys.Scripta, T36, 7

Pagel, B.E.J. 1993, Proc.Nat.Acad.Sci.USA, 90, 4789

Pagel, B.E.J. 1994a, The Light Element Abundances, P. Crane, Springer-Verlag

Pagel, B.E.J. 1994b, Cosmical Magnetism, D. Lynden-Bell, Dordrecht: Kluwer, p. 113

Pagel, B.E.J. & Kazlauskas, A. 1993, MNRAS, 256, 49P

Pagel, B.E.J., Simonson, E.A., Terlevich, R.J. & Edmunds, M.G. 1992, MNRAS, 255, 325

Pagel, B.E.J., Terlevich, R.J. & Melnick, J. 1986, PASP, 98, 1005

Peimbert, M. 1993, Rev.Mex.Astr.Astrofis., 27, 9

Peimbert, M. & Torres-Peimbert, S. 1974, ApJ, 193, 327

Perrin, M.-N., Hejlesen, P.M. Cayrel de Strobel, G. & Cayrel, R. 1977, AA, 54, 779

Pilachowski, C.A., Sneden, C. & Booth, J. 1993, ApJ, 407, 699

Pilyugin, L. 1993, AA, 277, 42

Pinsonneault, M.H., Deliyannis, C.P. & Demarque, P. 1992, ApJ Suppl., 78, 179

Prantzos, N. 1994, AA, in press

Prantzos, N., Cassé, M. & Vangioni-Flam, E. 1993, ApJ, 403, 630

Rebolo, R., Molaro, P. & Beckman, J.E. 1988, AA, 192, 192

Reeves, H. 1994a, Rev.Mod.Phys., 66, 193

Reeves, H. 1994b, The Light Element Abundances, P. Crane, Springer-Verlag

Roelfsema, P.R., Goss, W.M. & Mallik, D.C.V. 1992, ApJ, 394, 188

Rood, R.T., Bania, T.M. & Wilson, T.L. 1992, Nature, 355, 618

Ryan, S., Norris, J., Bessell, M.S. & Deliyannis, C. 1992, ApJ, 388, 184

Sasselov, D. & Goldwirth, D. 1994, Preprint

Searle, L. & Sargent, W.L.W. 1972, ApJ, 173, 25

Shaver, P.A., Kunth, D. & Kjär, K. (eds.) 1983, Primordial Helium, Garching: ESO

Simonson, E.A. 1990, PhD Thesis, Sussex University

Skillman, E.D. & Kennicutt, R.C., Jr. 1993, ApJ, 411, 655

Skillman, E.D., Terlevich, R.J., Kennicutt, R.C., Jr., Garnett, D.R. & Terlevich, E. 1994a, ApJ, in press; 1994b, in prep.

Smith, M.S., Kawano, L.H. & Malaney, R.A. 1993, ApJ Suppl., 85, 219

Smith, V., Lambert, D.L. & Nissen, P.E. 1993, ApJ, 408, 262

Smits, D.P. 1991ab, MNRAS, 248, 193, and 251, 316

Smits, D.P. 1994, MNRAS, in press

Songaila, A., Cowie, L.L., Hogan, C.J. & Rugers, M. 1994, Nature, 368, 599

Spite, F. & Spite, M. 1992, Nature, 297, 483

Steigman, G. 1993, ApJ Lett., 413, L73

Steigman, G. & Tosi, M. 1992, ApJ, 401, 15

Swenson, F.J. Faulkner, J., Rogers, F.J. & Iglesias, C.A. 1994, ApJ, 425, 286

Terlevich, R.J., Melnick, J., Masegosa, J. & Moles, M. 1991, AA Suppl., 91, 285

Thorburn, J.A. 1994, ApJ, 421, 318

Vangioni-Flam, E. & Cassé, M. 1994, ApJ, submitted

Vangioni-Flam, E., Olive, K. & Prantzos, N. 1994, ApJ, 427, 618

Walker, T.P., Mathews, G.J. & Viola, V.E. 1985, ApJ, 299, 745

Walker, T.P., Steigman, G., Schramm, D.N., Olive, K.A. & Kang, H. 1991, ApJ, 376, 51

Yang, J., Turner, M.S., Steigman, G., Schramm, D.N. & Olive, K. 1984, ApJ, 281, 493

Chemical Evolution of the Galaxy

J. W. Truran[1,2] and F. X. Timmes[1,3]

[1] Department of Astronomy and Astrophysics, Enrico Fermi Institute
University of Chicago, Chicago, IL 60637

[2] Max-Planck Institute for Astrophysics
D-85740 Garching, Germany

[3] General Studies Division, Lawrence Livermore National Laboratory
Livermore, CA 94550

Our knowledge of the early history of our Milky Way Galaxy remains woefully incomplete, shrouded as it is by the consequences of the intervening 10-15 billion years of halo and disk evolution. This early history of our Galaxy, it is now believed, was characterized by an epoch of star formation and supernova activity, occurring in a spheroidal protogalactic environment. The halo field stars, the globular clusters, and the Galactic bulge are all assumed to be remnants of this evolutionary phase. This halo stage presumably ended when the residual metal-enriched gas had dissipated its energy and collapsed into the equatorial plane, forming a disk-like substructure. The subsequent restriction of star formation to a thin Galactic disk (see, e.g., Burkert, Truran, & Hensler 1992) marked the onset of a stage of evolution that has persisted until today, during which the bulk of the star formation and nucleosynthesis activity in our Galaxy appears to have occurred. In this paper, we will be concerned with interesting abundance trends as a function of metallicity [Fe/H], for both the halo and the disk components of the Galaxy, and their possible implications for the star formation history and nucleosynthesis timescale.

Halo Star Abundance Trends. Reviews of interesting abundance patterns as a function of metallicity have recently been provided by Spite & Spite (1985), Lambert (1989), and Wheeler, Sneden & Truran (1989). A significant point emphasized in the latter review concerns the observed abundances, relative to iron, of the intermediate mass α-chain elements (nuclei from oxygen to titanium). These stellar abundance determinations generally show that these nuclei are enriched relative to iron by a factor of approximately three, viz. [(O,Mg,Si,S,Ca,Ti)/Fe] $\sim$ +0.5 dex in metal deficient halo dwarfs ([Fe/H] $\lesssim$ -1.0 dex). This trend is shown specifically for the [Ca/Fe] ratio in Figure 1. It is also important to recognize that similar abundance patterns are known to be exhibited by stars associated with the halo population of globular clusters (Wheeler *et al.* 1989; Brown, Burkert & Truran 1991).

High quality data regarding heavy element abundances in metal-poor stars is now also becoming increasingly available, and further significant abundance features are to be found here. It has been determined, specifically, that the heavy element abundance patterns in the most extreme metal-poor stars reflect a pure r-process contamination (Truran 1981; Gilroy *et al.* 1988).

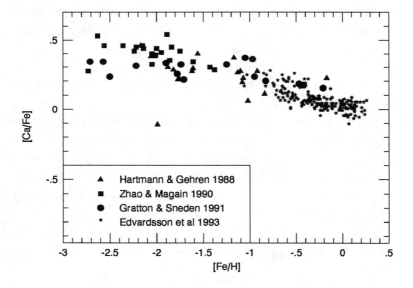

Fig. 1.— Evolution of the calcium to iron ratio [Ca/Fe] as a function of [Fe/H]. Many of the observed trends of calcium are typical of the α-chain nuclei – a factor of about three enhancement over solar in the halo, a slow decline due to mass and metallicity effects, followed by a drop down to the solar ratio. The star HD89499 from the Hartmann & Gehren (1988) survey, which has [Ca/Fe]=-0.11 dex and [Fe/H]=-2.0 dex, also possesses peculiar abundances of magnesium, aluminum and scandium.

As the metallicity increases, there is now evidence that the ratio s-process/r-process (as reflected in the Ba/Eu ratio) also increases (Mathews, Bazan, & Cowan 1992). These trends, as we shall see, suggest further constraints on the nucleosynthesis timescale.

A critical factor here is the recognition that the abundances that are expected to characterize the ejecta of Type II + Ib supernovae, whose initial masses are greater than 10 $M_\odot$, show similar distinguishing features (Tinsley 1980; Truran 1983, 1984; Woosley & Weaver 1986). The mass of iron ejected in such events, initially in the form of ^{56}Ni, has been found from observations of SNe 1987A and 1993J to be approximately 0.1 $M_\odot$. The α-chain nuclei in the mass region from oxygen to calcium are predicted, in recent theoretical studies, to be overproduced, relative to iron, by a factor ~ 3. This overproduction factor of three is uncertain, due to the intrinsic non-monoticity of the pre-supernova iron cores with stellar mass, and irresolution of the parameters that determine 1-dimensional explosion models (i.e the mass cut and the explosion energy). Uncertainties aside, the overproduction of the classical α-chain elements is thought to be a unique characteristic of nucleosynthesis in massive stars, which is then presumably reflected in the abundance patterns observed in metal-deficient field halo stars and globular cluster stars. That is, one would expect $[(O - Ca)/Fe] \sim 0.5$ dex for the α-chain elements.

As Figure 1 illustrates, the oldest stars do indeed exhibit signatures of massive star nucleosynthesis. A more interesting issue is the interpretation of the deviations from these early trends. These deviations seem to first appear at the point at which the metallicity, as reflected in [Fe/H], approaches $\sim 1/10\ Z_\odot$, which is approximately the metallicity of the oldest disk stars. Presumably, this change in behavior is a consequence of the first introduction of iron peak products from Type Ia supernovae. The fact that there exists a delay in the entry of the iron from Type I supernovae is perhaps not a surprise, however, as the timescale for the evolution of an intermediate-low mass star to a white dwarf, in the progenitor binary system, can range from $\sim 10^8$ to greater than 10^9 years. A further timescale constraint on nucleosynthesis is that provided by the gradual increase in the s-process/r-process ratio, as s-process nucleosynthesis is expected to occur in lower mass AGB stars, with lifetimes in excess of $\sim 10^9$ K. There is also some indication that s-process enrichment precedes the entry of the iron from SNe Ia.

History of Supernova Activity in the Galaxy. Recently, van den Bergh & Mcclure (1994) have reanalyzed the Evans *et al.* extragalactic supernova observations. Citing SN 1987A and SN 1991bg as good evidence, they relaxed the assumption that supernova display small intrinsic variations at maximum light by investigating the effects of different luminosity functions. They also dropped the $\sin i$ correction term, because the supernova survey of Muller *et al.* (1992) found no evidence for a dependence on inclination angle of the host galaxy. For a total Galactic blue luminosity of 2.3×10^{10} $L_\odot$, a Hubble constant of 75 km s^{-1} Mpc^{-1}, and a Sbc Galactic morphology, they estimate that the present Type II + Type Ib rate is 2.4 - 2.7 per century while the Type Ia rate is 0.3 – 0.6 per century. Thus, the the total rate is about 3 per century, and the ratio of core collapse to thermonuclear events is about 6.

While there clearly exist uncertainties associated with the history of Type Ia and Type II + Ib activity, there are several simple order of magnitude estimates that can aid in understanding these supernova ratios. In particular, one can estimate the *integrated* ratio of the total number of collapse events events to the total number of thermonuclear driven events. Assume (1) that each core collapse event produces 0.1 $M_\odot$ of iron and that they are responsible for a fraction F_{II+Ib} of the solar iron abundance, and (2) that each Type Ia produces 0.6 $M_\odot$ of iron and accounts for a fraction F_{Ia} of the solar iron abundance. These imply

$$\frac{N_{II+Ib}}{N_{Ia}} = 6 \left(\frac{F_{Ia}}{F_{II+Ib}}\right) \left(\frac{M(Fe)_{Ia}/0.6}{(M(Fe))_{II+Ib}/0.1}\right),$$

where the scaling with the fractions and iron production is explicitly shown. Thus, taking the canonical values of $F_{Ia} = 1/3$ and $F_{II+Ib} = 2/3$, the number ratio is 3. This is roughly consistent with the ratio of the frequencies observed in other (Sb or Sbc) spiral galaxies today, with perhaps some preference for the view that our Galaxy is type Sb instead of an Sbc. On the other hand, by taking $F_{Ia} = F_{II+Ib} = 1/2$, one duplicates the van den Bergh & Mcclure (1994) estimate, with the implied Sbc Galactic morphology.

Evolution of the Galactic supernova rates, calculated by integrating in radius across the model Galaxy, are shown in Figure 2. During the early stages of evolution there is a large amount of gas which gets turned into a large number of stars. The death of these massive stars gives a correspondingly large early Type II + Ib supernova rate. As the amount of gas decreases, due to formation of compact remnants and long-lived low mass stars, the Type II +Ib supernova rate decreases with time. The onset of Type Ia supernovae is delayed, due to the longer lifetime of the intermediate mass star which produces a white dwarf. After 15 billion years, the calculated Type II + Ib supernova rate is 3.07 per century and 0.53 for Type Ia supernova, in excellent agreement with the van den Bergh & McClure (1994) estimates. There is no inconsistency of the peaking of the Type II rate at early epochs with the order of magnitude estimate given above, since the order of magnitude estimate is concerned with the integrated number of events.

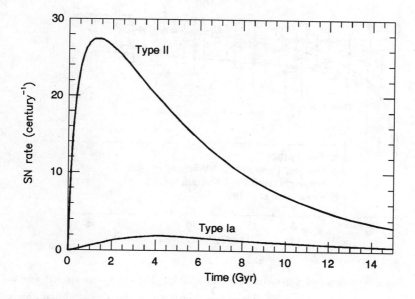

Fig. 2.— Number of supernova per century as a function of time. From Timmes *et al.* (1994).

The Age Metallicity Relation for the Disk. Abundance determinations for disk stars have also recently served further to constrain the nature of a possible "age-metallicity relation" for the Galactic disk (Twarog 1980; Carlberg *et al.* 1985; Meusinger, Reimann & Stecklum 1991; Edvardsson *et al.* 1993), whose nature contains important implications for the dynamical and chemical evolution of the disk. Employing a direct, spectroscopically calibrated age-metallicity relationship, Edvardsson *et al.* (1993) find evidence for only a relatively weak relation: the *mean* metallicity of the disk, as reflected it the iron abundance, was determined to vary by only a factor ≈ 3 over the approximately 10 to 12 billion year lifetime of the disk. In addition, the scatter in [Fe/H] values at any age was comparable to or exceeded the magnitude of this trend. This observed behavior is generally consistent with a relatively constant rate of nucleosynthesis over the history of the disk, at a level roughly comparable to, and never greatly exceeding, the present value.

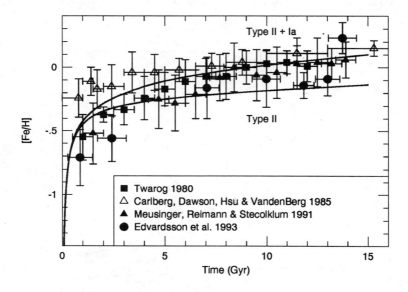

Fig. 3.— Solar neighborhood age-metallicity relationship. The error bars are not due to intrinsic uncertainties of a single observation, but represent the spread of many stars in the binned [Fe/H] data. From Timmes *et al.* (1994).

Type II and Type Ia supernovae provide two sources of iron production, each operating on distinctly different time scales, and each injecting very different amounts of iron (see the supernova ratio estimate given above). Figure 3 indicates that the inclusion of Type Ia events is important for reproducing the observed iron evolution, and suggests that about 2/3's of the solar iron abundance comes from Type II events and about 1/3 comes from Type Ia supernovae. Variations in these two fractions depend primarily on the iron yields of Type II supernovae, but the main-sequence lifetimes are also important. Fractions of 50% from Type Ia supernova and 50% from Type II supernova are well within the uncertainties associated with the iron yields from each type of supernovae. It may be that 2/3 of the solar iron abundance comes from Type Ia events and 1/3 from Type II events without doing grave injustice to the stellar physics.

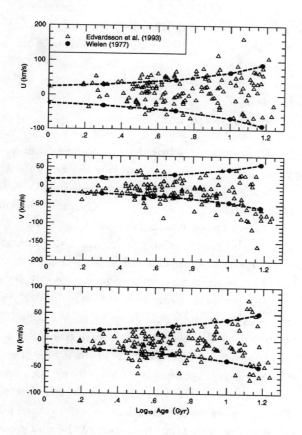

Fig. 4.— U, V and W components of the space motion as a function of stellar age.

There are two canonical mechanisms for achieving a real scatter in [Fe/H], or any other abundance ratio. The first explains the scatter by diffusion of stellar orbits or of the gas that becomes stars. Physical processes that may account for the diffusion are (1) that stellar velocities are randomized through chance encounters with other stars and interstelar cloud complexes, or (2) that stars are stochastically accelerated by scattering off the mass concentrations in a spiral density wave. The second mechanism relies on relaxing the assumption that the nucleosynthetic products of stars are mixed on an extremely short time scale. Models which invoke instantaneous mixing can only explain the average trends. Only a few models that relax the short mixing time scale assumption have been attempted (Malinie et al. 1993), but these tend to suffer from the addition of a few ad hoc parameters.

Figure 4 shows the space motion relative to the Sun, components U, V and W, as a function of stellar age, for the stars in the Edvardsson et al. (1993) survey. Also shown are the consequences of using Wielan's (1977) results for diffusion of stellar orbits for a velocity and time dependent diffusion coefficient. In general, the fit is very reasonable, even if the precise physical nature of the "diffusion" mechanism remains shrouded.

Halo and Disk Star Formation Timescales. Figure 5 shows the W space motion component as a function of [Fe/H] for the two oldest age bins of the Edvardsson et al. (193) survey. The observations show a distinct demarcation between stars with $1.0 \leq$ Log Age ≤ 1.1 Gyr and those with $1.1 \leq$ Log Age ≤ 1.2 Gyr. Stars in the younger age bin have a wide spead in W and [Fe/H], thought to be characteristic of an old disk population, while stars in the older age bin have a large spread in W over a quite small range in [Fe/H], which can be interpreted as evidence for a halo or thick disk population of stars. The solid dot is the mean values of W and [Fe/H] in the given age bin. The "error" bars are the result of applying Wielen's (1977) diffusion coefficients combined with the observed [Fe/H] gradients of -0.1 dex/kpc in the radial direction and -0.5 dex/kpc in the vertical z direction. Stars within the "error" bars can be explained by 'diffusion and the observed metallicity gradients. The scatter in [Fe/H] is examined in more detail by Timmes & Truran (1995). We interpret the change in behaviors noted above, occurring at $\sim 11 \times 10^9$ years, as indicating the onset of the disk phase of star formation activity in our Galaxy.

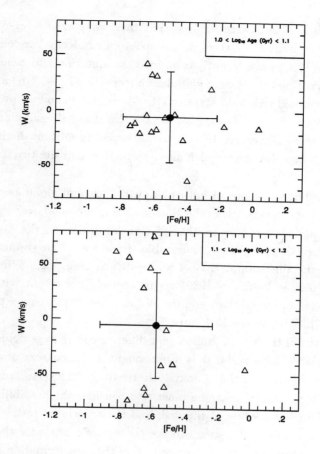

Fig. 5.— W component of the space motion as a function of [Fe/H] for the oldest stars in the Edvardsson *et al.* (1993) survey.

The age of the galactic halo population of globular clusters is currently estimated to lie in the range 13-18 Gyr (see e.g. the review by Cowan, Thielemann, & Truran 1991). This age is generally assumed to reflect the age of the Galaxy itself, as the spheroidal population of globular clusters certainly dates the time of halo collapse and early galactic chemical and dynamical evolution. In contrast, the disk component of the Galactic globular cluster system (Zinn 1985) is characterized by a very different kinematics and metallicity distribution, and appears to be 2-3 Gyr younger (Lee, Demarque,

& Zinn 1990). It is also now recognized that there may exist a significant spread in age of individual clusters in the spheroid. Indeed, VandenBerg, Bolte & Stetson (1991) have recently examined this question and concluded that, while the most metal deficient globular clusters ([Fe/H] $\sim$ -2.1) appear to be coeval, the metal rich halo systems ([Fe/H] $\sim$ -1.3) show a spread in age of perhaps 2-4 Gyr. Zinn (1993) has argued that the halo globular cluster population can be divided into two groups, based upon their distinctive horizontal branch morphologies, which also appear to constitute two distinct age populations.

The combined observations of both halo field stars and globular cluster stars suggest that star formation in the Galactic halo proceeded over a period of a few Gyr, and that the restriction of star formation to the Galactic disk was therefore delayed by a comparable timescale. The younger age determined for the disk population of white dwarfs (see, e.g., Winget *et al.* 1987 and Iben & Laughlin 1989) seems entirely consistent with this conclusion, as is the observed change in the W component of the space motion as a function of [Fe/H], shown in Figure 5.

Understanding of the several billion year difference in the age determinations for the Galactic halo and disk is often sought in the context of merger models for the formation of the Galaxy. Alternatively, Burkert, Truran, & Hensler (1992) have argued that one must also consider the possibility of a finite timescale for the cooling of the disk and the associated restriction of star formation to a thin disk region. In the context of models for the early chemodynamic evolution of the Galaxy, they find that the formation of such a thin disk structure can be delayed by 4-6 Gyr. It seems clear that an improved understanding of the star formation and nucleosynthesis histories of our Galaxy will require further and more detailed numerical studies of its combined chemical and dynamical evolution.

References

Brown, J. H., Burkert, A., & Truran, J. W. 1991, ApJ, 376, 115
Burkert, A., Truran, J. W., & Hensler, G. 1992, ApJ, 391, 651
Carlberg, R. G., Dawson, P. C., Hsu, T., VandenBerg, D. A. 1985, ApJ, 294, 674

Edvardsson B., Andersen J., Gustafsson B., Lambert D. L., Nissen P. E., & Tomkin J. 1993, A&A, 275, 101
Evans, R., van den Bergh, S., & McClure, R. D. 1989, ApJ, 345, 752
Gilroy, K. K., Sneden, C., Pilachowski, C. A., & Cowan, J. J. 1988, ApJ, 327, 298
Hartmann, K., & Gehren, T. 1988, A&A, 199, 269
Iben, I. Jr., & Laughlin, G. 1989, ApJ, 341, 312
Lacey, C. G. 1984, MNRAS, 208, 687
Lambert, D.L. 1989, in Cosmic Abundances of Matter, ed. C.J. Waddington (New York: AIP Conference 183), 168
Lee, Y.-W., Demarque, P., & Zinn, R. 1990, ApJ, 350, 155
Malinie, G., Hartmann, D. H., Clayton, D. D., & Mathews, G. J. 1993, ApJ, 413, 633
Mathews, G. J., Bazan, G., & Cowan, J. J. 1992, ApJ, 391, 719
Meusinger, H., Reimann, H.-G., & Stecklum, B. 1991, A&A, 245, 57
Muller, R. A., Newberg, H. J. M., Pennypacker, C. R., Perlmutter, S., Sassen, T. P., & Smith, C. K. 1992, ApJ, 384, L9
Spite, M., & Spite, F. 1985, ARA&A, 23, 225
Timmes, F. X., & Truran, J. W. 1995, ApJ, in preparation
Timmes, F. X., Woosley, S. E., & Weaver, T. A. 1995, ApJ Supp., in press
Tinsley, B. M. 1980, Fund. Cosmic. Phys., 5, 287
Truran, J. W. 1981, A&A, 97, 391
Truran, J. W. 1983, Mem. S. A. Italy, 54, 113
Truran, J. W. 1984, Ann. Rev. Nucl. Part. Sci., 34, 53
Twarog, B. A. 1980, ApJ, 242, 242
van den Bergh, S., McClure, R. D., & Evans, R. 1987, ApJ, 359, 277
VandenBergh, D. A., Bolte, M. & Stetson, P. B. 1990, AJ, 100, 445
Wheeler, J. C., Sneden, C., & Truran, J. W. 1989, ARA&A, 27, 279
Wielen, R. 1977, A&A, 60, 263
Winget, D. E., Hansen, C. J., Liebert, J., Van Horn, H., Fontaine, G., Nather, R. E., Kepler, S. O., & Lamb, D. Q. 1987, ApJ, 315, L77
Woosley, S. E., & Weaver, T. A. 1986, AAR&A, 24, 205
Zhao, G., & Magain, P. 1990, A&A, 238, 242
Zinn, R. 1985, ApJ, 293, 424

Dating methods from stellar evolution

Roberto Buonanno, Giacinto Iannicola

Osservatorio Astronomico di Roma

Abstract.
The principal relative dating methods from stellar evolution are reviewed and a new one, named $\Delta V_{0.05}$, is presented. This new parameter, based on the difference in magnitude between the Horizontal Branch and a point on the main sequence, is particularly robust and largely independent from the cluster metallicity (the only residual theoretical uncertainty being the slope $\Delta M_V(HB)/\Delta[Fe/H]$, i.e. the so-called Sandage effect). This new parameter is used to determine the empirical calibration $\Delta(B-V)_{TO-RGB}$ $vs[Fe/H]$ establishing a new valuable technique of calibration.

Key words: Globular Clusters Ages, Stellar Evolution.

1. INTRODUCTION

Stellar evolution offers one of the most direct methods to assess the age of the Universe. Sandage (1953) showed that this method is particularly strightforward when applied to chemically homogeneous samples of stars like those in globular clusters.

From the colour-magnitude diagram of the globular cluster M3 Sandage concluded that *"... if the turnoff-point from the main sequence in M3 and M92 are identified with the Schönberg-Chandrasekhar limit, ... computation of the rate of mass conversion at a given luminosity give 5×10^9 years as the time elapsed since all stars in M3 and M92 were on the main sequence"*.

The age derived for M3, so close to that of the Sun, is instructive insofar as it shows that a good CMD is only one of the ingredients needed to assess the age of glubolar clusters, the others being reliable models and realistic values of the input parameters.

Concerning the observation of globular clusters, an enormous progress has been made in the last ten years. This progress is due to the increased sensitivity of panoramic detectors and to the improved techniques of calibration.

A similar improvement has been obtained with regard to the models by many investigators (Ciardullo & Demarque, 1977, VandenBerg & Bell, 1985, Bertelli et al., 1990, Straniero & Chieffi, 1991,) so that it is possible, in principle, to determine detailed isochrones from grids of stellar evolutionary tracks and to compare such isochrones to the observed CMD.

2. ABSOLUTE DATING TECHNIQUES

To perform a direct comparison between observations and theory, the independent knowledge of several parameters is required. Among the others, the most important parameters are:
1) helium abundance and metallicity (respectively Y and Z)
2) distance
3) reddening

a) Metallicity

The heavy element abundance is usually represented by the observed iron abundance, $[Fe/H]$, determined through the observed lines along with models of atmosphere. Being the scale of Zinn & West (1984) now generally adopted, it is found that almost all the galactic clusters have been found to range from -1 to -2.3 in $[Fe/H]$.

The existing models are usually computed with a typical uncertainty of ± 0.15 *dex* <u>assuming</u> that the proportion of the heavy elements scale with that in the Sun.

However, there exists now a growing amount of data showing that low-metallicity stars present lines of all the α-elements enhanced by a factor of about 0.4, with respect to that in the Sun.

This result has far-reaching consequences as it puts into questions all the ages obtained so-far. In fact, Simoda & Iben (1968,1970) and Renzini (1977) found that the elements beyond the He contribute both to the opacity and to the nuclear burning and, therefore, alter the colors and the luminosity of the tracks.

In conclusion, the evidence of the enhancement of the α-elements in metal-poor stars, together with the above-mentioned theoretical results urges the revision of the models and of the ages obtained.

Although low-temperature opacities are not yet available untill now, Salaris *et al.* (1993) computed a set of evolutionary models with α-enanched chemical composition, having *"demonstrated that most of the evolutionary properties of Pop. II stellar models are not influenced by the metal content included in the opacity coefficient below 12.000 K"*.

b) Helium

Concerning the helium mass fraction of the stars in globular clusters, one usually adopts the value determined with the R-method (Buzzoni *et al.* 1983) or through the observations of the extragalactic HII regions (Davidson & Kinman 1985) or by the standard Big Bang nucleosynthesis (Boesgaard & Steigman 1985).

c) Distance

A major problem is the determination of the distance to the cluster. The standard candles commonly used, namely the absolute magnitude of the RR Lyrae or that of the horizontal branch, are still controversial at least as regards the zero point (in fact, the oft-stated controversy on the slope $\Delta M_V(HB)/\Delta[Fe/H]$ is probably no longer valid and will be the topic of a forthcoming paper).

In addition, new standard candles like the white dwarf cooling sequence seems now theoretically uncertain. It should be kept in mind, in any case, that in order to know the ages of GCs within the 10%, their distance moduli must be known within 0.1 mag.

d) Reddening

Concerning the reddening, we note that a "typical" error, $\sigma(E_{B-V})$ is of the order of $0.02 mag$ for low-reddening globular clusters. This immediately translates into an error of $1.5 Gyr$ in the estimate of the age.

The indeterminacies mentioned above naturally add to the systematic errors introduced by the use of theoretical models. Among the latter we may mention the errors determined by the present poor knowledge of convection theory and by the transformations from temperatures to colors which heavily affect model radii and temperatures. All in all, the uncertainties and the systematics mentioned above make the absolute ages currently quoted for globular clusters not yet fully reliable.

We will deal, therefore, for the rest of the paper, only with relative cluster ages where the models are used only differentially or not at all.

3. RELATIVE CLUSTER AGES

a) ΔV_{HB}^{TO}

This dating method has been thoroughly described by Iben and Renzini (1984) and is based on the property of the HB luminosity to remain unchanged as the cluster evolves while the luminosity of the main sequence turn-off becomes fainter.

ΔV_{HB}^{TO} is defined as the magnitude difference between the horizontal branch and the turnoff at the color of the latter and it is independent from the assumed reddening and distance. In addition, theoretical models show that it is only slightly affected by indeterminacies on the assumed metallicity of the cluster. In fact, the "typical" error of ± 0.15 dex in $[Fe/H]$ translates into a ± 0.4 Gyr in age.

There are two major problems with this method:

1) the region around V_{TO} is substantially vertical in the color-magnitude plane and

2) many clusters do not have stars in the horizontal part of the HB (which tends to assume a nearly vertical morphology in the plane V, B-V) making the estimate of ΔV_{HB}^{TO} extremely uncertain.

This latter is a particularly disappointing characteristic, as clusters with HB "blue tails" could all belong to the same class of very old globulars and are potentially interesting with respect to the problem of dating the Galaxy.

b) $\Delta V_{0.05}$ method

This technique is an improvement of the previous method as it avoids the first of the two major problems identified above, i.e. the difficulty to accurately estimate the turn-off luminosity.

Following VandenBerg et al. (1990), hereafter VBS, who defined a point on the upper main sequence that is $0.05 mag$ redder than the MSTO, Buonanno et al. (1994), hereafter BCPF, defined the parameter $\Delta V_{0.05}$ as the distance in magnitude between the HB and $V_{0.05}$, the point on the main sequence defined by VBS. Clearly $\Delta V_{0.05}$ shares with ΔV_{HB}^{TO} the advantage of being independent from reddening and distance to the cluster, and shows an almost identical sensitivity to age-variations. As a bonus, $\Delta V_{0.05}$ is even less sensitive than ΔV_{HB}^{TO} to indeterminacies in the estimate of $[Fe/H]$. This is shown in figure 1 where the run of $\Delta V_{0.05}$ with $[Fe/H]$ is shown for different absolute ages.

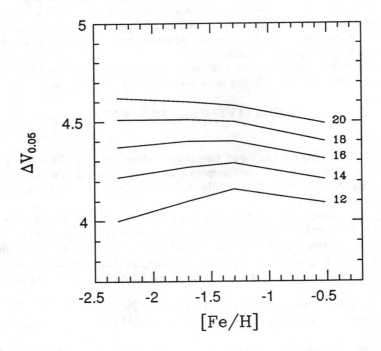

Fig. 1. The parameter $\Delta V_{0.05}$ vs $[Fe/H]$ for different ages. Absolute ages are labelled. The models used are those of Castellani et al. (1991) and Straniero and Chieffi (1991) for the HB and the TO, respectively.

c) The $\Delta(B-V)_{TO-RGB}$ method

The $\Delta(B-V)_{TO-RGB}$ has been independently invented by Chieffi & Straniero (1989), VBS, and Sarajedini & Demarque (1990) and is defined as the *color* difference between the turnoff and the base of the Red Giant Branch. The basis of the method is the fact that the temperature of the RGB is a weak function of age while the luminosity and the temperature of the MSTO decrease strongly with the age: the *difference* in temperature (and, then, in color) between the MSTO and RGB is therefore a good age-indicator. Since this parameter is independent of cluster reddening and distance, easy and straightforward to determine, it is observationally attractive. Naturally, $\Delta(B-V)_{TO-RGB}$ is totally independent from the HB morphology and it is ideally applicable to all globulars for which the CMD has been determined.

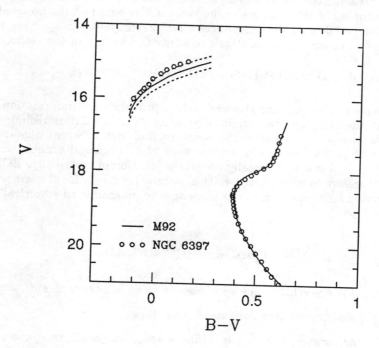

Fig. 2. The fiducial line of NGC 6397 (shifted by $2.17 mag$ in V and $-0.168 mag$ in $B-V$) is compared to that of M92. It turns out that as an artifact of the registration, the horizontal age-parameter $\Delta(B-V)_{TO-RGB}$ seems to indicate that two clusters are stricly coeval while the HB of NGC 6397 appears brighter than that of M92 (the dotted lines running parallel to the HB of M92 indicate the expected location of the HB of two clusters $2 Gyr$ respectively younger or older than M92).

However there are problems with this method. In particular, the RGB colors are so sensitive to the uncertainties in convection theory and in the temperature-color transformations that it is easily possible that the color-difference $\Delta(B-V)_{TO-RGB}$ should be reduced or increased relatively to that of an isochrone of different metallicity. For this reason VBS performed their analysis of relative ages, grouping galactic clusters in "metallicity boxes" which appear to be incommunicable one from the other. This is a substancial problem as shown in figure 2 where the fiducial sequence of M92 (Stetson & Harris 1988) and of NGC 6397 (Buonanno et al. 1989) has been superimposed following the precepts of VBS. Although the lower RGB and the MS of the two clusters have been aligned in figure 2, the two HBs show a clear displacement. We interpret this result as due to the difference in metallicity between M92 and NGC 6397 which, although small, nonetheless heavily affects the estimate of relative ages of the two clusters based on the horizontal age parameter. All in all this introduces a serious limitation in applying the $\Delta(B-V)_{TO-RGB}$ method as it stands to study the history of the Galaxy.

4. DISCUSSION AND CONCLUSIONS

The ages of the globular clusters system provide a key information to understand the story and the origin of galaxies. The age dating techniques outlined in the previous paragraph seem to find two different obstacles: $\Delta V_{0.05}$ cannot be applied to old clusters with blue horizontal branch, while $\Delta(B-V)_{TO-RGB}$ is a theoretically uncertain function of metallicity. BCPF proposed therefore to measure the vertical parameter $\Delta V_{0.05}$ for clusters with well-populated HBs and then to use these ages to determine an empirical relationship:

$$\Delta(B-V)_{TO-RGB} = f([Fe/H], \Delta t)$$

where Δt is the relative age with respect to an arbitrary cluster.

The proposed procedure consists in four steps:

1) Use the parameter $\Delta V_{0.05}$ to define a set of coeval clusters in a wide range of metallicity.
This step is easily accomplished selecting clusters with high-quality CMD and well-populated HB.

2) Check the helium abundance.
This is performed applying the R-method This requires well-populated CMDs along the RGB and the HB. As some clusters with $[Fe/H] > -1$ turn out to be relatively He-rich (for instance 47 Tuc and NGC 6838 give $Y = 0.27 \pm 0.02$ and $Y = 0.029 \pm 0.03$, respectively) and, consequently, could

present an anomalously bright HB, it is wise to limit the empirical calibration to clusters metal-poorer than $[Fe/H] = -1$.

3) Measure the parameter $\Delta(B-V)_{TO-RGB}$ for clusters known to be coeval

This step is accomplished following the procedure described by VBS and offers the opportunity to find empirically the function $\Delta(B-V)_{TO-RGB}$ vs $[Fe/H]$ without any need of the models.

BCPF in fact obtained

1) $\quad \Delta(B-V)_{TO-RGB} = 0.325(\pm 0.008) + 0.049(\pm 0.005)[Fe/H]$

with a linear best fit to the data available for 12 clusters (relation valid for $[Fe/H] < -1$).

4) Find the relative age of clusters with Blue HBs

The final step is to measure $\Delta(B-V)_{TO-RGB}$ for clusters for which the vertical age parameter, $\Delta V_{0.05}$, cannot be reliably estimated and compare the observed value of $\Delta(B-V)_{TO-RGB}$ with that given by equation 1).

REFERENCES

Bertelli, G., Betto, R., Bressan, A., Chiosi, C., Nasi, E. and Vallenari, A. 1990, A.A. Suppl., **85**, 845.

Boesgaard, A. and Steigman, G. 1985, Ann. Rev. AA **23**, 319

Buonanno, R., Corsi, C.E. and Fusi Pecci, F. 1989, A.A. **216**, 80

Buonanno, R., Corsi C.E., Fusi Pecci, F. and Pulone, L. 1994, in preparation (BCPF)

Buzzoni, A., Fusi Pecci, F., Buonanno, R., Corsi, C.E. 1983, A.A., **128**, 94

Castellani, V., Chieffi, A. Pulone, L. 1991, Ap.J. Suppl., **76**, 911

Chieffi, A. and Straniero, O. 1989, Ap.J. Suppl., **71**, 47

Ciardullo, R.B. and Demarque, P. 1977, "Tables of Isochrones", Yale Trans., Vol. 33

Davidson, K. and Kinman, T.D. 1985, Ap.J. Suppl., **58**, 321

Iben, I. and Renzini, A. 1984, Phys. Rep. **105**, 329

Lambert, D.L. 1989, in "Cosmic Abundances of Matter", ed. C.J. Weddington (N.Y.:AIP), 168

Renzini, A. 1977 in "Advenced Stages in Stellar Evolution, ed. P. Bouvier & A. Maeder, 152

Salaris, M., Chieffi, A., Straniero, O. 1993, Ap.J. **414**, 580

Sandage, A.R. 1953, A.J. **58**, 61

Sarajedini, A. and Demarque, P. 1990, Ap.J, **365**, 219

Simoda, M. & Iben, I. 1968 Ap.J., **152**, 509

Simoda, M. & Iben, I. 1970 Ap.J. Suppl., **22**, 81

Stetson, P.B. and Harris, W.E. 1988, A.J. **96**, 909

Straniero O. and Chieffi A., 1991, Ap.J. Suppl., **76**, 525
VandenBerg, D.A. and Bell, R.A. 1985, Ap.J. Suppl., **58**, 561
VandenBerg, D.A., Bolte, M. and Stetson, P.B. 1990, A.J., **100**, 445 (VBS)
Zinn, R.J. and West, M.J. 1984, Ap.J. Suppl. **55**, 45

Gamma-Ray Line Astronomy and Interstellar Radioactivity

Mark D. Leising and Donald D. Clayton

*Department of Physics and Astronomy,
Clemson University, Clemson, SC 29634-1911*

The interstellar medium contains a substantial amount of radioactive ^{26}Al, which is well known from its γ-ray line emission, and probably holds a similar quantity of ^{60}Fe, which has yet to be seen in γ-rays. The interstellar abundances of these place significant constraints on the current global Galactic nucleosynthesis rate and therefore on models of Galactic chemical evolution. We review the γ-ray observations to date and their implications. The nucleus ^{44}Ti might have been detected in the supernova remnant Cas A, but surprisingly has not been detected in any younger remnants. These observations offer considerable insight into supernova explosion dynamics and constrain the very recent Galactic supernova rate. Both ^{26}Al and ^{60}Fe, along with other radioactivities too rare to be detected anytime soon in γ-rays, are also known to have existed in significant quantities in the forming solar system. It is natural then to compare those abundances inferred for the presolar nebula with expected interstellar abundances. Our arguments suggest that most of the solar system extinct radioactivities can have been inherited from the standard interstellar condition; but others, especially ^{26}Al, require a special injection event or production processes in the star-forming cloud itself.

INTERSTELLAR ^{26}AL

Mahoney et al. (1984) discovered a strong 1.809 MeV ^{26}Al line from the central Galaxy. Their *HEAO-3* high energy-resolution measurement left no doubt as to its identification. Share et al. (1985) used the very broad field *Solar Maximum Mission (SMM)* spectrometer and its ten year database to determine with high precision the total inner Galaxy 1.809 MeV line flux, $4.0 \pm 0.4 \times 10^{-4}$ cm^{-2} s^{-1}. This flux corresponds to roughly 1.8 M$_\odot$ of ^{26}Al if it is distributed like supernova remnants in the Galaxy. Diehl et al. (1994) have added important new information with the imaging Compton telescope, COMPTEL, on the *Compton Gamma Ray Observatory (CGRO)*. Their images show a few significant point-like features dominating the emission. This in itself implicates young objects, therefore relatively massive stars, as the source of some if not all of the ^{26}Al. They argue that the bright spots are foreground regions which contribute a significant part of the total flux but do not represent a large mass of ^{26}Al. If they subtract these features, the diffuse flux left in the COMPTEL data require approximately 1 M$_\odot$ of ^{26}Al. One of these bright spots is coincident with the Vela supernova remnant. It might ultimately be demonstrable that the Vela supernova or the progenitor is the source of that ^{26}Al (Oberlack et al. 1994), although contributions from several supernova can not now be ruled out. The OSSE instrument also on the *CGRO* has also detected ^{26}Al emission but at surprisingly low flux levels (Leising et al. 1994). Those data are not entirely consistent with any models or with the COMPTEL maps.

Clayton (1984) argued that supernovae could not be the source of the ^{26}Al, without overproducing stable ^{27}Al. Yet clearly supernovae could be adequate sources of ^{26}Al if three per century eject 0.7×10^{-4} M$_\odot$ of ^{26}Al (Weaver and Woosley 1993). This apparent conflict between simple models of Galactic chemical evolution and the estimated rate of core-collapse supernovae has been resolved.

Weaver and Woosley's (1993) production ratio p(^{26}Al)/p(^{27}Al)=6×10^{-3} for Type II supernovae is three times larger than earlier values. Clayton (1984) also employed an overly restrictive closed-box model of Galactic chemical evolution, which sets the interstellar ratio ^{26}Al/^{27}Al equal to their production ratio times the ratio of the ^{26}Al lifetime to the age of the Galactic disk. That model is contrary to the growing belief that the mass of the Galactic disk grew for many Gyr while low-metallicity material fell onto it. Clayton (1988) presented many of the reasons for this belief and also its effect on the interstellar concentrations of radioactive nuclei. In a family of exact solutions (Clayton 1985) the infall is characterized by a parameter k; namely the infall rate $f(t)$ is given by $k/(t+\Delta)$ times the current mass of the ISM. The net effect is that all interstellar radioactivities have higher concentration with respect to a stable reference isotope by a factor $(k+1)$. Clayton et al. (1993) adopted these ideas and found a value $k \simeq 3$-4 brings the ^{26}Al mass into line with observations. If $k = 4$, today's steady-state ISM ratio becomes ^{26}Al/^{27}Al=3×10^{-6}, giving 1.5 M$_\odot$ of ^{26}Al from supernovae. They also pointed out that the best way to set the factor $(k+1)$ measuring the past infall is not by the classic tests but by a comparison of the ISM mass fraction of a supernova product (e.g. ^{16}O or ^{27}Al) with the value it would have had in a closed-box model using the same supernova yield.

If it proves correct that one solar mass of ^{26}Al should be attributed to the general ISM smaller values of k are allowed. Other factors might be important. One should account for the Galactic radial metallicity gradient if the ^{26}Al/^{27}Al production ratio increases in proportion to the initial metallicity of the massive stars involved. The importance of past infall is probably greater as one moves away from the Galactic center, at least as suppression of metallicity growth is concerned. Early infall, even very heavy, still allows the disk to evolve almost as a closed box subsequently; it is continuing infall that suppresses metallicity. If the matter of low angular momentum near the center was in place earlier than that of higher angular momentum farther out, the inner Galaxy may have higher metallicity because it has evolved more nearly as a closed-box. Then the effective enhancement factor $(k+1)$ in the inner regions may offset the metallicity effect. Timmes et al. (1994 and this volume) have presented a very useful chemical evolution model for the Galaxy which obtains almost 2 $M_\odot$ of ^{26}Al in bulk. Its (assumed exponential) infall rate is strong enough to hold the Al mass fraction down to the solar value; but its radial distribution of infall is arbitrarily chosen.

As we evaluate other interstellar radioactivities, we will need an assumption about past infall; namely, what value of $(k+1)$ should be used in applying

$$< \frac{X(A',Z)}{X(A,Z)} > = \frac{p(A',Z)}{p(A,Z)} * (\tau/T_{Gal}) * (k+1) \qquad (1)$$

to the mean interstellar medium near the Sun?

INTERSTELLAR ^{60}FE

The *SMM* Galactic plane 1.17 MeV flux limit of $<8\times10^{-5}$ cm^{-2} s^{-1} radian^{-1} (99.5%) corresponds to a limit of 1.7 $M_\odot$ in the present interstellar medium (Leising & Share 1994). Given the usual assumption of steady state between production and decay, the current rate of synthesis of ^{60}Fe is less than 1.7 $M_\odot$ per 2.2 Myr.

It has been suggested that a neutron rich NSE occurs in small regions in both Type Ia supernovae supernovae and in core collapse supernovae (Hartmann et al. 1985). Either type might eject significant quantities of ^{60}Fe. If we know the frequency of a particular type of ^{60}Fe-producing event then we can limit the mean ^{60}Fe mass ejected per event. Taking 1.7 $M_\odot$ /2.2 Myr, which is 8×10^{-5} $M_\odot$ century^{-1}, we have $M_{ej}(^{60}Fe) \leq \frac{8\times10^{-5}}{R_{SN}} M_\odot$, where R_{SN} is the frequency of the supernovae which eject ^{60}Fe, per century. Woosley (1991) estimates that Type Ia supernovae eject roughly 10^{-4} $M_\odot$ of ^{60}Fe, which is very close to this limit for rates near 1 century^{-1}. We can again compare the production of the radioactivity to stable nuclei. In particular ^{48}Ca and ^{50}Ti probably originate primarily in the neutron rich NSE. Estimating the current production of these stable isotopes from their solar abundances suggests that there should be about 0.45 $(k+1)$ $M_\odot$ of ^{60}Fe in the interstellar medium (Leising & Share 1994), limiting $k \leq 3$ in these models. ^{60}Fe could soon be detected with a slightly more sensitive instrument. This probably will not be accomplished by the *CGRO* because *SMM* already constrains its flux to be at least five times below that of ^{26}Al, which would itself be very difficult for OSSE or COMPTEL to detect if reduced five times, given its angular distribution.

^{44}TI IN SUPERNOVA REMNANTS

Titanium-44 is produced primarily in the α-rich freezeout from nuclear statistical equilibrium. The site for this can be Type Ia (Nomoto et al. 1984; Woosley et al. 1986) or Type II supernovae (Woosley et al. 1988; Thielemann et al. 1990). The Galactic recurrence time of these events is comparable to the ^{44}Ti lifetime, so with current sensitivities we expect to be able to see at most a few ^{44}Ti remnants at any given time. Like most Galactic supernovae, these events will have been undetected optically. Mahoney et al. (1992) used the γ-ray spectrometer on the *HEAO 3* spacecraft to set upper limits on Galactic emission from ^{44}Ti, and Leising & Share (1994) lowered those limits with *SMM* data.

It is likely that most ^{44}Ca is produced as ^{44}Ti in nature. Attesting to this may be supernova-produced meteoritic particles rich in ^{44}Ca (Amari et al. 1992). Therefore expectations of current ^{44}Ti production can be determined from the requirement that pre-solar nucleosynthesis produce the solar abundance of ^{44}Ca. The above models suggest that about $2\times10^{-4}(k+1)$ M$_\odot$ of ^{44}Ca are produced per century (Leising & Share 1994). The product of the supernova frequency (of types producing ^{44}Ti) times the ^{44}Ti yield per event must equal this. The upper limit on the γ-ray line flux, for example, 8×10^{-5} cm^{-2} s^{-1} from the Galactic center direction (Leising & Share 1994), gives a lower limit on the time since the latest supernova with a given ^{44}Ti yield at a given distance. Assuming a Poisson random process for the occurrence of supernovae, one can derive the probability that we have had to wait so long given the rate.

Even assuming that only the latest event would be seen, rates in excess of 2 century^{-1} are ruled out at $\geq 97\%$ confidence for $k=1$. Only rates less than 0.5 century^{-1} are acceptable at $>10\%$ confidence, and this means that the yield per event must be $\geq 10^{-3}$ M$_\odot$ to produce the requisite ^{44}Ca. Rates this low are incompatible with current estimates for Type II supernovae, and yields this high are also very difficult to understand. Perhaps some very rare type of event, such as helium detonation supernovae (Woosley et al. 1986) produces most of the ^{44}Ca. If the preceding considerations are on the right track at all, even a slightly more sensitive instrument has a very high probability of detecting ^{44}Ti in some young supernova remnants.

In this context it was surprising that ^{44}Ti was reportedly detected from the 300 year old Cas A remnant (Iyudin et al. 1994). For standard supernova rates, there should have been several more recent events, and for at least one of them its relative youth should overcome its (possibly) larger distance. Galaxy simulations substantiate this expectation. However the reported line flux, 7.0×10^{-5} cm^{-2} s^{-1}, corresponds to *at least* 3×10^{-4} M$_\odot$ of ^{44}Ti ejected, perhaps pointing to Cas A as a rare type of event. The et al. (1994) were unable to confirm this detection and set an upper limit (99% confidence) to the flux of 5.5×10^{-5} cm^{-2} s^{-1}. These two measurements are consistent at only 2% confidence. Future measurements should resolve the discrepancy.

TABLE 1. Initial Solar System Supernova Radioactivities

(A, Z)/(A', Z)	$\frac{^{26}Al}{^{27}Al}$	$\frac{^{129}I}{^{127}I}$	$\frac{^{107}Pd}{^{108}Pd}$	$\frac{^{244}Pu}{^{238}U}$	$\frac{^{53}Mn}{^{55}Mn}$	$\frac{^{146}Sm}{^{144}Sm}$
SN process	expl-Ne	r	r, s	r	expl-Si	expl-Ne
$p_A/p_{A'}$	6(-3)	1.4	0.65	0.7	0.13	0.1
τ (Myr)	1.05	23.1	9.4	118	5.3	150
$\langle ISM \rangle^1$	3(-6)	0.018	3(-3)	0.07	4(-4)	8(-3)
$\langle MC/ISM \rangle^2$	1.0(-3)	0.16	0.049	0.59	0.020	0.65
$\langle MC \rangle$	3(-9)	3(-3)	2(-4)	0.04	7(-6)	5(-3)
Meteorites	4(-5)	1(-4)	2(-5)	7(-3)	8(-6)	0.015
Meteor./$\langle MC \rangle$	1(4)	0.03	0.1	0.2	1	3
$\langle MC'/ISM \rangle^3$	2.3(-5)	8.8(-3)	1.7(-3)	0.11	5.5(-4)	0.15
$\langle MC' \rangle$	8(-11)	1.5(-4)	6(-6)	0.007	2(-7)	1.2(-3)
Meteor./$\langle MC' \rangle$	6(5)	0.7	3.6	1.0	42	12

^{1} $\langle ISM \rangle = p_A/p_{A'}$ $(k+1)$ $(\tau/7400$ Myr$)$, with $k=3$, except for the Pu/U ratio, which is increased by factor 1.4 owing to U decay
^{2} ratio from eq. (10) of Clayton (1983) using $T_1 = T_2 = 50$ Myr.
^{3} ratio from eq. (10) of Clayton (1983) using $T_1 = 300, T_2 = 400$ Myr

EXTINCT RADIOACTIVITY IN THE ISM AND IN METEORITES

A number of radioactive nuclei are thought to have been alive in the early solar system (Wasserburg, this volume). For example, meteorite observations suggest that the solar system formed from matter containing an abundance ratio $^{26}Al/^{27}Al = 5 \times 10^{-5}$, much greater than the value observed in the ISM. This meteoritic evidence is also telling that ^{26}Al can be enhanced by more than an order of magnitude in some locations, presumably also at places today in the ISM, unless the circumstances of solar birth were almost unique. It also suggests that some ^{26}Al ejecta can be slowed and assimilated by a molecular cloud core before great dilution. For many other isotopes, such conclusions are not so obvious, so we try to place them in some astrophysical context.

Table 1 lists key facts and expectations for well documented cases: production ratios, anticipated interstellar mean abundance ratios at the time of solar birth using $k = 3$, expected associated mean ratios in molecular clouds, and the meteoritic abundances. The calculation of the relationship of the radioactive concentrations in average molecular clouds to that in the mean ISM is described below.

^{129}I: This oldest known extinct radioactivity (Reynolds 1960; Podosek and Swindle 1988) had early solar system abundance about $^{129}I/^{127}I = 0.8$–1.5×10^{-4}, with 40% variations from one meteorite to another. It is noteworthy that the amount expected in the mean ISM, $\langle ^{129}I/^{127}I \rangle = 0.018$, is 180 times greater than the meteoritic value, in the opposite sense of ^{26}Al.

Although the mean ISM ratio 0.018 is theoretical, it can hardly be wrong (except for the exact choice of k), because the very successful r-process theory (Seeger, Fowler and Clayton 1965; Meyer et al. 1992) fits well the expectation that stable ^{129}Xe owes its high abundance to r-process nucleosynthesis of ^{129}I, and 1.4 times more abundantly than ^{127}I.

^{53}Mn: This 5.3 Myr activity reveals itself in meteorites as excess ^{53}Cr associated with manganese. Virtually all of ^{53}Cr was synthesized as radioactive ^{53}Fe, which decays quickly to ^{53}Mn. Regrettably, the ^{53}Mn decay emits no γ-ray, for if it did it would be twice as bright as ^{26}Al! Lugmair et al. (1992) have convincingly shown that angrite meteorites, dated by Pb-Pb chronology to have formed 4.5578 Gyr ago, contained the ratio ^{53}Mn/^{55}Mn=1.3×10^{-6}. This is the first clear mineral isochron, and the meaning of the high ^{53}Cr and high Mn/Cr in the olivine separates can not be in doubt. Those meteorite ages are 10 Myr younger than the most primitive solar system samples, so allowing for two mean-decay lifetimes, the initial abundance was ^{53}Mn/^{55}Mn=8.6×10^{-6}. The larger values $4-7\times10^{-5}$ found by Birck and Allegre (1985) might be attributed to mixing lines between a ^{53}Cr-rich Mn-bearing component and a ^{53}Cr-poor Mn-free component—in other words, to fossil ^{53}Cr. The Galactic evolution models anticipate 40 times more than the meteoritic concentration, but approximately the mean expectation in molecular clouds (see below).

^{244}Pu: Evidence from both etched fission tracks and from Xe fission fragments indicate an abundance 0.007 that of ^{238}U (Hudson et al. 1989; Podosek and Swindle 1988). The mean ISM would carry the ratio $<^{244}$Pu/^{238}U$>$ = 0.045 in steady state if ^{238}U were itself stable. Because the ^{238}U remainder is about 70% (Clayton 1985), we increase the calculated ISM ratio by the factor 1.4 to 0.065 to account for ^{238}U decay. Thus ^{244}Pu was tenfold less abundant in the early solar system than it was expected to be in the mean ISM.

^{60}Fe: In the meteorite Chervony Kut the isotopic excess of ^{60}Ni correlates with the Fe/Ni element abundance ratio, requiring at the time the meteorite solidified ^{60}Fe/^{56}Fe = 3.9×10^{-9} (Shokulyukov and Lugmair 1993). Its solidification probably required about 10 ± 2 Myr (Shukolyukov and Lugmair 1993), which requires an initial ratio near 3.5×10^{-6}, with an uncertainty of a factor of 3.9 owing to the 2 Myr age uncertainty. If this value were typical of the mean ISM it would correspond to 10–140 $M_\odot$ of ^{60}Fe, which is excluded by the γ-ray data described above (Leising & Share 1994), limiting the present interstellar ratio ^{60}Fe/^{56}Fe to be less than 1.7 $M_\odot$ /1×10^7 $M_\odot$. Although greater than the amount measured in Chervony Kut at the time it solidified, that amount is considerably less than the inferred solar initial value if that solidification required 10 Myr! If that solidification time, which now assumes great astrophysical importance, is correct, we may already conclude that the early solar system would have to have been far richer than the mean ISM in ^{60}Fe, and therefore require either an injection event or cosmic ray production.

^{107}Pd: From the ratio ^{107}Ag/^{108}Pd=0.68 and the s-process theory, we know that almost all of ^{107}Ag was produced as ^{107}Pd parent, with 80% of ^{107}Ag produced as ^{107}Pd in the r-process, and that the total production ratio owing to both r-and s-processes production of ^{107}Pd was $p(^{107}Pd)/p(^{108}Pd) = 0.65$, allowing only a few % of direct ^{107}Ag production by the s-process. Thus we can securely expect the mean ISM to contain the steady-state ratio $^{107}Pd/^{108}Pd= 3.3\times10^{-3}$. Continuing the trend of other r-process products, this is 100 times more than has been found in meteorites; but the expected molecular cloud concentration <MC> (Table 1) is only tenfold less than <ISM>. The correlation of excess ^{107}Ag with the Pd/Ag elemental ratio in samples from iron meteorites shows that they solidified with $^{107}Pd/^{108}Pd = 2\times10^{-5}$ (Kelly and Wasserburg 1978).

^{146}Sm: The convincing mineral isochron measured in differentiated meteorites shows $^{146}Sm/^{144}Sm$=0.008 when they solidified, and that value would have been 0.015 in the early solar system based on the solidification ages measured by the associated ^{147}Sm decay to ^{143}Nd (Prinzhofer, Papanastassiou & Wasserburg 1989). Taking Woosley & Howard's (1990) best nuclear argument for the production ratio $p(146)/p(144)=0.1$, eq.(1) yields <ISM>=<$^{146}Sm/^{144}Sm$>=0.008, sufficiently close to the solar system value to suggest that this relatively long-lived extinct radioactivity is a residual of continuous Galactic nucleosynthesis.

The solar system is believed to have formed near a molecular-cloud core (Levy and Lunine 1993), and so might be expected to have had the radioactive concentrations found there. Those activities are surely smaller (on average) than in the mean ISM, because most supernova products will be violently ejected into a badly disrupted and heated local ISM. Considerable time will be required for that matter to undergo a cooling transition, join a cloud, and then mix into a cloud core. To address this Clayton (1983) introduced a mixing scheme built on three phases: dilute ISM plus diffuse clouds small enough to be disrupted by supernova shock waves (M_3=40%); larger diffuse clouds (M_2=30%); and molecular clouds (M_1=30%). Supernova ejecta are placed in the dilute phase 3, whereas stars form in molecular phase 1. He postulated that on average phases 3 and 2 mix on a mass-exchange timescale T_2, and that phases 2 and 1 mix on the timescale T_1. This mixing is not physical mixing, like diffusion, but rather processes that cause mass elements initially in one phase to transform to another phase owing to thermal instabilities. On average those mass transformations balance (see his Fig. 1). With these assumptions the ratio of a radioactive concentration in the molecular phase, $<MC>$, to its concentration in the average over all phases, given by chemical evolution, is fixed. Because of the continuous mixing, the dependence upon the nuclear lifetime τ is quite different than the exponential dependence of a simple free-decay epoch. In Table 1 the mass-exchange timescales are chosen to be $T_1=T_2=50$ Myr. That choice seems reasonable on the grounds that molecular clouds are believed to live about 50 Myr, which is also the e-folding time required to grow molecular clouds from collisions of diffuse clouds and the time for incorporating supernova ejecta into diffuse clouds. The fifth line of Table 1 shows the resulting value of <MC/ISM> as calculated from eq.(2); and the sixth line gives the resulting ratio <MC> = <X(A',Z)/X(A,Z)> in the molecular phases. There follows the ratio of the observed meteoritic ratio to the one expected in the average molecular cloud.

The r-process activities ^{107}Pd, ^{129}I and ^{244}Pu are all expected to be a factor 10 more abundant in the average molecular cloud than they are observed to have been in the initial solar system; the explosive products ^{53}Mn and ^{146}Sm are expected to have concentrations near the meteoritic values; the ^{26}Al has negligible expected molecular-cloud abundance (as would ^{41}Ca and ^{60}Fe if we included them). This remarkable result calls for several speculations. The first concerns the relatively uniform underabundance of r-process nuclei with respect to others (^{53}Mn and ^{146}Sm) that are also supernova products. This suggests that not all supernovae eject r-process matter, so that the solar cloud was deficient in ejecta from recent r-type events. One possibility is that the ^{53}Mn and ^{146}Sm in the solar cloud may have been primarily Type Ia products, whereas the r-ejecta are from Type II. Another is that the massive Type IIs may not eject r-nuclei (Mathews and Cowan 1990). Another is that the high entropy of the r-ejecta and their central location in SN remnants may unduly slow their assimilation by clouds. Final lines of Table 1 show that 300 Myr for model mixing times are needed to interpret ^{129}I and ^{244}Pu as average interstellar concentrations; but then special events are required for *all* of the others.

There is simply so much meteoritic ^{26}Al that one is forced to assume some event that synthesized it and mixed it quickly into the collapsing solar cloud. The only possible escape would be fossil ^{26}Mg produced by prior ^{26}Al decay within Al-rich ISM grains (Clayton 1986; Podosek and Swindle 1988). Cameron (1993) attributes the live ^{26}Al to a nearby AGB star at the time of solar birth, as do Wasserburg et al. (1994). None of the other activities listed in Table 1 with $T_1=T_2=50$ require a special event in the solar cloud, although ^{60}Fe and ^{41}Ca may ultimately. Quite to the contrary, the nuclei tabulated are expected more abundantly in <MC> than found to be in the solar system. We inherited too little, not too much. Nonetheless, Cameron (1993) attributes many to that injection event by choice. If a special event is needed for ^{26}Al, then that same event may indeed have been responsible for some of the other activities, in which case our comparison to the mean molecular clouds in the ISM would be irrelevant. And therein lies the point; that the relevance to the early solar system lies not in the radioactive abundances themselves but rather in their deviations from those expected in a mean-ISM picture. Equation (2), even without questioning its basis, describes only the average molecular cloud, not a specific molecular cloud such as the one of solar birth. Our comparison of the observations with the mean expected molecular cloud is the correct thing to do if we hope to identify what has happened by what lies out of the anticipated. Precisely this same reasoning has traditionally been invoked to compare to the mean bulk <ISM> expected from nucleosynthesis, as in eq. (1). What we have demonstrated by this approach is a clear separation in solar material of the Galactic r-process products (^{107}Pd, ^{129}I, and ^{244}Pu) from the Galactic explosive-shell products (^{53}Mn and ^{146}Sm), even though each of those five nuclei are ample in the average molecular cloud for explaining their meteoritic abundances.

LOW-ENERGY COSMIC RAYS IN ORION CLOUDS

One attempt to understand the natural occurrence of ^{26}Al in star forming regions has been inspired by the recent discovery (Bloemen et al. 1994) of 4.43 and 6.13 MeV γ-rays from the Orion molecular clouds. Because this suggests a large flux of low-energy (~ 10 MeV/nucleon) cosmic rays in the Orion clouds, Clayton (1994) calculated that a high ^{26}Al/^{27}Al in Orion clouds naturally follows. It is almost surely not enough ^{26}Al to provide 1 $M_\odot$ Galaxy-wide, because most molecular clouds are, at any given time, dormant with respect to star formation and particle acceleration. Furthermore, this large production would require the 4.43 MeV line to be brighter in the diffuse emission than is the 1.809 MeV line by at least a factor 10. This is not the case, but if only about 1/10 of the Orion cloud mass is stopping the low-energy cosmic rays, the ^{26}Al/^{27}Al ratio in that portion falls near 4×10^{-5}. And if that portion of the mass is where stars commonly form, the coincidence with the activity in the early solar system would be normal.

Other nuclei of particular relevance for the cosmic-ray irradiation model are ^{60}Fe and ^{53}Mn. If that model is not able to account for meteoritic ^{60}Fe, so that an "injection event" is needed for it, that same event may as well inject ^{26}Al too, as in the atypical-AGB calculated by Wasserburg et al. (1994). And unless the cosmic rays are restricted to rather low energy (near 10 MeV/nucleon) they can easily overproduce ^{53}Mn, for which meteoritic abundance estimates have recently declined greatly (Lugmair et al. 1992). On the other hand, the new evidence for live ^{41}Ca/^{40}Ca $= 1.5 \times 10^{-8}$ (Srinivasan et al. 1994) supports the irradiation model, for which ^{41}K(p,n)^{41}Ca also works at low energy. If that cross section is taken equal to ^{26}Mg(p,n), one immediately obtains from simple scaling that ^{41}Ca/^{41}K$=\tau_{41}/\tau_{26}$ ^{26}Al/^{26}Mg, giving ^{41}Ca/^{40}Ca$=1.1 \times 10^{-8}$ if ^{26}Al/^{26}Mg$=3.6 \times 10^{-5}$ from the irradiation. On the other hand, the short halflife (0.1 Myr) requires incorporation of ^{41}Ca from stellar ejecta to have been very prompt (Srinivasan et al. 1994) if its presence in CAIs is to be attributed to stellar nucleosynthesis.

REFERENCES

Amari, S., Hoppe, P., Zinner, E. & Lewis, R. S. 1992, ApJ, 394, L43.
Birck, J.-L. & Allegre, C. J., 1985, Geophys. Res. Lett., 12, 745.
Bloemen, H. et al. 1994, A&A, 281, L5.
Cameron A. G. W. 1993, in Protostars & Planets III, eds. E. Levy and J. Lunine (U. Arizona: Tucson), p. 47.
Clayton, D. D., 1983, ApJ, 268, 381.
Clayton, D. D., 1984, ApJ, 285, 411.

Clayton, D. D., 1985, in Nucleosynthesis, eds. W. D. Arnett & J. W. Truran (Univ. Chicago Press) p. 65.
Clayton, D. D., 1986, ApJ 310, 490.
Clayton, D. D., 1988, MNRAS 234, 1.
Clayton, D. D., 1994, Nature, 368, 222.
Clayton, D. D., Hartmann, D. H. & Leising, M. D. 1993, ApJ, 415, L25.
Diehl, R. et al, 1994, A&A (submitted).
Hartmann, D. H., Woosley, S. E., & El Eid, M. F. 1985, ApJ, 297, 837.
Hudson, G.B., Kennedy, B.M., Podosek, F.A. & Hohenberg, C.M., 1989, Proc. Lunar Planet. Sci. 19, 547.
Iyudin, A. F. et al. 1994, A&A, in press.
Lee, T., Papanastassiou, D. A. & Wasserburg, G. J. 1977, ApJ (Lett.)
Levy, E. & Lunine, J. 1993, Protostars & Planets III (U. Arizona:Tucson).
Leising, M. D. & Share, G. H. 1994, ApJ 424, 200.
Leising, M. D. et al. 1994, in preparation.
Lugmair, G. W., MacIsaac, C. & Shukolyukov, A. 1992, Lunar Planet. Sci., 23, 823.
Mahoney, W. A., Ling, J. C., Wheaton, W. A., & Jacobson, A. S. 1984, ApJ, 286, 578.
Mahoney, W. A., Ling, J. C., Wheaton, W. A., & Higdon, J. C. 1992, ApJ, 387, 314.
Mathews, G. J. & Cowan, J. J. 1990, Nature, 345, 491.
Meyer, B. S., Mathews, G. J., Howard, W. M., Woosley, S. E. & Hoffman, R. D. 1992, ApJ, 399, 656.
Nomoto, K., Thielemann, F. K., & Yokoi, K. 1984, ApJ, 286, 644.
Oberlack, U., Diehl, R., Montmerle, T., Prantzos, N. & Von Ballmoos, P. 1994, ApJ supp., 92, 433.
Podosek, F. A. & Swindle, T. D. 1988, in Meteorites & the Early Solar System, eds. J. Kerridge & M. Mathews (U. Arizona:Tucson), p. 1093.
Prinzhofer, A., Papanastassiou, D.A. & Wasserburg, G. J. 1989, ApJ, 344, L81.
Reynolds, J. H. 1960, Phys. Rev. Lett., 4, 8.
Seeger, P. A., Fowler, W. A. & Clayton, D. D. 1965, ApJ Supp., 11, 121.
Share, G. H., Kinzer, R. L., Kurfess, J. D., Forrest, D. J., Chupp, E. L., & Rieger, E. 1985, ApJ , 292, L61.
Shukolyukov, A. & Lugmair, G. W. 1993, Science, 259, 1138.
Srinivasan, G., Ulyanov, A.A. & Goswami, J. N. 1994, ApJ(Lett.) 431, L67.
The, L.-S., et al. 1994, ApJ, submitted.
Thielemann, F. -K., Hashimoto, M. & Nomoto, K. 1990, ApJ 349, 222.
Thielemann, F. -K., Nomoto, K. & Yokoi (1986) A&A 158, 17.
Timmes, F.X., Woosley, S. E. & Weaver, T.A., 1994, ApJ, submitted.
Wasserburg, G.J., Busso, M., Gallino, R. & Raiteri, C.M. 1994, ApJ 424, 412.
Weaver, T. A. & Woosley, S. E. 1993, Phys. Rep., 227, 65.
Woosley, S. E. 1991, in Gamma-Ray Line Astrophysics, ed. P. Durouchoux & N. Prantzos, AIP Conference Proceedings, No. 232, (New York: AIP), 270.
Woosley, S. E., Pinto, P. A., & Weaver, T. A. 1988, Proc. Astron. Soc. Aust., 7, 355.
Woosley, S. E., Taam, R. E., & Weaver, T. A. 1986, ApJ, 301, 601.

Galactic Evolution of D,^{3}He,Li,Be,B

N. Prantzos

Institut d'Astrophysique de Paris

Abstract. In this work the evolution of the light elements in the Galaxy is investigated in the framework of consistent models for the solar neighborhood and the galactic disk. The following points are made:

1- It is shown, as already suggested in (19), that a consistent model of the solar neighborhood (Figs. 1 to 3) fails to explain the observed primary behaviour of Be vs. Fe *if* Be is produced as a secondary element, as usually assumed. Such a problem is already identified in the halo (for Be and B), where it can be marginally solved by various, more or less plausible, assumptions. The difficulties encountered (independently) in both the halo and the disk can be "naturally" solved with the assumption that *Be and B are produced as primary elements during the whole galactic evolution*. The advantages and problems of a recently proposed site for such a primary production are presented.

2- The abundance gradients of D and ^{3}He in the disk are evaluated in the framework of a model that reproduces most of disk properties. It is found that the D gradient is larger than the one of oxygen in the inner Galaxy and smaller than that in the outer Galaxy (the two gradients are, of course, anticorrelated); such a gradient is already observable in the inner Galaxy (through very uncertain radio-measurements) and would eventually be observable in the outer Galaxy through UV observations by the FUSE-LYMAN mission. The anticorrelated ^{3}He gradient is found to be considerably smaller but still exceeding observations in the inner Galaxy; however, the current uncertainties in the stellar production of ^{3}He prevent from definite conclusions.

Key words: galactic evolution — cosmic rays

1. THE EVOLUTION OF Be AND B

Since the works of Reeves et al. (Ref. 26) and Meneguzzi et al. (Ref. 15) it has been established that the presolar ^{6}Li, a part of ^{7}Li, ^{9}Be, ^{10}B and (most of) ^{11}B have been produced by the spallation of the CNO nuclei during the propagation of GCR in the interstellar medium (ISM). However, the meteoritic ^{11}B/^{10}B=4 ratio cannot be reproduced by this scenario, which gives only ^{11}B/^{10}B~2.5. Obviously, other sources of ^{11}B are required, like an intense low-energy GCR flux (15) or neutrino-induced nucleosynthesis in SNII (18), but none has been found satisfactory yet.

The galactic evolution of Be and B in the early Galaxy has received considerable attention since the first detection of these elements in low-metallicity halo stars (6, 11, 28). The quasi-linear relationship of their abundances w.r.t. Fe is difficult to understand in the framework of the "standard" model, i.e. if the abundances of the LiBeB producing CNO nuclei increase with time. In that case Be and B are produced as *secondary* elements and a slope of ~2 is expected in the Be (or B) vs. Fe relationship. Several potential solutions to

© 1995 American Institute of Physics

that problem have been proposed (6, 7, 19, 21), the last one having the less serious problems.

Despite a few investigations of the evolution of Be (23, 31) the problem of the observed *linearity of the Be vs. Fe relationship in the disk* has not drawn attention up to now, with the exception of (19). In closed box models of the disk this linear relationship can be understood within the classical scenario of Be and B production as secondary elements. Indeed, the production rate of those species is proportional to the ∼const. product of two oppositely varying factors : the abundance of CNO nuclei (*increasing* by a factor of ∼10 during the disk evolution) and the GCR flux (*decreasing* by a factor of ∼10, since it follows the SN rate, itself proportional to the gas mass).

This "fortuitous" agreement with observations no longer holds in the framework of infall models for the disk, as recently remarked (19). Indeed, in such models the production rate of Be is the product of two factors *increasing* with time (at least until [Fe/H]∼-0.5) and its slope w.r.t. Fe is larger than 2, clearly failing to satisfy the observations (Fig. 4.) Combined to the difficulty of reproducing the halo data on Be and B this new difficulty strongly suggests that a *primary production of Be and B during the whole galactic evolution* should be seriously considered as a "natural" explanation.

A primary GCR production of Be and B results if they are predominantly made by fast CNO nuclei having always the ∼same composition, and the only conceivable way to do this is to assume that SNII accelerate their own ejecta. As discussed above this possibility is observationally excluded for high energy GCR (E>100 MeV/nucleon), but not for lower energy ones. An indication of the existence of such low energy CR fluxes, at least in one site, has been recently provided by the detection of 4.4 and 6.1 MeV gamma-ray lines in the Orion molecular cloud (2), attributed mostly to the de-excitation of fast C and O nuclei hitting the protons and alphas of Orion. Assuming a low-energy (E∼5-50 MeV/nucleon) power-law spectrum one sees that the derived emissivity (L∼2 10^{39} s^{-1}) should be accompanied by a production rate of ^{11}B, ^{10}B and ^{9}Be of ∼4 10^{38} s^{-1}, ∼10^{38} s^{-1} and ∼2.5 10^{37} s^{-1}, respectively. It has been suggested (4, 25) that this mechanism, operating galaxywise during the halo phase, could explain the observed linearity of Be and B vs. Fe in metal poor halo stars. On the other hand, if ∼100 Orion-like clouds operate permanently in the Galaxy at the above rates during the last 10^{10} years, they could manufacture the total galactic Be and B, producing them as primaries and with a ∼solar ratio of ^{11}B/^{10}B (for a slope x∼4-5 of the power-law CR spectrum).

The above scenario is somewhat different from the traditional picture of Be and B production as secondaries, i.e. with the parent CNO abundances increasing with time. It is attractive, since it explains at one stroke the primary behaviour of Be and B in both the halo and the disk (Fig. 5) and the solar ^{11}B/^{10}B ratio (=4), allowing for a large B/Be>20 ratio in the early Galaxy; it also produces most of the disk Li by low-energy CR (Fig. 6), barely avoiding an overproduction around [Fe/H]∼-1. On the other hand it has, in my opinion, several weak (but, perhaps, not lethal) points:

- The duration of the active period in Orion is completely unknown; it could well be a transient phenomenon, lasting much less than the molecular cloud lifetime of a few 10^6 years required by the scenario described above.

- A restricted activity in only a few halo regions would produce locally very large enhancements of Be and B abundances w.r.t. the ones of the general ISM, their dilution requiring timescales of several 10^7 years. Since low mass stars are born in both irradiated and non-irradiated regions, there should be a considerable dispersion in the observed abundances of Be in halo stars, which does not seem to be the case.

2. THE EVOLUTION OF D AND ^{3}He

There is currently considerable interest in the galactic evolution of D and ^{3}He, especially after the presumed detection of high primordial $D/H \sim 2\ 10^{-4}$ in a high red-shift (z=3.32) hydrogen cloud (29). Such a high primordial D/H corresponds to a Universe of low baryonic density, with almost no room for baryonic dark matter. However, the pre-solar $((D/H)_\odot = 2.6 \pm 1.\ 10^{-5}$, Ref. 10) and current ISM $((D/H)_O = 1.5\ 10^{-5}$, Ref. 13) values of D suggest that a high primordial value is difficult to accomodate in "standard" models of galactic chemical evolution (i.e 32).

Since the discovery of high primordial D is still controversial, I simply ignore it in this work and I try rather to determine depletion factors within "standard" best-fit models of chemical evolution. I am mostly interested in the depletion of D during the evolution of the disk, since the evolution of the halo has obviously processed only a small amount of gas (because of the small mass and metal content of that site). Notice that this point is not always made in chemical evolution models, some of them destroying most of the D before [Fe/H]$\sim$-1. The results of the calculations appear in Fig. 7 for a closed box and an infall model. The former can deplete D by $D_P/D_\odot \sim 3$ until solar system formation and by a factor of $D_P/D_O \sim 5$ until today, leading to a pre-disk D/H$\sim 8\ 10^{-5}$. In the latter, the corresponding factors are $D_P/D_\odot \sim 2$ and $D_P/D_O \sim 3$ only. Similar results have also been obtained in several recent calculations (30, 32, 9).

The situation is much more complicated for ^{3}He. Standard stellar evolution models give a net production of ^{3}He in low mass stars (9, 12), supported by observations in planetary nebulae (27), and this leads inevitably to a large overproduction of ^{3}He during galactic evolution. Since the situation with the stellar production of ^{3}He is not yet quite settled, I adopt here the prescrition of (32), where production by D+p, not the p-p chains, is considered. Depletion factors $g_3 = {}^3\text{He}_{final}/(D+{}^3\text{He})_{initial} = 1, 0.7, 0.7$. are adopted for stars with mass 1, 2 and 3 $M_\odot$, respectively. Even with this "minimal" assumption the abundance of ^{3}He increases during galactic evolution, the increase been larger (and incompatible with the presolar value) in the closed box model than in the infall model (Fig. 8).

The aim of this work is to make a preliminary investigation of the galactic gradients of D and ^{3}He, in the framework of a model that reproduces most of the observational constraints of the galactic disk (Figs. 9 to 14). A SFR dependent on galactic radius r is adopted: SFR$\propto \sigma_{gas}^2/$r, after (34). Contrary to that work, instantaneous recycling approximation is not assumed here and the disk is not a closed box, but slowly built by infall.

The results on D and ^{3}He appear in Figs. 15 and 16, respectively. Because of important astration in the inner Galaxy D is largely depleted there, its (positive) gradient being larger in absolute value than the (negative) one of oxygen in that region; in the outer Galaxy, the D gradient is smaller but still significant. Radio observations of deuterated molecules show indeed an important depletion of D in the Galactic Center, but the uncertainties due to chemical fractionation are quite large (33). UV observations with the future FUSE-LYMAN mission could possibly detect the D gradient in the outer Galaxy. Finally, there is a small negative ^{3}He gradient in the inner Galaxy (even with the adopted here "minimal" production), that makes difficult to understand the recent observations of a flat ^{3}He profile in the galactic disk (1).

REFERENCES

1. Balser D. et al. (1994), preprint
2. Bloemen H. et al. (1994), A&A, 281, L5
3. Bykov A. and Bloemen H. (1994), A&A, 283, L1
4. Cassé M., Lehoucq R., Vangioni-Flam E. (1994), Nature, submitted
5. Clayton D. (1984), ApJ, 285, 411
6. Duncan D., Lambert D., Lemke M. (1992), ApJ, 401, 584
7. Feltzing M. and Gustaffson B. (1994), ApJ, 423, 68
8. Edvardsson et al. (1993), A&A, 275, 101
9. Galli D., Palla F., Straniero O., Ferrini F. (1994), ApJ Letters, in press
10. Geiss J. (1993) in *Origin and Evolution of the Elements*, Eds. N. Prantzos, E. Vangioni-Flam and M. Cassé (Cambridge: CUP), 89
11. Gilmore G., Gustafsson B., Edvradsson B., Nissen P. (1992), Nature, 357, 379
12. Iben I. and Truran J. (1978), ApJ, 220, 980
13. Linsky J. et al. (1992), ApJ, 402, 694
14. Matteucci F. and Greggio L. (1986), A&A, 154, 279
15. Meneguzzi M., Audouze J., Reeves H. (1971), A&A, 15, 337
16. Meneguzzi M. and Reeves H. (1975), A&A, 40, 99
17. Norris J. and Ryan S. (1991), ApJ, 380, 403
18. Olive K., Prantzos N., Skully S, Vangioni-Flam E. (1994), ApJ, 424, 666
19. Pagel B. (1993) in *Cosmical Magnetism*, Ed. D. Lynden-Bell (Kluwer), p. 113
20. Pagel B. (1994) this volume
21. Prantzos N., Cassé M., Vangioni-Flam E. (1993), ApJ, 403, 630
22. Rebolo R., Molaro P., Abia C., Beckman J. (1988), A&A, 193, 93
23. Reeves H. (1994), Rev. Mod. Phys., 66, 193
24. Reeves H. and Meyer J.P. (1978), ApJ, 266, 613
25. Reeves H. and Prantzos N. (1994), preprint
26. Reeves H., Fowler W. and Hoyle F. (1970), Nature, 227, 727
27. Rood R., Bania T., Wilson T. (1992), Nature, 355, 658
28. Ryan S. et al. (1992), ApJ, 388, 184
29. Songaila A., Cowie L., Hohan C., Rugers M. (1994), Nature, 368, 599
30. Steigmann G. and Tosi M. (1992), ApJ, 401, 150
31. Vangioni-Flam E., Cassé M., Audouze J, Oberto Y. (1990), ApJ, 364, 568
32. Vangioni-Flam E., Olive K., Prantzos N (1994), ApJ, 427, 618
33. Walmsley C. and Jacq T. (1991), in *Atoms, Ions and Molecules*, Eds. A. Haschick and P. Ho, p. 305
34. Wang B. and Silk J. (1994), ApJ, 427, 759

FIGURE CAPTIONS

FIG. 1: Age-metallicity relationship in the solar neighborhood. Data from Edvardsson et al. (1993). *Dashed line:* Closed box; *Solid line:* infall model.

FIG. 2: Oxygen vs. Fe relationship in the solar neighborhood. Data from Edvardsson et al. (1993). The decline is due to SNIa, injecting 0.7 $M_\odot$ of Fe at a rate SNIa = 0.2 SNII with a delay of 2 Gyr.

FIG. 3: Differential metallicity distribution in the galactic disk. Data from Norris and Ryan (1991, *histrogram*). *Dashed line:* Closed box model; *Solid line:* Clayton's (1984) infall model with k=2 and Δ=1.

FIG. 4: Be vs. Fe in the galactic disk. Data from Rebolo et al. (1988). *Dashed line:* Closed box model; *Solid line:* Infall model.

FIG. 5: Be vs. Fe in the galactic halo and disk. The discontinuity in the model curve is due to the assumption of outflow during the halo phase, from which the disk is subsequently built by infall (both outflow and infall assumptions are made to reproduce the metallicity distributions in the halo and the disk). Be is produced by the low-energy Orion-like CR with SNII composition, i.e. it is always produced as a primary.

FIG. 6: Li vs. Fe in the galactic halo and disk, in the scenario of Fig. 5. Only the upper envelope of the Li data is drawn. Li is almost exclusively produced by low-energy $\alpha + \alpha$ reactions, assuming a SNII composition for low-energy CR. An overproduction is barely avoided around [Fe/H]$\sim$-1.

FIG. 7: D/H as a function of time in models with no instantaneous recycling. *Dashed line:* Closed box; *Solid line:* Infall with primordial composition. Data: Geiss (1993) and Linsky et al. (1992).

FIG. 8: ^{3}He/H as a function of time in models with no instantaneous recycling. "Minimal" production of ^{3}He is assumed (see text). *Dashed line:* Closed box; *Solid line:* Infall with primordial composition.

FIG. 9: Snapshots of the total surface density profile during the evolution of the galactic disk, built by infall of primordial composition at a rate: $f(t,r) \propto e^{-t/\tau(r)} e^{-r/R}$. Curves are at 2 Gyr intervals, the solid line corresponding to the final configuration, i.e. an exponential disk with R=3. kpc scalelength and a total mass of $\sim$5 10^{10} $M_\odot$.

FIG. 10: Snapshots of the gass surface density profile of the galactic disk. The molecular ring at $\sim$4 kpc is reproduced if a SFR$\propto \sigma_{gas}^2/r$ is adopted (see text). *Solid and dashed lines:* as in Fig. 9; *Long-dashed lines:* Observational limits (from Wang and Silk 1994).

FIG. 11: Snapshots of the star formation rate profile of the galactic disk. Curves as in Fig. 9. Data from Wang and Silk (1994).

FIG. 12: Snapshots of the Oxygen profile during disk evolution. Curves as in Fig. 9. Data from Shaver et al. (1983), adapted to $R_\odot$=8.5 kpc.

FIG. 13 and 14: Fe galactic gradient as a function of stellar age. Data from Edvardsson et al. (1993). Solid lines correspond to model ages of 2 and 4 Gyr (Fig. 13) or 6, 8 and 10 Gyr (Fig. 14).

FIG. 15 and 16: Snapshots (every 2 Gyr) of the D and ^{3}He abundance profiles. *Solid lines:* Current profiles. ^{3}He data from Balser et al. (1994).

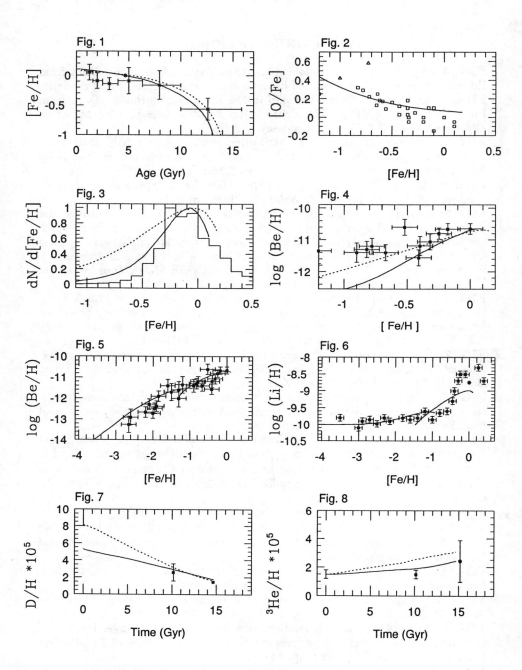

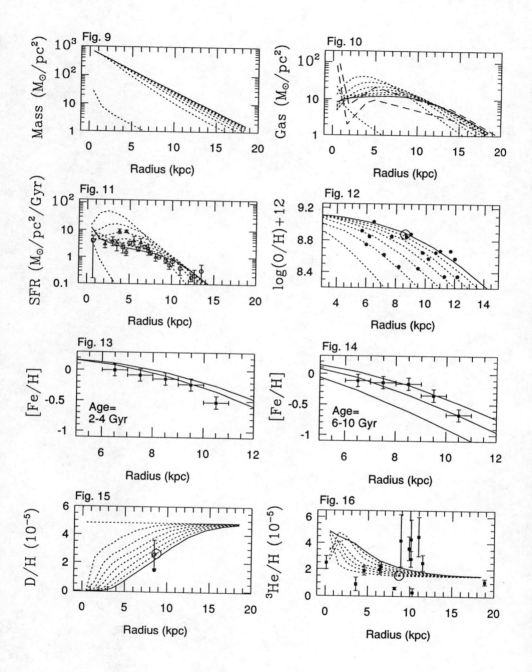

Genesis And Evolution of LiBeB Isotopes II: Galactic Evolution

M. Cassé[1], E. Vangioni-Flam[2], R. Lehoucq[1] and Y. Oberto[2]

[1]*DSM/DAPNIA/Service d'Astrophysique, CE-Saclay, F91191 Gif sur Yvette cedex*
[2]*Institut d'Astrophysique de Paris, 98 bis boulevard Arago, F75014 Paris*

Abstract. In previous contributions we have shown that copious amounts of LiBeB are likely to be produced in Orion at the present time, as witnessed by the high flux of gamma-rays it radiates (1, 2). It is shown here that spallation of fast nuclei within molecular clouds surrounding the regions of formation of massive stars, alone, could govern the behaviour of light species in the early galaxy (up to [Fe/H] = -1). Afterwards two other processes, already identified, combine with the first one to built the abundances observed in the Solar System and the present interstellar medium : i) standard galactic cosmic rays interactions with ISM and ii) ^{7}Li formation by slowly evolving stars or binary systems (low mass AGB and/or novae). This new model reproduces simply and elegantly all the elemental and isotopic abundances of light elements as observed in the Solar System, including ^{11}B/^{10}B, together with their evolution.

INTRODUCTION

Recent gamma-ray observations by COMPTEL (3) have revealed that the Orion nebula is internally irradiated by fast particles. The interaction of rapid nuclei with the ambient cloud medium offers an unsuspected source of LiBeB. More generally, regions where multiple supernova (SN) remnants and numerous OB stars are concentrated should produce local enhancements of fast particle density, possibly of SN composition (α and O enriched).

Competition between gain and losses during acceleration would lead to a power law source spectrum (with slope -n) flattened below a critical energy E_c (4). The source spectrum is modified by energy losses in the course of propagation, giving rise to an equilibrium spectrum with slopes 1.8 up to E_c and (-n+1.8) above. All cross sections are integrated over this equilibrium spectrum to get the relevant production rates. The composition of the accelerated beam is assumed to reflect that ejected by rather massive supernovae. The mass of the progenitor is varied between 15 and 35 $M_\odot$.

In the following, we will explore the hypothesis that the processes experienced by Orion apply to the whole Galaxy at all times. Our choice of spectral parameters (E_c, n) is based on the following criteria : 1. avoiding to overproduce Li/Be at very low metallicity. This translates to Li/Be < 100 ; 2. reproducing the

B/Be ratio in both extreme pop. II stars (5), i.e. B/Be ≈ 20-35 ; 3. getting $^{11}B/^{10}B$ = 4 at solar birth after mixing with standard galactic cosmic rays (GCR) spallation products ; 4. limiting the luminosity of the galactic disk at 4.4 and 6.1 MeV (deexcitation lines of ^{12}C and ^{16}O) since no positive detection has been announced ; 5. getting a $^{12}C*/^{16}O*$ ratio close to 1 (3).

PRODUCTION RATES AND GALACTIC EVOLUTION

All these constraints can be fulfilled with a limited choice of parameters. A 35 $M_\odot$ SN composition (6), together with E_c = 7 MeV/n and n = 9 is satisfactory. Indeed, it is reasonable to consider that only the most massive stars explode within giant molecular clouds due to their relatively limited lifetime. The production ratios corresponding to this case are presented in Table 1. The associated production rates monitored on the gamma-ray line emission are displayed in Table 2. We have calculated two cases : a solar composition for the ISM to simulate the present evolution and a zero metallicity for the early phase of the Galaxy.

In a first step, to integrate this process in a classical galactic evolutionary model (8) we affect the same LiBeB yield to stars more massive than 35 $M_\odot$ and assume an irradiation time of 10^5 yr (7). The SN rate, itself regulated by the star formation rate, is proportional to the gas mass fraction. The initial mass function is parametrized as $\phi(m)$ dm $\propto m^{-2.7}$. We get the results presented in Figure 1.

DISCUSSION

The evolution of Li, Be and B is well reproduced, considering the uncertainty affecting the B/Be ratio (5, 9). To disentangle the contribution of the different components, we have suppressed the effect of GCR and stars. It appears that the low energy component alone has produced, about 60% of ^6Li, 45% of ^7Li, 1/3 of ^9Be, 40% ^{10}B and 70% of ^{11}B. The remaining ^7Li comes from stars (40%) and GCR (15%). The $^{11}B/^{10}B$ ratio evolves from about 11, initially, to about 4 at solar birth. The B/Be varies from about 35 up to [Fe/H] = -1, to about 20 presently. The transition between the epochs where the low energy component

TABLE 1. Production ratios			
Ratios	Z = 0	Z = $Z_\odot$	GCR
$^6Li/^9Be$	13.7	11.5	5
$^7Li/^9Be$	95.6	71.4	7
$^{10}B/^9Be$	2.9	4.6	5
$^{11}B/^9Be$	33.3	36.5	12
$^7Li/^6Li$	6.0	6.2	1.4
$^{11}B/^{10}B$	11.0	8.0	2.5
Li/Be	109.4	82.9	12
B/Be	35.2	41.1	17

TABLE 2. Production rates (10^{36} s^{-1})		
Isotopes	Z = 0	Z = $Z_\odot$
^6Li	24.0	27.0
^7Li	167.0	167.8
^9Be	1.7	2.5
^{10}B	4.9	10.8
^{11}B	43.7	85.9

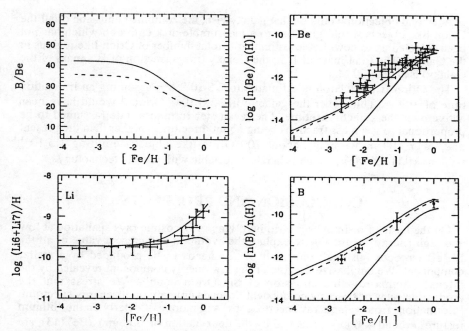

FIGURE 1. B/Be, Li/H, Be/H and B/H vs [Fe/H]. Solid line : $Z = Z_\odot$. Dashed line : $Z = 0$. Both including the three components (Orion-like, stars and GCR). Lower solid line : GCR only. Data points are referenced in (9) and (14).

governs the production of ^{9}Be and that of the GCR dominance takes place at [Fe/H] = -1, which correspond to the beginning of the disk evolution, as shown by the evolution of B/Be versus [Fe/H] in figure 1. Note the little difference introduced by the composition of the ISM ($Z = 0$ or $Z = Z_\odot$) on the evolution of the light elements.

Due to the high amount of light elements produced in a single stellar generation, one may object to the whole scenario that presently observed extremely metal poor stars should exhibit a wide range of Be and B abundances, contrary to observations. This difficulty can bee circumvented, assuming that the bulk of low-mass stars formation occurs in a single event after efficient mixing by turbulence has occured (10, 11). More specifically, the formation of low mass stars observed presently in the halo is thought to occur at the fringe of their parent clouds, behind the supersonic supershell induced by SN explosions in the middle of the cloud, in a zone thoroughly mixed by the passage of this supershell. A clue to the very high mixing efficiency in proto-globular cluster clouds is that the abundance of even the rarest species, like r-process elements (for instance Europium) suffer only moderate abundance dispersion (12). Note that in our case the problem of dilution and mixing is somewhat alleviated due to the fact that fast particles permeate, in all likelihood, a significant fraction of clouds. Thus spallative products are inherently scattered in the medium, contrary to products of stellar nucleosynthesis.

A second objection would be that galaxywise, the combined emission of all the Orion-like objects would give rise to a measurable disk emission which has not been detected up to now. Let us estimate the total number of Orion-like regions in the Galaxy. The total mass of B in the Galaxy (mass fraction of B × mass of the Galaxy) is about 100 $M_\odot$.

The Orion B production is estimated to 2.5 10^{-6} $M_\odot$ assuming an irradiation time of 10^5 yr. Thus, over the galactic life 3.5 10^7 "Orions" would have been active over the galactic lifetime. The mean star formation rate (assumed to be proportional to gas mass fraction) being about three times higher than the present one, over 10^5 yr, we count about 100 Orion-like objects, emitting 1.5 10^{41} photons s^{-1} at 4.4 MeV, which is barely detectable with present technology.

CONCLUSION AND PROSPECTS

On the basis of a simple scenario involving only cosmic rays spallation at low and high energy we are able to explain the evolution and present value of all the LiBeB isotopes but one ; only half of ^{7}Li needs to be produced by a stellar component. We emphasize the role of the low energy component revealed by the intense gamma-ray line emission of the Orion nebula. We stress that the production rates of this new component depends crucialy on its energy spectrum. Informations from gamma ray spectroscopy, and more particularly on the "lithium feature" around 450 keV produced by the deexcitation of ^{7}Li* and ^{7}Be* (13), are eagerly awaited to define precisely its form.

ACKNOWLEDGMENTS

We thank Martin Lemoine and Nicolas Prantzos for stimulating discussions and numerical help. We are endebted to Philippe Ferrando for providing us with up to date cosmic ray spectra.

REFERENCES

1. Vangioni-Flam E., Lehoucq R. and Cassé M., ESO/EIPC Workshop on "The light elememts abundances", Isola d'Elba, Italy, 1994, to appear
2. Cassé M., Lehoucq R. and Vangioni-Flam E., Nature, 1994, accepted
3. Bloemen et al., A&A 281, L5, 1994
4. Ramaty R., Koslovsky B. and Lingenfelter R., ApJ Suppl. 40, 487, 1979
5. Edwardson et al., A&A, 1994, in press
6. Weaver T.A. and Woosley S.E., Phys. Rep. 227, 65, 1993
7. Marti K. and Lingenfelter R., in "Nuclei in the Cosmos", L'Aquila, Italy, 1994, to appear
8. Vangioni-Flam E., Cassé M., Audouze J. and Oberto Y., ApJ 364, 568, 1990
9. Olive K.A., Prantzos N., Scully S. and Vangioni--Flam E., ApJ 424, 666, 1993
10. Cayrel R., A&A 168, 81, 1986
11. Brown J.H., Burkert A. and Truran J.W., ApJ 376, 115, 1991
12. Gilroy K.K. et al., ApJ 327, 298, 1988
13. Koslovsky B. and Ramaty R., ApJ Lett. 19, 19, 1977
14. Abia C., Isern J. and Canal R., preprint, 1994

Galactic Chemical Evolution: Neutrino-Process Contributions

F. X. Timmes[1,2,3], S. E. Woosley[2,3], and Thomas A. Weaver[3]

[1] Laboratory for Astrophysics and Space Research, Enrico Fermi Institute
University of Chicago Chicago, IL 60637

[2] Board of Studies in Astronomy and Astrophysics, UCO/Lick Observatory
University of California at Santa Cruz, Santa Cruz, CA 95064

[3] General Studies Division, Lawrence Livermore National Laboratory
Livermore, CA 94550

Using the output from a grid of 52 Type II supernova models (Woosley & Weaver 1995) of varying mass ($11 \lesssim M/M_\odot \lesssim 40$) and metallicity (0, 10^{-4}, 0.01, 0.1, and 1 $Z_\odot$), the chemical evolution of 76 stable isotopes, from hydrogen to zinc, was calculated (Timmes, Woosley & Weaver 1995). The models used to represent the dynamical and isotopic evolution are simple and relatively standard. Each radial zone in the exponential disk begins with zero gas and accretes primordial or near-primordial material over a 4 Gyr e-folding time scale. The isotopic evolution at each radial coordinate is calculated using "zone" models (as opposed to hydrodynamic models) of chemical evolution. Standard auxiliary quantities such as a Salpeter initial mass function and a Schmidt birth rate function were used. In this brief note, the theoretical results are compared with observed stellar abundances of ^{7}Li, ^{11}B and ^{19}F.

The evolution of ^{7}Li as a function of [Fe/H] is shown in Figure 1. The calculated ^{7}Li abundance is shown as the solid line, and the dashed lines show factors of two variation in the ν-process yields. Also shown are abundance determinations of ^{7}Li in disk and halo dwarfs that have $T_{\text{eff}} > 5500$ K, and the severely depleted solar photospheric value. It is common practice to assume that the lithium abundance of very metal-poor stars ([Fe/H] < -1.4 dex) represents the primordial value. Agreement of the calculations with the Spite plateau is due strictly to the infall of primordial material with the homogeneous Big Bang composition of Walker *et al.* (1991). Beyond the Spite plateau, the upper envelope of the observations rises smoothly up to the maximum value found in Population I objects. Below the upper envelope are dwarfs that span the entire range of measurable lithium abundances because it is very easy to destroy lithium in a star. At metallicities larger that [Fe/H] $\simeq$ -1.0 dex, injections of freshly synthesized ^{7}Li (produced chiefly in the helium shell) into the interstellar medium by the ν-process causes the lithium abundance to rise above the Spite plateau. The exploded massive star models are producing lithium prior to [Fe/H] $\simeq$ -1.0 dex, but the contributions are small compared to the infall values. Figure 1 shows that these ^{7}Li contributions can account for a significant fraction of the shape and amplitude of the upper envelope, and that Type II supernova contribute about 1/2 of the solar ^{7}Li abundance.

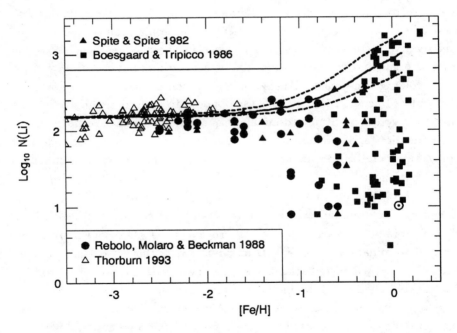

Fig. 1.— Evolution of ^{7}Li relative to hydrogen as a function of the metallicity [Fe/H]. The Thorburn (1994) observations have been "corrected" by -0.2 dex due to differences in the stellar atmosphere models.

The evolution of elemental boron (whose dominant isotope is ^{11}B) as a function of [Fe/H] is shown in Figure 2. The calculated boron abundance is shown as the solid line, while the dashed lines depict factors of two variation in the ν-process yields. Boron is very difficult to measure in stars because it is a trace element, and all of its usable transitions are in the ultraviolet. Linearity of the boron abundances with [Fe/H] strongly suggests that boron is produced in lockstep along with the other metals. The ν-process produces ^{11}B in the carbon shell. The fit to the observations, which span over 2 orders of magnitude, and include the solar abundance, is encouraging. Using some of our preliminary ^{11}B yields, Olive et al. (1993) found similar results. In addition, they found that by including a standard cosmic ray model, without the unobserved low-energy component to the cosmic ray spectrum, that the solar ^{11}B/ ^{10}B ratio was reproduced.

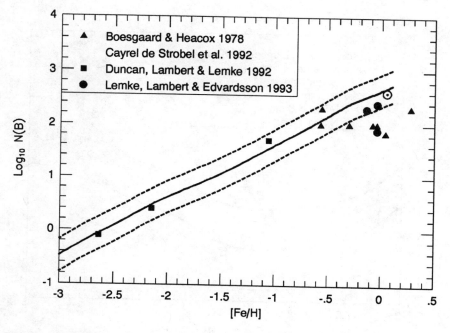

Fig. 2.— Evolution of boron relative to hydrogen as a function of the metallicity [Fe/H]. Recent non-LTE investigations suggest that the boron abundances of the halo dwarfs and Procyon, should have their boron abundances increased by 0.2 dex.

The evolution of fluorine to oxygen ratio [F/O] as a function of [Fe/H] is shown in Figure 3. The calculated history is shown as the solid line, and the dashed lines depict factors of two variation in the ν-process yields. Fluorine is synthesized in the oxygen shell by the ν-process. Fluorine is very difficult to measure in stars primarily because it is the least abundant of all the stable $12 \leq A \leq 38$ nuclides. All abundances determinations of F are based on analysis of the hydrogen fluoride molecule. The significant deviation from a solar [F/O] ratio at low values of [Fe/H] is a direct reflection of the enhanced oxygen abundance at these metallicities. Thus, our model makes the prediction that metal-poor dwarfs will have a subsolar [F/O] ratio. At larger metallicities the model fits the chemically normal K giant observations. The fit to red giants stars that are chemically peculiar is not good, nor should the model be expected to fit these spectral classes as they are dominated by the effects of low mass stellar evolution rather than chemical evolution.

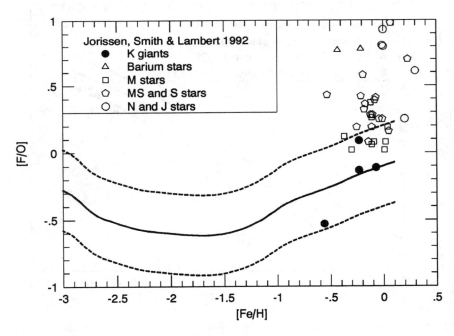

Fig. 3.— Evolution of the fluorine to oxygen [F/O] ratio as a function of the metallicity [Fe/H].

Although the choice of the μ and τ neutrino temperature strongly influences the results, the sum of Figures 1, 2, 3 suggest that the ν-process provides an an attractive site for the nucleosynthetic origin of the rare and fragile ^{7}Li, ^{11}B, and ^{19}F isotopes. The major evolutionary features and a significant fraction of their solar abundances are accounted for. The ν-process, cosmic ray spallation, and homogeneous Big Bang nucleosynthesis are complimentary, and together they yield a comprehensive set of prescriptions for the evolution of the light ($A \leq 11$) elements.

This work has been supported by the National Science Foundation under grant number AST91-15367; NASA under grant number NAGW 2525, the US Department of Energy by the Lawrence Livermore National Laboratory under contract number W-7405-ENG-48, the California Space Institute under grant number CS86-92, and an Enrico Fermi Postdoctoral Fellowship (FXT).

REFERENCES

Boesgaard, A. M., & Heacox, W. D. 1978, ApJ, 226, 888
Boesgaard, A. M., & Tripico, M. J. 1986, ApJ, 303, 724
Cayrel de Strobel, G., Hauck, B., Francois, P., Thevenin, F., Friel, E., Mermilliod, M., & Borde, S. 1992, A&A Supp., 95, 273
Duncan, D. K., Lambert, D. L., & Lemke, M. 1992, ApJ, 401, 584
Jorissen, A., Smith, V. V., & Lambert, D. L. 1992, A&A, 261, 164
Lemke, M., Lambert, D. L., & Edvardsson, B. 1993, PASP, 105, 468
Olive, K. A., Prantzos, N., Scully, S., & Vangioni-Flam, E. 1993, ApJ, 424, 666
Rebolo, R., Molaro, P., & Beckman, J. E. 1988, A&A, 192, 192
Spite, F., & Spite, M. 1982, A&A, 115, 357
Thorburn, J. A. 1994, ApJ, 421, 318
Timmes, F. X., Woosley, S. E., & Weaver, T. A. 1995, ApJ, submitted
Walker, T. P., Steigman, G., Schramm, D. N., Olive, K. A., & Kang, H. 1991, ApJ, 376, 51
Woosley, S. E., & Weaver, T. A. 1995, ApJ, in preparation

The Orion Phenomenon: Particle Fluences in the Solar Nebula

K. Marti* and R. E. Lingenfelter

CASS and *Department of Chemistry,
University of California, San Diego, La Jolla, California 92093

Abstract. The detection of intense γ-ray line emission in the star-forming Orion complex suggests a new option for particle radiation during accretion in the solar nebula. These observations also offer a unique measure of the particle irradiation environment in a star formation region. The discovery of ^{12}C and ^{16}O nuclear deexcitation lines in the Orion molecular cloud provides the first direct "live" measure of nuclear processes and nuclide production rates in an active star-forming region. The effects of such an "Orion phenomenon" 4.56 Ga ago on the evolving solar nebula can be compared with the "fossil" records observed in meteorites.

INTRODUCTION

The discovery by COMPTEL (1) of ^{12}C and ^{16}O nuclear deexcitation lines from the Orion giant molecular cloud complex of recent OB star formation, provides the first direct measure of high energy nuclear processes in a star formation region. Bloemen et al. (1) observed nuclear deexcitation lines at 4.443 MeV from ^{12}C and at 6.129 MeV from ^{16}O with a combined intensity of 1.0 (±0.15) x 10^{-4} photon cm^{-2} s^{-1}. The bulk of the line emission comes from a region of ~3° radius, centered at $l = 208°$ and $b = -18°$ in the middle of the ORI A and B complex.

The observed intensity of the lines is much greater (>10^2 x) than that predicted (2) simply from low energy cosmic ray interactions with the molecular clouds. Possible sources of the accelerated particles are either T-Tauri stars, flarings on massive stars, or a supernova. The observed intensity has prompted suggestions that the very high abundances of extinct radioactive nuclides observed in meteoritic materials may be related to this phenomenon (3,4). Circumstantial evidence has been presented (1) that the γ-rays may be due to strongly enhanced abundances of C and O in low energy cosmic rays.

Although the specifics of the COMPTEL phenomenon are still not clear, the detected γ-radiation implies that enormous rates of nuclear excitations and reactions occur in the Orion nebula. If this represents a typical, even transient, environment in star-forming regions, it is possible that 4.56 Ga ago the evolving solar nebula was subjected to similar external particle radiations. The observed pre-irradiation

© 1995 American Institute of Physics

effects in meteorites could be the result of reactions due to an external particle radiation, rather than an active, early Sun.

By comparing calculated with observed relative intensities and widths of the 4.443 and 6.129 MeV lines, and the limits on other lines, it may be possible to constrain the total number, energy spectrum and composition of the accelerated particles, the number and energy spectrum of the accompanying secondary neutrons, the resulting nuclide production rates, as well as the thermal energy deposition rate, associated with the γ-ray emission region in the Orion region. We can also try to discriminate between such sources as supernovae and flares on T-Tauri or massive stars.

THE ORION PHENOMENON

Although the origin of the accelerated particles interacting in the Orion complex is not yet known, a comparison of the observed (1) intensities of the nuclear deexcitation lines with new calculations (5) of both the nuclear line production and the accompanying energy dissipation by accelerated particles, using updated cross sections, already constrains some models. These calculations show (5) that if the emission comes predominantly from heavy ions, produced and accelerated by roughly 10 supernovae within the last 3×10^6 years, as recently suggested by Bykov and Bloemen (6), the inferred energy deposition by the accelerated particles, estimated (2,5) to be >1 erg per deexcitation line photon, is $>2 \times 10^{32}$ J s^{-1}, which approaches the estimated bolometric luminosity of the entire Orion OB association. Moreover, if the energy of the accelerated particles is supplied by the $\sim 10^{44}$ J of kinetic energy in each supernova, then the present rate of energy dissipation could not be sustained by 10 supernovae for more than 10^5 years, or <3% of the estimated 3×10^6 year age of the Orion complex. The energy dissipation per photon depends on both the assumed energy spectrum and composition of the accelerated particles, as well as the ambient medium.

We clearly need to make more detailed calculations to explore the full range of possibilities. Nonetheless, the overall energetic problem suggests that the observed gamma-ray line emission comes not from a steady state cosmic ray acceleration on the scale of the whole association, but instead from a more localized, rather transient emission resulting from particle acceleration in flaring on T-Tauri stars or young, massive stars. If this interpretation is correct, this could imply that observed irradiation effects in solar system matter, which have been assigned to the T-Tauri phase of our Sun, may in fact reflect such effects but from a different source, namely some neighbors in the star-forming cluster, possibly at high latitudes relative to the disc-shaped solar nebula.

Comparing the observed intensities of the nuclear deexcitation lines at 4.443 and 6.129 MeV, and the observational limits set on the intensities of other deexcitation lines, especially those from ^{14}N, ^{20}Ne, ^{24}Mg, ^{28}Si, ^{32}S, ^{56}Fe, and other heavy nuclei, in the 1 to 3 MeV band, with new calculations (5) of the production of these lines and other nuclei can also constrain the production of these other nuclei. For example, very significant ^{26}Al production from cosmic ray interactions in the Orion complex has recently been suggested (3), based on the observations of the ^{12}C and ^{16}O deexcitation lines. However, the yield of ^{26}Al nuclei per photon in the 1 to 3 MeV band obtained (5), combined with the COMPTEL upper limits on the 1 to 3 MeV flux, set a 3σ limit on ^{26}Al production that is <7% of the value estimated (3) from the 3 to 7 MeV band. The further suggestion (3) that all of the galactic ^{26}Al, inferred from the observed (7) diffuse galactic 1.809 MeV ^{26}Al decay line emission, may result from cosmic ray production in starformation regions is also inconsistent with the upper limits set by SMM(8) on the accompanying diffuse galactic 4.4 and 6.1 MeV fluxes which limit (5) such cosmic ray production to <10% of the required ^{26}Al production.

COMPTEL PHOTON FLUX VS. METEORITIC EFFECTS

Observed neutron effects in solar system matter are due to locally produced neutrons, as primary free neutrons will decay in transit. Such secondary neutron fluxes are comparable to proton fluxes after one interaction length in precompacted matter with densities $>10^{16}$ cm^{-3}. Neutrons are also produced in fission processes, with the expected fluences depending upon the abundances of fissioning nuclei.

We can estimate the range of implied irradiation volumes and time scales in starforming regions. The observations (1) of 1.0×10^{-4} photon cm^{-2} s^{-1} in the 3-7 MeV range scaled to the Orion distance (~1500 Ly) yield a photon production rate of 2.3×10^{39} photon s^{-1} at the source. Assuming comparable photon and neutron production rates (5) in a source region of radius r, we calculate neutron fluxes ranging from 10^{10} cm^{-2} a^{-1} for r = 10^5 AU, to 10^{16} cm^{-2} a^{-1} for r = 100 AU. A neutron fluence of 10^{16} n cm^{-2} inferred (see below) from the meteoritic particle recorders 4.56 Ga ago, could be obtained anywhere from 1 Ma to 1a. The COMPTEL photon flux has been observed for more than 1 yr so far, and since the Orion region is a few Ma old, the observed radiation environment can extend for some time.

It is possible, in principle, that neutron effects may be observed, while spallation effects due to higher energy particles may be difficult to detect, especially since meteorites also recorded the recent exposure to GCR during their exposure ages. An irradiation of evolving solar disk matter by a high latitude COMPTEL-type environment might produce a situation where low energy reactions due to secondaries are favored.

FOSSIL RECORDS IN METEORITES

Caffee et al. (9) observed that track-rich grains reveal considerably more spallogenic ^{21}Ne than track-free grains. If the excesses ^{21}Ne$_c$ are due to enhanced solar-flare activity, as implied by Caffee et al. (9), the minimal required solar-flare proton fluence was $>10^{17}$ p cm^{-2}. Marti et al. (10) and Kim and Marti (11) reported the identification of a distinct Xe component (FVM-Xe) in chondritic metal and concluded that this signature may be a relict of a neutron irradiation in the early solar system. A neutron fluence of $\geq 1 \times 10^{16}$ n cm^{-2} is required.

Evidence for unexplained neutron effects on ^{127}I was reported by Nichols et al. (12) in a residue of the Inman chondrite, where significant and correlated excesses were observed for ^{128}Xe and ^{129}Xe. The ^{129}Xe excess is explained by the decay of extinct ^{129}I, while the ^{128}Xe excess requires a $\geq 2 \times 10^{16}$ n cm^{-2} epithermal neutron fluence.

The reported excess of ^{41}K in an Efremovka inclusion (13) was assigned to nebular live ^{41}Ca, which decayed in situ to ^{41}K. Srinivasan et al. (13) exclude the possibility of a neutron irradiation, since a fluence of 3×10^{16} n cm^{-2} would be required. The observed strong correlation with Ca has to be expected, whether ^{41}Ca was incorporated live or produced in situ.

Kaiser and Wasserburg (14) invoked a neutron irradiation and an early active Sun based on exotic Ag isotopic systematics in iron meteorites. These authors found no evidence in terms of other spallation products for the presence of a high energy bombardment of sufficient fluence to produce ^{107}Pd and excess ^{107}Ag, but require an epithermal neutron fluence of $\sim 10^{17}$ cm^{-2}. Wasserburg (15) pointed out the constancy of inferred relative abundances of extinct radionuclides in the solar system. The abundance ratios ^{129}I/^{127}I, ^{107}Pd/^{108}Pd and ^{26}Al/^{27}Al, which are close to 10^{-4}, irrespective of their different half-lives, require an explanation.

The origin of the rare odd-odd nuclide ^{138}La is also not resolved. Shen et al. (16) studied its relative abundance in meteorites in order to search for evidence for proton fluence variations and obtained a limit $<10^{19}$ cm^{-2}. They note that the only possible excesses of ~ 1.2‰ (at the 3σ level) were observed in two carbonaceous chondrites. Murchison also shows excess ^{21}Ne$_c$ which was assigned to enhanced solar flare activity (9).

We conclude that many of the observed fossil neutron effects in meteorites require fluences of 10^{16} to 10^{17} n cm^{-2}. The COMPTEL photon flux which was observed (1) during the past year suggests that the particle fluence in an Orion-complex environment only requires exposure times of <1 Ma. This environment needs to be considered as an alternative to the postulated T-Tauri activity by our Sun. The irradiation geometry and disk opacity may be constrained by an evaluation of the relative importance of neutron and charged particle effects in solar system matter.

REFERENCES

1. Bloemen, H., et al., *Astr. Astrophys.* **281**, L5-L8 (1994).
2. Ramaty, R., Kozlovsky, B., and Lingenfelter, R. E., *Astrophys. J. Supp.* **40**, 487 (1979).
3. Clayton, D. D., *Nature* **368**, 222-224 (1994).
4. Cameron, A. G. W., *Nature* **368**, 192 (1994).
5. Ramaty, R., Lingenfelter, R. E., and Kozlovsky, B., *Nature* submitted (1994).
6. Bykov, A., and Bloemen, H., *Astr. Astrophys.* **283**, L1-L4 (1994).
7. Mahoney, W. A., et al., *Astrophys. J.* **286**, 578 (1984).
8. Share, G., et al., *Adv. Space Res.* **11**(8), 85 (1991).
9. Caffee, M. W., et al., *Astrophys. J.* **313**, L-31-L35 (1987).
10. Marti, K., et al., *Z. Naturforsch.* **44a**, 963-967 (1989).
11. Kim, J. S., and Marti, K., *LPSC* **XXIII**, 689 (1992).
12. Nichols, R., et al., *Geochim. Cosmochim. Acta* **55**, 2921 (1991).
13. Srinivasan, G., et al., *LPSC* **XXV**, 1325 (1994).
14. Kaiser, T., and Wasserburg, G. J., *Geochim. Cosmochim. Acta* **47**, 43-58 (1983).
15. Wasserburg, G. J., *Protostars and Planets II*, Tucson: Univ. of Arizona Press, 1985, pp. 703-737.
16. Shen, J. J-S., Lee, T., and Chang, C-T., *Geochim. Cosmochim. Acta* **58**, 1499-1506 (1994).

Distribution of Al-26 in the Galaxy

N. Prantzos

Institut d'Astrophysique de Paris

Abstract. The expected spatial profile of the 1.8 MeV emission from the radioactive decay of galactic ^{26}Al is calculated, assuming plausible distributions for the ^{26}Al sources and taking into account the spiral structure of our Galaxy. The resulting sky maps, compared to the recent one obtained by COMPTEL, suggest that the ^{26}Al sources are massive stars embedded in the spiral arms.

Key words: gamma-ray line astronomy — galactic structure

The COMPTEL instrument aboard the Compton Gamma-Ray Observatory definitively showed that the 1.8 MeV emission of the radioactive nucleus ^{26}Al is diffuse in the galactic plane (3). The detected flux implies that ~ 2 M$_\odot$ of ^{26}Al are ejected in our Galaxy every $\tau_{26} \sim 10^6$ years. Several astrophysical sites can produce considerable amounts of ^{26}Al, but the uncertainties in their modelling do not allow definite conclusions yet (see reviews in 2 and 5).

The first map of the galactic ^{26}Al has been recently obtained by COMPTEL (4). It shows a structured emission over a wide longitude range and a marked asymmetry relatively to the Galactic Centre. Those results can be tentatively interpreted as evidence for an underlying young population with lifetime $<10^8$ years and corresponding mass M>5 M$_\odot$. To that class belong the heavy AGB stars ($5<M/M_\odot<9$), SNII ($10<M/M_\odot<35$) and WR stars (M>35 M$_\odot$). It has been suggested a few years ago that the population of all those objects should trace the (still poorly known!) spiral structure of our Galaxy, leading to an asymmetric profile with several features in the directions that are tangent to the spiral arms (5, 6).

The distribution of all those objects in the Galaxy is not well known. It could follow the one of giant H$_2$ clouds (presumed site of massive star formation), of giant HII regions (ionised by the light of O stars) or the one of SNII, as observed in external spiral galaxies (Fig. 2). Theoretical sky maps corresponding to those radial distributions and with the spiral pattern of Fig. 1 are presented in Fig. 3a,b,c and d; they compare favourably (at first sight!) to the COMPTEL results and seem to support the idea that the 1.8 MeV flux traces indeed the spiral structure of the Galaxy.

REFERENCES

1. Bartunov O., Makarova J., Tsvetkov D. 1992, A&A, 264, 428
2. Clayton D., Leising M. 1987, Phys. Rep., 144, 1
3. Diehl et al. 1993 A&ASuppl., 97, 181
4. Diehl et al. 1994 A&A, submitted
5. Prantzos N. 1991, in "Gamma-ray Line Astrophysics", Eds. Ph. Durouchoux and N. Prantzos (AIP), p. 129
6. Prantzos N. 1993, ApJ Letters, 405, L55
7. Scoville N., Sanders D. 1987, in "Interstellar Processes", Ed. H. Thronson (Reidel), p. 21
8. Taylor J., Cordes J. 1993, ApJ, 411, 674

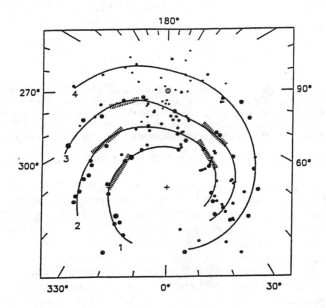

FIG. 1 Spiral structure of our Galaxy (from Taylor and Cordes 1993). Enhanced emission is expected from the directions tangent to the spiral arms.

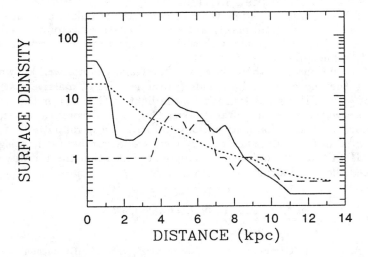

FIG. 2 Surface density as a function of galactocentric radius for: giant H_2 clouds (dotted line, from Scoville and Sanders 1987); giant HII regions (dashed line, from Scoville and Sanders 1987); and SNII in spiral galaxies (dashed-dotted line from Fig. 6 of Bartunov et al. 1992). Those three distributions and a flat one (i.e. $\sigma(R)$=const., not appearing in Fig. 2) are used for the construction of the theoretical sky maps of Fig. 3.

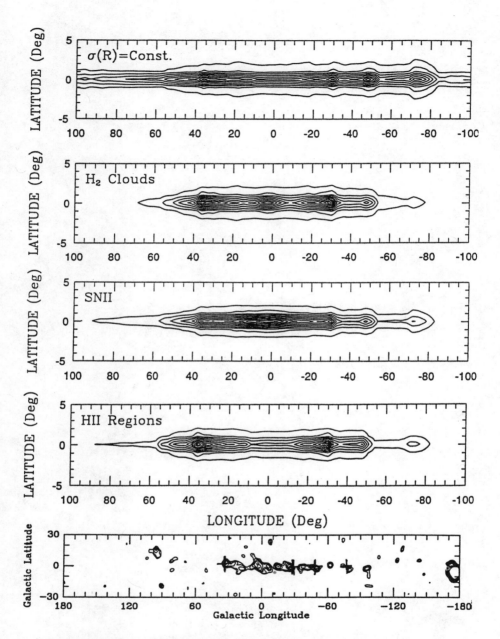

FIG 3 Theoretical sky maps, corresponding to the radial distributions of Fig. 2; they can be compared to the COMPTEL results (bottom).

The Main s-Component as a Superposition of Mean Neutron Exposures by TP-AGB Stars

R. Gallino(1), M. Busso(2), and C.M. Raiteri(2)

(1) Istituto di Fisica Generale dell'Università - Torino (Italy)
(2) Osservatorio Astronomico di Torino (Italy)

Abstract. We present the results of a series of s-process calculations in Thermally Pulsing Asymptotic Giant Branch stars of different metallicities, made on the assumption that the ^{13}C source is at the origin of the neutron capture episodes. We assume that the amount of ^{13}C made available by partial mixing of protons at the interface of the convective envelope with the ^{12}C-enriched region is constant over the metallicity range spanned by galactic disc stars. In this way we show that a very simple arithmetic mean of the production factors of s-nuclei by stars of different metallicity, from one tenth to the solar value as characteristic of the disc population, is a good approximation to the solar system *main s-process component*. This last is therefore likely to be a normal outcome of the chemical enrichment of the galactic disc.

Key words: Stars: evolution of — Nucleosynthesis — s-process

1. INTRODUCTION

The s-enhancements shown by peculiar Red Giants are characterized by a large spread of mean neutron exposures, from 0.1 to 0.6 mbarn^{-1} [1]. On the other hand, it is known from the phenomenological analysis that it is formally possible to reproduce the main s-component in the solar system, i.e. the solar distribution of the heavy s-isotopes from Zr to Pb, with an exponential distribution of neutron exposures $\rho(\tau) = exp[-\tau/\tau_0]$ provided that the mean neutron exposure is $\tau_0 \sim 0.30$ mbarn^{-1} [2].

From s-process nucleosynthesis calculations in thermally pulsing asymptotic giant branch stars (TP-AGB) of low mass, we showed that the solar system main s-component can be well reproduced by stars of metallicity $Z \sim 1/3\ Z_\odot$ [3].

In this kind of models, a fixed amount (some $10^{-6}\ M_\odot$) of ^{13}C was assumed to be produced after each thermal instability at the interface between the hydrogen-rich envelope and the He shell; neutrons are then released by the activation of the ^{13}C(α,n)^{16}O reaction. In Fig. 1 the best fit distribution of heavy s-isotopes, reproducing the main component is shown for a metallicity $Z = 0.006$.

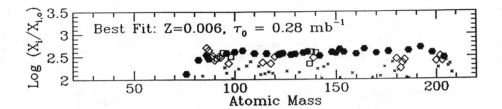

Fig. 1. The overabundances of the best fit to the main component

In the Figure, the enhancement factors of the s-process isotopes with respect to solar abundances are plotted as a function of atomic mass. The heavy exagons represent the s-only isotopes, the open squares and the diamonds represent isotopes with a major s-process contribution, and the small crosses show those isotopes with a minor s-contribution.

2. s-PROCESSING AT DIFFERENT METALLICITIES

In Fig. 2 we show s-process calculations repeated with the same assumptions as in Fig. 1 (in particular assuming that the same amount of ^{13}C is created in the interpulse phases) but changing the stellar metallicity. As is evident from the various panels, this yields non solar s-process distributions.

This differential trend of s-abundances is a well known effect (see [4], [5], [6]): the Fe seeds for the s-process and the light neutron poisons scale with metallicity, whereas the ^{13}C neutron source, deriving from captures of protons on the newly synthesized ^{12}C in the He shell, behaves as a primary nucleus.

A non solar distribution of s-elements, as resulting in low mass TP-AGB stars of initial metallicity close to solar, has been suggested as the key to interpret the isotopic anomalies in carbon-rich interstellar grains recovered in pristine meteorites and most likely condensed in the mass-losing extended envelopes of C stars.

Here we note that the main s-component, i.e. the average composition of the s-elements in the interstellar medium from which the solar system condensed, can more realistically be obtained by properly averaging different s-process distributions within the observed spread of mean neutron exposures. This process of averaging is actually what is expected to result from the continuous recycling of matter through subsequent stellar generations, of

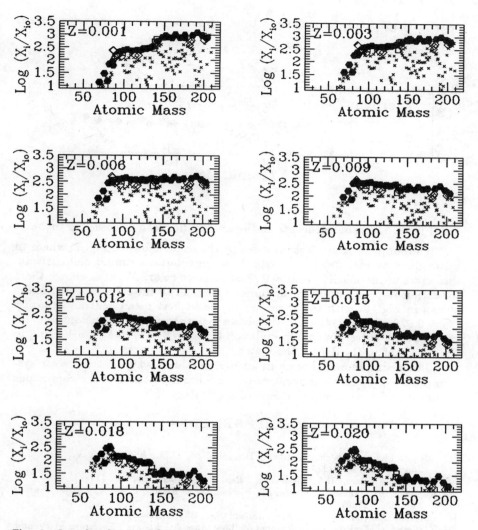

Fig. 2. Overabundances with respect to the solar composition obtained by allowing the same amount of ^{13}C per pulse to burn in stars of different metallicity.

metallicity gradually increasing in time, up to the epoch of the solar system formation.

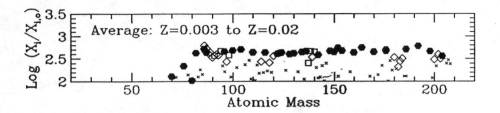

Fig. 3. Overabundances from the arithmetic mean discussed in the text.

The simplest example is provided by the result shown in Fig. 3, where an unweighted average has been made of the contributions coming from galactic disc stars of metallicity from $\sim 1/7$ of solar to solar. What is shown there for each isotope is simply the arithmetic mean of all the enhancement factors shown in Fig. 2, with the exclusion of the highest mean neutron exposure ($\tau_0 = 0.926$ mbarn^{-1}) that, in the above hypotheses for the constant amount of ^{13}C burnt, corresponds to population II stars ($Z = 1/20\ Z_\odot$), belonging to the halo, not to the disc of the Galaxy.

The result shows, though in an oversimplified way, that the solar system main s-process component can be a normal outcome of the continuous chemical enrichment in s-elements of the Galaxy.

REFERENCES

[1]. Busso, M., Gallino, R., Lambert, D.L., Raiteri, C.M., & Smith, V.V., *Astrophys. Journal* **399**, 218, (1992).
[2]. Käppeler, F., Beer, H.,& Wisshak, K., *Rep. Progr. Phys.*, **52**, 945, (1989).
[3]. Käppeler, F., Gallino, R., Busso, M., Picchio, G., & Raiteri, C.M., *Astrophys. Journal*, **354**, 630, (1990).
[4]. Clayton, D.D., *Month. Not. Roy. Astron. Soc.*, **234**, 1, (1988).
[5]. Gallino, R., "s-Process Enrichment in Low-Mass AGB Stars" *Evolution of Peculiar Red Giant Stars*, IAU Colloquium No. 106, ed. H.R. Johnson & B. Zuckerman, (Cambridge: Cambridge Univ. Press), 176 (1989).
[6]. Gallino, R., Busso, M., Picchio, G., Raiteri, C.M., & Renzini, A. *Astrophys. Journal*, **334**, L45, (1988).

In Early Universe
Primordial Molecules and Thermal Effects

Denis Puy[1,2]

[1]International School of Advanced Studies, I.S.A.S.-S.I.S.S.A
Via Beirut, Trieste (Italy)

[2]Observatoire de Paris-Meudon, D.A.E.C.
5, Place Jules Janssen, 92195 Meudon (France)

Abstract. The elements available from primordial nucleosynthesis are H, H^+, D, D^+, He and Li forming primordial molecules such as H_2, HD, LiH. A primordial chemistry yielding to the formation of these molecules is presented. Since the reaction rates depending on temperature, the density and the temperature evolution equations must be solved simultaneously, which needs in turn a determination of molecular cooling and heating rates. The primordial molecules seem to play a moderate role in the thermal history of the Universe. Molecules could play a more decisive role in a situation of gravitational collapse. I shall describe the physics and chemistry of a simple collapsing protocloud, and present the abundances of these molecules and the consequences.

INTRODUCTION

It is now well-known that the conventional models of nucleosynthesis can lead to the elements H, H^+, D, D^+, He and Li. Since the work of Saslaw and Zipoy (1), Lepp and Shull (2) and more recently Puy *et al* (3) have described the post-recombination chemistry. Two main routes lead to H_2 formation (via H^- and H_2^+). Concerning the molecule HD, similar processes are possible, however its formation proceeds mainly through the molecule H_2. The molecule LiH is supposed to be formed through radiative association. All these reactions are coupled with others reactions which are listed in Puy *et al* (3). The numerical integration of the coupled chemical equations is an initial value problem for stiff differential equation. In the standard Big Bang model, the composition of the Universe remains essentially unchanged as it cools - from the nucleosynthesis period - toward the epoch of hydrogen recombination at $T_r \sim 4000^\circ K$ or $z_r \sim 1000$. Below $3000^\circ K$, only excitation of rotational levels is mainly due to collisional excitation and de-excitation

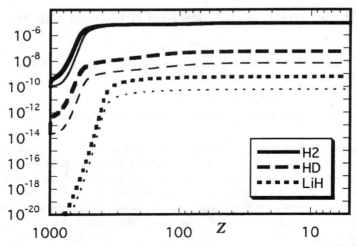

Figure 1: Fractional abundances of H_2, HD, LiH for models A (thin line) and B (bold line).

with H, H_2 and He. Molecular cooling corresponds to collisional excitation followed by radiative radiation. Molecular heating is due to radiative excitation from the Cosmic Microwave Background Radiation (CMBR) followed by collisional de-excitation.

POST-RECOMBINATION EVOLUTION

As first shown by Peebles (4), coupling between matter and radiation via the Thomson diffusion during the phase of recombination leads to a thermal evolution of the Universe. Once the molecules form, the thermal evolution must also take into account molecular processes. These contributions are molecular (cooling and heating rates) and chemical due to the fact that some chemical reactions can be exothermical or endothermical. I have considered two different models (A and B). Model A corresponds to typical values of the parameters given by the standard nucleosynthesis. The parameters of model B are given by the inhomogeneous nucleosynthesis. Relative abundances of H_2, HD and LiH are shown on Fig.1 for both models. H_2 abundances rises through two processes (H^- and H_2^+). The HD abundance rises also in two steps with a transition around $z = 100$, due to D^+ and HD^+ (see Puy al (3). LiH formation through radiative association becomes effective when the photodestruction decreases. The Compton scattering is important and the molecular contribution remains small, Puy et al (3). So molecular cooling and heating functions have a slight influence on the decrease of the temperature of the matter. Moreover the diffuse radiative background due to radiative transitions of the primordial molecules is found to be presently undetectable.

GRAVITATIONAL COLLAPSE

In a context of gravitational collapse, the chemistry seems to be different and leads to new abundances. In fact, fluctuations that survive decoupling are subject to gravitational instabilities if they are of sufficiently large scale. Their growth and eventual collapse may lead to the formation of structures. The balance between pressure and gravity is determined by the Jeans criterion. Moreover thermal processes may be important in particular cooling effects produced by molecules and could lead to fragmentation. Hutchins (5) analyzed this point but did consider only the hydrogen molecules in their collapse. The initial perturbations are described by the spectrum $\frac{\delta \rho}{\rho} = (\frac{M}{M_o})^\alpha$ with $\alpha = -\frac{2}{3}$ and $M_o = 10^{15} M_\odot$ as Gott and Rees (6). I assume that the perturbation region is a spherical homogeneous non-rotating cloud without opacity and magnetic field. The behaviour of the *cloud's* radius is analogous to that of the scale factor in Friedmann's solution, Lahav (7). Hence, the radius reaches a maximum value, *the turn-around point*, and then collapses. I choose a cloud with a mass $M = 10^{10} M_\odot$ which corresponds to a turn around redshift $z_{ta} = 25$, and consider the baryonic density parameter and the initial relative elemental abundance of deuterium and lithium in the case of model B. Fig. 2 shows the evolution of abundances. Thus we obtain an increase of H_2, LiH abundances (if the density increases, the collisions begin efficient). Because the formation is only due to collisional formation, the abundance for these molecules are going to increase naturally (Destruction for these molecules are radiative). For HD the situation is different as Figure 2 shows. This molecule is destructed by collision with H^+, see Puy et al (8).
Maoli et al (9) and Signore et al (10) analyzed a strategy for the detection of primordial LiH molecules through experiments devoted to the observations of the CMBR anisotropies. The chemical approach of a gravitational collapse suggests that LiH abundances could induce anisotropies at a detectable level, because abundance increases. Nevertheless this simple model of collapse do not consider the opacity effect, necessary for a valuable estimation of these secondary anisotropies. The very promising result of this study warrants future detailed investigations Puy et al (11).

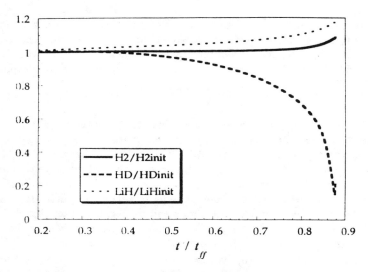

Figure 2: Evolution of abundances of primordial molecules (relative to initial abundances) as a function of time (in unit of free-fall time $t_{ff} = \sqrt{3\pi}/\sqrt{32G\rho}$).

Acknowledgements. The author gratefully acknowledge Monique Signore and Francesco Melchiorri for the interesting discussions and the supervision of this work.

References.
(1) W.C. Saslaw, D. Zipoy, *Nature* 216, 1967, 976
(2) S. Lepp, J. M. Shull, *Ap. J.* 280, 1984, 465.
(3) D. Puy et al, *Astron. Astroph.* 267, 1993, 337.
(4) P. J. E. Peebles, *Ap. J.* 153, 1968, 1.
(5) J.B. Hutchins J. B., *Ap. J.* 205, 1976, 103.
(6) R. J. Gott III, M. J. Rees, *Astron. Astrop.* 45, 1975, 365.
(7) O. Lahav, *Mon. Not. R. astr. Soc.* 220, 1988, 259.
(8) D. Puy, F. Melchiorri, M. Signore, Proceeding of the International *Birth of the Universe and Fundamental Physics-Roma 1994*, in press
(9) R. Maoli, F. Melchiorri, D. Tosti, *Ap. J.* 425, 1994, 372.
(10) M. Signore et al, *Ap. J. Supp.* 92, 1994, 535.
(11) D. Puy, F. Melchiorri, M. Signore, *in preparation*

SECTION V

ISOTOPIC COMPOSITION OF METEORITES AND THE EARLY SOLAR SYSTEM

Interstellar Grains from Primitive Meteorites: New Constraints on Nucleosynthesis Theory and Stellar Evolution Models

Ernst Zinner

McDonnell Center for the Space Sciences and the Physics Department
Washington University, Saint Louis, Missouri 63130-4899 USA

Abstract. Primitive meteorites contain interstellar grains that formed in stellar outflows or from supernova ejecta, survived interstellar travel and the formation of the solar system, and were incorporated into meteorites. To date diamond, silicon carbide, graphite, corundum and possibly silicon nitride have been identified. Their isotopic compositions provide new information on nucleosynthesis and stellar evolution and on the numbers and types of stars that contributed material to the solar system. Examples are the C-, N- and Si-isotopic ratios of SiC grains from AGB stars, the C-, N-, O-, Mg- and Si-isotopic ratios of graphite grains from massive stars (WR or SN), and the O-isotopic ratios of corundum grains from red giants.

1. INTRODUCTION

The seminal papers by Burbidge *et al.* (1) and Cameron (2) established the field of nucleosynthesis theory by proposing a set of nuclear processes for the production of all elements from C on in stars. The theoretical efforts of these authors were essentially guided by the goal to explain the elemental and isotopic abundances of the solar system (3). The solar system abundances underwent many refinements (4-7), but they remained the most important constraint for the theory of nucleosynthesis (e.g., 8). The obvious advantage provided by the solar system abundances is that they encompass all stable (and a fair number of unstable) isotopes of almost all elements. A disadvantage is that the composition of the solar system is a mixture from many different stellar sources and does not give information on the contribution from individual stars. The decomposition of the solar system abundances into contributions from different nucleosynthetic processes was most successful for the heavy (>Fe) elements, whose abundances can be explained as a superposition of s-, r- and p-process material (9-11). For the lighter elements, however, there is much more ambiguity as to their stellar sources (e.g., 12,13).

Another important set of constraints for nucleosynthesis and stellar evolution theory is provided by astronomical measurements of elemental and isotopic abundances in stars (14). The advantage of this type of information is that it is obtained from individual stars. One of the most important discoveries was the detection of the unstable element Tc in the spectra of S-stars (15), which provided direct evidence for stellar synthesis of the elements. One major shortcoming of astronomical observations of isotopic abundances is that the measurements usually are limited to a few elements such as carbon and oxygen (e.g., 16-20).

In the last few years a new source of information about the isotopic composition of stellar matter has become available in the form of interstellar grains from primitive meteorites. These grains are believed to have condensed in stellar outflows and thus reflect the elemental and isotopic composition of stellar atmospheres. In comparison to the two types of constraints for nucleosynthesis theory mentioned above, in favorable cases the isotopic compositions of **several** elements can be obtained on individual grains (and thus on individual stellar sources). Furthermore, interstellar grains can, in principle, provide isotopic information on stars from which no isotopic information can be obtained by astronomical observations. These advantages are contrasted by the fact that the stellar source is unknown for any particular grain or type of grains and must be inferred from the elemental and isotopic compositions. Since the first isolation of interstellar grains (21) their study has grown to an extent that a detailed review is beyond the scope of this article. The interested reader will find more comprehensive treatments in the papers by Anders and Zinner (22) and Ott (23). Here, after a short survey of the interstellar grains identified so far I shall concentrate on a few topics to demonstrate that the study of interstellar grains yields information that has not been obtained in any other way.

2. INTERSTELLAR GRAINS IN METEORITES

2.1. Isolation and identification

The most important clue that led to the discovery of interstellar grains was the presence of isotopic anomalies in the noble gases Ne and Xe in primitive meteorites. The search for isotopic anomalies in meteorites with the goal of finding extrasolar material goes back to the twenties (24-26), but it wasn't until the fifties that isotopic anomalies in meteorites were found, first of the D/H ratio (27). However, in spite of the finding of further anomalies in the noble gases Xe (28) and Ne (29), it took almost 20 years and the discovery of ^{16}O excesses in refractory inclusions in 1973 (30) before the notion of the survival of relict interstellar material in primitive meteorites was widely accepted. The isolation and identification of three types of carbonaceous interstellar grains, all labeled by "exotic" (i.e., isotopically anomalous) noble gases, took another 14 years. These types of grains are: **diamond** (21), carrier of Xe-HL (28); **silicon carbide** (31,32), carrier of Ne-E(H) (29) and Xe-S (33); **graphite** (34), carrier of Ne-E(L) (35). In addition, SiC and graphite were found to contain tiny subgrains of Ti, Zr and Mo carbides identified in the transmission electron microscope (36-38). Interstellar oxide grains apparently do not carry any noble gases (39) and their presence in primitive meteorites was established by ion microprobe isotopic analysis of single grains (40-43). A possible additional type of interstellar grains, silicon nitride or silicon oxynitride, has also been identified in the ion microprobe (44).

The identification of the interstellar origin of these grains is based on their highly anomalous isotopic compositions that, in general, agree with those expected for fresh stardust. However, since meteoritic material, believed to be of solar system origin, also can carry anomalous isotopic signatures inherited from interstellar grains that were reprocessed and diluted in the solar nebula (e.g., 45, 46), a careful distinction between this material and intact interstellar grains is necessary. One criterion is the presence of **correlated** isotopic anomalies in the **major** elements. An example is shown in Figs. 1 and 2. Many refractory objects ("rocks") in primitive meteorites exhibit ^{26}Mg excesses that have been interpreted as being due to the decay of short-lived ^{26}Al. The inferred ^{26}Al/^{27}Al ratios show a wide distribution with a sharp upper limit of 5×10^{-5}, indicating the presence of

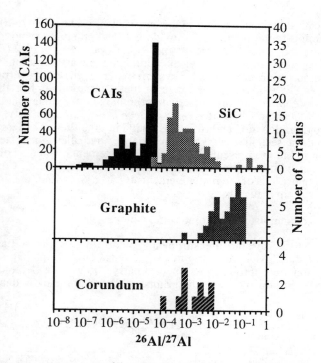

Figure 1. Distributions of $^{26}Al/^{27}Al$ ratios in CAIs and interstellar grains.

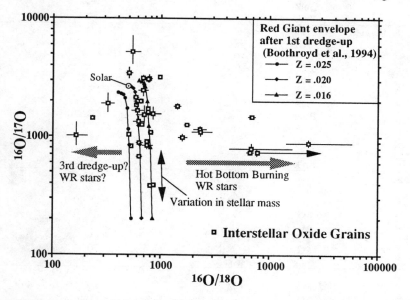

Figure 2. Oxygen isotopic compositions of the solar system and of interstellar corundum grains.

live ^{26}Al with this concentration in the early solar system (45,47,48). In contrast, ^{26}Al/^{27}Al ratios found in interstellar oxide, SiC and graphite extend to many orders of magnitude above this "canonical" solar system value of 5×10^{-5} (Fig. 1).

The situation is similar for oxygen. Although there are variations in the oxygen isotopic composition of early solar system material (including refractory objects with ^{26}Al/^{27}Al $\leq 5 \times 10^{-5}$), the range of these variations is less than 10%. The range of O-isotopic ratios in interstellar corundum, on the other hand, is a factor of 10 for ^{16}O/^{17}O and a factor of 100 for ^{16}O/^{18}O (Fig. 2). Similarly, the variations in the isotopic ratios of C and Si in SiC and C in graphite are orders of magnitude larger than those seen in solar system materials. Anomalies in interstellar grains are not only large but are present in essentially all elements, whereas objects of solar system origin with isotopic anomalies in certain elements (e.g., Ca, Ti, Cr) have isotopic compositions of the major elements (O, Mg, Si) within the limited range of other solar system materials.

2.2. Types of stardust and isotopic compositions

Table 1 lists isotopic measurements made on the interstellar grain types identified to date. Many of these measurements could be made on single grains, mostly by ion microprobe mass spectrometry. For more details the reader is referred to the review by Anders and Zinner (22) and to the individual papers listed in the table.

Table 1. Isotopic Analyses of Stardust

Grain Type	Analysis Type	Element	References†
Diamond	Bulk	Noble gases	21,49-53
	Bulk	C	54-56
	Bulk	N	56
	Bulk	Sr,Ba	57
Silicon Carbide	SG*	C	41,58-65
	SG	N	41,58,60,62-64
	SG	Si	41,58-65
	SG	Mg-Al	41,60,63,65,66
	SG	Ti	63,65,67,68
	SG	Ca	65,68
	SG	He,Ne	69-71
	Bulk	Noble gases	58,72-76
	Bulk	Sr	77-80
	Bulk	Ba	81-83
	Bulk	Nd,Sm	82,84
	Bulk	Dy,Er	85
Graphite	SG	C	34,86-88
	SG	N	34,86-88
	SG	Mg-Al	66,88,90,91
	SG	O	88-90
	SG	Si	88,90-92
	SG	Ca,Ti	88,90,92
	SG	He,Ne	70,93
	Bulk	Noble gases	94-96
Oxide	SG	O	43,97-99
	SG	Al-Mg	40,97-99
	SG	Ti	100
Silicon nitride	SG	C,N,Si	44

* Single grain analysis † The numbers refer to the reference list at the end of the paper

Diamond grains are too small for single grain analysis and shall not be discussed any further.

Silicon carbide grains are believed to originate mostly from thermally pulsing AGB stars because: 1) the distribution of $^{12}C/^{13}C$ ratios in single SiC grains (61, 63,64) is similar to that in carbon stars (16,101); 2) AGB stars are the main contributor of carbonaceous dust to the interstellar medium (102,103); 3) there is evidence for SiC in the dusty envelopes of carbon stars from their 11.2 μm emission feature (e.g., 104) and 4) AGB stars are believed to be the main source of s-process elements (105). Most isotopic compositions conform to the notion of an AGB origin, some deviations will be discussed below.

Graphite grains show the most complexity of all the interstellar grains identified so far. They vary in size, density, chemical compositions and morphology and their isotopic compositions depend on these properties (86,87,90, 96) and indicate several types of stellar sources. The C-isotopic ratios found in the grains indicate that they come from stellar sources dominated by H-burning as well as from sources dominated by He-burning.

Corundum grains show a large range in their O-isotopic ratios and $^{26}Al/^{27}Al$ ratios (42,98,99). Some must have formed in red giant atmospheres before the third dredge-up, most during the TP-AGB phase; some show evidence for hot-bottom burning.

3. NEW INFORMATION FROM INTERSTELLAR GRAINS

In the rest of this paper I shall concentrate on a few topics with two goals in mind: 1) Show that existing theoretical models of nucleosynthesis and stellar evolution cannot satisfactorily explain all isotopic data obtained on interstellar grains, i.e. show that the data on I.S. grains present new challenges, and 2) Show that, under the assumption of the validity of certain stellar evolution models, interstellar grains provide information on the numbers and types of stellar sources that have contributed material to the solar system.

3.1. C and N isotopes in SiC

With the exception of grains X, the C- and N-isotopic compositions of SiC grains generally agree with the expectations for carbon stars: heavy to moderately light C and light N (Fig. 3). However, neither the C- nor the N-isotopic ratios of certain grains can be explained by existing stellar models. In these models $^{12}C/^{13}C$ ratios after the first dredge-up are ~20 or higher (106-108) and become even higher during the TP-AGB phase when the star turns into a carbon star. This problem has been known before for $^{12}C/^{13}C$ ratios in carbon stars (16), but the SiC grains provide additional constraints from the N-isotopic ratios (Fig. 3) and $^{26}Al/^{27}Al$ ratios (63). The $^{14}N/^{15}N$ ratios of most grains with $^{12}C/^{13}C < 20$ are significantly lower than the ratios expected to exist after the first dredge-up (107,108). Hot bottom burning (HBB) in more massive AGB stars ($M \geq 5\ M_\odot$) is expected to result in low $^{12}C/^{13}C$ ratios (109-113), but not in low $^{14}N/^{15}N$ ratios. Furthermore, HBB is believed to prevent AGB stars from becoming carbon stars. More detailed model calculations for C-, N- and Al-isotopic ratios in stars, where HBB takes place, are needed. The situation is similar for J-type carbon stars, which are observed to dominate carbon stars with $^{12}C/^{13}C < 5$ (16). A major theoretical effort is needed to understand the nucleosynthesis of C, N and ^{26}Al in such stars and to determine whether J-stars can possibly be sources of SiC grains with small $^{12}C/^{13}C$ ratios.

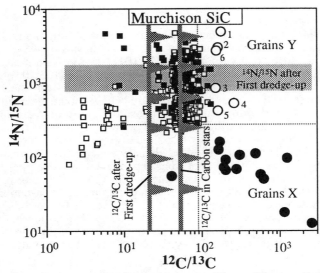

Figure 3. Carbon- and N-isotopic compositions of interstellar SiC.

3.2. Graphite grains from massive stars

Another challenge for nuclear astrophysics is posed by the isotopic compositions of graphite grains. A more detailed presentation of the isotopic

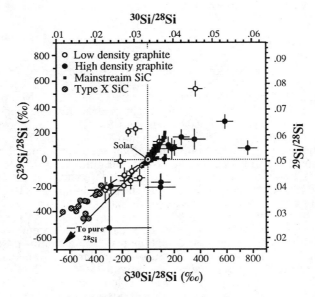

Figure 4. Silicon isotopic compositions of interstellar SiC and graphite grains.

compositions of these grains is given by Amari *et al.* (88). The Si-isotopic ratios of two density fractions of such grains, the low density fraction on KE3 (1.60 to 1.90 g cm^{-3}) and the high density fraction KFC1 (2.15 to 2.20 g cm^{-3}) are shown in Figure 4, together with those of SiC grains (41,63). Most KE3 grains show depletions in ^{29}Si and ^{30}Si relative to ^{28}Si, similar to the SiC grains of type X, which are characterized by excesses in ^{12}C (in most grains), ^{15}N and large ^{26}Al/^{27}Al ratios and for which a supernova origin has been proposed (65). The most diagnostic signatures of the KE3 grains with regard to their stellar origin are the oxygen isotopic compositions (Fig. 5). Most grains have large ^{18}O excesses (up to 100×, corresponding to ^{16}O/^{18}O = 5). Such high ^{18}O enrichments are expected to

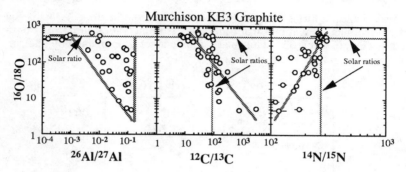

Figure 5. Carbon-, N-, O- and ^{26}Al/^{26}Al ratios of low density interstellar graphite grains.

be found in the He-burning zone of massive stars at the end of their evolution just before their explosion as supernovae (114,115). In this layer ^{14}N has been transformed to ^{18}O by He-burning via ^{14}N$(\alpha,\gamma)^{18}$F$\rightarrow^{18}$O, but temperatures are not high enough to destroy ^{18}O by the ^{18}O$(\alpha,\gamma)^{22}$Ne reaction. KE3 grains have also high ^{26}Al/^{27}Al ratios (Fig. 5). While ^{26}Al is low in the He-burning shell of massive stars, a large amount of it is produced in the overlying ^{14}N-rich zone. Table 2 shows the C, N and O isotopic ratios in these two zones of a 25M$_\odot$ star (115).

Table 2. Isotopic ratios in two zones of 25 M$_\odot$ supernova

Zone	C/O	^{12}C/^{13}C	^{14}N/^{15}N	^{16}O/^{18}O	^{26}Al/^{27}Al
He/N	1.1	4.8	30,000	73	0.19
He/C	4.5	10^6	1.8	5.8	2.7 × 10^{-3}

In both zones C>O and graphite grains could thus form from the ejecta of these layers. Mixing between them can at least qualitatively account for the sign of the correlations between the ^{16}O/^{18}O, ^{12}C/^{13}C and ^{14}N/^{15}N ratios (Fig. 5), but fails to explain the distribution of data points in the ^{16}O/^{18}O versus ^{26}Al/^{27}Al scatter plot, where grains with the highest ^{18}O excesses show high ^{26}Al/^{27}Al ratios while grains with normal ^{18}O/^{16}O have much lower ^{26}Al/^{27}Al ratios. There are some other detailed features that remain unexplained, for example the lack of large ^{14}N excesses in grains with small ^{12}C/^{13}C ratios. Work by Langer (116) and Langer and Henkel (117) has shown that the surface of massive mass-losing stars (Wolf-Rayet stars) can be ^{18}O rich at the WN-WC transition as long as semiconvection dominates the mixing processes in such stars. Because the stellar evolution models of Weaver and Woosley, from which Table 2 is derived, use semiconvective mixing and because the application of the Schwarzschild criterion for convective instability would not result in large $^{18}O/^{16}O$ ratios, either in the He/C shell (Woosley, private communication) or on the surface of WR stars (117), the O-isotopic composition of the graphite grains favor semiconvection as mixing mechanism in massive stars. However, much more theoretical work is needed for a thorough understanding of the graphite data (see also 88).

3.3. Silicon and titanium isotopes in SiC: evidence for multiple stellar sources

The Si-isotopic ratios of most SiC grains fall along a slope 1.34 line when plotted as δ-values (deviations from the solar isotopic ratios in permil) on a 3-isotope plot (Fig. 6 – Ref. 63). This distribution cannot be explained by nuclear

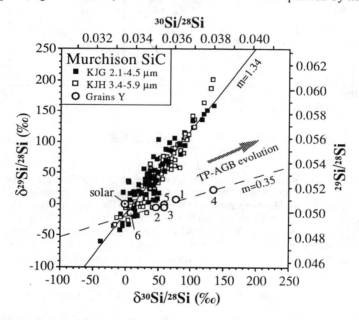

Figure 6. Silicon isotopic compositions of mainstream SiC grains.

reactions taking place during H- and He-burning, either in a single massive star or in a single TP-AGB star (most of the SiC grains are believed to originate from AGB stars). At the temperatures reached by these processes, charged particle reactions do not affect the Si isotopes. Neutron capture in the He-shell affects the Si isotopes by (n,γ), and (n,α) on ^{33}S (118), resulting in a shift of the Si-isotopic ratios along the thick arrow ("TP-AGB evolution") in Fig. 6 (119,120). However, only a few SiC grains (named "grains Y") lie on a line parallel to the expected TP-AGB evolution. The n-capture effects are not sensitive to the nature of the neutron source and the same Si-isotopic evolution is expected in WR stars. It thus has to be concluded that the distribution of the Si-isotopic ratios in SiC grains must reflect contributions from different stars with different isotopic compositions. Gallino and coworkers (119) have used a galactic evolution model to relate different initial isotopic compositions to different initial metallicities. The observed range in Si-isotopic compositions can be explained as originating from stars with [Fe/H] values between −0.35 and +0.10. This conclusion is remarkable in that it agrees with the observation by Edvardssen et al. (121) that stars of any given epoch show a range of metallicities of approximately a factor of four.

The conclusions reached from the Si isotopes are made even stronger by the Ti-isotopic compositions of SiC grains (63,67). Data points do not lie along mixing lines between an isotopically normal component and a Ti-isotopic component representing neutron capture in the He-burning zones of TP-AGB and WR stars. The most satisfying explanation is again that multiple stellar sources with different initial Ti-isotopic compositions (corresponding to different metallicities) contributed SiC grains to the solar nebula that were subsequently incorporated into primitive meteorites (119).

3.4. Oxygen isotopes and ^{26}Al in oxide grains

A situation similar to that for interstellar SiC exists for interstellar oxide grains: their O-isotopic compositions cannot be explained by nucleosynthetic processes taking place in a single star. A more detailed discussion of the oxide data is given by Nittler et al. (42,98), here I wish to concentrate only on the question of multiple sources. Previous work (108,122,123) has shown that mixing into the envelope (first dredge-up; also second dredge-up in more massive stars) of material processed in the star's interior during core H-burning significantly decreases the ^{16}O/^{18}O ratio (see Fig. 2). The exact predicted O-isotopic compositions depend on the still very uncertain ^{17}O(p,γ)^{18}F, ^{17}O(p,α)^{14}N and ^{18}O(p,α)^{15}N reaction rates. The essential point, however, is, that the spread in O-isotopic compositions found in meteoritic oxide grains (Fig. 2) can only be explained by assuming that different stars with different masses (variations in ^{16}O/^{17}O) and different initial isotopic compositions (variations in ^{16}O/^{18}O) contributed oxide grains to the solar system. Huss et al. (99) and Boothroyd et al. (123) related variations of initial ^{16}O/^{18}O ratios to variations in metallicity (see Fig. 2) according to a galactic evolution model of the oxygen isotopes (124). Although this relationship has been revised (125), their basic conclusions remains the same as that reached from the Si- and Ti-isotopic compositions in SiC: several stars with a range in metallicities must have contributed stardust to the solar system.

However, there are several potential complications with this interpretation of the O-isotopic compositions in oxide grains. Some grains have high ^{16}O/^{18}O ratios (Fig. 2), indicating hot bottom burning. Others have ^{18}O excesses (low ^{16}O/^{18}O). Whereas the calculations of Boothroyd et al. (123) indicate that the third dredge-up does not change the O-isotopic composition of the envelope, Mowlavi and Arnould (pers. commun.) obtain lower temperatures in the He-shell of a 2.5 M$_\odot$ TP-AGB star, with the consequence that ^{18}O produced by ^{14}N(α,γ)^{18}O survives during the first few thermal pulses. The question is whether a limited degree of HBB and the dredge-up of ^{18}O from the He-shell can cause variations in the ^{16}O/^{18}O ratio at the surface of a given star, relieving the need for stars with different metallicities. Additional theoretical models and better determinations of nuclear reaction rates, including that of the crucial ^{18}O(α,γ)^{22}Ne reaction (126) are needed to answer this question.

ACKNOWLEDGMENTS

E. Anders developed the chemical and physical separation methods that led to the identification of interstellar grains. R. M. Walker played a crucial role in establishing microanalytical facilities at Washington University that made the analysis of single grains of stardust and the discovery of interstellar oxide grains possible. I thank P. Eberhardt and the University of Bern for their hospitality during a visit when part of this paper was written and P. Hoppe for his help with problems of computer communication. I also thank C. Alexander, S. Amari, M. Arnould, M. El Eid, R. Gallino, P. Hoppe, N. Langer, N. Mowlavi, L. Nittler, and S. Woosley for comments and discussions and E. Koenig for help with the manuscript. This work was supported by NASA.

REFERENCES

1. Burbidge, E. M., Burbidge, G. R., Fowler, W. A., and Hoyle, F., *Rev. Mod. Phys.* **29**, 547-650, 1957.
2. Cameron, A. G. W., AELC (Atomic Energy of Canada, Ltd.), 1957, CRL-41.
3. Suess, H., and Urey, H. C., *Rev. Mod. Phys.* **28**, 53-74, 1956.
4. Cameron, A. G. W., *Space Sci. Rev.* **15**, 121-146, 1973.
5. Cameron, A. G. W., in *Essays in Nuclear Astrophysics*, C. A. Barnes, D. D. Clayton and D. N. Schramm (eds.), Cambridge: Cambridge University Press, 1982, 23-43.
6. Anders, E., and Ebihara, M., *Geochim. Cosmochim. Acta* **46**, 2363-2380, 1982.
7. Anders, E., and Grevesse, N., *Geochim. Cosmochim. Acta* **53**, 197-214, 1989.
8. Trimble, V., *Astron. Astrophys. Rev.* **3**, 1-46, 1991.
9. Woolum, D. S., in *Meteorites and the Early Solar System*, J. F. Kerridge and M. S. Matthews (eds.), Tucson: Univ. Arizona Press, 1988, 995-1020.
10. Käppeler, F., Beer, H., and Wisshak, K., *Rep. Prog. Phys.* **52**, 945-1013, 1989.
11. Lambert, D. L., *Astron. Astrophys. Rev.* **3**, 201-256, 1992.
12. Woosley, S. E., in *Nucleosynthesis and Chemical Evolution*, B. Hauck, A. Maeder and G. Meynet (eds.), Observatoire de Genève, 1986, 1-195.
13. Clayton, D. D., in *Meteorites and the Early Solar System*, J. F. Kerridge and M. S. Matthews (eds.), Tucson: Univ. of Arizona Press, 1988, 1021-1062.
14. Michaud, G., and Tutukov, A. (Eds.), *Evolution of Stars: The Photospheric Abundance Connection*, Dordrecht: Kluwer, 1991, 487 pp.
15. Merrill, P. W., *Astrophys. J.* **116**, 21-26, 1952.
16. Smith, V. V., and Lambert, D. L., *Astrophys. J. Suppl.* **72**, 387-416, 1990.
17. Lambert, D. L., Gustafsson, B., Eriksson, K., and Hinkle, K. H., *Astrophys. J. Suppl.* **62**, 373-425, 1986.
18. Harris, M. J., Lambert, D. L., Hinkle, K. H., Gustafsson, B., and Eriksson, K., *Astrophy. J.* **316**, 294-304, 1987.
19. Kahane, C., Cernicharo, J., Gomez-Gonzalez, J., and Guelin, M., *Astron. Astrophys.* **256**, 235-250, 1992.
20. Kahane, C., in *Nuclei in the Cosmos*, this volume, 1994.
21. Lewis, R. S., Tang, M., Wacker, J. F., Anders, E., and Steel, E., *Nature* **326**, 160-162, 1987.
22. Anders, E., and Zinner, E., *Meteoritics* **28**, 490-514, 1993.
23. Ott, U., *Nature* **364**, 25-33, 1993.
24. Briscoe, H. V. A., and Robinson, P. L., *J. Chem. Soc.* **127**, 696-720, 1925.
25. Bradley, C. A., and Urey, H. C., *Phys. Rev.* **40**, 889-890, 1932.
26. Manian, S. H., Urey, H. C., and Bleakney, W., *J. Am. Chem. Soc.* **56**, 2601-2609, 1934.
27. Boato, G., *Geochim. Cosmochim. Acta* **6**, 209-220, 1954.
28. Reynolds, J. H., and Turner, G., *J. Geophys. Res.* **69**, 3263-3281, 1964.
29. Black, D. C., and Pepin, R. O., *Earth Planet. Sci. Lett.* **6**, 395-405, 1969.

30. Clayton, R. N., Grossman, L., and Mayeda, T. K., *Science* **182**, 485-488, 1973.
31. Tang, M., and Anders, E., *Geochim. Cosmochim. Acta* **52**, 1235-1244, 1988.
32. Bernatowicz, T., Fraundorf, G., Tang, M., Anders, E., Wopenka, B., Zinner, E., and Fraundorf, P., *Nature* **330**, 728-730, 1987.
33. Srinivasan, B., and Anders, E., *Science* **201**, 51-56, 1978.
34. Amari, S., Anders, E., Virag, A., and Zinner, E., *Nature* **345**, 238-240, 1990.
35. Jungck, M. H. A., and Eberhardt, P., *Meteoritics* **14**, 439-441, 1979.
36. Bernatowicz, T. J., Amari, S., Zinner, E. K., and Lewis, R. S., *Astrophys. J.* **373**, L73-L76, 1991.
37. Bernatowicz, T. J., Amari, S., and Lewis, R. S., *Lunar Planet. Sci. XXIII*, 91-92, 1992.
38. Bernatowicz, T. J., Amari, S., and Lewis, R. S., *Lunar Planet. Sci. XXV*, 103-104, 1994.
39. Lewis, R. S., and Srinivasan, B., *Lunar Planet. Sci. XXIV*, 873-874, 1993.
40. Huss, G. R., Hutcheon, I. D., Wasserburg, G. J., and Stone, J., *Lunar Planet. Sci. XXIII*, 563-564, 1992.
41. Nittler, L. R., Amari, S., Walker, R. M., Zinner, E. K., and Lewis, R. S., *Meteoritics* **28**, 413, 1993.
42. Nittler, L., Alexander, C. M. O., Gao, X., Walker, R. M., and Zinner, E., *Nature*, in press, 1994.
43. Hutcheon, I. D., Huss, G. R., Fahey, A. J., and Wasserburg, G. J., *Astrophys. J.* **425**, L97-L100, 1994.
44. Hoppe, P., Strebel, R., Eberhardt, P., Amari, S., and Lewis, R. S., *Lunar Planet. Sci. XXV*, 563-564, 1994.
45. Wasserburg, G. J., and Papanastassiou, D. A., in *Essays in Nuclear Astrophysics*, C. A. Barnes et al. (ed.) Cambridge Univ. Press, 1982, 77-140.
46. Lee, T., in *Meteorites and the Early Solar System*, J. F. Kerridge and M. S. Matthews (eds.), Tucson: University of Arizona Press, 1988, 1063-1089.
47. Podosek, F. A., Zinner, E. K., MacPherson, G. J., Lundberg, L. L., Brannon, J. C., and Fahey, A. J., *Geochim. Cosmochim. Acta* **55**, 1083-1110, 1991.
48. MacPherson, G. J., Davis, A. M., and Zinner, E. K., in preparation, 1994.
49. Lewis, R. S., and Anders, E., *Lunar Planet. Sci. XIX*, 679-680, 1988.
50. Nichols, R. H., Jr., Hohenberg, C. M., Alexander, C. M., O'D., Olinger, C. T., and Arden, J. W., *Geochim. Cosmochim. Acta* **55**, 2921-2936, 1991.
51. Verchovsky, A. B., Russell, S. S., Pillinger, C. T., Fisenko, A. V., and Shukolyukov, Y. A., *Meteoritics* **27**, 301-302, 1992.
52. Huss, G. R., and Lewis, R. S., *Geochim. Cosmochim. Acta*, in press, 1994.
53. Huss, G. R., and Lewis, R. S., *Meteoritics*, in press, 1994.
54. Virag, A., Zinner, E., Lewis, R., and Tang, M., *Lunar Planet. Sci. XX*, 1158-1159, 1989.
55. Alexander, C. M. O., Arden, J. W., Ash, R. D., and Pillinger, C. T., *Earth & Planet. Sci. Letters* **99**, 220-229, 1990.
56. Russell, S. S., Arden, J. W., and Pillinger, C. T., *Science* **254**, 1188-1191, 1991.
57. Lewis, R. S., Huss, G. R., and Lugmair, G., *Lunar Planet. Sci. XXII*, 807-808, 1991.
58. Zinner, E., Tang, M., and Anders, E., *Geochim. Cosmochim. Acta* **53**, 3273-3290, 1989.
59. Stone, J., Hutcheon, I. D., Epstein, S., and Wasserburg, G. J., *Earth Planet. Sci. Lett.* **107**, 570-581, 1991.
60. Virag, A., Wopenka, B., Amari, S., Zinner, E. K., Anders, E., and Lewis, R. S., *Geochim. Cosmochim. Acta* **56**, 1715-1733, 1992.
61. Alexander, C. M., O'D., *Geochim. Cosmochim. Acta* **57**, 2869-2888, 1993.
62. Hoppe, P., Geiss, J., Bühler, F., Neuenschwander, J., Amari, S., and Lewis, R. S., *Geochim. Cosmochim. Acta* **57**, 4059-4068, 1993.
63. Hoppe, P., Amari, S., Zinner, E., Ireland, T., and Lewis, R. S., *Astrophys. J.* **430**, 870-890, 1994.

64. Huss, G. R., Hutcheon, I. D., and Wasserburg, G. J., *Lunar Planet. Sci. XXV*, 687-688, 1994.
65. Amari, S., Hoppe, P., Zinner, E., and Lewis, R. S., *Astrophys. J.* **394**, L43-L46, 1992.
66. Zinner, E., Amari, S., Anders, E., and Lewis, R. S., *Nature* **349**, 51-54, 1991.
67. Ireland, T. R., Zinner, E. K., and Amari, S., *Astrophys. J.* **376**, L53-L56, 1991.
68. Ireland, T. R., Zinner, E. K., Amari, S., and Anders, E., *Meteoritics* **26**, 350-351, 1991.
69. Nichols, R. H., Jr., Hohenberg, C. M., Amari, S., and Lewis, R. S., *Meteoritics* **26**, 377-378, 1991.
70. Nichols, R. H., Jr., Hohenberg, C. M., Hoppe, P., Amari, S., and Lewis, R. S., *Lunar Planet. Sci. XXIII*, 989-990, 1992.
71. Nichols, R. H., Jr., Amari, S., Hoppe, P., and Lewis, R. S., *Meteoritics* **28**, 410-411, 1993.
72. Lewis, R. S., Amari, S., and Anders, E., *Nature* **348**, 293-298, 1990.
73. Lewis, R. S., Amari, S., and Anders, E., *Geochim. Cosmochim. Acta* **58**, 471-494, 1994.
74. Tang, M., and Anders, E., *Geochim. Cosmochim. Acta* **52**, 1245-1254, 1988.
75. Tang, M., and Anders, E., *Astrophys. J.* **335**, L31-L34, 1988.
76. Ott, U., Begemann, F., Yang, J., and Epstein, S., *Nature* **332**, 700-702, 1988.
77. Ott, U., and Begemann, F., *Lunar Planet. Sci. XXI*, 920-921, 1990.
78. Richter, S., Ott, U., and Begemann, F., *Lunar Planet. Sci. XXIII*, 1147-1148, 1992.
79. Prombo, C. A., Podosek, F. A., Amari, S., and Lewis, R. S., *Lunar Planet. Sci. XXIII*, 1111-1112, 1992.
80. Podosek, F. A., Prombo, C. A., Amari, S., and Lewis, R. S., *Astrophys. J.*, submitted, 1994.
81. Ott, U., and Begemann, F., *Astrophys. J.* **353**, L57-L60, 1990.
82. Zinner, E., Amari, S., and Lewis, R. S., *Astrophys. J.* **382**, L47-L50, 1991.
83. Prombo, C. A., Podosek, F. A., Amari, S., and Lewis, R. S., *Astrophys. J.* **410**, 393-399, 1993.
84. Richter, S., Ott, U., and Begemann, F., in *Nuclei in the Cosmos*, F. Käppeler and K. Wisshak (eds.), Philadelphia: IOP Publishing, 1993, 127-136.
85. Richter, S., Ott, U., and Begemann, F., in *Nuclei in the Cosmos*, this volume, 1994.
86. Amari, S., Hoppe, P., Zinner, E., and Lewis, R. S., *Nature* **365**, 806-809, 1993.
87. Zinner, E., Amari, S., Wopenka, B., and Lewis, R. S., *Meteoritics*, submitted, 1994.
88. Amari, S., Zinner, E., and Lewis, R. S., in *Nuclei in the Cosmos*, this volume, 1994,.
89. Hoppe, P., Amari, S., Zinner, E., and Lewis, R. S., *Meteoritics* **27**, 235, 1992.
90. Amari, S., Zinner, E., and Lewis, R. S., *Lunar Planet. Sci. XXV*, 27-28, 1994.
91. Hoppe, P., Amari, S., Zinner, E., and Lewis, R. S., *Lunar Planet. Sci. XXIII*, 553-554, 1992.
92. Ireland, T. R., Amari, S., Hoppe, P., Zinner, E., and Lewis, R. S., *Meteoritics* **27**, 237-238, 1992.
93. Nichols, R. H., Jr., Kehm, K., Brazzle, R., Amari, S., Hohenberg, C. M., and Lewis, R. S., *Meteoritics* **29**, 510-511, 1994.
94. Amari, S., Lewis, R. S., and Anders, E., *Lunar Planet. Sci. XXI*, 19-20, 1990.
95. Lewis, R. S., and Amari, S., *Lunar Planet. Sci. XXIII*, 775-776, 1992.
96. Amari, S., Lewis, R. S., and Anders, E., *Meteoritics*, in press, 1994.
97. Nittler, L. R., Walker, R. M., Zinner, E., Hoppe, P., and Lewis, R. S., *Lunar Planet. Sci. XXIV*, 1087-1088, 1993.
98. Nittler, L., Alexander, C., Gao, X., Walker, R., and Zinner, E., in *Nuclei in the Cosmos*, this volume, 1994.
99. Huss, G. R., Fahey, A. J., Gallino, R., and Wasserburg, G. J., *Astrophys. J.*, **430**, L81-L84, 1994.

100. Huss, G. R., Fahey, A. J., and Wasserburg, G. J., *Meteoritics* **29**, 475-476, 1994.
101. Dominy, J. F., and Wallerstein, G., *Astrophys. J.* **317**, 810-818, 1987.
102. Gehrz, R. D., in *Interstellar Dust*, L. J. Allamandola and A. G. G. Tielens (eds.), Dordrecht: Kluwer Academic Publishers, 1989, 445-453.
103. Whittet, D. C. B. *Dust in the Galactic Environment*. R. J. Taylor and R. E. White (Eds.), New York: Institute of Physics, 1992, 295 pp.
104. Little-Marenin, I. R., *Astrophys. J.* **307**, L15-L19, 1986.
105. Gallino, R., Busso, M., Picchio, G., and Raiteri, C. M., *Nature* **348**, 298-302, 1990.
106. Bazan, G., Ph.D. Thesis, University of Illinois, Urbana-Champaign, 1991.
107. Bressan, A., Fagotto, F., Bertelli, G., and Chiose, C., *Astron. Astrophys. Suppl. Ser.* **100**, 647-664, 1993.
108. El Eid, M., *Astron. Astrophys.* **285**, 915-928, 1994.
109. Scalo, J. M., Despain, K. H., and Ulrich, R. K., *Astrophys. J.* **196**, 805-817, 1975.
110. Renzini, A., and Voli, M., *Astron. Astrophys.* **94**, 175-193, 1981.
111. Sackmann, I.-J., and Boothroyd, A. I., *Astrophys. J.* **392**, L71-L74, 1992.
112. Boothroyd, A. I., Sackmann, I.-J., and Ahern, S. C., *Astrophys. J.* **416**, 762-768, 1993.
113. Cannon, R. C., Frost, C. A., Lattanzio, J. C., and Wood, P. R., in *Nuclei in the Cosmos*, this volume, 1994.
114. Weaver, T. A., and Woosley, S. E., *Phys. Reports* **227**, 65-96, 1993.
115. Meyer, B. S., Weaver, T. A., and Woosley, S. E., *Astrophys. J.*, submitted, 1994.
116. Langer, N., *Astron. Astrophys.* **248**, 531-537, 1991.
117. Langer, N., and Henkel, C., in *Nuclei in the Cosmos*, 1994, this volume.
118. Wagemans, C., D'hondt, P., and Brissot, R., in *Nuclei in the Cosmos*, F. Käppeler and Wisshak (eds.), Bristol and Philadelphia: Inst. Physics. Publ., 1992, 247-252.
119. Gallino, R., Raiteri, C. M., Busso, M., and Matteucci, F., *Astrophys. J.* **430**, 858-869, 1994.
120. Brown, L. E., and Clayton, D. D., *Astrophys. J.* **392**, L79-L82, 1992.
121. Edvardssen, B., Anderson, J., Gustaffson, B., Lambert, D. L., Nissen, P. E., and Tomkin, J., *Astron. Astrophys.* **275**, 101-152, 1993.
122. Dearborn, D. S. P., *Phys. Rep.* **210**, 367-382, 1992.
123. Boothroyd, A. I., Sackmann, I.-J., and Wasserburg, G. J., *Astrophys. J.*, **430**, L77-L80, 1994.
124. Timmes, F. X., Woosley, S. E., and Weaver, T. A., in *Proc. of the VI Advanced School of Astrophysics in São Paulo, Brazil*, B. Barbui, J. A. Freites Pacheco and E. Janot Pacheco (eds.), São Paulo: IAGUSP, 1993, in press.
125. Timmes, F. X., Woosley, S. E., and Weaver, T. A., *Astrophys. J. Suppl.*, submitted, 1994.
126. Wiescher, M., Giesen, U., Görres, J., Azuna, R. E., Kratz, K. L., and Trautvetter, H.-P., in *Nuclei in the Cosmos*, F. Käppeler and K. Wisshak (eds.), IOP, 1992, 191-196.

Interstellar Graphite from the Murchison Meteorite

Sachiko Amari[1,2], Ernst Zinner[1] and Roy S. Lewis[2]

[1]McDonnell Center for the Space Sciences and the Physics Department, Washington University, One Brookings Drive, St. Louis, MO 63130-4899, USA. [2]Enrico Fermi Institute, University of Chicago, 5630 Ellis Avenue, Chicago IL 60637-1433, USA.

Abstract. Graphite grains from two density fractions isolated from the Murchison carbonaceous chondrite have been measured for their C, N, O, Mg, Si, Ca, and Ti isotopic compositions. Many graphite grains from the low density fraction (1.65-1.72g/cm^3) have enormous ^{18}O excesses (^{18}O/^{16}O ranges up to 100x solar), ^{28}Si excesses, and ^{26}Al/^{27}Al ratios from 10^{-2} to 10^{-1}. Some of the grains also have ^{42}Ca, ^{43}Ca, and ^{49}Ti excesses. These isotopic ratios indicate that the low density grains originated from massive stars. Type II supernovae can qualitatively explain the correlations between several isotopic ratios. Graphite grains from the high density fraction (2.15-2.20g/cm^3), on the other hand, have no ^{18}O and ^{28}Si excesses, but have ^{29}Si and ^{30}Si excesses. None of them have ^{26}Mg excesses originating from the decay of ^{26}Al. Several grains also have ^{42}Ca, ^{43}Ca, and ^{49}Ti excesses. These and the Kr isotopic ratios suggest AGB stars as probable stellar sources.

1. INTRODUCTION

Graphite is the third type of carbonaceous interstellar dust isolated from primitive meteorites (1). Most studies of graphite have been done on density separates, with densities ranging from 1.6 to 2.2g/cm^3, extracted from the Murchison meteorite (2). These studies have shown that isotopic and other properties depend on density, and have indicated that graphite grains come from several types of stellar sources (e.g., refs. 3,4).

We have measured isotopic ratios of several elements in grains from the two most extreme fractions in terms of density, the lowest density fraction KE3 (1.65-1.72g/cm^3) and the highest density fraction KFC1 (2.15-2.20g/cm^3). We shall summarize these isotopic measurements and then shall discuss the stellar sources of the grains.

2. RESULTS

Table 1 summarizes the isotopic data obtained on the two separates. Carbon and N isotopic ratios are shown in Fig. 1. ^{12}C/^{13}C ratios in both fractions range from 5 to 3000, although the distributions are very different; grains from the low density fraction tend to have isotopically heavy C, while those from the high density fraction isotopically light C. In contrast to the large range in C isotopic ratios, N isotopic ratios of most grains are surprisingly normal, especially those of the high density graphite grains. Stellar sources are expected to produce N with non-solar isotopic ratios; the almost normal N isotopic ratios in many grains may be due to the exchange or dilution of indigenous nitrogen with isotopically nor-

TABLE 1. Comparison of analyses on separates KE3 and KFC1

	KE3	KFC1
Density	1.65-1.72g/cm^3	2.15-2.20g/cm^3
Grains measured (C and N)	39	88
C	Isotopically heavy	Isotopically light
Fraction of grains with heavy N (^{14}N/^{15}N < 250 within 2σ)	41%	3%
^{26}Al/^{27}Al	10^{-2} - 10^{-1}	< 8x10^{-3} (2σ)
^{16}O/^{18}O	Down to 5	Normal (499)
Si (enriched in)	^{28}Si	^{29}Si and ^{30}Si
^{42}Ca and/or ^{43}Ca excesses	2 out of 7 (2σ)	3 out of 14 (2σ)
^{49}Ti excess	6 out of 7 (2σ)	6 out of 14 (2σ)

mal nitrogen, either in the solar system or in the laboratory. However, since not all grains show normal N, at least some of the grains must have retained indigenous N. Actually, in the low density fraction, 41% of all measured grains show isotopically heavy N (^{14}N/^{15}N<250 within 2σ errors). Furthermore, isotopically heavy N seems to be correlated with isotopically light C (Fig. 1).

In fraction KE3 about 70% of all grains have ^{18}O excesses, with ^{18}O/^{16}O ratios ranging up to 100 times solar (Fig. 2). In contrast, none of the KFC1 grains show such large anomalies. All are normal within 2σ except for one grain that has a small ^{18}O excess (^{16}O/^{18}O =441±28).

Differences also exist between the ^{26}Al/^{27}Al ratios of the two fractions. Grains from KE3 have ratios ranging from 10^{-2} to 10^{-1} (see Fig. 5 in ref. 5), while none of the KFC1 grains have ^{26}Mg excesses originating from the decay of ^{26}Al;

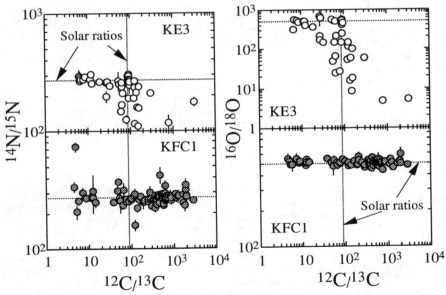

FIGURE 1. C and N isotopic ratios   **FIGURE 2.** C and O isotopic ratios

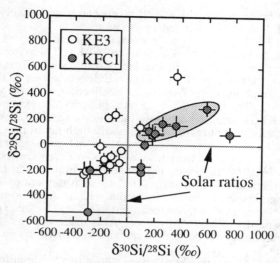

FIGURE 3. Si isotopic ratios

the 2σ upper limit for the $^{26}Al/^{27}Al$ ratio is 8×10^{-3}. Five KE3 grains show ^{25}Mg excesses with $^{25}Mg/^{24}Mg$ ratios ranging up to 3 times solar.

Silicon isotopic compositions are shown in Fig. 3; only grains with Si isotopic compositions deviating from normal by more than 2σ are plotted (39% of the KE3 grains and 24% of the KFC1 grains). Low density graphite grains are, in general, enriched in ^{28}Si, while high density graphite grains are enriched in ^{29}Si and ^{30}Si.

Calcium and Ti isotopic ratios were measured in a limited number of grains from both fractions. Two out of 7 grains from KE3 have ^{42}Ca and/or ^{43}Ca excesses ($\delta^{42}Ca=343\pm122$‰, and 175 ± 36‰; $\delta^{43}Ca=429\pm66$‰) and 6 of them have ^{49}Ti excesses up to 1720 ± 486 ‰. Similarly, 3 out of 14 grains from KFC1 have excesses of up to 941 ± 145‰ in ^{42}Ca and 1569 ± 557‰ in ^{43}Ca. Six out of 14 grains have ^{49}Ti excesses of up to 3629 ± 1019 ‰. In addition, two of them have ^{46}Ti excesses of up to 1115 ± 437‰.

3. DISCUSSION

The large ^{18}O excesses ($^{18}O/^{16}O$ ratios of up to 100x solar) in KE3 indicate massive-stars -- Wolf-Rayet stars (6,7) or supernovae (8) -- as stellar sources. The $^{16}O/^{18}O$ ratios correlate with the $^{26}Al/^{27}Al$, $^{12}C/^{13}C$, and $^{14}N/^{15}N$ ratios (see Fig. 5 in ref. 5). Qualitatively, these correlations can be explained by mixing of material from the He/C shell and He/N shell in Type II pre-SN stars, as discussed in detail in Section 3.2 of ref. 5. Excesses of ^{25}Mg, ^{42}Ca, ^{43}Ca, and ^{49}Ti in some of the KE3 grains indicate neutron capture, which could have taken place in the He/C shell, where He-burning produces neutrons by the $^{22}Ne(\alpha,n)^{25}Mg$ reaction (9). However, mixing with inner layers is required to account for the Si isotopic ratios; since the He/C shell is expected to be enriched in ^{29}Si and ^{30}Si, addition of ^{28}Si-rich material has to come from mixing with the O-shell after explosive nucleosynthesis. Whether mixing of several layers can quantitatively explain the isotopic ratios of KE3 graphite grains remains to be seen.

Some of the KFC1 grains show excesses of ^{29}Si and ^{30}Si, ^{42}Ca and ^{43}Ca, and ^{49}Ti (δ^{29}Si/^{28}Si up to 285‰, δ^{30}Si/^{28}Si up to 761‰). These isotopic features suggest neutron capture reactions in the He-burning shell. This could have happened in massive stars but also in AGB stars. Evidence for an AGB star origin of the high density graphite comes from bulk (≡collection of many grains) noble gas measurements (4). The high ^{86}Kr/^{82}Kr ratio (4.80±0.52), extrapolated to the He-shell from the KFC1 data, indicate low metallicity (Z≤0.002) AGB stars or stars with other metallicities but with with similarly high neutron exposures (τ_0≥ 0.55mb^{-1}) as possible stellar sources (4). The lack of ^{18}O excesses in KFC1 grains indicates negligible contributions from massive stars, which is consistent with the Kr data. Although ^{16}O/^{18}O ratios are expected to be higher than normal in the envelopes of low-metallicity AGB stars (10), no such depletions in ^{18}O were observed in KFC1 grains. Furthermore, for grains with an AGB origin, ^{26}Al/^{27}Al ratios of about 10^{-3} - the same as those measured in SiC (11) - are expected, since ^{26}Al/^{27}Al depends on the ^{25}Mg/^{27}Al ratio (^{26}Al is made by the ^{25}Mg(p,γ)^{26}Al reaction) and this ratio does not depend on metallicity. However, no evidence for ^{26}Al is found in KFC1.

Six grains with ^{29}Si and ^{30}Si excesses (those inside the ellipse in Fig. 3) have ^{12}C/^{13}C ratios ranging from 55 to 2007, 4 of them have ratios between 255 to 755. These high ratios can be achieved in low metallicity AGB stars (12, 13). However, initial Si in the envelope of such stars is expected to be enriched in ^{29}Si and ^{30}Si; in stars with Z=0.002, the excesses should be about 60% for both isotopes (13). As such low-metallicity AGB stars evolve, the Si isotopic ratios are expected to move along a line of slope 0.5 in a δ^{29}Si versus δ^{29}Si 3-isotope plot, starting from δ^{29}Si=δ^{30}Si=600‰. This is not the case for these grains.

From the overall isotopic features, it is certain that many low density graphite grains come from massive stars, possibly Type II supernovae. Some of the high density graphite grains most probably come from AGB stars. However, if we look at the details, many questions remain to be resolved.

REFERENCES

1. Amari S., Anders E., Virag A., and Zinner E. *Nature* **345**, 238-240 (1990).
2. Amari S., Lewis R. S., and Anders E. *Geochim. Cosmochim. Acta* **58**, 459-470 (1994).
3. Amari S., Hoppe P., Zinner E, and Lewis R. S. *Nature* **365**, 806-809 (1993).
4. Amari S., Lewis R. S., and Anders E. *Meteoritics*, in press (1994).
5. Zinner E. this volume.
6. Meynet G. and Arnould M. in *Proc. Europ. Workshop on Heavy Element Nucleosynthesis*, Budapest, p. 52-56 (1994).
7. Langer N. and Henkel C. this volume.
8. Weaver T. A. and Woosley S. E. *Physics Reports* **227**, 65-96 (1993).
9. Thielemann F-K. and Arnett W. D. Nucleosynthesis, University of Chicago Press, 1985, Chicago and London, Ch. 11, p.151-189.
10. Boothroyd A. I., Sackmann I.-J., and Wasserburg G. J. *Astrophys. J.* **430**, L77-L80 (1994).
11. Hoppe P., Amari S., Zinner E, Ireland T., and Lewis R. S. *Astrophys. J.* **430**, 870-890 (1994).
12. Tsuji T., Iye M., Tomioka K., Okada T., and Sato H. *Astron. Astrophys.* **252**, L1-L5 (1991).
13. Gallino R., private communication.

Oxygen-rich Stardust in Meteorites

Larry R. Nittler, Conel M. O'D. Alexander, Xia Gao, Robert M. Walker, and Ernst K. Zinner

McDonnell Center for the Space Sciences and Dept. of Physics, Washington University, St. Louis, Mo. 63130, USA.

Abstract. Isotopic analyses of forty circumstellar oxide grains isolated from primitive meteorites have revealed a large range of O-isotopic ratios and evidence for the prior presence of the short-lived radionuclide ^{26}Al. One group of grains is characterized by slight ^{18}O depletions, and a large range of ^{16}O/^{17}O and initial ^{26}Al/^{27}Al ratios: ^{16}O/^{18}O=502 − 850 (Solar ratio=499), ^{16}O/^{17}O= 385 − 5000 (Solar ratio=2610), and ^{26}Al/^{27}Al=(0 − 8) × 10^{-3}. These grains probably formed around red giant stars. Another group has extreme ^{18}O depletions, as well as enrichments in ^{26}Al and ^{17}O: ^{16}O/^{18}O= (.016 − 2.3) × 10^5, ^{16}O/^{17}O= 735 − 1500, initial ^{26}Al/^{27}Al= (2 − 8) × 10^{-3}. These grains may have formed in hot-bottom-burning TP-AGB (thermally pulsing asymptotic giant branch) stars or in Wolf-Rayet stars. Three grains have enrichments in ^{18}O (^{16}O/^{18}O= 164 − 320); one of these grains also has ^{25}Mg and ^{26}Mg excesses, suggestive of He-burning.

1. Introduction

The first individual circumstellar dust grains to be discovered in meteorites and extensively studied were carbon-rich (1). The dominating presence of solar-system-produced oxygen-rich material in meteorites has made the isolation of presolar oxide grains much more difficult than that of carbonaceous grains, but forty such grains have now been found: 39 corundums (Al$_2$O$_3$) and one spinel (MgAl$_2$O$_4$) (2-5). The isotopic compositions of these grains are extremely unusual, compared to solar system material. Isotopic analyses of these circumstellar grains give new constraints on theoretical models of nucleosynthesis within O-rich stars, and provide information about the stellar sources which contributed material to the solar system.

2. Oxygen Isotopes

The O-isotopic ratios of the forty interstellar oxide grains found to date (2-5) are shown in Figure 1, along with those measured spectroscopically in red giant stars (6-9). For the purposes of discussion, we have divided the grains into four groups (Groups 1-3 were treated in an earlier paper (5)). Group 1 grains have compositions similar to O-rich red giants, *i.e.*, significant enrichments in ^{17}O and modest depletions in ^{18}O (500 ≲ ^{16}O/^{18}O ≲ 1000), and such stars are thus likely stellar sources for these grains. The O-isotopic compositions of red giants are explained by theoretical models of the first and second dredge-ups, where the convective envelope of the star extends deep into the stellar interior and brings the ashes of main-sequence H-burning

© 1995 American Institute of Physics

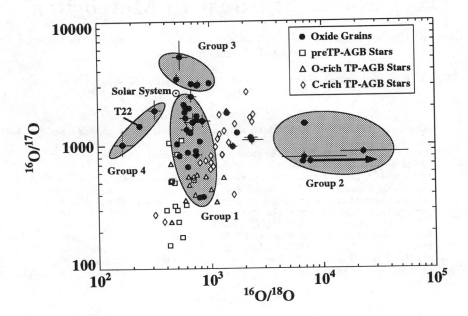

FIGURE 1.— O-isotopic ratios in forty interstellar oxide grains (2-5) and in red giant stars of different types (6-9).

to the surface (10-13). This CNO-cycled material is highly enriched in ^{17}O and has essentially no ^{18}O. Later, when a red giant reaches the TP-AGB phase, it periodically mixes material from the H- and He-burning shells to the surface (10). In most, but not all, models, this third dredge-up is not predicted to significantly change the envelope's O-isotopic ratios. The range of O-isotopic ratios observed in Group 1 grains requires that they formed in several distinct red giants, with different masses and initial O-isotopic ratios, probably reflecting a spread in metallicities (5,11-13). Low-mass and/or low-metallicity red-giants are likely sources for the grains with slightly depleted ^{17}O and ^{18}O (Group 3).

The newly-discovered Group 4 grains have large ^{18}O enrichments, in addition to ^{17}O enrichments. ^{18}O is both produced and destroyed by He-burning, and some models predict that early thermal pulses in some TP-AGB stars produce large amounts of ^{18}O, which could be mixed to the surface during the third dredge-up (12,14-15). The magnitude of any such enrichments depends on the temperature of the thermal pulses and on the uncertain ^{18}O$(\alpha,\gamma)^{22}$Ne rate (12,15-16). Although observations of AGB stars have not yet shown such ^{18}O enrichments, one barium giant has been observed with ^{16}O/^{18}O $\approx$ 100 (17). Wolf-Rayet stars are also predicted to have large ^{18}O enrichments at the

WN-WC transition (18).

Group 2 grains have $^{16}O/^{18}O$ ratios much larger than those that have been observed in stellar atmospheres, as well as ^{17}O enrichments. Hot-bottom burning (HBB), where nuclear burning occurs at the base of the convective envelope of a TP-AGB star, is predicted to produce such compositions at the surface of stars of $M \geq 5M_\odot$(12). Wolf-Rayet stars in the Of and WN phases are also predicted to have large surface ^{18}O depletions (19). Both HBB-AGB stars and Wolf-Rayet stars are possible stellar sources for these grains. A few grains have ^{18}O depletions intermediate between Group 1 and Group 2 ($^{16}O/^{18}O \approx 1000 - 2000$), similar to those measured in some C-star atmospheres (8). Corundum may form in C-rich atmospheres (20). On the other hand, these grains may have formed in O-rich atmospheres that have undergone a small amount of hot-bottom burning.

3. Magnesium Isotopes

In addition to their unusual O-isotopic compositions, sixteen interstellar oxide grains, out of 22 measured for Al-Mg, have large excesses of ^{26}Mg, but normal $^{25}Mg/^{24}Mg$ ratios. Such compositions are most likely due to the in situ

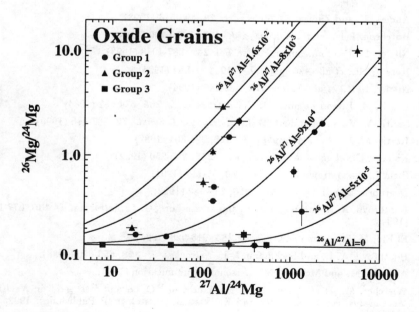

FIGURE 2.— Mg-Al isotopic ratios in 21 interstellar oxide grains. Curves correspond to the initial $^{26}Al/^{27}Al$ ratios indicated.

decay of ^{26}Al and the inferred initial ^{26}Al/^{27}Al ratios are between 1.2×10^{-4} and 1.6×10^{-2} (Figure 2), much higher than the "canonical" early solar system value of 5×10^{-5}. ^{26}Al can be produced by shell H-burning and hot-bottom burning in TP-AGB stars at temperatures higher than those reached during main-sequence evolution (21-23). Since the ^{16}O/^{18}O ratios of Group 1 grains do not indicate hot-bottom burning, Group 1 grains with ^{26}Al probably formed in TP-AGB stars where the third dredge-up had mixed ^{26}Al from the H-shell into the envelope, and those without ^{26}Al originated in red giants before they reached the TP-AGB phase. All of the Group 2 grains have large initial ^{26}Al/^{27}Al ratios. This does not constrain their stellar sources very much, however, since ^{26}Al is predicted to be found at the surface of both Wolf-Rayet stars and hot-bottom-burning TP-AGB stars.

One ^{18}O-enriched grain, T22, has excesses of both ^{25}Mg and ^{26}Mg. Since the anomalies are comparable in size, we cannot ascribe the ^{26}Mg to the radioactive decay of ^{26}Al. However, ^{25}Mg and ^{26}Mg are predicted to be produced by n-capture in the He-burning shell of TP-AGB stars (24), suggesting that the ^{18}O, ^{25}Mg, and ^{26}Mg excesses in this grain are all the result of third dredge-up in an unusual star.

References

1. Anders, E. and Zinner, E., *Meteoritics* **28**, 490-514 (1993).
2. Hutcheon, I. D., et al., *Astrophys. J.* **425**, L97-L100 (1994).
3. Nittler, L. R., et al., *Lunar Planet. Sci.* **24**, 1087-1088 (1993).
4. Huss, G. R., et al., *Astrophys. J.* **430**, L81-L84 (1994).
5. Nittler, L. R., et al., *Nature* **370**, 443-446 (1994).
6. Harris, M. J., and Lambert, D. L., *Astrophys. J.* **285**, 674-682 (1984).
7. Smith, V. V., and Lambert, D. L., *Astrophys. J. Suppl.* **72**, 387-416 (1990).
8. Harris, M. J., et al., *Astrophys. J.* **316**, 294-304 (1987).
9. Kahane, C., et al., *Astron. Astrophys.* **256**, 235-250 (1992).
10. Iben, I., Jr., *Astrophys. J. Suppl.* **76**, 55-114 (1991).
11. Dearborn, D. S. P., *Phys. Rep.* **210**, 367-382 (1992).
12. Boothroyd, A. I., Sackmann, I.-J., and Wasserburg, G. J., *Astrophys. J.* **430**, L77-L80 (1994).
13. El Eid, M. F., *Astron. Astrophys.* **285**, 915-928 (1994).
14. Boothroyd, A. I., and Sackmann, I.-J., *Astrophys. J.* **328**, 653-670 (1988).
15. Arnould, M. and Mohlavi, N. , personal communication (1994).
16. Wiescher, M., et al., "Low energy resonances in ^{18}O $+\alpha$ and ^{22}Ne $+\alpha$," in *Nuclei in the Cosmos*, ed. F. Käppeler and K. Wisshak (Bristol: IOP Publishing), 1992, pp. 191-196.
17. Harris, M. J., Lambert, D. L., and Smith, V. V., *Astrophys. J.* **292**, 620-627 (1985).
18. Maeder, A., *Astron. Astrophys.* **173**, 247-262 (1987).
19. Prantzos, N., et al., *Astrophys. J.* **304**, 695-712 (1986).

20. Lattimer, J. M., Schramm, D. N., and Grossman, L., *Astrophys. J.* **219**, 230-249 (1978).
21. Forestini, M., Paulus, G., and Arnould, M., *Astron. Astrophys.* **252**, 597-604 (1991).
22. Gallino, R., *et al.*, *Astrophys. J.* **430**, 858-869 (1994).
23. Nørgaard, H., *Astrophys. J.* **236**, 895-898 (1980).
24. Gallino, R., *et al.*, *Nature* **348**, 298-302 (1990).

The Abundance of ^{60}Fe in the Early Solar System.

G. W. Lugmair, A. Shukolyukov[1], and Ch. MacIsaac,

University of California, San Diego, La Jolla, CA 92093-0212, USA.
[1]*Now at the Max-Planck-Institute für Chemie, Mainz.*

Abstract. A reliable estimate of the abundance of ^{60}Fe in the early solar system is critically dependent on the absolute ages of the meteorites in which its decay product, ^{60}Ni, was first discovered. Because of their violent early history the ages of these meteorites are difficult to determine with the required resolution and accuracy. The long-lived U – Pb chronometers reflect these disturbances and do not yield the primary formation ages of these meteorites. An indirect approach had to be taken using the short-lived ^{53}Mn – ^{53}Cr chronometer. The results indicate that the ^{60}Fe abundance in the early solar system may have been lower than originally assumed. This lower estimate brings it closer to coincide with recent results obtained for stellar production rates.

Evidence for the existence in the early solar system of live ^{60}Fe ($T_{½} = 1.5$ My (1)) has recently been discovered by Shukolyukov and Lugmair (2, 3). This evidence consists of excess ^{60}Ni, the decay product of ^{60}Fe, which was found in differentiated meteorites — the eucrites. The eucrites are magmatic products of a differentiated planetesimal. In the silicate portion of this planetesimal Ni is extremely depleted relative to Fe. This depletion occurred either before or during accretion of proto-planetesimals or during planetary differentiation and core formation or both. It is this very low Ni content in eucrites which made it possible to detect the small excesses in ^{60}Ni (^{60}Ni*). However, a high Fe/Ni ratio is not the only prerequisite — the meteorite had to solidify or, at the very least, the relative depletion of Ni had to occur at a time when ^{60}Fe was still alive. Because a correlation of ^{60}Ni* with the Fe/Ni ratio was shown to exist in various bulk samples from these meteorites it is apparent that the ^{60}Fe – ^{60}Ni system was closed before all the ^{60}Fe had decayed. The value obtained for the relative abundance of ^{60}Fe in one of these meteorites, Chervony Kut (CK), is ^{60}Fe/^{56}Fe $= (3.9\pm0.6) \times 10^{-9}$, while the ^{60}Fe/^{56}Fe ratio in another eucrite, Juvinas (JUV), is one order of magnitude lower: $\sim 4.3 \times 10^{-10}$. The simplest explanation for the difference in these ^{60}Fe/^{56}Fe ratios is that the ^{60}Fe – ^{60}Ni system in JUV closed ~ 4.7 My later than that in CK.

Although the existence of extant ^{60}Fe in the early solar system and in an early formed planetary body is now clearly established there is considerable uncertainty with regard to its abundance at the onset of planetary accretion. This is due to the fact that no sufficiently precise absolute age is known for the meteorites in which ^{60}Fe was found. Previous estimates (2) were based on arguments using the very low initial ^{87}Sr/^{86}Sr ratios measured in eucrites and another class of differentiated meteorites, the angrites. For the angrites a very precise Pb – Pb age of 4.5578 ± 0.0005 Ga was recently obtained (4). Since both the angrites and eucrites have the same initial ^{87}Sr/^{86}Sr ratios (4) the absolute age of the eucrites, at least that of CK, was estimated to be within 2 My of that of the angrites (2). If this estimate were correct then a value for the ^{60}Fe/^{56}Fe ratio of close to ~1 x 10^{-6} is calculated for the time when the first condensates formed within the solar nebula. An abundance of this magnitude would be sufficient to lead to complete melting of relatively small planetesimals shortly after their accretion. On the other hand, current estimates of ^{60}Fe production within various stellar settings (cf. (5, 6)) cannot account for a high abundance of this kind. Obviously, a directly measured precise and absolute age for the eucrites is very important.

We are currently attempting to measure absolute U – Pb ages for these eucrites. Unfortunately, however, the results we have obtained so far (7) do not hold much promise that meaningful primary ages for these meteorites can be obtained. The U – Pb systems in both eucrites apparently have undergone severe resetting at a time much later than would be consistent with the life time of ^{60}Fe. This thermal processing was probably also responsible for the *mm* to *cm* scale migration of ^{60}Ni* that was originally observed (2, 3). Thus, since the U – Pb system cannot solve the 'absolute age' problem, and other long-term chronometers do not offer the required time resolution, we have to rely on another short-term chronometer which is based on an extinct radioisotope, the ^{53}Mn – ^{53}Cr system.

If effects from the decay of ^{53}Mn ($T_{½}$ = 3.7 My) can be found in mineral separates from these eucrites and if the radiogenic excesses of ^{53}Cr correlate with the Mn/Cr ratios in these samples then we could calculate the relative abundance of ^{53}Mn at the time when the ^{53}Mn – ^{53}Cr system had closed in these meteorites. We then can relate these ^{53}Mn/^{55}Mn ratios for the eucrites via the value of ^{53}Mn/^{55}Mn = (1.29 ± 0.07) x 10^{-6} obtained for the angrites to the angrite absolute age of 4.5578 Ga. The only major assumption entering this indirect way of obtaining absolute ages for the eucrites is that ^{53}Mn was homogeneously distributed in the early solar system, at least where the parent planetesimals of the angrites and eucrites were formed.

We have obtained good quality ^{53}Mn – ^{53}Cr isochrons for Chervony Kut and Juvinas. For CK the ^{53}Mn/^{55}Mn ratio at the time of ^{53}Mn – ^{53}Cr system closure was (3.6 ± 0.5) x 10^{-6}. When compared to the ^{53}Mn/^{55}Mn ratio obtained for the angrites then CK is (5.5 ± 0.7) My older than the angrites resulting in an absolute age of 4.563 ± 0.001 Ga. For JUV we found a ^{53}Mn/^{55}Mn ratio of (3.1±0.5) x 10^{-6}, within error the same as that of CK. Thus, both eucrites have also the same age within about 1 My.

The ^{53}Mn – ^{53}Cr system clearly does not reflect the 4.7 My age difference between the two eucrites as indicated by the ^{60}Fe – ^{60}Ni system. There is only a hint of a slightly lower ^{53}Mn/^{55}Mn ratio for JUV as compared to CK. Even at the extremes of the error limits this difference would only amount to less than half of the time difference indicated by the ^{60}Fe – ^{60}Ni system. This, therefore, suggests that the lower ^{60}Fe/^{56}Fe ratio in JUV is most likely due to partial re-equilibration and prolonged migration of ^{60}Ni* — JUV is also much more brecciated than is CK. However, there is no way to tell at present how much re-distribution of Ni has occurred in CK itself. Thus, the ^{60}Fe/^{56}Fe ratio obtained from CK has to be also regarded as a lower limit to the true ^{60}Fe/^{56}Fe ratio at 4.563 Ga, the time when the ^{53}Mn – ^{53}Cr system closed.

However, if we take the ^{60}Fe/^{56}Fe ratio from CK at face value we can calculate the ^{60}Fe/^{56}Fe ratio for the time when the first condensates were formed in the solar nebula by using the Pb – Pb age of 4.566 ± 0.002 Ga obtained for Allende meteorite inclusions (8). Accounting for all known uncertainties the range of values for ^{60}Fe/^{56}Fe is then 4 x 10^{-9} to 6 x 10^{-8} with a most probable value of about 1.6 x 10^{-8}. This value is much lower than our previous estimate which was obtained without the benefit of the new ^{53}Mn – ^{53}Cr data. We note, however, that this estimate is still a lower limit for the reasons discussed above for CK. Nevertheless, it is in much closer agreement with recent calculations of ^{60}Fe production in AGB stars (5).

ACKNOWLEDGMENTS

This work was supported by NASA grant NAGW-3285.

REFERENCES

1. Kutschera, W., Billquist, P. J., Frekers, D., Henning, W., Jensen, K. J., Xiuzeng, M., Pardo, R., Paul, M., Rehm, K. E., Smither, R. K., and Yntema, J. L., *Nucl. Inst. Meth. Phys. Res.* **B5**, 430 – 435 (1984).
2. Shukolyukov, A., and Lugmair, G. W., *Science* **259**, 1138 – 1142 (1993).

3. Shukolyukov, A., and Lugmair, G. W., *Earth and Planet. Science Letters* **119,** 159–166 (1993).
4. Lugmair, G. W., and Galer, S. J. G., *Geochim. Cosmochim. Acta* **56,** 1673–1694 (1992).
5. Wasserburg, G. J., Busso, M., Gallino, R., Raiteri, C. M., *Ap. J.* **424,** 412–428 (1994).
6. Meyer, B., *Meteoritics* **27,** 261 (1992).
7. Galer, S. J. G., and Lugmair, G. W., *work in progress and unpublished data.*
8. Göpel, C., Manhès, G., and Allègre, C. J., *Meteoritics* **26,** 338 (1991).

Evidence in Meteorites for the Presence of ^{41}Ca in the Early Solar System

J.N. Goswami and G. Srinivasan

Physical Research Laboratory, Ahmedabad 380009, India

Abstract. ^{41}K excess that can be attributed to *in-situ* decay of ^{41}Ca has been found in several Ca-Al-rich refractory inclusions (CAIs) of the Efremovka carbonaceous chondrite. Our observation shows that ^{41}Ca ($\tau \sim 0.15$ Myr) was present in the early solar system. We have inferred an initial ^{41}Ca/^{40}Ca ratio of $(1.6\pm0.3)\times10^{-8}$ at the time of formation of the Efremovka CAIs based on our data. Observation of excess ^{41}K in these objects constrains the time interval between the cessation of fresh nucleosynthetic input to the solar nebula and the formation of some of the first solar system solids (CAIs) to less than a million years. Several possibilities exist for the stellar nucleosynthesis of ^{41}Ca and its addition to the early solar system.

The presence of several extinct radionuclides with meanlife ≥ 1 Myr in the early solar system has been established from isotopic studies of samples from primitive meteorites (see e.g. 1). Their presence constrains the time interval between the last addition of freshly synthesized material to the solar nebula and the formation of the first solar system solids. Obviously, the observation of the radionuclide with the shortest meanlife will put the most stringent constraint on this time interval. At present, ^{26}Al ($\tau \sim 1$ Myr) is the shortest-lived radionuclide whose presence in the early solar system has been established conclusively. We have now found evidence for the presence of a much shorter-lived radionuclide ^{41}Ca ($\tau \sim 0.15$ Myr) in the early solar system.

A hint for the possible presence of ^{41}Ca in the early solar system was initially provided by the data of Hutcheon et al.(2), who analysed a couple of CAIs from the Allende meteorite for their Ca-K isotopic systematics and observed a marginal excess of ^{41}K in them. This study also highlighted the experimental difficulties one encounters during K isotopic measurements in phases with high Ca/K ratio using an ion microprobe. In the present study, we have analysed several CAIs from the Efremovka meteorite whose petrography, trace element and Mg-isotope data suggest them to be less altered and more pristine than the Allende CAIs (see e.g. 3). The Ca-K isotopic measurements were carried out using a ion microprobe (Cameca ims-4f) at a mass resolution (M/ΔM) of $\sim$ 5000, sufficient to resolve the ^{40}CaH interference on ^{41}K. The unresolved (^{40}Ca^{42}Ca)$^{++}$ interference was corrected by monitoring the $(^{40}\text{Ca}^{43}\text{Ca})^{++}$ signal at mass 41.5 and the measured ratio of $^{42}\text{Ca}/^{43}\text{Ca}$ in the analysed samples. All the instrument parameters including the dynamic background of the pulse counting system was also checked routinely [see Srinivasan et al. (4) for details]. Terrestrial samples with Ca/K ratio varying over several orders of magnitudes (10^{-3} to 10^6) were also analysed. In Fig. 1 we show the measured K isotopic composition in these samples; all of them yielded ($^{41}\text{K}/^{39}\text{K}$) ratios close to the reference value of 0.072. Note that the magnitude of $(^{40}\text{Ca}^{42}\text{Ca})^{++}$ interference increases with increasing Ca/K ratio, and account for $\sim 55\%$ of the total signal at mass ^{41}K for the sample with the highest Ca/K ratio (4×10^6). These results demonstrate that the doubly charged Ca interference is properly accounted for and that the mass fractionation effect is small in the case of K isotopes.

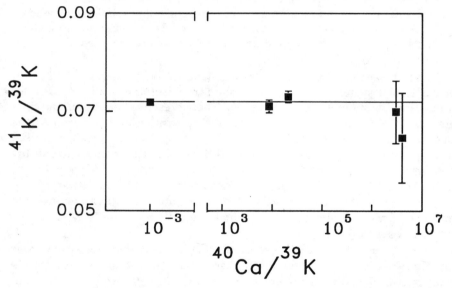

Figure 1: Measured $^{41}\text{K}/^{39}\text{K}$ ratios in terrestrial minerals. The solid line represents the reference value (0.072).

We have so far analysed three Efremovka CAIs (E44, E50, E65) for their Ca-K isotopic compositions. Here we discuss the data obtained from analysis of Ca-rich pyroxenes in E44 and E65. The small sizes of Ca-rich perovskite (< 10 micron) in E50 made accurate measurement difficult due to contributions to the observed Ca and K ion signals from adjacent mineral phases. The results obtained by us are displayed in Fig. 2. It is clear that the meteoritic pyroxenes, particularly those with Ca/K $> 3\times 10^5$, have clearly resolvable ^{41}K excess with $^{41}\text{K}/^{39}\text{K}$ ratios much above the reference value

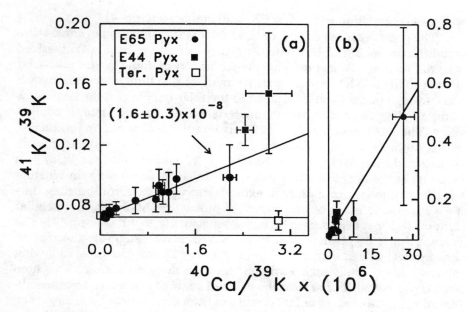

Figure 2: Measured ^{41}K/^{39}K ratios in terrestrial and meteoritic pyroxenes (pyx), corrected for (^{40}Ca^{42}Ca)$^{++}$ interference, plotted against their ^{40}Ca/^{39}K. The horizontal line represents the value of reference ^{41}K/^{39}K (0.072); the slope of the best-fit line through the meteorite (Efremovka CAI) data is also given.

of 0.072. The excess ^{41}K also correlates well with the ^{40}Ca/^{39}K ratios of the analysed phases. If we attribute this excess to *in-situ* decay of ^{41}Ca within the Efremovka CAIs, the slope of the correlation line yields the initial value of ^{41}Ca/^{40}Ca at the time of formation of these objects. A value of $(1.6\pm0.3)\times10^{-8}$ for this parameter can be inferred from our data. An upper limit of $(8\pm3)\times10^{-9}$ was suggested by Hutcheon et al.(2) based on their data for Allende CAIs. We may note here that both the observed excess in ^{41}K and its correlation with ^{40}Ca/^{39}K could also result from secondary neutron induced effect during the cosmic ray exposure of the Efremovka meteorite via the reaction: ^{40}Ca (n,γ) ^{41}Ca $\rightarrow$ ^{41}K. However, the thermal neutron fluence needed to explain the excess ^{41}K is more than an order of magnitude higher than the estimated value of neutron fluence experienced by the Efremovka meteorite (4). The possibility that the observed ^{41}K excess is of 'fossil' origin with 'stardust' as its carrier (5,6) can be ruled out as it is difficult to accommodate both the magnitude of the observed ^{41}K excess as well as its correlation with $^{40}Ca/^{39}K$ ratio in this scenario (4).

The presence of excess ^{41}K in early solar system objects put some stringent limit on the time interval between the last input of freshly synthesized material to the solar nebula and the formation of some of the first solar system solids (CAIs). An accurate estimate of this time interval demands a knowledge of the fresh ^{41}Ca input to the solar nebula and the level of its dilution with pre-existing nebular material with normal K-isotopic composition. The short meanlife of ^{41}Ca effectively rules out long-term production of this isotope and points towards a very late event that introduced freshly synthesized material into the solar nebula. The input value for $^{41}Ca/^{40}Ca$ could be in the range of a few times 10^{-3} to 10^{-2}, depending on whether the production is taking place in explosive oxygen or silicon burning during a supernova event or in a thermally pulsating AGB star. The nebular dilution factor is expected to be in the range of 10^3 to 10^4 (see e.g. 7). The value of 1.6×10^{-8} for the initial $^{41}Ca/^{40}Ca$, therefore, suggests a formation time interval of less than a million year for the Efremovka CAIs following the cessation of fresh nucleosynthetic input to the solar nebula. This time scale is somewhat shorter than the value of a few Myr inferred from the observation of ^{26}Mg excess in CAIs resulting from the decay of ^{26}Al. However, the input of ^{26}Al is not well constrained and could vary over two orders of magnitude depending on the choice of the stellar nucleosynthesis site(s). The short time scale inferred by us is consistent with the observation of excess ^{60}Ni in differentiated meteorites, resulting from the decay of ^{60}Fe ($\tau \sim 2$ Myr), that suggests the formation of large meteorite parent bodies and their subsequent melting, cooling and recrystallization within a short span of $\leq$ 5 Myr (8,9).

REFERENCES

1. Cameron, A.G.W., *Protostars and Planets III*, Tucson, Arizona Univ. Press, 1993, pp 47-73.
2. Hutcheon, I.D., Armstrong, J.T., and Wasserburg, G.J., *Meteoritics* 19, 243 (1984).
3. Goswami, J.N., Srinivasan, G., and Ulyanov, A.A., *Geochim. Cosmochim. Acta* 58, 431-447 (1994).
4. Srinivasan, G., Ulyanov, A.A., and Goswami, J.N., *Ap.J. (Letters)*, 1994 (In Press).
5. Clayton, D.D., *Earth Planet. Sci. Lett.* 36, 381-390 (1977).
6. Clayton, D.D., *Quar. J. Royal Astron. Soc..* 23, 174-212 (1982).
7. Wasserburg, G.J., Busso, M., Gallino, R., and Raiteri, C.M., *Ap. J.* 424, 412-428 (1994).
8. Shukolyukov, A., and Lugmair, G.W., *Science* 259, 1138-1142 (1993).
9. Lugmair, G.W., Shukolyukov, A., and MacIsaac, Ch., (1994) (This volume).

The Astrophysical Site of the Origin of the Solar System Inferred from Extinct Radionuclides

Charles L. Harper Jr.

Department of Earth & Planetary Sciences, Harvard University, 20 Oxford Street, Cambridge, MA, 02138, USA (E-mail: harper@geochemistry.harvard.edu; FAX: USA 617-496-3439)

Abstract. *Ab initio* abundances of extinct radionuclides in the solar abundance distribution (SAD) provide clues sufficient to characterize the molecular cloud environment within which the solar system formed 4566±2 Ma ago. Key observations are: (i) the low abundance of the longer-lived r-process radionuclide ^{129}I indicates an ~10^2 Ma isolation time from freshly-synthesized Type II supernova products; (ii) abundances of the shorter-lived species (^{26}Al, ^{60}Fe, ^{53}Mn) are consistent with late "injection" of freshly-synthesized Type II supernova products. This apparent contradiction is resolved in a simple two time scale molecular cloud self-contamination model consistent with formation of the sun in an evolved star-forming region in the vicinity of an OB association. Admixture of an ~10^{-5} to ~10^{-6} mass fraction of Type II supernova ejecta into the cloud dominates the shorter-lived species and ^{107}Pd, but contributes only ~1/2 of the ^{129}I budget. ^{129}I consequently preserves the longer time scale information constraining the mean isolation/condensation/accretion model age of the cool molecular mass reservoir with respect to ^{129}I/^{127}I in the global ISM. All extinct radionuclide abundances except ^{26}Al are reproduced in this model. An alternate hypothesis involving late admixture of matter from a mass-losing low mass AGB star can account for ^{26}Al, ^{60}Fe and ^{107}Pd, but fails for ^{53}Mn, requires unusual s-process conditions, and otherwise appears *ad hoc* and unlikely. Cosmic ray spallation in an OB association environment may contribute significantly to ^{53}Mn in the protosolar reservoir, but is limited as a putative ^{26}Al source by overproduction of ^{53}Mn, ^{92g}Nb and Li.

INTRODUCTION

Astronomers observe star formation proceeding in a variety of environments. Protostars are seen deep within quiescent giant molecular cloud complexes, and in hot and violent OB associations. They occur in giant bound clusters and in tiny Bok globules. They form in compressed zones in spiral arms and in clouds in inter-arm regions. The question of the astrophysical site of the Sun's birth 4566±2 Ma ago is that of type of astrophysical environment: in an old or young cloud, deep within a quiet complex, or near some heated interface involving massive stars?

Extinct radionuclides are tracers of nucleosynthetic activity and their abundances yield information about the nature of their sources as well as about time scales for matter transport between synthesis sites and the solar birthplace. Evidence of extinct radionuclides is preserved in meteorites. Initial abundances are inferred from isotopic systematics. *Ab initio* (4566 Ma) estimates are given in table 1 and incorporate age constraints on relevant cosmochemical element fractionations.

RADIONUCLIDES IN STAR-FORMING REGIONS

A star-forming region is a particular "local" sub-reservoir within the "global" ISM. Meteorite data on initial radionuclide abundances in the solar system may be compared to global ISM estimates based on:

$$^{ij}\zeta_{<ISM>} = (P_i/P_j)_x (N_{x,j}/N_{SAD,j})[(1+\kappa)\tau_i/T] \quad (1)$$

where $^{ij}\zeta$ is the ratio of radionuclide i to stable index j, (P_i/P_j) is the relative production ratio of the i-j pair, τ_i/T is ratio of the mean life of the radionuclide to the presolar duration of nucleosynthesis (~7 Ga), $(N_{x,j}/N_{SAD,j})$ is the fraction of the index nuclide in the SAD attributable to dominant production process x, and κ is Clayton's (11) infall parameter (~1.5). The difference between the initial SAD ratio $^{ij}\zeta_{SAD}$ and $^{ij}\zeta_{<ISM>}$ can be interpreted as a model "free decay interval", δ (Table 2):

$$^{ij}\zeta_{SAD} / \,^{ij}\zeta_{<ISM>} = e^{-\delta/\tau_i}. \quad (2)$$

Bulk self-contamination (SC) is described by:

$$^{ij}\zeta_{SC} = (P_i/P_j)_x D_j \quad (3)$$

where D_j is the self-contamination dilution factor (j*/j) for the index species:

$$D_j = [X_{j,SC} / X_{j,SAD}][D_{SC}], \quad (4)$$

where X's denote mass or atom fractions and D_{SC} is the ejecta-cloud mass dilution.

General expectations for radionuclide abundances in star-forming regions can be inferred from consideration of the schematic ISM cycle for molecular star-forming reservoirs shown in figure 1. In terms of roughly-defined cloud age characteristics, these are (in evolutionary sequence: young & small ... older & large ... oldest & large-in-process-of-self-destruction):

(1) Abundances of longer-lived species (*e.g.*, ^{129}I) will be close to the global ISM in "young" clouds, but substantially reduced in "older" clouds by decay.

(2) In the case of the shorter-lived supernova-produced species (*e.g.*, ^{60}Fe), abundances in "young" clouds should be variable due to spatial inhomogeneities of nucleosynthesis, whereas in "older" clouds (*viz.*, without self-contamination) their abundances will be negligible.

(3) In the "oldest" clouds, self-contamination due to massive star evolution and supernovae may admix short-lived supernova-produced species in excess abundances relative to that expected for free-decay during the isolation time of the star-forming reservoir (*e.g.*, as estimated from ^{129}I).

The rough pattern of initial extinct radionuclide abundances in the solar system (Table 2) suggests the Sun formed in an old star-forming region, likely containing one or more OB associations old enough for supernovae to have admixed ejecta into compressed peripheral star-forming shell regions.

TABLE 1. INITIAL SOLAR SYSTEM ABUNDANCES OF EXTINCT RADIONUCLIDES

NUCLIDE	$T_{1/2}$ (Ma)	RATIO	AB INITIO VALUE	ΔT-CORR.	REFERENCES
^{26}Al	0.73	^{26}Al/^{27}Al	~5 × 10^{-5}	none	(1)
^{60}Fe	1.5	^{60}Fe/^{56}Fe	~1 × 10^{-8}	~3 Ma	(2)
^{135}Cs	2.3	^{135}Cs/^{133}Cs	≤1 × 10^{-5}	none	(3)
^{53}Mn	3.7	^{53}Mn/^{55}Mn	~6 × 10^{-6}	8 Ma	(4)
^{107}Pd	6.5	^{107}Pd/^{110}Pd	~9 × 10^{-5}	5 Ma	(5)
^{182}Hf	9	^{182}Hf/^{180}Hf	≥5 × 10^{-6}	uncertain	(6)
^{129}I	15.7	^{129}I/^{127}I	~1.5 × 10^{-4}	none	(7)
^{92g}Nb	36	^{92}Nb/^{93}Nb	(1.5±0.3) × 10^{-5}	none	(8)
^{244}Pu	80	^{244}Pu/^{238}U	~ 0.007	none	(9)
^{146}Sm	103	^{146}Sm/^{144}Sm	= 0.008±1	8 Ma	(10)

TABLE 2. COMPARISON OF "GLOBAL" AND "LOCAL" RADIONUCLIDE ABUNDANCES

RATIO i/j	PROD.RATIO $(P_i/P_j)_x$	GLOBAL ISM $^{ij}\zeta_{<ISM>}$	SOLAR SYSTEM $^{ij}\zeta_{SAD}$	FREE DECAY INTERVAL δ (Ma)
r-process input:				
^{107}Pd/^{110}Pd	1.5	5×10^{-3}	$\sim 9 \times 10^{-5}$	~ 38
^{129}I/^{127}I	1.3	9×10^{-3}	$\sim 1.5 \times 10^{-4}$	~ 93
Lower mass species:*				
^{26}Al/^{27}Al	~ 0.006	$\sim 2 \times 10^{-6}$	5×10^{-5}	local > global
^{53}Mn/^{55}Mn	0.13	$\sim 2 \times 10^{-4}$	6×10^{-6}	~ 19
^{60}Fe/^{56}Fe	~ 0.0004	$\sim 3 \times 10^{-7}$	$\sim 1 \times 10^{-8}$	~ 7

* IMF-integrated production in Type II SNe from (14), and assuming $(P_{53}P_{55}) = (^{53}Cr/^{55}Mn)_{SAD}$.

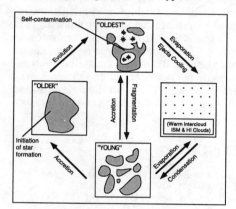

FIGURE 1. Schematic ISM Cycle for Molecular Star-forming Reservoirs.

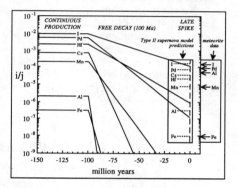

FIGURE 2. Isotopic evolution in a two time scale scenario. For a 10^2 Ma isolation time, the late spike dominates the budgets of all species except ^{129}I.

SELF-CONTAMINATION SCENARIO

For further evaluation of this hypothesis, a simple Type II SN self-contamination model is shown in figure 2. The model is normalized by an ^{127}I dilution factor $D_i = 5 \times 10^{-5}$ to contribute $\sim 50\%$ of the initial SAD ^{129}I budget. The remaining $\sim 50\%$ is attributed to an old galactic component remaining after ~ 100 Ma of free decay. This normalization essentially fits both ^{129}I and ^{107}Pd using: (i) r-process ^{107}Pd/^{110}Pd and ^{129}I/^{127}I production ratios (1.5 and 1.3, respectively) and (ii) decompositions for ^{110}Pd (100% r-process) and ^{127}I (94% r-process) obtained from residuals after subtraction of the s-process main component distribution in the SAD (12). Additional yields are based on the SAD (13) for ^{27}Al ($\sim 100\%$ from Type II SNe), ^{55}Mn ($\sim 100\%$ from Type II SNe), and ^{56}Fe ($\sim 50\%$ from Type II SNe), using Weaver & Woosley's (14) IMF-integrated model production ratios for ^{26}Al/^{27}Al (0.006) and ^{60}Fe/^{56}Fe (0.0004). The production ratio for ^{53}Mn/^{55}Mn is assumed equal to $(^{53}Cr/^{55}Mn)_{SAD} = 0.13$. It can be seen that the fit is excellent for all extinct radionuclide species except ^{26}Al. As ^{26}Al is predominantly and variably produced in the presupernova evolution in massive stars, and is present in the envelope and in winds, disagreement is not unexpected between one particular SN and an IMF-integrated SN yields model assuming homogenized ejecta. The ejecta mass dilution (D_{SC}) for this scenario is $\sim 10^{-5}$ to $\sim 10^{-6}$ for a 25 M_O SN, assuming an ^{56}Fe yield range of 0.07 to 0.7 M_O, with $D_{56Fe} = 3 \times 10^{-5}$ to fit $(^{60}Fe/^{56}Fe)_{SAD}$.

ALTERNATE HYPOTHESES

Two alternate sources recently proposed to explain extinct radionuclide abundances are a mass-losing AGB star (15), and cosmic ray (CR) spallation with enhanced flux due to one or more supernovae in the vicinity of the protosolar environment (16). The AGB star model cannot explain ^{53}Mn and has been criticized as *ad hoc* and statistically unlikely (17). To fit ^{60}Fe and ^{107}Pd, it also requires conditions of high neutron density and low exposure not observed either in the spectral signatures of AGB stars, or in isotopic compositions of meteoritic SiC.

Clayton proposed the CR spallogenesis scenario as a dominant ^{26}Al source in both the early solar system and present day ISM in response to the COMPTEL observation of unexpectedly high ^{12}C* and ^{16}O* γ-ray de-excitation lines from Orion (18). It has been shown, however, that ^{53}Mn is strongly overproduced (by ~two orders of magnitude) in spallogenic models with high enough fluence to generate ^{26}Al/^{27}Al ~5 x 10^{-5} (19). (Li and ^{92g}Nb are also overproduced.) The initial SAD abundance of ^{53}Mn strongly constrains the CR fluence on the protosolar reservoir at the time of the Sun's formation. Consequently, cosmic rays cannot account for ^{26}Al, ^{60}Fe and ^{107}Pd and do not alleviate the need for a stellar source.

On balance, Type II SN self-contamination on the periphery of an OB association seems the most plausible scenario, in agreement with (20). Studies of additional radioactivities (^{10}Be, ^{97,98,99}Tc, ^{135}Cs, ^{182}Hf & ^{205}Pb) should provide key tests.

REFERENCES

1. Podosek, F. A., Zinner, E. K., MacPherson, G. J., Lundberg, L. L., Brannon, J. C., and Fahey, A. J., *Geochim. Cosmochim. Acta* **55**, 1083-1110 (1991).
2. Shukolyukov, A., and Lugmair, G. W., *Earth Planet. Sci. Letts* **119**, 159-166 (1993).
3. Harper, C. L. Jr., *J. Phys. G (Nucl. Part. Phys.)* **19**, S81-S94 (1993).
4. Harper, C. L. Jr., and Wiesmann, H., *Lunar Planet Sci.* **XXIII** (1992), 489-490; Lugmair, G. W. (pers. comm., 1994).
5. Chen, J. H., and Wasserburg, G. J., *Geochim. Cosmochim. Acta* **54**, 1729-1743 (1990).
6. Harper, C. L. Jr., and Jacobsen, S. B., *Lunar Planet. Sci.* **XXV**, 509-510 (1994); (and in prep).
7. Swindle, T. D., and Podosek, F. A., "Iodine-xenon dating," in eds. J. F. Kerridge & M. S. Mathews, *Meteorites and the Early Solar System*, (Tucson, U. of Arizona), 1127-1146 (1988).
8. Harper, C. L. Jr., submitted to the *Astrophysical Journal*.
9. Hudson, G. B., Kennedy, B. M., Podosek, F. A., and Hohenberg, C. M., *Proc. Lunar Planet. Sci. Conf.* **19**, 547-557 (1989).
10. Lugmair, G. W., and Galer, S. J. G., *Geochim. Cosmochim. Acta* **56**, 1673-1694 (1992).
11. Clayton, D. D. *MNRAS* **234**, 1-36 (1988).
12. Beer, H. *ApJ* **375**, 823-832 (1991); Käppeler, F., Beer, H., and Wisshak, K., *Rep. Prog. Phys.* **52**, 945-1013 (1989).
13. Anders, E., and Grevesse, N., *Geochim. Cosmochim. Acta* **53**, 197-214 (1989).
14. Weaver, T. A., and Woosley, S. E., *Phys. Reps.* **227**, 65-96 (1993).
15. Wasserburg, G. J., Busso, M., Gallino, R., and Raiteri, C. M., *ApJ*, **424**, 412-428 (1994).
16. Clayton, D. D. *Nature* **368**, 222-224 (1994).
17. Kastner, J. L., and Myers, P. C. 1994, *ApJ*, **421**, 605-614 (1994).
18. Bloemen, H., Wijnands, R., Bennett, K., Diehl, R., Hermsen, W., Lichti, G., Morris, D., Ryan, J., Schönfelder, V., Strong, A. W., Swanenburg, B. N., de Vries, C., and Winkler, C. *Astron. Astrophys.* **281**, L5-L8 (1994).
19. Clayton, D. D., Dwek, E., and Woosley, S. E. *ApJ* **214**, 300-315 (1977); Lee, T. *ApJ* **224**, 217-226 (1978); Wasserburg, G. J., and Arnould, M., in ed. W. Hillebrandt, *Nuclear Astrophysics*, (Lect. Not. Phys. v. **287**), 262-276 (1987); Harper, C. L. Jr., Wiesmann, H., Nyquist, L. E., Howard, W. M., Meyer, B., Yokoyama, Y., Rayet, M.,Arnould, M., Palme, H., Spettel, B., and Jochum, K. P. *Lunar Planet. Sci.* **XXII**, 519-520 (1991).
20. Reeves, H. in ed. T. Gehrels, *Protostars and Planets*, (Tucson, U. of Arizona), 399-426 (1978); Reeves, H. in ed. D. Lal, *Early Solar System Processes and the Present Solar System*, (Proc. Int. Sch. Phys. "Enrico Fermi" LXXIII, New York, North Holland), 54-57 (1980).

Chemically Fractionated Fission Xenon (CFF-Xe) on the Earth and in Meteorites

Alexander P. Meshik[1], Yuri A. Shukolyukov[1] and Elmar K. Jeβberger[2]

[1] *Vernadsky Institute of Geochem. and Analyt. Chemistry, 117975 Moscow, Russia,*
[2] *Max-Planck-Institute fur Kernphysik, 103980 Heidelberg, Germany*

Abstract. A new explanation of Xe isotopic composition in the terrestrial mantle has been suggested on the basis of Xe isotopic analyses in the n-irradiated pitchblende, in the natural nuclear reactor Oklo, in Colorado type deposit, in the sandstone from the nuclear test epicentre. Xe isotopic anomalies in these samples and in some meteorites as well as in mantle rocks and well gases can be explained by the migration processes of β^--active xenon precursors in the isobaric chains formed during heavy nuclei fission.

INTRODUCTION

The excess ^{129}Xe have been observed in a great variety of rocks and natural well gases of presumably mantle origin (1-11). It was interpreted as a result of β-decay of extinct primordial ^{129}I initially presented in the Earth's mantle. In mantle rocks and well gases excess ^{129}Xe correlates with the excess of ^{136}Xe which is a typical product of heavy nuclei fission. (4, 8-10). This correlation is interpreted as a result of the mixing of ^{129}I-derived Xe with products of spontaneous fission of ^{238}U (and ^{244}Pu) and atmospheric Xe. On this basis a number of models of the Earth's degassing have been suggested, with the ratio ^{129}Xe/^{136}Xe being a very important parameter in the modelling (4, 7, 10, 12).

The purpose of this paper is to check the validity of the wide-spread opinion that ^{129}Xe originates from primordial ^{129}I and ^{136}Xe derives directly from the spontaneous fission of ^{238}U and ^{244}Pu.

All available published data on Xe in samples of mantle origin are plotted on the Fig. 1. Actually, the correlation between ^{129}Xe/^{130}Xe and ^{136}Xe/^{130}Xe (Fig. 1a) looks like the result of the mixing of atmospheric, solar, ^{129}I-derived and fissiogenic Xe components. If it is correct, after replacement ^{129}Xe/^{130}Xe by ^{134,132,131}Xe/^{130}Xe ratios experimental points should lie in the mixing area between atmospheric, solar and fissiogenic Xe. But it is not the case (Fig. 1b, c, d). Mass-fractionation and experimental uncertainties can not be the reason of these enrichments.

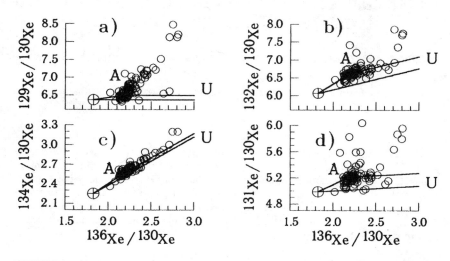

FIGURE 1. Isotopic composition of Xe in mantle samples (1–12). ⊕ - solar Xe, A - atmosperic Xe, U - direction to Xe produced by spontaneous fission of U.

Thus, some *alien* Xe component enriched in $^{136-131}$Xe and probably ^{129}Xe undoubtedly exists in samples of mantle origin. A large portion (probably, all) of excess ^{129}Xe is a part of this mysterious component, but not the β-decay product of primordial ^{129}I. To understand the origin of this *alien* Xe the following experiments have been carried out.

During step-wise heating of the pitchblende from ordinary uranium deposit no Xe isotopic anomalies were found. Then the sample evacuated in ampoule was irradiated by thermal neutron. A month after the irradiation we measured Xe isotopic composition both in the ampoule and in the sample in step-wise heating experiment. At low temperature Xe released was enriched in $^{134-131}$Xe. Opposite, Xe in the ampoule was depleted in these isotopes (Fig 2a). No doubts, the ampoule Xe came from the sample, its isotopic pattern is a mirror image of low temperature gas fraction. Their sum gives a normal neutron-induced xenon derived from the ^{235}U fission. At the first time this effect was observed by Kennet and Thode (14).

The second experiment was the Xe isotopic analyses in partially molten sandstone from the epicentre of the first nuclear test, Alamogordo, USA. During step-wise annealing of this sample a huge excesses of ^{132}Xe and ^{131}Xe were found (16) in comparison with Xe produced from the n_{th}-induced fission of U in the sandstone (Fig. 2b).

In both experiments one condition was the same: the elevated temperature during uranium fission. Such temperatures may result in the migration of β-radioactive Xe precursors (Sn, Sb, Te, I) before they decayed into stable Xe isotopes. The migration rates vary because of their different geochemical properties. This results in the chemical fractionation of the precursors. Generally, the probability of precursor's migration depends also

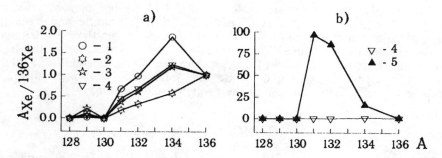

FIGURE 2. Isotopic composition of: a - Xe released from the pitchblende #1-822 (Siberia) at low temperature (460°C) - 1 and in the ampoule - 2. Their sum - 3 is very close to Xe produced by n-induced fission of ^{235}U - 4 (13); b - Xe extracted from the sandstone from the epicentre of the nuclear test in Alamogordo, NM - 5.

on the half-life period of the precursor. That is why maximal Xe enrichments were observed in ^{132}Xe and ^{131}Xe, their precursors ^{132}I and ^{131}I have maximal half-life periods. Xe produced by the process was called Chemically Fractionated Fission Xenon or CFF-Xe (13).

To understand whether or not the excess ^{129}Xe observed in the mantle belongs to CFF-Xe two natural samples whose ages were sufficient for complete ^{129}I ($T_{1/2} = 17 \cdot 10^6$ y) decay into ^{129}Xe have been studied: Oklo natural nuclear reactor in Gabon and Colorado type deposit in Kazakhstan (16). In both cases excesses of ^{129}Xe, ^{134}Xe, ^{132}Xe and ^{131}Xe were found. Some of the results as well as Xe in ancient Greenland anorthosite (2) and average *alien* Xe from the mantle are plotted in Fig 3a, b, c. The isotopic signature of CFF-Xe is clearly pronounced in all examples.

Thus, the *alien* Xe observed in mantle rocks is Chemically Fractionated Fission Xenon. The excess ^{129}Xe is a part of the CFF-Xe originated from the migration of xenon precursors. The value of the excess may vary depending on the migration conditions.

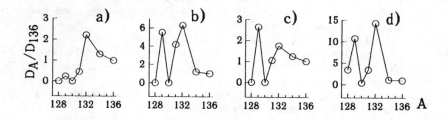

FIGURE 3. CFF-Xe in a - Oklo reactor (16), b - Greenland anorthosite (2), c - the mantle (calculated from Fig. 1), d - Allende CAIs (13) expressed as values $D_A/D_{136} = [(^AXe/^{130}Xe) - (^AXe/^{130}Xe)_{atm.}] / [(^{136}Xe/^{130}Xe) - (^{136}Xe/^{130}Xe)_{atm.}]$.

Surprisingly, the characteristic features of CFF-Xe are present also in meteorites e.g. sample 1C5 of Olivenza LL5 (18), chromite and px-plag separate samples of Tzarev L5-6 (19), in an Apollo 16 anorthosite (20) and in CAI inclusions of Allende C3V (13). The isotopic composition of the latter is shown on the Fig. 3d. Probably, due to the "Orion phenomenon" particle fluence 4.56 Ga ago could produce neutron-induced fission xenon (15) and, hence, CFF-Xe in the dust of the evolving solar nebula.

CONCLUSION

Under definite conditions Chemically Fractionated Fission Xe could be produced due to the migration of β-active precursors in fission chains. The isotopic composition of CFF-Xe is not fixed, it may vary depending on migration conditions. Enrichment in CFF-Xe might be very local. CFF-Xe is not unique. It found in various solar system materials. The ^{129}Xe portion in CFF-Xe could be large. The care is needed to interpret the ^{129}Xe-excess in mantle rocks and natural gases as a result of the primordial ^{129}I decay.

REFERENCES

1. Butler, W. A., Jeffery, P. M., Reinolds, J. H., and Wasserburg, G. J., *Journ. Geophys. Res.* **68**, 3283–3291 (1963).
2. Jeffety, P. M., *Nature*, **223**, 260–261 (1971).
3. Boulos, M. S., and Manuel, O. K., *Science*, **174**, 1334–1337 (1971).
4. Staudacher, T., and Allegre, C. J., *Earth Planet. Sci. Lett.*, **60**, 389–406 (1982).
5. Phinney, D., Tennyson, J., and Frick, U., *Jour. Geophys. Res.* **83**, 2313–2319 (1978).
6. Staudacher, T., Kurz, M. D., and Allegre, C. J., *Chem. Geology*, **56**, 193–205 (1986).
7. Marty, B., *Earth Planet. Sci. Lett.*, **94**, 45–56 (1989).
8. Poreda, R. J., and Farley, K. A., *Earth Planet. Sci. Lett.*, **113**, 129–144 (1992).
9. Ozima, M., and Zashu, S., *Earth Planet. Sci. Lett.*, **105**, 13–27 (1991).
10. Hiyagon, H., Ozima, M., Marty, B., Zashu, S., and Sakai, H., *Geochim. et Cosmochim. Acta*, **56**, 1301–1316 (1992).
11. Allegre, C. J., Staudacher, T., Sarda, P., and Kurz, M., *Nature*, **303**, 762–766 (1983).
12. Azbel, J. Ya., and Tolstikhin, I. N., *Meteoritics*, **28**, 609–621 (1993).
13. Shukolyukov, Yu. A., Jeßberger, E. K., Meshik, A. P., Dang Vu Minh, and Jordan, J. L., *Geochim. et Cosmochim. Acta*, (in press).
14. Kennett, T. J., Thode, H. G., *Can. Journ. Phys.*, **38**, 945–954 (1960).
15. Marti, K., and Lingenfelter, R. E., "The Orion Phenomenon: Particle Fluences in the Solar Nebula", presented at the Conference on "Nuclei in the Cosmos", L'Aquila, Italy, July 8–13, 1994.
16. Shukolyukov, Yu. A., Meshik, A. P., Pravdivtseva, O. V., Verkhovsky, A. B., and Kryuchkova, O. I., *Geochemistry International*, **7**, 27–37 (1988).
17. Shukolyukov, Yu. A., and Meshik, A. P., *Isotopenpraxis*, **26**, 364–370 (1990).
18. Alaerts, L. et al, *Geochim. et Cosmochim. Acta*, **43**, 183–198 (1978).
19. Shukolyukov, Yu. A. et al, *Geochimia*, **10**, 1379–1389 (1982) (in Russian).
20. Leich, D. A., and Niemeyer, S., "Apollo 16..." in *Proceeding of Lunar and Planetary Science Conference.* **VI**, 1953–1965 (1975).

List of Participants

C. M. O'D. Alexander
Mc Donnell Center for the Space Sciences
Washington University
Campus Box 1105
St. Louis, MO 63130
USA

M. Aliotta
Istituto di Astronomia
Citta' Universitaria
V.le Doria 6
95125 Catania
ITALY

S. Amari
Mc Donnell Center for the Space Sciences
Campus Box 1105
Washington University
One Brookings Drive
St. Louis, Missouri 63130-4899
USA

A. C. Andersen
The Niels Bohr Institute
Blegdamsvej 17
DK-2100 Copenhagen
DENMARK

O. Aubert
IAP
98bis Bd. Arago
75014 Paris
FRANCE

R. E. Azuma
Dept. of Physics
University of Toronto
60 St. George St.
Toronto, Ontario
M5S 1A7
CANADA

H. Beer
Kernforschugszentrum Karlsruhe
IK III
Postfach 3649
D-76021 Karlsruhe
GERMANY

G. F. Bignami
C.N.R.
Istituto di Fisica Cosmica
via Bassini 15
20133 Milano
ITALY

T. Blöcker
Astrophysikalisches Institut
Potsdam
Telegraphenberg
D-14473 Potsdam
GERMANY

R. N. Boyd
Dept. of Physics
Ohio State University
174 W. 18th Ave.
Columbus, OH 43210
USA

E. Bravo
Universitat Politecnica
de Catalunya
E.T.S.E.I.B
Av. Diagonal 647
E-08028 Barcelona
SPAIN

E. Brocato
Osservatorio Astronomico
di Collurania
64100 Teramo
ITALY

C. Broggini
Laboratori Nazionali dell'INFN
del Gran Sasso
Strada Statale 17 bis
67100 Assergi (AQ)
ITALY

R. Buonanno
Osservatorio Astronomico
Via del Parco Mellini 84
00136 Roma
ITALY

M. Busso
Osservatorio Astronomico di Torino
Strada Osservatorio 20
10025 Pino Torinese
ITALY

V. Caloi
Istituto di Astrofisica Spaziale
C.P. 67
I-00044 Frascati (Rm)
ITALY

M. Casse'
Service d'Astrophysique
CE Saclay
Gemes des Merisiers
91191 Gif-sur-Yvette Cedex
FRANCE

V. Castellani
Istituto di Astronomia
Piazza Torricelli 2
56100 Pisa
ITALY

C. Charbonnel
Observatoire Midi-Pyrenees
14 Avenue Edouard-Belin
31400 Toulouse
FRANCE

A. Chieffi
Istituto di Astrofisica Spaziale
C.P. 67
I-00044 Frascati (Rm)
ITALY

C. Chiosi
Dipartimento di Astronomia
Vicolo dell' Osservatorio 5
I-35122 Padova
ITALY

A. Coc
CSNSM, Bat. 104
91405 Orsay Campus
FRANCE

F. Corvi
Central Bureau for Nuclear
Measurements, EURATOM
Steenweg op Retie
B-2440 Geel
BELGIUM

K. Czerski
Technische Universitaet Berlin
Institut fuer Strahlungs-
und Kernphysik
Hardenbergstr. 36, Sekr. PN3-1
D-10623 Berlin
GERMANY

P. Decrock
I.K.S. Katholieke Universiteit
Leuven
Celestijnenlaan 200D
B-3001 Leuven (Heverlee)
BELGIUM

S. degli Innocenti
Universita' di Pisa
Sezione di Astronomia
Piazza Torricelli 2
56100 Pisa
ITALY

A. Denker
Institut fuer Strahlenphysik
Universitaet Stuttgart
Allmandring 3
D-70569 Stuttgart
GERMANY

P. Descouvemont
Universite Libre de Bruxelles
CP 229, Boulevard du Triomphe
B-1050 Bruxelles
BELGIUM

P. Doll
Kernforschungszentrum Karlsruhe
Postfach 3640
D-76021 Karlsruhe 1
GERMANY

M. El Eid
Universitaets-Sternwarte
Geismarlandstr. 11
D-37083 Goettingen
GERMANY

A. Ferrari
Osservatorio Astronomico
di Torino
Strada Osservatorio 20
10025 Pino Torinese
ITALY

G. Fioni
Institut Laue-Langevin - BP 156
F-38042 Grenoble Cedex 9
FRANCE

G. Fiorentini
Universita' di Ferrara
Dipartimento di Fisica
Via del Paradiso 1
44100 Ferrara
ITALY

E. Fiorini
Dipartimento di fisica
Universita' di Milano
Via G. Celoria 16
20133 Milano
ITALY

C. Frost
Dept. of Mathematics
Monash University
Clayton, Victoria 3168
AUSTRALIA

A. Fubini
ENEA INN FIS.
P.O. Box 65
I-00044 Frascati (Roma)
ITALY

Z. Fülop
Institute of Nuclear Research
ATOMKI
H-4001 Debrecen PF. 51.
HUNGARY

M. Gaelens
Katholieke Universiteit Leuven
Instituut voor
Kern-en Stralingsfysika
Celestijnenlaan 200D
B-3001 Leuven (Heverlee)
BELGIUM

P. Galeotti
Istituto di Cosmogeofisica del CNR
Corso Fiume 4
I-10125 Torino
ITALY

R. Gallino
Istituto di Fisica Generale
Universita' di Torino
Via P. Giuria 1
10125 Torino
ITALY

W. Galster
Inst. de Phys. Nucl.
Universite Cath. de Louvain
Chemin du Cyclotron 2
B-1348 Louvain-La-Neuve
BELGIUM

Yu. M. Gledenov
Frank Laboratory of Neutron
Physics
JINR, Moscow Region,
141980 Dubna,
RUSSIA

J. N. Goswami
Physical Research Laboratory
Ahmedabad-380 009
INDIA

C. Grama
Institute of Atomic Physics
P.O.B.MG-6
RO-7690 Bucharest
ROMANIA

R. Gratton
Dipartimento di Astronomia
Vicolo dell' Osservatorio 5
I-35122 Padova
ITALY

U. Greife
Ruhr-Universitaet Bochum
Fakultaet fuer Physik and Astronomie
Inst. f. Experimentalphysik III
D-44780 Bochum
GERMANY

W. J. Hammer
Institut fuer Strahlenphysik
Universitaet Stuttgart
Allmandring 3
D-70569 Stuttgart
GERMANY

W. Hampel
MPI fuer Kernphysik
P.O. Box 103980
D-69029 Heidelberg
GERMANY

C. L. Harper
Dept. of Earth and
Planetary Sciences
Harvard University
20 Oxford Street
Cambridge, MA 02138
USA

D. Hartmann
Dept. of Physics and Astronomy
Clemson University
Clemson, SC 29634-1911
USA

M. Hashimoto
Dept. of Physics
College of General Education
Kyushu University 01
Ropponmatsu, Fukuoka 810
JAPAN

P. Heide
Technische Universitaet Berlin
Institut fuer Strahlungs-
und Kernphysik
Hardenbergstr. 36, Sekr. PN3-1
D-10623 Berlin 12
GERMANY

R. D. Hoffman
Board of Studies in Astronomy
and Astrophysics
University of California
Santa Cruz, CA 95064
USA

M. A. Hofstee
Triangle Universities
Nuclear Laboratory
Duke University
Box 90308
Durham, NC 27708-00398
USA

H. Holweger
Universitat Kiel
Institut fuer Theoretische
Physik und Sternwarte
D-24098 Kiel
GERMANY

H. Huber
Institut fuer Kernphysik
Technische Universitaet Wien
Wiedner Haudtstrasse 8110
1100 Wien
AUSTRIA

I. Iben Jr.
Department of Astronomy
University of Illinois
341 Astronomy Building
Urbana, IL 61801
USA

S. Jaag
Kernforschugszentrum Karlsruhe
IK III
Postfach 3640
D-76021 Karlsruhe
GERMANY

A. Jorissen
Institut d'Astronomie
Universite' Libre de Bruxelles
C.P. 226 Bd. du Triomphe
B-1050 Bruxelles
BELGIUM

List of Participants

M. Junker
Ruhr-Universitaet Bochum
Fakultaet fuer Physik und Astronomie
Inst. f. Experimentalphysik III
D-44780 Bochum
GERMANY

F. Käppeler
Kernforschugszentrum Karlsruhe
IK III
Postfach 3640
D-76021 Karlsruhe
GERMANY

C. Kahane
Laboratoire d'Astrophysique
de l'Observatoire de Grenoble
BP 53
38041 Grenoble
Cedex 9
FRANCE

G. Kaniadakis
Politecnico di Torino
Corso Duca degli Abruzzi 24
I-10129 Torino
ITALY

R. W. Kavanagh
Kellogg Radiation Laboratory
California Institute of Technology
Pasadena, CA 91125
USA

R. Kingsburgh
IAUNAM
P.O. Box 439027
San Diego, CA 92143-9027
USA

V. Kobychev
Institute for Nuclear Research
of Ukrainian Academy of Science
Prospect Nauki 47
Kiev 252028
UKRAINE

P. Koehler
Oak Ridge National Laboratory
Building 3500, Mail stop 6010
Oak Ridge, TN 37831-6010
USA

E. Kolbe
Institut fuer Theoretische Physik
Klingelbergstrasse 82
Universitaet Basel
CH-4056 Basel
SWITZERLAND

K.-L. Kratz
Johannes Gutenberg-Universitat
Institut fuer Kernchemie
Fritz-Strassmann-Weg 2
D-55099 Mainz
GERMANY

J. M. Lambert
Georgetown University
Department of Physics
Washington, D.C. 20816
USA

N. Langer
MPI fur Physik und Astrophysik
Postfach 1523
D-85740 Garching
GERMANY

List of Participants   613

J. C. Lattanzio
Dept. of Mathematics
Monash University
Clayton, Victoria 3168
AUSTRALIA

M. D. Leising
Dept. of Physics and Astronomy
Clemson University
Clemson, SC 29634
USA

P. Leleux
Inst. de Phys. Nucl.
Universite Cath. de Louvain
Chemin du Cyclotron 2
B-1348 Louvain-La-Neuve
BELGIUM

M. Liebendoerfer
Institut fuer Theoretische Physik
Klingelbergstrasse 82
Universitaet Basel
CH-4056 Basel
SWITZERLAND

M. Limongi
Istituto di Astrofisica Spaziale
C.P. 67
I-00044 Frascati (Rm)
ITALY

G. Lobov
Inst. of Theor. and
Exp. Physics
117259 Moscow M-259
RUSSIA

G. W. Lugmair
Scripps Institution of Oceanography
University of California, San Diego
La Jolla, CA 92093
USA

M. Marengo
Istituto di Fisica Generale
Universita' di Torino
Via P. Giuria 1
10125 Torino
ITALY

K. Marti
Department of Chemistry, 0317
University of California, San Diego
La Jolla, CA 92093-0317
USA

A. Mengoni
E.N.E.A.
Via Mazzini 2
I-40138 Bologna
ITALY

A. P. Meshik
Vernadsky Institute of
Geochemistry and Analytical
Chemistry
Kosygin St. 19
SU-117975 Moscow
RUSSIA

B. S. Meyer
Department of Physics and
Astronomy
Clemson University
Clemson, SC 29634-1911
USA

List of Participants

G. Mezzorani
Dipartimento di Scienze Fisiche
Universita' degli Studi
Via Ospedale 72
09124 Cagliari
ITALY

P. Mohr
Physikalisches Institut
Universitaet Tuebingen
Auf der Morgenstelle 14
D-72076 Tuebingen
GERMANY

P. Monacelli
Laboratori Nazionali dell'INFN
del Gran Sasso
Strada Statale 17 bis
67100 Assergi (AQ)
ITALY

N. Mowlavi
Institut d'Astronomie
Universite' Libre de Bruxelles
C.P. 226 Bd. du Triomphe
B-1050 Bruxelles
BELGIUM

F. Münnich
Inst. f. Metallphysik
Mendelssohnstr. 3
D-38106 Braunschweig
GERMANY

P. Mutti
Central Bureau for Nuclear
Measurements, EURATOM
Steenweg op Retie
B-2440 Geel
BELGIUM

Y. Nagai
Tokio Institute of Technology
Department of Applied Physics
Oh-okayama 2-12-1
Meguro
Tokyo 152
JAPAN

G. Navarra
Istituto di Fisica
Universita' di Torino
Via P. Giuria 1
10125 Torino
ITALY

L. R. Nittler
Mc Donnell Center for the Space Sciences
Washington University
Campus Box 1105
St. Louis, MO 63130
USA

H. Oberhummer
Institut fuer Kernphysik
Technische Universitaet Wien
Wiedner Hauptstrasse 8-10/142
A-1040 Wien
AUSTRIA

U. Oberlack
Max-Planck-Institut fuer
Extraterrestrische Physik
85740 Garching
GERMANY

K. Oyamatsu
Dept. of Energy Engineering
and Science
Furo-Cho, Chikusa-Ku
Nagoya 464-01
JAPAN

List of Participants   615

B. Pagel
NORDITA
Blegdamsvej 17
DK-2100 Copenhagen 0
DENMARK

T. Paradellis
NRC Democritos
Tandem Accelerator Laboratory
GR-15341 Aghia Paraskevi
GREECE

L. Paterno'
Istituto di Astronomia
V.le A. Doria 6
95125 Catania
ITALY

M. Perinotto
Dipartimento di Astronomia e
Scienza dello Spazio
Largo E. Fermi 5
50125 Firenze
ITALY

A. Piechaczek
GSI Darmstadt
Planckstrasse 1
D-64220 Darmstadt
GERMANY

N. Prantzos
IAP
98bis Bd. Arago
75014 Paris
FRANCE

D. Puy
Observatoire de Paris-Meudon
DAEC, 5 Place Jules Janssen
92195 Meudon
FRANCE

P. Quarati
Politecnico di Torino
Corso Duca degli Abruzzi 24
I-10129 Torino
ITALY

G. Raimann
The Ohio State University
174 W. 18th Avenue
Columbus, Ohio 43210
USA

C. M. Raiteri
Osservatorio Astronomico di Torino
Strada Osservatorio 20
10025 Pino Torinese
ITALY

T. Rauscher
Institut fuer Kernchemie
Universitaet Mainz
Fritz-Strassmann-Weg2
D-55099 Mainz
GERMANY

S. Richter
MPI fuer Chemie
Abt. Isotopenkosmologie
Becherweg 27
D-55128 Mainz
GERMANY

W. S. Rodney
Department of Physics
Georgetown University
Washington, DC 20057
USA

C. Rolfs
Inst. f. experimentelle Physik
NB3
Ruhr-Universitt Bochum
D-44780 Bochum
GERMANY

S. Rosswog
Institut fuer Theoretische Physik
Klingelbergstrasse 82
Universitaet Basel
CH-4056 Basel
SWITZERLAND

G. Roters
Inst. fr Experimentelle Physik
Ruhr-Universitt Bochum
D-44780 Bochum
GERMANY

S. Sandrelli
Dipartimento di Astronomia
Universita' di Bologna
Via Zamboni 33
40100 Bologna
ITALY

A. Scalia
Dipartimento di Fisica
Universita' di Catania
Corso Italia 57
95129 Catania
ITALY

E. Schmid
Institut fuer Theoretische Physik
Universitaet Tuebingen
Auf der Morgenstelle 14
D-72076 Tuebingen
GERMANY

Michael Schumann
Kernforschungszentrum Karlsruhe
Institut f. Kernphysik III
Postfach 3640
76021 Karlsruhe
GERMANY

P. V. Sedyshev
Frank Laboratory of Neutron
Physics, Joint Institute for
Nuclear Research
Moscow Region
141980 Dubna
RUSSIA

T. Shima
Tokio Institute of Technology
Department of Applied Physics
Oh-okayama 2-12-1
Meguro
Tokyo 152
JAPAN

T. D. Shoppa
California Institute of
Technology
Kellogg Radiation Lab. 106-38
Pasadena, California 91125
USA

G. Silvestro
Istituto di Fisica Generale
Universita' di Torino
Via P. Giuria 1
10125 Torino
ITALY

C. Sneden
Department of Astronomy
The University of Texas at Austin
78712-1083 Austin (Texas)
USA

E. Somorjai
ATOMKI
Institute of Nuclear Research
of the Hungarian Academy
of Sciences
P.O. Box 51
H-4001 Debrecen
HUNGARY

O. Sorlin
Institut de Physique Nucleaire
Universite de Paris
B. P. 1
F-91406 Orsay Cedex
FRANCE

G. Staudt
Physikalisches Institut der
Universitaet Tuebingen
Auf der Morgenstelle 14
D-72076 Tuebingen
GERMANY

O. Straniero
Osservatorio Astronomico
di Collurania
64100 Teramo
ITALY

E. Sugarbaker
OSU Van de Graaff Lab.
1302 Kinnear Road
Columbus, OH 43212
USA

F. K. Sutaria
Theoretical Astrophysics
Tata Institute of Fundamental
Research
Homi Bhabha Road
Navy Nagar
Bombay 400 005
INDIA

J. Szabo'
Institute of Experimental Physics
Kossuth University
H-4001 Debrecen, Pf.105
HUNGARY

C. Theis
Kernforschungszentrum Karlsruhe
Institut fr kernphysik
Postfach 3640
D-76021 Karlsruhe
GERMANY

F.-K. Thielemann
Institut fuer Theoretische Physik
Klingelbergstrasse 82
Universitaet Basel
CH-4056 Basel
SWITZERLAND

F. X. Timmes
University of Chicago
Laboratory for Astrophysics
and Space Research
933 East 56th Street
Chicago, IL 60637
USA

W. Tkaczyk
University of Lodz
Department Experimental Physics
ul. Pomorska 149/153
90-236 Lodz
POLAND

A. Tornambe'
Istituto di Astrofisica Spaziale
C.P. 67
Prof. Dr. Tornamb
I-00044 Frascati (Rm)
ITALY

H.-P. Trautvetter
Inst. f. experimentelle Physik
NB3
Ruhr-Universitt Bochum
D-44780 Bochum
GERMANY

C. Travaglio
Istituto di Fisica Generale
Universita' di Torino
via P. Giuria 1
10125 Torino
ITALY

J. W. Truran
Dept. of Astronomy and
Astrophysics
University of Chicago-LASR
933 East 56th Street
Chicago, Illinois 60637
USA

G. Vancraeynest
I.K.S. Katholieke Universiteit
Leuven
Celestijnenlaan 200D
B-3001 Leuven (Heverlee)
BELGIUM

L. Van Wormer
Department of Physics
Hiram College
P.O. Box 1778
Hiram, Ohio 44234-1778
USA

M. Villata
Osservatorio Astronomico di Torino
Strada Osservatorio 20
10125 Pino Torinese
ITALY

F. Voss
Kernforschungszentrum Karlsruhe
Institut fr Kernphysik
Postfach 3640
D-76021 Karlsruhe
GERMANY

S. Vouzoukas
Department of Physics
University of Notre Dame
Notre Dame, IN 46556
USA

C. Wagemans
University of Gent
Proeftuinstraat 86
B-9000 Gent
BELGIUM

W. Walters
Department of Chemistry
University of Maryland
College Park, MD 20742
USA

S. Wanajoh
Department of Astronomy
School of Science
University of Tokio
Bunkyo-ku, Tokio 113
JAPAN

G. J. Wasserburg
Caltech 170-25
Geology Division
Pasadena, CA 91125
USA

F. Weber
Universitt Mnchen
Institut fr Theoretische Kernphysik
Theresienstr. 37/III
D-80333 Munich
GERMANY

K. Werner
Institut fuer Theoretische Physik
und Sternwarte der Universitaet
D-24098 Kiel
GERMANY

M. Werner
MPI fr Chemie
F.-F.-Becher-Weg 27
D-55128 Mainz
GERMANY

A. Wierling
MPG-AG Vielteilchenphysik
Universitatsplatz 1
D-18051 Rostock
GERMANY

M. Wiescher
Department of Physics
University of Notre Dame
Notre Dame, IN 46556
USA

K. Wisshak
Kernforschungszentrum Karlsruhe
Institut fr Kernphysik
Postfach 3640
D-76021 Karlsruhe
GERMANY

S. E. Woosley
Astronomy Department
University of California
Santa Cruz, CA 95064
USA

D. Zahnow
Ruhr-Universitaet Bochum
Fakultaet fuer Physik und Astronomie
Inst. f. Experimentalphysik III
D-44780 Bochum
GERMANY

E. K. Zinner
Mc Donnell Center for the Space
Sciences and Physics Department
Washington University
St. Louis, MO 63130-4899
USA

Author Index

A

Abele, H., 299
Aguer, P., 259, 311
Amari, S., 581
Andersen, A. C., 481
Andersen, M.-L., 481
Angulo, C., 311
Anne, R., 191
Arnould, M., 403
Arpesella, C., 31
Athanassopulos, K., 165
Atzrott, U., 243
Aubert, O., 473
Axelsson, L., 191

B

Bain, C. R., 253
Balbes, M. J., 331
Baraffe, I., 473
Barnes, C. A., 31
Barthélémy, R., 169
Bartolucci, F., 31
Bateman, N. P. T., 195
Bazin, D., 191, 303
Beer, H., 165, 247, 307
Bellotti, E., 31
Bieber, R., 183
Binon, F., 253
Blackmon, J. C., 195
Blöcker, T., 399
Bogaert, G., 311
Böhmer, W., 191
Borrel, V., 191
Boyd, R. N., 235, 303, 331
Brandt, U., 79
Bravo, E., 429, 437
Bressan, A., 337
Broggini, C., 31
Brown, B. A., 263
Brown, J. A., 303
Browne, C. P., 259
Buchmann, L., 303
Bucka, H., 271

Buonanno, R., 513
Busso, M., 407, 557
Butt, Y., 195

C

Cannon, R. C., 469
Cassé, M., 539
Champagne, A. E., 195
Charbonnel, C., 433
Chieffi, A., 407
Chiosi, C., 337
Chloupek, F. R., 235, 331
Clayton, D. D., 447, 521
Coc, A., 259, 311, 453
Corvi, F., 165
Corvisiero, P., 31
Coszach, R., 253
Cowan, J. J., 37
Crawford, H. J., 209
Czerski, K., 271

D

Danevich, F. A., 285
Daumiller, K., 79
Davinson, T., 253
Debus, K., 291
Decrock, P., 253
Delbar, Th., 253
Deng, L., 337
Denker, A., 255
de Oliveira, F., 311
Descouvemont, P., 149
Doll, P., 79, 209
Dreizler, S., 45
Drotleff, H. W., 255
Druyts, S., 169
Duhamel, P., 253

E

El Eid, M. F., 393
Engelmann, C., 243
Erdas, A., 319, 323
Eriguchi, Y., 425

F

Fauerbach, M., 331
Fioni, G., 153
Fiorentini, G., 31
Forestini, M., 403
Fortier, S., 311
Frost, C. A., 469
Fubini, A., 31
Fujimaki, M., 235
Fülöp, Zs., 277

G

Gaelens, M., 253
Gallino, R., 407, 557
Galster, W., 91, 253
Gao, X., 585
García, D., 437
Georgadze, A. Sh., 285
Gervino, G., 31
Gledenov, Yu. M., 173
Glejbøl, K., 481
Glendenning, N. K., 485
Görres, J., 259, 263, 303
Gorris, F., 31
Goswami, J. N., 595
Grama, C., 223
Grama, N., 223
Gratton, R. G., 3
Greife, U., 31
Grollmuss, H., 243
Grosse, M., 255
Grün, K., 299, 315
Guber, K., 227
Guillemaud-Mueller, D., 191
Gumbsheimer, R., 79
Gustavino, C., 31

H

Hammer, J. W., 255
Hampel, W., 59
Harper, C. L., Jr., 599
Hartmann, D., 447
Hashimoto, M., 379, 419, 425, 457
Heber, U., 45
Heide, P., 271
Hellström, M., 303, 331
Hencheck, M., 331
Henkel, C., 413
Herndl, H., 259, 263, 299, 307
Hoffman, R. D., 441, 463
Hofstee, M. A., 195
Holweger, H., 41
Howard, A. J., 195
Hoyler, F., 243
Huber, H., 315
Hucker, H., 79
Huke, A., 271
Huyse, M., 253

I

Iannicola, G., 513
Igashira, M., 145, 201, 205
Iliadis, C., 259
Isern, J., 429

J

Jaag, S., 231
Jading, Y., 191
Jank, J., 315
Jessberger, E. K., 603
Jørgensen, U. G., 481
Jorissen, A., 67, 403
Junker, M., 31

K

Kahane, C., 19
Kan, M., 425
Kaniadakis, G., 319, 323, 327

Käppeler, F., 101, 157, 161, 227, 231, 291, 307
Kavanagh, R. W., 31
Kazakov, L., 227
Keller, H., 191
Kelley, J. H., 303
Kettner, Ch., 485
Kiener, J., 311
Kii, T., 145, 205
Kikuchi, T., 145
Kimura, K., 235
Kingsburgh, R. L., 51
Kiss, Á. Z., 277
Kitazawa, H., 201
Klages, H. O., 79
Kleinwächter, P., 79
Knee, H., 255
Kobayashi, T., 235
Kobychev, V. V., 281, 285
Koehler, P. E., 157, 173
Kolata, J. J., 235
Kolb, G., 79
Kölle, V., 243
Kornilov, N., 227
Kratz, K.-L., 113, 191
Krauss, H., 299
Krishnan, T., 447
Kropivyansky, B. N., 285
Kryger, R. A., 303
Kubono, S., 235
Kunz, R., 255
Kuts, V. N., 285

L

Lang, J., 79
Langer, N., 413
Lanza, A., 31
Lattanzio, J. C., 353, 469
Leeb, H., 315
Lefébvre, A., 259, 311
Lehoucq, R., 539

Leising, M. D., 521
Leleux, P., 253
Lemke, M., 41
Lewis, R. S., 581
Lewitowicz, M., 191
Licot, I., 253
Liénard, E., 253
Limongi, M., 407
Lingenfelter, R. E., 549
Lingner, S., 183
Lipnik, P., 253
Lugmair, G. W., 591
Lukyanov, S. M., 191
Luo, N., 447

M

MacIsaac, Ch., 591
Maino, G., 179
Maison, J. M., 311
Marengo, M., 63, 71
Marigo, P., 477
Marti, K., 549
Masuda, K., 201
Mathews, G. J., 441
Mayer, A., 255
McWilliam, A., 37
Mehren, T., 191
Meissner, J., 259, 303, 307
Mengoni, A., 179
Meshik, A. P., 603
Meyer, B., 331, 441, 447
Mezzorani, G., 31, 319, 323
Michotte, C., 253
Mitchell, C. A., 331
Mochkovitch, R., 453
Möhle, M., 83
Mohr, P., 243, 299
Morrissey, D. J., 303, 331
Mowlavi, N., 403
Mueller, A. C., 191
Mutti, P., 165

N

Nagai, Y., 145, 201, 205
Nakajima, Y., 179
Navarra, G., 55
Nikolaiko, A. S., 285
Ninane, A., 253
Nittler, L. R., 585
Nomoto, K., 379, 419, 457

O

Oberhummer, H., 183, 243, 277, 299, 315
Oberto, Y., 453, 539
O'D. Alexander, C. M., 585
Ohsaki, T., 201
Okunev, I. S., 173
Origlia, L., 63
Oyamatsu, K., 239, 425
Ozawa, A., 235

P

Page, R. D., 253
Pagel, B. E. J., 491
Palme, H., 247
Paravicini Bagliani, E., 71
Parker, P. D., 195
Penionzhkevich, Yu. E., 191
Persi, P., 63
Pfaff, R., 331
Pfeiffer, B., 191
PIAFE Collaboration, 153
Pougheon, F., 191
Powell, C. F., 331
Prantzos, N., 473, 531, 553
Prati, P., 31
Preston, G. W., 37
Puy, D., 561

Q

Quarati, P., 31, 319, 323, 327

R

Raimann, G., 235, 331
Raiteri, C. M., 407, 557
Rauch, T., 45
Rauscher, T., 183, 277
Ray, A., 187
Rentzsch-Holm, I., 41
Rodney, W. S., 31
Rolfs, C. E., 31, 217, 277
Rosier, L., 311
Ross, J. G., 259
Rotbard, G., 311
Roters, G., 217
Ruprecht, G., 271

S

Saint-Laurent, M. G., 191
Salamatin, V. S., 191
Salaris, M., 407
Salatski, V. I., 173
Sato, H., 145, 205
Scalia, A., 213
Schatz, H., 303, 307
Scheller, K. W., 303
Schmidt, S., 83
Schoedder, S., 191
Schönberner, D., 399
Schulte, W. H., 31
Searle, L., 37
Sedyshev, P. V., 173
Sedysheva, M. V., 173
Seidel, R., 255
Sellin, P. J., 253
Serichol, N., 437
Sherrill, B. M., 331
Shima, T., 145, 201, 205
Shoppa, T. D., 267
Shotter, A. C., 253
Shukolyukov, A., 591
Shukolyukov, Yu. A., 603
Silvestro, G., 63, 71
Smith, M. S., 195
Sneden, C., 37
Soiné, M., 255
Somorjai, E., 277

Sorlin, O., 191
Spettel, B., 247
Srinivasan, G., 595
Staudt, G., 243, 299
Steiner, M., 303, 331
Straniero, O., 407
Stürenburg, S., 41
Sugarbaker, E., 135
Sükösd, Cs., 253
Sutaria, F. K., 187
Suzuki, T. S., 145
Szabó, J., 295

T

Tanihata, I., 235
Tatischeff, V., 311
Theis, Ch., 227
Thibaud, J.-P., 259, 311, 453
Thielemann, F.-K., 263, 379, 419
Timmes, F. X., 501, 543
Tkaczyk, W., 75
Tornambé, A., 429
Toukan, K. A., 291
Trautvetter, H.-P., 31, 83, 217, 277
Tretyak, V. I., 281, 285
Truran, J. W., 501

U

Unrau, B., 271
Utku, S., 195

V

Vancraeynest, G., 253
Vandegriff, J., 331
Van Duppen, P., 253
Vangioni-Flam, E., 453, 539
Vanhorenbeeck, J., 253
Van Wormer, L., 263
van Wormer, L., 259

Vernotte, J., 311
Vervier, J., 253
Vesna, V. A., 173
Vogelaar, R. B., 195
Voss, F., 161, 227
Vouzoukas, S., 259, 303

W

Wagemans, C., 169
Walker, R. M., 585
Wanajoh, S., 457
Watanabe, K., 205
Watanabe, Y., 235
Weaver, T. A., 365, 463, 543
Weber, F., 485
Werner, K., 45
Wiescher, M., 125, 253, 259, 263, 303, 307
Wilmes, S., 243
Wilson, J. R., 441
Wisshak, K., 161, 227
Wöhr, A., 191, 255
Wolf, G., 255
Wood, P. R., 469
Woods, P. J., 253
Woosley, S. E., 365, 441, 463, 543

Y

Yamada, M., 239
Yennello, S. J., 331
Yildiz, K., 195
Yoshida, K., 235

Z

Zahnow, D., 31
Zamfirescu, I., 223
Zdesenko, Yu., 285
Zinner, E., 567, 581, 585

AIP Conference Proceedings

		L.C. Number	ISBN
No. 130	Laser Acceleration of Particles (Malibu, CA, 1985)	85-48028	0-88318-329-3
No. 131	Workshop on Polarized ^{3}He Beams and Targets (Princeton, NJ, 1984)	85-48026	0-88318-330-7
No. 132	Hadron Spectroscopy – 1985 (International Conference, Univ. of Maryland)	85-72537	0-88318-331-5
No. 133	Hadronic Probes and Nuclear Interactions (Arizona State University, 1985)	85-72638	0-88318-332-3
No. 134	The State of High Energy Physics (BNL/SUNY Summer School, 1983)	85-73170	0-88318-333-1
No. 135	Energy Sources: Conservation and Renewables (APS, Washington, DC, 1985)	85-73019	0-88318-334-X
No. 136	Atomic Theory Workshop on Relativistic and QED Effects in Heavy Atoms (Gaithersburg, MD, 1985)	85-73790	0-88318-335-8
No. 137	Polymer-Flow Interaction (La Jolla Institute, 1985)	85-73915	0-88318-336-6
No. 138	Frontiers in Electronic Materials and Processing (Houston, TX, 1985)	86-70108	0-88318-337-4
No. 139	High-Current, High-Brightness, and High-Duty Factor Ion Injectors (La Jolla Institute, 1985)	86-70245	0-88318-338-2
No. 140	Boron-Rich Solids (Albuquerque, NM, 1985)	86-70246	0-88318-339-0
No. 141	Gamma-Ray Bursts (Stanford, CA, 1984)	86-70761	0-88318-340-4
No. 142	Nuclear Structure at High Spin, Excitation, and Momentum Transfer (Indiana University, 1985)	86-70837	0-88318-341-2
No. 143	Mexican School of Particles and Fields (Oaxtepec, México, 1984)	86-81187	0-88318-342-0
No. 144	Magnetospheric Phenomena in Astrophysics (Los Alamos, NM, 1984)	86-71149	0-88318-343-9
No. 145	Polarized Beams at SSC & Polarized Antiprotons (Ann Arbor, MI & Bodega Bay, CA, 1985)	86-71343	0-88318-344-7
No. 146	Advances in Laser Science—I (Dallas, TX, 1985)	86-71536	0-88318-345-5
No. 147	Short Wavelength Coherent Radiation: Generation and Applications (Monterey, CA, 1986)	86-71674	0-88318-346-3

No. 148	Space Colonization: Technology and The Liberal Arts (Geneva, NY, 1985)	86-71675	0-88318-347-1
No. 149	Physics and Chemistry of Protective Coatings (Universal City, CA, 1985)	86-72019	0-88318-348-X
No. 150	Intersections Between Particle and Nuclear Physics (Lake Louise, Canada, 1986)	86-72018	0-88318-349-8
No. 151	Neural Networks for Computing (Snowbird, UT, 1986)	86-72481	0-88318-351-X
No. 152	Heavy Ion Inertial Fusion (Washington, DC, 1986)	86-73185	0-88318-352-8
No. 153	Physics of Particle Accelerators (SLAC Summer School, 1985) (Fermilab Summer School, 1984)	87-70103	0-88318-353-6
No. 154	Physics and Chemistry of Porous Media—II (Ridgefield, CT, 1986)	83-73640	0-88318-354-4
No. 155	The Galactic Center: Proceedings of the Symposium Honoring C. H. Townes (Berkeley, CA, 1986)	86-73186	0-88318-355-2
No. 156	Advanced Accelerator Concepts (Madison, WI, 1986)	87-70635	0-88318-358-0
No. 157	Stability of Amorphous Silicon Alloy Materials and Devices (Palo Alto, CA, 1987)	87-70990	0-88318-359-9
No. 158	Production and Neutralization of Negative Ions and Beams (Brookhaven, NY, 1986)	87-71695	0-88318-358-7
No. 159	Applications of Radio-Frequency Power to Plasma: Seventh Topical Conference (Kissimmee, FL, 1987)	87-71812	0-88318-359-5
No. 160	Advances in Laser Science—II (Seattle, WA, 1986)	87-71962	0-88318-360-9
No. 161	Electron Scattering in Nuclear and Particle Science: In Commemoration of the 35th Anniversary of the Lyman-Hanson-Scott Experiment (Urbana, IL, 1986)	87-72403	0-88318-361-7
No. 162	Few-Body Systems and Multiparticle Dynamics (Crystal City, VA, 1987)	87-72594	0-88318-362-5
No. 163	Pion–Nucleus Physics: Future Directions and New Facilities at LAMPF (Los Alamos, NM, 1987)	87-72961	0-88318-363-3
No. 164	Nuclei Far from Stability: Fifth International Conference (Rosseau Lake, ON, 1987)	87-73214	0-88318-364-1

No. 165	Thin Film Processing and Characterization of High-Temperature Superconductors (Anaheim, CA, 1987)	87-73420	0-88318-365-X
No. 166	Photovoltaic Safety (Denver, CO, 1988)	88-42854	0-88318-366-8
No. 167	Deposition and Growth: Limits for Microelectronics (Anaheim, CA, 1987)	88-71432	0-88318-367-6
No. 168	Atomic Processes in Plasmas (Santa Fe, NM, 1987)	88-71273	0-88318-368-4
No. 169	Modern Physics in America: A Michelson-Morley Centennial Symposium (Cleveland, OH, 1987)	88-71348	0-88318-369-2
No. 170	Nuclear Spectroscopy of Astrophysical Sources (Washington, DC, 1987)	88-71625	0-88318-370-6
No. 171	Vacuum Design of Advanced and Compact Synchrotron Light Sources (Upton, NY, 1988)	88-71824	0-88318-371-4
No. 172	Advances in Laser Science—III: Proceedings of the International Laser Science Conference (Atlantic City, NJ, 1987)	88-71879	0-88318-372-2
No. 173	Cooperative Networks in Physics Education (Oaxtepec, Mexico, 1987)	88-72091	0-88318-373-0
No. 174	Radio Wave Scattering in the Interstellar Medium (San Diego, CA, 1988)	88-72092	0-88318-374-9
No. 175	Non-neutral Plasma Physics (Washington, DC, 1988)	88-72275	0-88318-375-7
No. 176	Intersections Between Particle and Nuclear Physics (Third International Conference) (Rockport, ME, 1988)	88-62535	0-88318-376-5
No. 177	Linear Accelerator and Beam Optics Codes (La Jolla, CA, 1988)	88-46074	0-88318-377-3
No. 178	Nuclear Arms Technologies in the 1990s (Washington, DC, 1988)	88-83262	0-88318-378-1
No. 179	The Michelson Era in American Science: 1870–1930 (Cleveland, OH, 1987)	88-83369	0-88318-379-X
No. 180	Frontiers in Science: International Symposium (Urbana, IL, 1987)	88-83526	0-88318-380-3
No. 181	Muon-Catalyzed Fusion (Sanibel Island, FL, 1988)	88-83636	0-88318-381-1
No. 182	High T_c Superconducting Thin Films, Devices, and Applications (Atlanta, GA, 1988)	88-03947	0-88318-382-X

No. 183	Cosmic Abundances of Matter (Minneapolis, MN, 1988)	89-80147	0-88318-383-8
No. 184	Physics of Particle Accelerators (Ithaca, NY, 1988)	89-83575	0-88318-384-6
No. 185	Glueballs, Hybrids, and Exotic Hadrons (Upton, NY, 1988)	89-83513	0-88318-385-4
No. 186	High-Energy Radiation Background in Space (Sanibel Island, FL, 1987)	89-83833	0-88318-386-2
No. 187	High-Energy Spin Physics (Minneapolis, MN, 1988)	89-83948	0-88318-387-0
No. 188	International Symposium on Electron Beam Ion Sources and their Applications (Upton, NY, 1988)	89-84343	0-88318-388-9
No. 189	Relativistic, Quantum Electrodynamic, and Weak Interaction Effects in Atoms (Santa Barbara, CA, 1988)	89-84431	0-88318-389-7
No. 190	Radio-frequency Power in Plasmas (Irvine, CA, 1989)	89-45805	0-88318-397-8
No. 191	Advances in Laser Science—IV (Atlanta, GA, 1988)	89-85595	0-88318-391-9
No. 192	Vacuum Mechatronics (First International Workshop) (Santa Barbara, CA, 1989)	89-45905	0-88318-394-3
No. 193	Advanced Accelerator Concepts (Lake Arrowhead, CA, 1989)	89-45914	0-88318-393-5
No. 194	Quantum Fluids and Solids—1989 (Gainesville, FL, 1989)	89-81079	0-88318-395-1
No. 195	Dense Z-Pinches (Laguna Beach, CA, 1989)	89-46212	0-88318-396-X
No. 196	Heavy Quark Physics (Ithaca, NY, 1989)	89-81583	0-88318-644-6
No. 197	Drops and Bubbles (Monterey, CA, 1988)	89-46360	0-88318-392-7
No. 198	Astrophysics in Antarctica (Newark, DE, 1989)	89-46421	0-88318-398-6
No. 199	Surface Conditioning of Vacuum Systems (Los Angeles, CA, 1989)	89-82542	0-88318-756-6
No. 200	High T_c Superconducting Thin Films: Processing, Characterization, and Applications (Boston, MA, 1989)	90-80006	0-88318-759-0
No. 201	QED Structure Functions (Ann Arbor, MI, 1989)	90-80229	0-88318-671-3
No. 202	NASA Workshop on Physics From a Lunar Base (Stanford, CA, 1989)	90-55073	0-88318-646-2

No. 203	Particle Astrophysics: The NASA Cosmic Ray Program for the 1990s and Beyond (Greenbelt, MD, 1989)	90-55077	0-88318-763-9
No. 204	Aspects of Electron-Molecule Scattering and Photoionization (New Haven, CT, 1989)	90-55175	0-88318-764-7
No. 205	The Physics of Electronic and Atomic Collisions (XVI International Conference) (New York, NY, 1989)	90-53183	0-88318-390-0
No. 206	Atomic Processes in Plasmas (Gaithersburg, MD, 1989)	90-55265	0-88318-769-8
No. 207	Astrophysics from the Moon (Annapolis, MD, 1990)	90-55582	0-88318-770-1
No. 208	Current Topics in Shock Waves (Bethlehem, PA, 1989)	90-55617	0-88318-776-0
No. 209	Computing for High Luminosity and High Intensity Facilities (Santa Fe, NM, 1990)	90-55634	0-88318-786-8
No. 210	Production and Neutralization of Negative Ions and Beams (Brookhaven, NY, 1990)	90-55316	0-88318-786-8
No. 211	High-Energy Astrophysics in the 21st Century (Taos, NM, 1989)	90-55644	0-88318-803-1
No. 212	Accelerator Instrumentation (Brookhaven, NY, 1989)	90-55838	0-88318-645-4
No. 213	Frontiers in Condensed Matter Theory (New York, NY, 1989)	90-6421	0-88318-771-X 0-88318-772-8 (pbk.)
No. 214	Beam Dynamics Issues of High-Luminosity Asymmetric Collider Rings (Berkeley, CA, 1990)	90-55857	0-88318-767-1
No. 215	X-Ray and Inner-Shell Processes (Knoxville, TN, 1990)	90-84700	0-88318-790-6
No. 216	Spectral Line Shapes, Vol. 6 (Austin, TX, 1990)	90-06278	0-88318-791-4
No. 217	Space Nuclear Power Systems (Albuquerque, NM, 1991)	90-56220	0-88318-838-4
No. 218	Positron Beams for Solids and Surfaces (London, Canada, 1990)	90-56407	0-88318-842-2
No. 219	Superconductivity and Its Applications (Buffalo, NY, 1990)	91-55020	0-88318-835-X
No. 220	High Energy Gamma-Ray Astronomy (Ann Arbor, MI, 1990)	91-70876	0-88318-812-0

No. 221	Particle Production Near Threshold (Nashville, IN, 1990)	91-55134	0-88318-829-5
No. 222	After the First Three Minutes (College Park, MD, 1990)	91-55214	0-88318-828-7
No. 223	Polarized Collider Workshop (University Park, PA, 1990)	91-71303	0-88318-826-0
No. 224	LAMPF Workshop on (π, K) Physics (Los Alamos, NM, 1990)	91-71304	0-88318-825-2
No. 225	Half Collision Resonance Phenomena in Molecules (Caracas, Venezuela, 1990)	91-55210	0-88318-840-6
No. 226	The Living Cell in Four Dimensions (Gif sur Yvette, France, 1990)	91-55209	0-88318-794-9
No. 227	Advanced Processing and Characterization Technologies (Clearwater, FL, 1991)	91-55194	0-88318-910-0
No. 228	Anomalous Nuclear Effects in Deuterium/Solid Systems (Provo, UT, 1990)	91-55245	0-88318-833-3
No. 229	Accelerator Instrumentation (Batavia, IL, 1990)	91-55347	0-88318-832-1
No. 230	Nonlinear Dynamics and Particle Acceleration (Tsukuba, Japan, 1990)	91-55348	0-88318-824-4
No. 231	Boron-Rich Solids (Albuquerque, NM, 1990)	91-53024	0-88318-793-4
No. 232	Gamma-Ray Line Astrophysics (Paris-Saclay, France, 1990)	91-55492	0-88318-875-9
No. 233	Atomic Physics 12 (Ann Arbor, MI, 1990)	91-55595	088318-811-2
No. 234	Amorphous Silicon Materials and Solar Cells (Denver, CO, 1991)	91-55575	088318-831-7
No. 235	Physics and Chemistry of MCT and Novel IR Detector Materials (San Francisco, CA, 1990)	91-55493	0-88318-931-3
No. 236	Vacuum Design of Synchrotron Light Sources (Argonne, IL, 1990)	91-55527	0-88318-873-2
No. 237	Kent M. Terwilliger Memorial Symposium (Ann Arbor, MI, 1989)	91-55576	0-88318-788-4
No. 238	Capture Gamma-Ray Spectroscopy (Pacific Grove, CA, 1990)	91-57923	0-88318-830-9
No. 239	Advances in Biomolecular Simulations (Obernai, France, 1991)	91-58106	0-88318-940-2

No. 240	Joint Soviet-American Workshop on the Physics of Semiconductor Lasers (Leningrad, USSR, 1991)	91-58537	0-88318-936-4
No. 241	Scanned Probe Microscopy (Santa Barbara, CA, 1991)	91-76758	0-88318-816-3
No. 242	Strong, Weak, and Electromagnetic Interactions in Nuclei, Atoms, and Astrophysics: A Workshop in Honor of Stewart D. Bloom's Retirement (Livermore, CA, 1991)	91-76876	0-88318-943-7
No. 243	Intersections Between Particle and Nuclear Physics (Tucson, AZ, 1991)	91-77580	0-88318-950-X
No. 244	Radio Frequency Power in Plasmas (Charleston, SC, 1991)	91-77853	0-88318-937-2
No. 245	Basic Space Science (Bangalore, India, 1991)	91-78379	0-88318-951-8
No. 246	Space Nuclear Power Systems (Albuquerque, NM, 1992)	91-58793	1-56396-027-3 1-56396-026-5 (pbk.)
No. 247	Global Warming: Physics and Facts (Washington, DC, 1991)	91-78423	0-88318-932-1
No. 248	Computer-Aided Statistical Physics (Taipei, Taiwan, 1991)	91-78378	0-88318-942-9
No. 249	The Physics of Particle Accelerators (Upton, NY, 1989, 1990)	92-52843	0-88318-789-2
No. 250	Towards a Unified Picture of Nuclear Dynamics (Nikko, Japan, 1991)	92-70143	0-88318-951-8
No. 251	Superconductivity and its Applications (Buffalo, NY, 1991)	92-52726	1-56396-016-8
No. 252	Accelerator Instrumentation (Newport News, VA, 1991)	92-70356	0-88318-934-8
No. 253	High-Brightness Beams for Advanced Accelerator Applications (College Park, MD, 1991)	92-52705	0-88318-947-X
No. 254	Testing the AGN Paradigm (College Park, MD, 1991)	92-52780	1-56396-009-5
No. 255	Advanced Beam Dynamics Workshop on Effects of Errors in Accelerators, Their Diagnosis and Corrections (Corpus Christi, TX, 1991)	92-52842	1-56396-006-0
No. 256	Slow Dynamics in Condensed Matter (Fukuoka, Japan, 1991)	92-53120	0-88318-938-0

No. 257	Atomic Processes in Plasmas (Portland, ME, 1991)	91-08105	0-88318-939-9
No. 258	Synchrotron Radiation and Dynamic Phenomena (Grenoble, France, 1991)	92-53790	1-56396-008-7
No. 259	Future Directions in Nuclear Physics with 4π Gamma Detection Systems of the New Generation (Strasbourg, France, 1991)	92-53222	0-88318-952-6
No. 260	Computational Quantum Physics (Nashville, TN, 1991)	92-71777	0-88318-933-X
No. 261	Rare and Exclusive B&K Decays and Novel Flavor Factories (Santa Monica, CA, 1991)	92-71873	1-56396-055-9
No. 262	Molecular Electronics—Science and Technology (St. Thomas, Virgin Islands, 1991)	92-72210	1-56396-041-9
No. 263	Stress-Induced Phenomena in Metallization: First International Workshop (Ithaca, NY, 1991)	92-72292	1-56396-082-6
No. 264	Particle Acceleration in Cosmic Plasmas (Newark, DE, 1991)	92-73316	0-88318-948-8
No. 265	Gamma-Ray Bursts (Huntsville, AL, 1991)	92-73456	1-56396-018-4
No. 266	Group Theory in Physics (Cocoyoc, Morelos, Mexico, 1991)	92-73457	1-56396-101-6
No. 267	Electromechanical Coupling of the Solar Atmosphere (Capri, Italy, 1991)	92-82717	1-56396-110-5
No. 268	Photovoltaic Advanced Research & Development Project (Denver, CO, 1992)	92-74159	1-56396-056-7
No. 269	CEBAF 1992 Summer Workshop (Newport News, VA, 1992)	92-75403	1-56396-067-2
No. 270	Time Reversal—The Arthur Rich Memorial Symposium (Ann Arbor, MI, 1991)	92-83852	1-56396-105-9
No. 271	Tenth Symposium Space Nuclear Power and Propulsion (Vols. I–III) (Albuquerque, NM, 1993)	92-75162	1-56396-137-7 (set)
No. 272	Proceedings of the XXVI International Conference on High Energy Physics (Vols. I and II) (Dallas, TX, 1992)	93-70412	1-56396-127-X (set)

No. 273	Superconductivity and Its Applications (Buffalo, NY, 1992)	93-70502	1-56396-189-X
No. 274	VIth International Conference on the Physics of Highly Charged Ions (Manhattan, KS, 1992)	93-70577	1-56396-102-4
No. 275	Atomic Physics 13 (Munich, Germany, 1992)	93-70826	1-56396-057-5
No. 276	Very High Energy Cosmic-Ray Interactions: VIIth International Symposium (Ann Arbor, MI, 1992)	93-71342	1-56396-038-9
No. 277	The World at Risk: Natural Hazards and Climate Change (Cambridge, MA, 1992)	93-71333	1-56396-066-4
No. 278	Back to the Galaxy (College Park, MD, 1992)	93-71543	1-56396-227-6
No. 279	Advanced Accelerator Concepts (Port Jefferson, NY, 1992)	93-71773	1-56396-191-1
No. 280	Compton Gamma-Ray Observatory (St. Louis, MO, 1992)	93-71830	1-56396-104-0
No. 281	Accelerator Instrumentation Fourth Annual Workshop (Berkeley, CA, 1992)	93-072110	1-56396-190-3
No. 282	Quantum 1/f Noise & Other Low Frequency Fluctuations in Electronic Devices (St. Louis, MO, 1992)	93-072366	1-56396-252-7
No. 283	Earth and Space Science Information Systems (Pasadena, CA, 1992)	93-072360	1-56396-094-X
No. 284	US-Japan Workshop on Ion Temperature Gradient-Driven Turbulent Transport (Austin, TX, 1993)	93-72460	1-56396-221-7
No. 285	Noise in Physical Systems and 1/f Fluctuations (St. Louis, MO, 1993)	93-72575	1-56396-270-5
No. 286	Ordering Disorder: Prospect and Retrospect in Condensed Matter Physics: Proceedings of the Indo-U.S. Workshop (Hyderabad, India, 1993)	93-072549	1-56396-255-1
No. 287	Production and Neutralization of Negative Ions and Beams: Sixth International Symposium (Upton, NY, 1992)	93-72821	1-56396-103-2

No. 288	Laser Ablation: Mechanismas and Applications-II: Second International Conference (Knoxville, TN, 1993)	93-73040	1-56396-226-8
No. 289	Radio Frequency Power in Plasmas: Tenth Topical Conference (Boston, MA, 1993)	93-72964	1-56396-264-0
No. 290	Laser Spectroscopy: XIth International Conference (Hot Springs, VA, 1993)	93-73050	1-56396-262-4
No. 291	Prairie View Summer Science Academy (Prairie View, TX, 1992)	93-73081	1-56396-133-4
No. 292	Stability of Particle Motion in Storage Rings (Upton, NY, 1992)	93-73534	1-56396-225-X
No. 293	Polarized Ion Sources and Polarized Gas Targets (Madison, WI, 1993)	93-74102	1-56396-220-9
No. 294	High-Energy Solar Phenomena A New Era of Spacecraft Measurements (Waterville Valley, NH, 1993)	93-74147	1-56396-291-8
No. 295	The Physics of Electronic and Atomic Collisions: XVIII International Conference (Aarhus, Denmark, 1993)	93-74103	1-56396-290-X
No. 296	The Chaos Paradigm: Developments an Applications in Engineering and Science (Mystic, CT, 1993)	93-74146	1-56396-254-3
No. 297	Computational Accelerator Physics (Los Alamos, NM, 1993)	93-74205	1-56396-222-5
No. 298	Ultrafast Reaction Dynamics and Solvent Effects (Royaumont, France, 1993)	93-074354	1-56396-280-2
No. 299	Dense Z-Pinches: Third International Conference (London, 1993)	93-074569	1-56396-297-7
No. 300	Discovery of Weak Neutral Currents: The Weak Interaction Before and After (Santa Monica, CA, 1993)	94-70515	1-56396-306-X
No. 301	Eleventh Symposium Space Nuclear Power and Propulsion (3 Vols.) (Albuquerque, NM, 1994)	92-75162	1-56396-305-1 (Set) 156396-301-9 (pbk. set)
No. 302	Lepton and Photon Interactions/ XVI International Symposium (Ithaca, NY, 1993)	94-70079	1-56396-106-7

No. 303	Slow Positron Beam Techniques for Solids and Surfaces Fifth International Workshop (Jackson Hole, WY 1992)	94-71036	1-56396-267-5
No. 304	The Second Compton Symposium (College Park, MD, 1993)	94-70742	1-56396-261-6
No. 305	Stress-Induced Phenomena in Metallization Second International Workshop (Austin, TX, 1993)	94-70650	1-56396-251-9
No. 306	12th NREL Photovoltaic Program Review (Denver, CO, 1993)	94-70748	1-56396-315-9
No. 307	Gamma-Ray Bursts Second Workshop (Huntsville, AL 1993)	94-71317	1-56396-336-1
No. 308	The Evolution of X-Ray Binaries (College Park, MD 1993)	94-76853	1-56396-329-9
No. 309	High-Pressure Science and Technology—1993 (Colorado Springs, CO 1993)	93-72821	1-56396-219-5 (Set)
No. 310	Analysis of Interplanetary Dust (Houston, TX 1993)	94-71292	1-56396-341-8
No. 311	Physics of High Energy Particles in Toroidal Systems (Irvine, CA 1993)	94-72098	1-56396-364-7
No. 312	Molecules and Grains in Space (Mont Sainte-Odile, France 1993)	94-72615	1-56396-355-8
No. 313	The Soft X-Ray Cosmos ROSAT Science Symposium (College Park, MD 1993)	94-72499	1-56396-327-2
No. 314	Advances in Plasma Physics Thomas H. Stix Symposium (Princeton, NJ 1992)	94-72721	1-56396-372-8
No. 315	Orbit Correction and Analysis in Circular Accelerators (Upton, NY 1993)	94-72257	1-56396-373-6
No. 316	Thirteenth International Conference on Thermoelectrics (Kansas City, Missouri 1994)	95-75634	1-56396-444-9
No. 317	Fifth Mexican School of Particles and Fields (Guanajuato, Mexico 1992)	94-72720	1-56396-378-7
No. 318	Laser Interaction and Related Plasma Phenomena 11th International Workshop (Monterey, CA 1993)	94-78097	1-56396-324-8

| No. 319 | Beam Instrumentation Workshop (Santa Fe, NM 1993) | 94-78279 | 1-56396-389-2 |

| No. 320 | Basic Space Science (Lagos, Nigeria 1993) | 94-79350 | 1-56396-328-0 |

| No. 321 | The First NREL Conference on Thermophotovoltaic Generation of Electricity (Copper Mountain, CO 1994) | 94-72792 | 1-56396-353-1 |

| No. 322 | Atomic Processes in Plasmas Ninth APS Topical Conference (San Antonio, TX) | 94-72923 | 1-56396-411-2 |

| No. 323 | Atomic Physics 14 Fourteenth International Conference on Atomic Physics (Boulder, CO 1994) | 94-73219 | 1-56396-348-5 |

| No. 324 | Twelfth Symposium on Space Nuclear Power and Propulsion (Albuquerque, NM 1995) | 94-73603 | 1-56396-427-9 |

| No. 325 | Conference on NASA Centers for Commercial Development of Space (Albuquerque, NM 1995) | 94-73604 | 1-56396-431-7 |

| No. 326 | Accelerator Physics at the Superconducting Super Collider (Dallas, TX 1992-1993) | 94-73609 | 1-56396-354-X |

| No. 327 | Nuclei in the Cosmos III Third International Symposium on Nuclear Astrophysics (Assergi, Italy 1994) | 95-75492 | 1-56396-436-8 |

| No. 328 | Spectral Line Shapes, Volume 8 12th ICSLS (Toronto, Canada 1994) | 94-74309 | 1-56396-326-4 |

| No. 329 | Resonance Ionization Spectroscopy 1994 Seventh International Symposium (Bernkastel-Kues, Germany 1994) | 95-75077 | 1-56396-437-6 |